Marine Biology and Ecology

Marine Biology and Ecology

Edited by Jonas Bailey

SYRAWOOD
PUBLISHING HOUSE
New York

Published by Syrawood Publishing House,
750 Third Avenue, 9th Floor,
New York, NY 10017, USA
www.syrawoodpublishinghouse.com

Marine Biology and Ecology
Edited by Jonas Bailey

International Standard Book Number: 978-1-68286-680-1 (Hardback)

Cataloging-in-Publication Data

Marine biology and ecology / edited by Jonas Bailey.
p. cm.
Includes bibliographical references and index.
ISBN 978-1-68286-680-1
1. Marine biology. 2. Marine ecology. 3. Marine sciences. 4. Aquatic biology.
5. Aquatic ecology. I. Bailey, Jonas.
QH91 .M37 2019
578.77--dc21

Printed in China.

TABLE OF CONTENTS

PREFACE

I am honored to present to you this unique book which encompasses the most up-to-date data in the field. I was extremely pleased to get this opportunity of editing the work of experts from across the globe. I have also written papers in this field and researched the various aspects revolving around the progress of the discipline. I have tried to unify my knowledge along with that of stalwarts from every corner of the world, to produce a text which not only benefits the readers but also facilitates the growth of the field.

The field of marine biology and ecology includes the study of marine animals and plants in relation to their ecosystems. It also focuses on the influence of biotic and abiotic factors on marine life populations. The field also studies changes in topological, ecological and atmospheric factors that affect the survival and reproduction of marine life. Our oceans are changing due to a complex combination of ocean eddies and atmospheric winds as well as human activities like fossil fuel burning, industrial waste disposal and fertilizer use. This has adversely affected the marine ecosystem which has major impacts in climate change, changes in coastlines and coastal environment, etc. This book is a compilation of chapters that discuss the most vital concepts and emerging trends in the field of marine ecology and biology. The various advancements in theoretical and applied marine science are also glanced at in this book. It will be vital for students and researchers looking for an elaborate reference text.

Finally, I would like to thank all the contributing authors for their valuable time and contributions. This book would not have been possible without their efforts. I would also like to thank my friends and family for their constant support.

Editor

Acoustic Characterization of Fish and Seabed Using Underwater Acoustic Technology in Seribu Island Indonesia

Henry M Manik*

Department of Marine Science and Technology, Faculty of Fisheries and Marine Sciences, Bogor Agricultural University (IPB) Kampus IPB, Darmaga, Bogor-16680, Indonesia

Abstract

Remote sensing technique using underwater acoustics are one of the most cost-effective methods of resource detection and mapping, particularly in the coastal zone. We measured the acoustic backscatter using single beam echosounder from water column target and sea bottom. The increasing of fish size is followed by target strength value. For sea bottom, the highest acoustic backscattering was sand followed by silt and clay. This was caused by the acoustic impedance difference and also the presence of the organisms in sediment. By this study, classification and characterization of underwater target such as fish and seabed using acoustic technology was possible.

Keywords: Acoustic; Sensing; Backscatter; Target; Seabed; Mapping

Introduction

Underwater acoustic technologies are one of the most cost-effective methods of resource detection and mapping. The use of acoustic techniques in seabed mapping and monitoring has proven to be a useful tool in marine resource management. Underwater acoustic as a remote sensing tool are recognized as one of the most effective tools available to detect, map and characterize the seafloor [1]. Depth measurement or bathymetry gives the relief of the seafloor and acoustic backscattering value can relate to study the morphology and composition of the seabed. Bathymetry maps produced from acoustic instrument is well developed, but the processing and analysis of acoustic backscatter data has not yet reached its full potential. Hence, the need for enhanced techniques of acoustic backscattering strength analysis was evident. The primary objective of this study was to examine and develop new methodologies for using acoustic backscattering data for fish, sea bottom and benthic habitat.

Methodology

Underwater acoustic technology used in this research is CruzPro single-beam echosounders. The acoustic transducer transmits one single vertical beam towards the seafloor for determining the water depth. A transmission of acoustic wave travels through water by displacement of water particles. Seawater has low acoustic impedance which results in a low resistance to the propagation of the acoustic wave. A part of the incident wave is reflected in symmetrical direction, a part is scattered in all directions, and another part penetrates to the seabed. The scattering of the acoustic energy back towards the sonar is called backscatter. This backscattered energy is received by the transducer echosounder and used for depth or bathymetry and echo strength measurements [2].

Acoustic data acquisition

Data acquisition at Seribu Island Indonesia was conducted on 7-9 June 2014 (Figure 1). The three-frequencies (50, 120, and 200 kHz) of CruzPro single beam transducer were acquired over substrates ranging from clayey silt to sand in the Seribu Island of Jakarta, using a hull-mounted normal-incidence underwater acoustics instrument. The beam width of the echosounder transducer for 50, 120, and 200 kHz is 20°, 15° and 9°, respectively, with respective pulse lengths of 0.85, 0.75, and 0.5 ms. The raw analog output on the receiver circuit board was tapped and connected to a 12-bit A/D converter with a sampling frequency of 1 MHz. The echoes were stored together with the information of the echosounder adjustments and ship position obtained from the GPS system.

The recorded echo data were converted from binary to ASCII format. The shape of the sea bottom echo envelope is generally influenced by various factors including natural variability of the underwater target, seafloor, transducer motion, and due to electrical noise of echosounder [3]. Echo alignment and echo averaging were performed to obtain good averaged echo envelopes.

Fish, zooplankton, and oceanography sampling

Fish and zooplankton sampling were conducted in order to verify the type of each species and relate to acoustic backscatter value. Data collections are simultaneously with acoustic transmission. Fish species data were collected using underwater camera, zooplankton sample were obtained using plankton net. Sample plankton were treated for further analysis in laboratory. Ocean temperature and salinity were collected using CTD instrument.

Sediment sampling

Sediment data were collected using a Van Veen grab, covering an area of 0.04 m² and penetration of 10 cm. About 50 g of sediment were taken from each grab sample to carry out the textural analyses using a 4.0 cm diameter core tube. The sediment was repeatedly washed in distilled water until all the chloride ions detectable with 5% silver nitrate were removed. These samples were treated with 10% sodium hexametaphosphate. The acquired sediment samples were processed to wet sieving using a 62 μm sieve to separate the sand from the mud

***Corresponding author:** Henry M Manik, Department of Marine Science and Technology, Faculty of Fisheries and Marine Sciences, Bogor Agricultural University (IPB) Kampus IPB, Darmaga, Bogor-16680, Indonesia
E-mail: henrymanik@ipb.ac.id

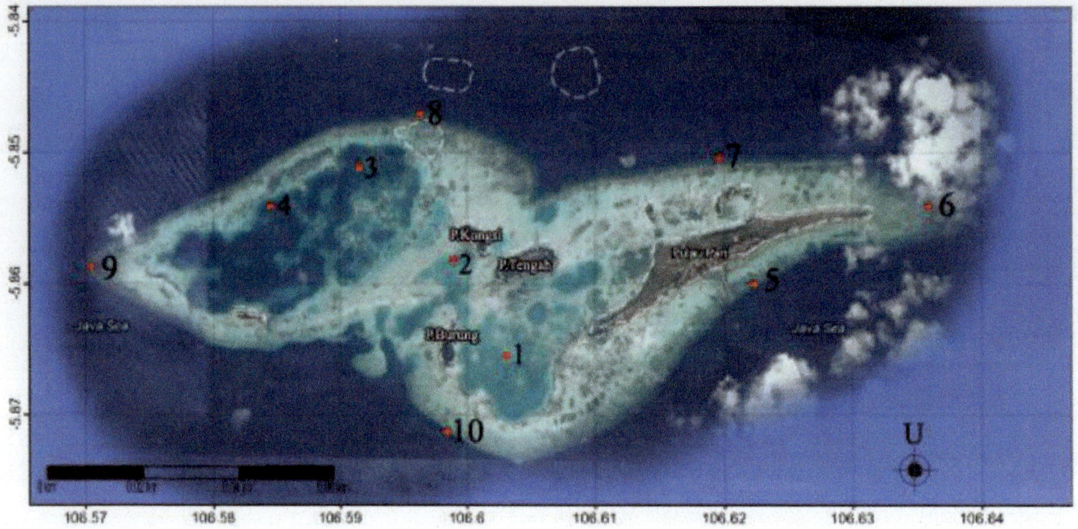

Figure 1: Ocean data acquisition in Pari Island. Acoustic track performed around the island with (•) are sampling points for plankton, diving, biota, coral reef, underwater vegetation, temperature and salinity.

No	Parameters	Value
1	Source Level: SL = 20 log (V_{TX})+TVR	220 dB
2	Directivity Index: DI_T=DI_R	30 dB
3	Transmit Voltage Response: TVR	170 dB
4	Open Circuit Voltage Response: OCVR	-178 dB
5	Detection Threshold: DT	-71 dB (V_{RX} = ± 3 mV rms)
6	Ambient Noise Level: NL	30 dB
7	Gain	60 dB (variable from 20 dB to 60 dB)
8	Absorption coefficient: α	0.0457 dB/m
9	Operating frequency: f	200 kHz

Table 1: Acoustic parameter for measuring detection range.

fraction. The size distribution of the sand fraction was determined using a dry sieving method [4].

Benthos sampling

The sediment samples for benthos identification were washed through a 0.5 mm mesh sieve, and all organisms retained on the sieve were collected and preserved in 15% seawater formalin. Benthos samples were washed through 0.5 mm mesh in running water in the laboratory to clear adhering sediments. All organisms were sorted into major groups, preserved in 90% alcohol for further identification and counted group-wise. The average number of organisms from the samples was then converted to number per m^2 (no. m^2). Biomass was determined by using the wet weight method. The obtained biomass was converted to gm^2 (wet weight). The species identification was based on manual book.

Result and Discussions

Use of underwater acoustic instrument as a primary tool to explore oceans has many advantages compared to conventional biological sampling, such as trawls and nets. First, underwater sound propagates at about 1500 m/s and can travel a much larger distance, making it possible to sample a much larger volume in a relatively shorter period of time. Secondly, acoustic measurements are remote, less invasive, and non-extractive. Thirdly, it can provide higher spatial resolution in both horizontal and vertical (or range for down-looking echosounders) directions [2]. Fisheries acoustics is the use of sound to measure the distribution and abundance of fish and other aquatic organisms. To gain a high quality of acoustic data, calibration was performed as shown in Table 1. Figure 2 shows the measurement method for detection range of acoustics instrument.

Figure 3 shows the target strength (TS) of zooplankton. The distribution of TS is ranged from -110 to -70 dB. Fish school distribution is ranged from -85 to -60 dB shown in Figures 4 and 5.

Figure 6 shows the Target Strength histogram of underwater target. This figure shows that target strength distribution is -70.0 dB to -47.0 dB. Figure 7 shows the relationship between fish size and acoustic target strength. By this figure, the increasing of fish size is followed by the increasing of target strength value.

Table 2 shows the acoustic density of detected target using Sonar. Number of single echo detector (SED) is shown in the second row, followed by volume density and area density. This table show the acoustic density is highest at 3433 target /ha and the lowest density at 28146 target/ha.

Figure 8 shows the hardness (red line) and roughness (blue line) of the sea bottom. The hardness of sea bottom is depending on acoustic impedance value, while roughness is depend on bottom surface morphology. From ping 1 until 100, the roughness is higher than hardness and from ping 100 until 200; the hardness is higher than roughness. Table 3 shows the ground truth data for seabed.

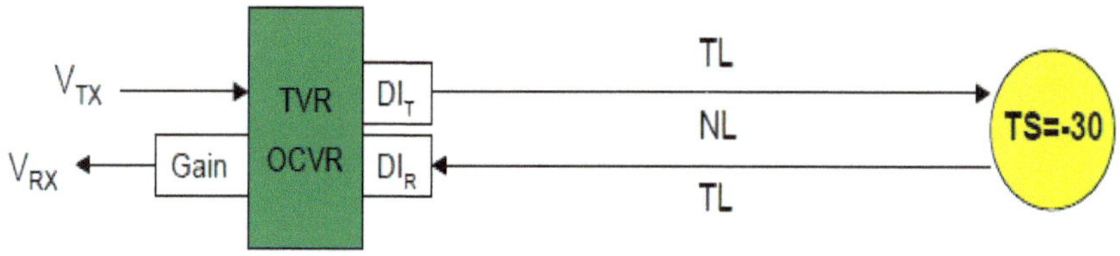

Figure 2: Measuring detection range of acoustic instrument.
Detection threshold (DT) was computed using the equation below:
$DT = SL\text{-}2\ TL+TS+DI_T+DI_R\ \text{-}NL+OCVR$
$2\ TL \leq SL\text{-}DT+TS+DI_T+DI_R\text{-}NL+\ OCVR$
$2\ TL \leq 220+70\text{-}30+30+30\text{-}30\text{-}178=122$
$20 \log R + \alpha R \leq 55 \quad R \leq 180\ m$
Where R is maximum detection range of acoustic instrument, SL is Source Level, TL is Transmission Loss, TS is Target Strength, DIT, DIR is directivity index for transmitting and receiving, respectively, NL is Noise Level; OCVR is open circuit voltage response. For data processing, the acoustic files contain run length packed and coded voltage samples.

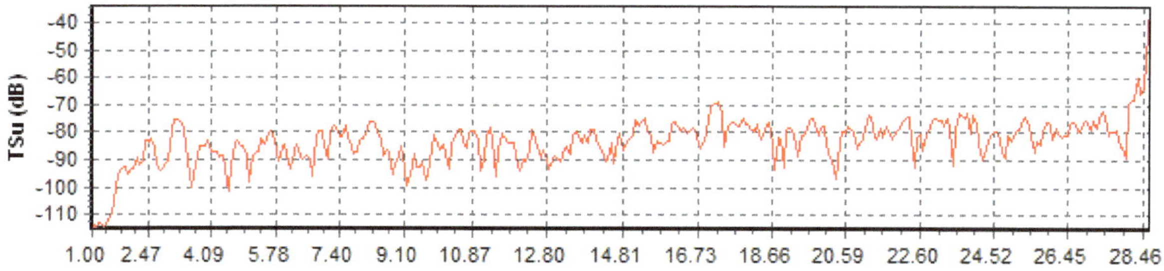

Figure 3: Acoustic target strength (TS) of zooplankton.

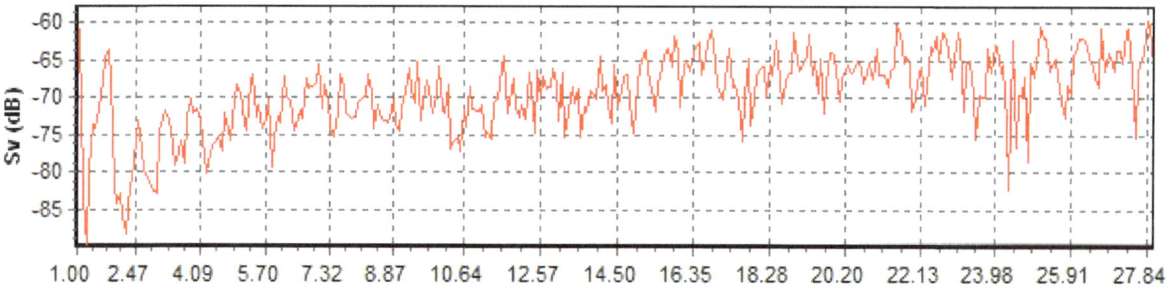

Figure 4: Acoustic volume backscattering strength (SV) of fish schools.

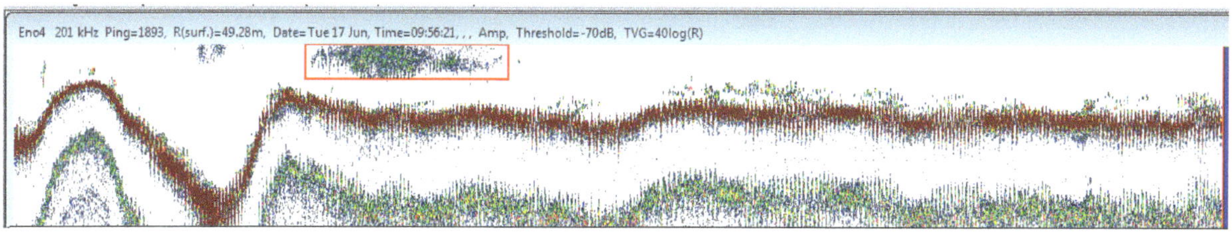

Figure 5: Acoustic image of underwater target and seabed profile.

Acoustic backscattering by the seafloor had long been studied in order to either predict the performance of instrument systems or to use sound to quantitatively detect and map the seafloor. The scattering is influenced by the roughness of the interfaces between the water and bottom and sub bottom layers as well as inhomogeneities [5-9]. There are both continuously varying inhomogeneities and discrete ones. Rocks, shells, and gas pockets are among the discrete inhomogeneities.

In this section, we measured backscatter strengths using underwater acoustic instrument with weight percentage of the sediment fraction and the number density of benthic macro-fauna present in the sea bottom. The establishment of traditional statistical techniques for the understanding of spatial and temporal dependence of benthic habitat on the acoustic backscatter has been proposed in the past [10].

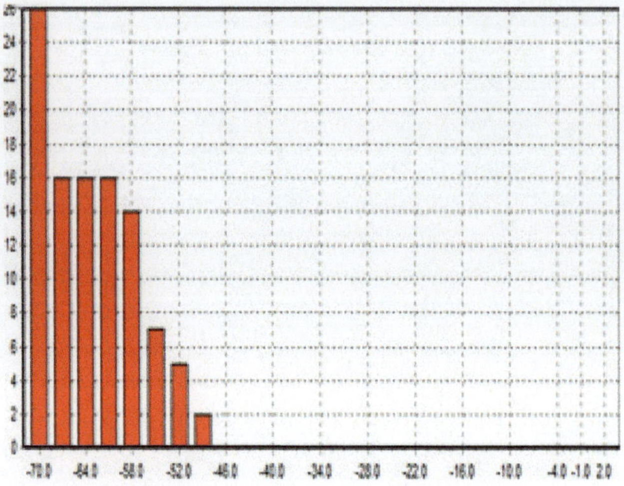

Figure 6: Histogram of Acoustic Target Strength.

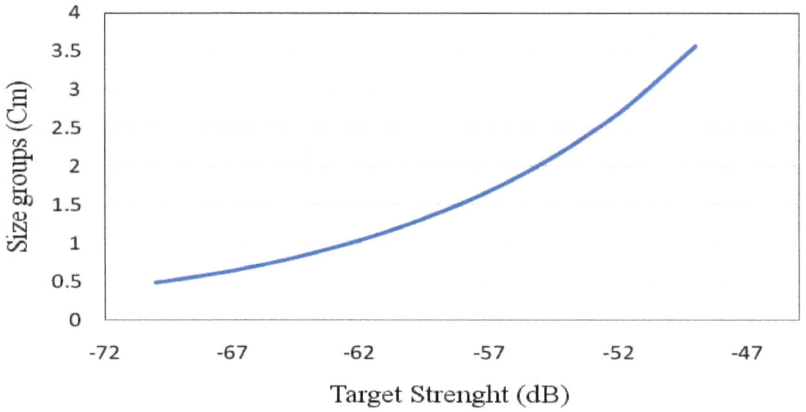

Figure 7: Acoustic target strength and fish size.

Size group (TS dB)	-70.0- 43.0	-70.0- -67.0	-67.0- -64.0	-64.0 - -61.0	-61.0- -58.0	-58.0- -55.0	-55.0- -52.0	-52.0- -49.0	-49.0- -46.0	-46.0- -43.0
SED number	56	5	4	12	13	4	4	8	3	3
Volume Density SED (/1000m³)	28	3	3	6	3	3	4	2	2	2
Volume Density Total (/1000m³)	235	19	24	52	57	15	15	33	10	10
Area density SED (/ha)	2896	255	333	689	752	189	189	239	125	125
Area density total (/ha)	28146	2889	2583	4884	5608	3150	1150	2016	2433	3433

Table 2: Acoustic density in the study area.

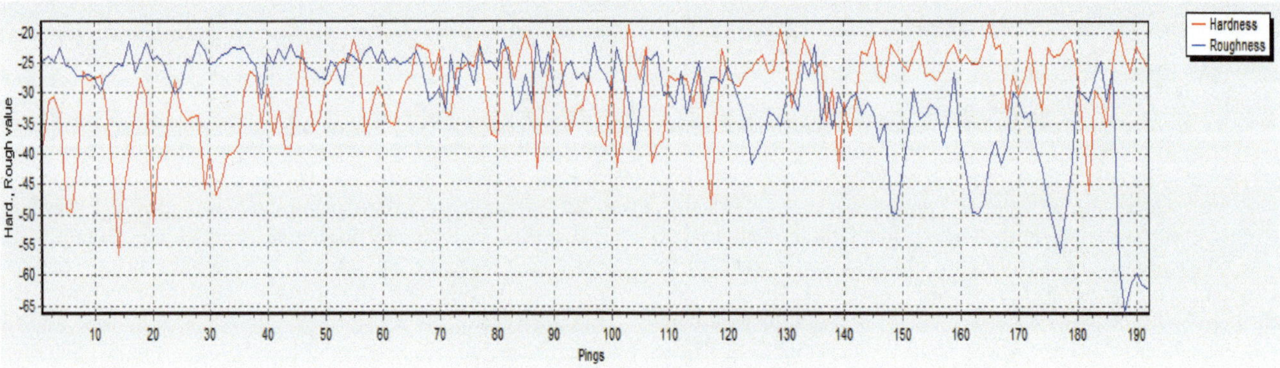

Figure 8: Hardness and roughness of sea bottom.

Station No.	Water Depth (m)	Laboratory measured of Grain size	Sand (%)	Silt (%)	Clay (%)	Sediment type
1	20	2-4 μm	0.75	20.98	78.27	Clay
2	25	4-63 μm	0.55	75.85	23.60	Silt
3	18	4-63 μm	0.65	75.90	23.45	Silt
4	32	1.5 mm	55.45	32.40	12.15	Silty sand
5	34	60 μm	0.18	75.53	24.29	Clayey silt
6	15	4-63 μm	0.95	79.08	19.97	Silt
7	18	63 μm - 2 mm	89.13	8.87	2.00	Sand
8	22	1.8 mm	73.65	22.45	3.90	Silty sand
9	24	1.9 mm	55.85	38.54	5.61	Silty sand
10	25	1.5 mm	80.75	15.01	4.24	Sand

Table 3: Summary of the ground-truth data with the percentage composition of each sediment type.

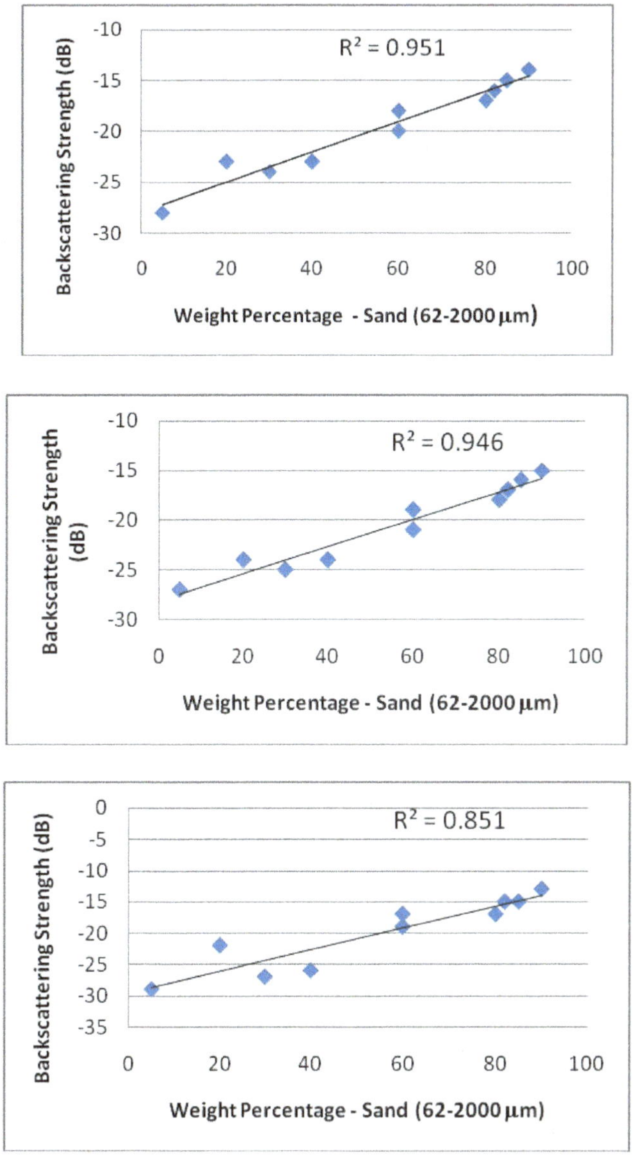

Figure 9: Acoustic backscatter of sand sea bottom for 50, 120, 200 kHz.

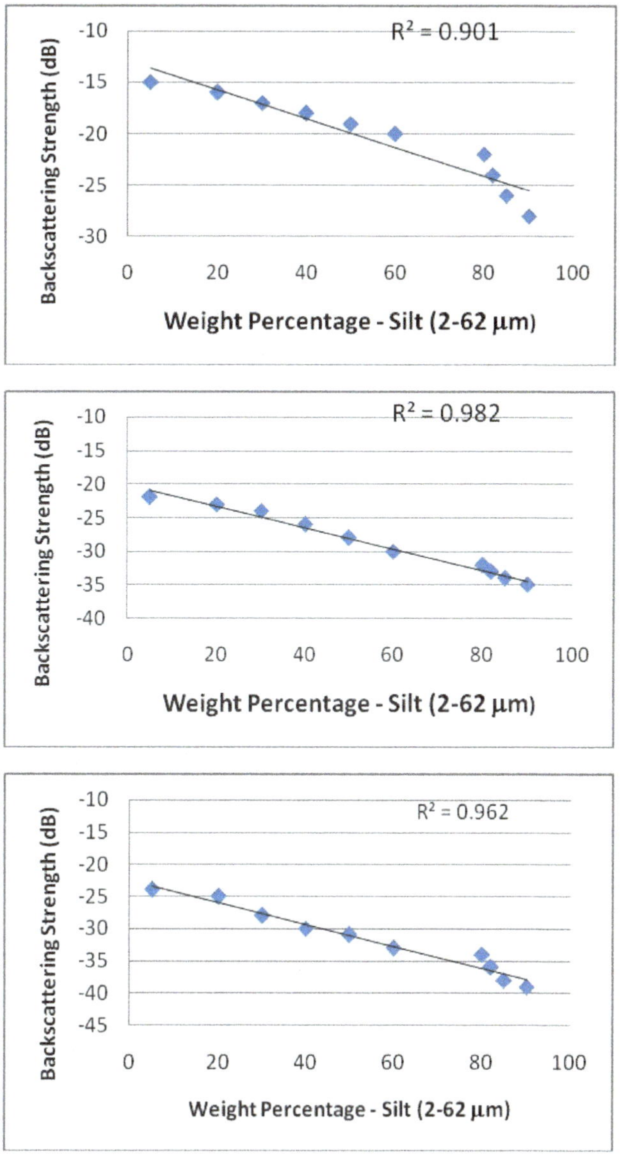

Figure 10: Backscattering strength for silt sea bottom for 50, 120, 200 kHz.

The percentage distribution of sediment compositions based on Shepard's classification [11] shows the presence of four seafloor sediment types: clayey-silt, silt, silty-sand and sand with varied levels of mixing of three textural grades of sand, silt, and clay. Sediment texture was relatively coarse in the deeper depths (20-35 m). Fine-grained sediment was located at the shallow depth region (15-25 m). Silty-sand and sand sediments will be referred to as coarse sediments; and clayey-silt and silt, and clay sediments will be referred to as fine sediments.

Backscatter strength data for three frequencies were compared with weight percentage of the individual grain size classes namely, sand, silt, and clay. The correlation coefficient R^2 between bottom backscattering strength and grain size classes is shown in Figures 9-11. For three frequencies, the backscattering strength is directly correlated with coarse fractions (sand within the range 62-2000 mm) and inversely correlated with finer fractions namely silt (2-62 mm) and clay (62 mm). The relationship between backscattering strength and the weight percentage of the sand fraction is shown in Figure 9. As the percentage of the sand fraction increases, the backscattering strength also increases linearly. The relationship between backscatter strength and the weight percentage of the silt and clay is expressed in Figures 10 and 11, and shows that the backscattering strength decreases with increasing weight percentage of both silt and clay fraction.

Sound wave interaction with the seabed depends partly on the impedance contrast between two layers. Acoustic impedance is a medium characteristic equal to the product of the density and sound speed. Large acoustic impedance contrast between water and rocky seabed with a considerable smooth surface means that the seabed surface behaves as perfect reflector. The value of acoustic impedance at softer sediments, mismatch is much less which means that larger energy will be able to penetrate to sub bottom layer. The signal encounters a different material and a portion of the acoustic energy is reflected and recorded by the system. The percentage of the acoustic energy reflected at each layer surface is a function of the relative densities, sound speeds, bottom material type, and the angle of incidence at the two layers.

Several studies have compared backscatter responses to ground-truth sediment data in order to assess the ability of different acoustic technologies to classify seafloor types [7,12-16]. The backscatter strength from a muddy seafloor has been shown to be inversely linearly related to the percentage content of silt and clay [17,18]. Fine sediments generally exhibit low backscatter strength due to low density and sound velocity [15,19]. However, the spatial variability of backscatter intensity along the seafloor characterized by coarse sediments has been shown to be mainly driven by the weight percent of coarse grains (sand) [10,20]. Coarse sediments are more likely to result in higher backscatter intensity due to scattering from coarse particles, lower porosity, higher density and sound velocity, and greater roughness of the water- sediment interface. The results of this study agree with those of Anderson in that backscatter strength and the proportion of the coarse fraction in sediments are strongly related [21,22]. In this study, a linear relationship between weight percentage of sand and backscatter intensity is also evident.

The present study for fine sea bottom, indicates an inverse linear relationship between percentage content of silt and clay. This findings were agreed with a previous study [23,24]. Benthic macro fauna present in the seafloor can affect backscatter in several ways. Hard-bodied fauna may individually be a discrete scatterer [25]. The benthic organism can influence seafloor roughness, and density or sound speed in the sediment are fluctuated. Benthic organism will contribute to acoustic backscattering by the seafloor. These animals can influence dominate the volume and seafloor reverberations [8,26]. The acoustic energy can penetrate into the sand sediment and scattered depends on the sizes of these buried inhomogeneities. For higher backscattering strength in coarse sediment where the benthic organism such as hard body organisms dominate. This is not a frequency related issue, as the single beam acoustic measurements are at 50, 120, and 200 kHz. At high frequencies acoustic, the backscattering from the seabed can generally be attributed to two contributing factors. Part of the energy is scattered by the interface relief and by bottom roughness. The other part of the energy penetrates to the sand and muddy sediment were reflected back by volume heterogeneities [2,21,22].

The number density of both hard and soft body organisms were compared with weight percentage of the individual grain size classes. Species groups analyzed by location is depicted in Table 4. The correlation coefficient R^2, which assesses the animal-sediment relationship, is shown in Figures 12 and 13.

The number density of hard body organism is directly correlated with coarse fractions and inversely correlated with finer fractions namely silt and clay bottom. The relationship between number density of hard body organism and the weight percentage of the sand fraction is shown in Figure 12. As the percentage of the sand fraction increases, the number density of hard body organism also increases linearly. Meanwhile, for fine region the number density of hardbody organism decreases with increasing weight percentage of both silt

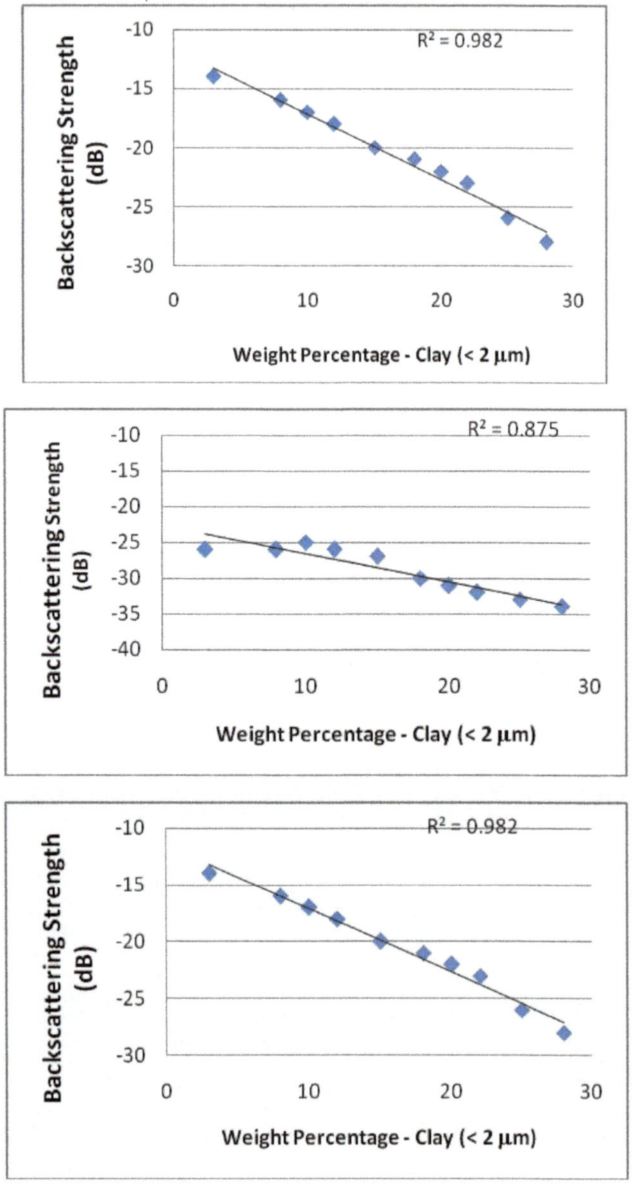

Figure 11: Backscattering strength for clay sea bottom for 50, 120, 200 kHz.

Location	Species Group						Density (number)
	Crustacea	Echinodermata	Mollusca	Polychaeta	Bivalvia	Gastropoda	
1	10	15	22	18	16	9	90
2	3	18	43	163	187	201	615
3	5	21	49	89	102	204	470
4	28	182	80	132	194	253	869
5	8	165	179	198	203	302	1055
6	12	198	152	203	246	198	1009
7	23	43	167	235	287	378	1133
8	15	46	98	327	23	14	523
9	18	126	23	213	312	341	1033
10	12	187	123	213	224	225	984

Table 4: Species groups analyzed by location.

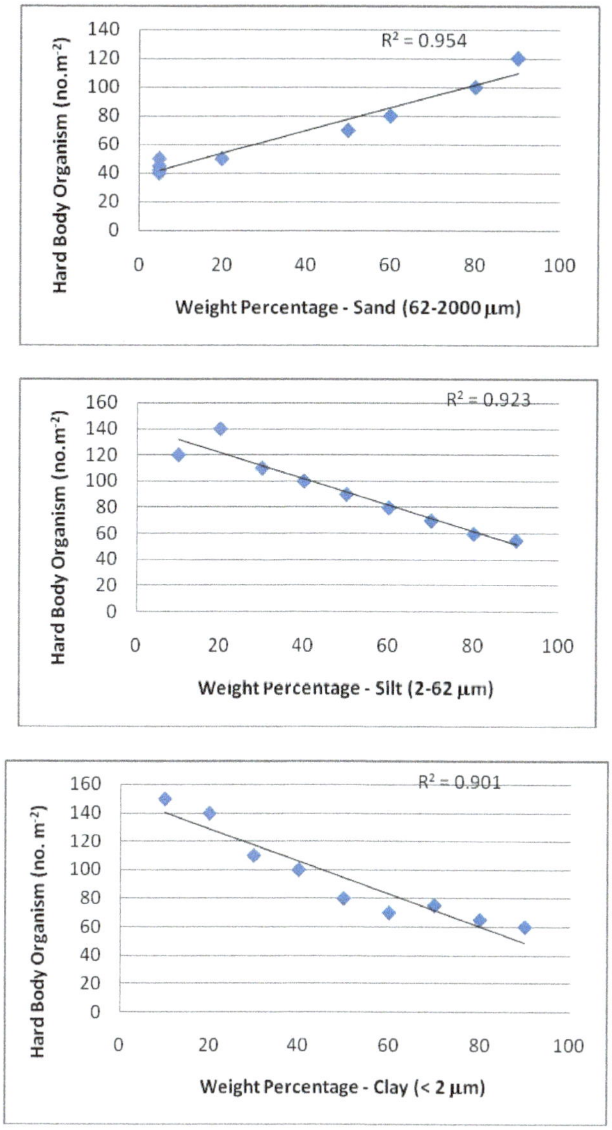

Figure 12: Correlations showing the relationship between the benthic macro-fauna and weight percentage of sediment types for hard body organism.

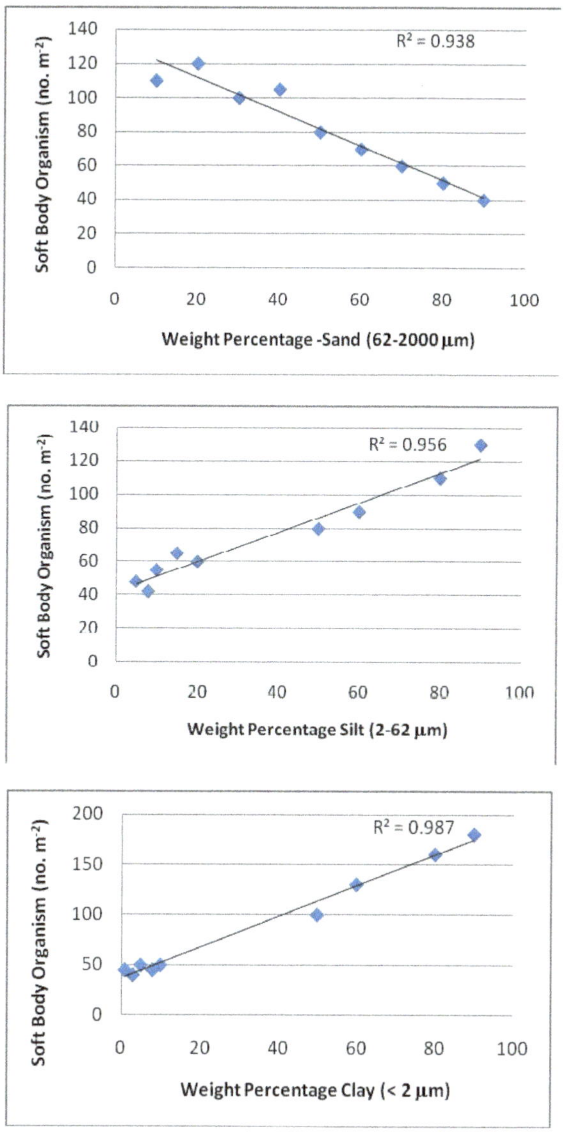

Figure 13: Correlations showing the relationship between the benthic macro-fauna and weight percentage of sediment types for hard body organism.

and clay fraction (Figure 13). Conversely, the number density of soft body organism is directly correlated with both silt and clay fraction sand inversely correlated with coarse fractions. The correlations show that there could be many factors in combination that influence the distribution of benthic organism. There were five groups of benthic organism found like crustacean, echinodermata, mollusca, polychaeta, bivalvia, and gastropoda. Hardbody organism such as bivalves and gastropods, while the soft body such as crustacean, echinodermata,

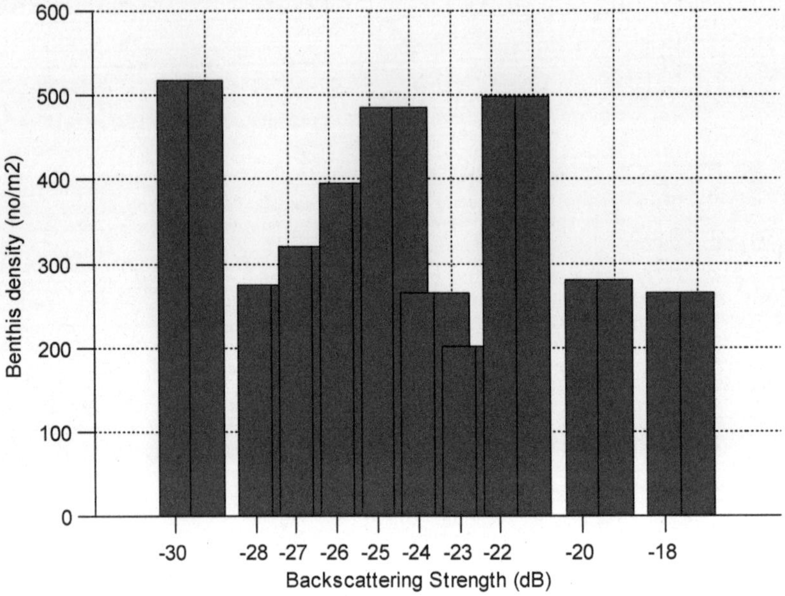

Figure 14: Relation backscattering strength and benthic density.

mollusca, and polychaeta. Jackson et al. and Jumars et al. have reported that fine-sediment such as clayey regions are not a favorable substratum for filter feeders [10,20]. Bloom et al and Bax et al had suggested that the sand bottom where the filter feeders dominate, reflects the more pronounced under water current activity and brings more potential food to filter feeding organisms than would weaker currents [27,28]. Conversely, fine sediment reflects the environment with feeble currents, which allow the fine particles to settle, so that only a small amount of organic matter in suspension is available as food for filter feeders, which in turn prevents them from inhabiting such environment and providing an adequate source of nutrition for deposit feeders only [29]. Relation of backscattering strength of sea bottom and benthic habitat is shown in Figure 14. From this figure, the value of backscatter was fluctuated in the presence of benthic organism. There have been very few controlled experiments involving acoustic backscattering by the seafloor in regions where there is a significant presence of benthic organism. Two such studies were published by Jackson et al. and Stanic et al. [10,30]. In the Jackson et al. study, the acoustic scattering by the seafloor was measured as a function of grazing angle, acoustic frequency, and seafloor type. One of the seafloor types involved a bottom material that consisted of very fine sand with a dense covering of live shellfish [8]. The scattering by the bed that contained the shellfish was elevated relative to the section of seafloor that contained sandy silt and no shellfish, indicating that the shellfish played a significant role in the scattering. In the studies by Stanic et al. the studies were focused entirely on a region where the seafloor was covered with shells and the acoustic scattering was measured as a function of grazing angle and acoustic frequency. Characterization of the shells was made possible through the use of samples collected at the site [30]. In the future, the knowledge on macrobenthic diversity is necessary to identify priority areas for conservation and manage human activities in such coastal zones using underwater acoustic method.

Conclusions

Remote sensing using underwater acoustic technology is useful for detection and quantification of marine fish and seabed characterization. From the sea bottom profile of the region, the depth of the neighboring sea of Seribu Islands particularly in Pari Island and Pramuka Island ranged approximately 15-55 m. Coral reefs in these areas were deep-rooted in the muddy bottom. The number of marine biota such as fish and zooplankton were measured using underwater acoustic effectively. We found the increasing of fish size is followed by the increasing of Target Strength.

The roughness and hardness were measured to characterize the sea bottom. The interrelationship between backscatter, grain size, and benthic macro-fauna abundance were demonstrated based on the acoustic data from three frequencies (50, 120, and 200 kHz) of single beam acoustic in conjunction with the sediment grain size and benthic macro-fauna information. The sediment textural properties and benthic macro-fauna information were collected using a grab for ground truth data. The findings of this study show that the different seabed types of the Pramuka Island of Jakarta can be mapped using a combination of data, including acoustic backscatter strength, biomass analysis, target strength, zooplankton, and ground-truth sediment information. The acoustic backscattering data presented here provide information about the sediment type and also give information about the occupancy of the organisms present in the seafloor. It can be deduced that a given location is suitable for occupancy by a given type of organism.

Acknowledgment

The author would like to thank SEAMEO BIOTROP Indonesia for financial support for this research project. I finally express my profound gratitude to editor and two anonymous reviewers of for their suggestions to improve the clarity of this manuscript.

References

1. Holliday DV (2007) Theory of sound scattering from the seabed, ICES Cooperative Research Report 286: 7-28.
2. Lurton X (2010) An Introduction to Underwater Acoustics: Principles and Applications, Springer Praxis Books.
3. Simmonds J (2007) Survey design for acoustic seabed classification. ICES Cooperative Research Report, 286: 145-152.
4. Folk RL (1968) Petrology of Sediment Rocks. Hemphills, Austin, Texas, pp.177

5. Medwin H, Clay C (1998) Fundamental of Acoustical Oceanography. Academic Press.

6. Ogilvy JA (1991) Theory of Wave Scattering from Random Rough Surfaces, Institute of Physics Publishing, Bristol, UK.

7. Urick RJ (1983) Principles of underwater sound. New York, McGraw-Hill, pp.423

8. Jackson DR, Winebrenner DP, Ishimaru A (1986) Application of the composite roughness model to high-frequency bottom backscattering. J. Acoust. Soc. Am 79: 1410-1422.

9. Jackson DR, Briggs KB, "High-frequency bottom backscattering: Roughness vs. sediment volume scattering," J. Acoust. Soc. Am 92: 962-977.

10. Jackson DR, Williams KL, Briggs KB (1996) high-frequency acoustic observations of benthic spatial and temporal variability. Geo-Marine Letters 16: 212-218.

11. Shepard FP (1954) Nomenclature based on sand-silt- clay ratios. Journal of Sedimentary Petrology 24: 151-158.

12. Stewart WK, Chu D, Mali S, Lerner S, Singh H, (1994) Quantitative seafloor characterization using a bathymetric sidescan sonar. IEEE Journal of Oceanic Engineering 19: 599-610.

13. Fonseca L, Mayer L, Orange D, Driscoll N (2002) The high frequency backscattering angular response of gassy sediments: model/data comparison from the Eel River Margin, California. Journal of the Acoustical Society of America 111: 2621-2631.

14. Parnum IM, Siwabessy PJW, Gavrilov AN (2004) Identification of seafloor habitats in Coastal Shelf waters using a multibeam echosounder. In: Acoustics. Gold Coast: Proceedings of the Annual Conference of the Australian Acoustical Society.

15. Manik HM, Furusawa M, Amakasu K (2006) Measurement of Sea Bottom Surface Backscattering Strength by Quantitative Echo Sounder. Fisheries Science 72: 503-512.

16. Manik HM (2012) Seabed identification and characterization using sonar. Advances in Acoustics and Vibration 2012: 1-5.

17. Briggs KB, Richardson MD (1997) Small-scale fluctuations in acoustic and physical properties in surficial carbonate sediments and their relationships to bioturbation. Geo-Marine Letters 17:306-315.

18. Briggs KB, Williams KL, Jackson DR, Jones C D, Ivakin AN et al. (2002) Fine-scale sedimentary structure: implications for acoustic remote sensing. Marine Geology 182: 141-159.

19. Greenlaw CF, Holliday DV, McGehee DE (2004) High-frequency scattering from saturated sand sediments. JASA 115: 2818-2823.

20. Jumars PA, Jackson DR, Gross TF, Sherwood C (1996) Acoustical remote sensing of benthic activity: A statistical approach. Limnol Oceanogr 41: 1220-1241.

21. Anderson JT (2006) Report of the study group on acoustic seabed classification (SGASC).ICES Fisheries Technology Committee, pp.7.

22. Anderson JT, Holliday V, Kloser R, Reid D, Simard Y (2007) Acoustic seabed classification of marine physical and biological landscapes. ICES Cooperative Research Report, 286: 198.

23. Boyle FA, Chotiros NP (1995a) A model for high-frequency acoustic backscatter from muddy sediments. J. Acoust. Soc. Am 98: 525-530.

24. Boyle FA, Chotiros NP (1995b) A model for high-frequency acoustic backscatter from gas bubbles in sandy sediments at shallow grazing angles. J. Acoust. Soc. Am 98: 531-541.

25. Stanton TK (2000) On acoustic scattering by a shell-covered seafloor. See comment in PubMed Commons below. J Acoust Soc Am 108: 551-555.

26. Ferrini VL, Flood RD (2006) The effects of fine-scale surface roughness and grain size on 300 kHz multibeam backscatter intensity in sandy marine sedimentary environments. Marine Geology 228: 153-172.

27. Bloom SA, Simon J, Hunter VD (1972) Animal sediment relationship and community analysis of a Florida estuary. Marine Biology 13: 43-56.

28. Bax N, Kloser R, Williams A, Gowlett-Holmes K, Ryan, T (1999) Seafloor habitat definition for spatial management in fisheries: a case study on the continental shelf of the southeast Australia. Oceanologica Acta 22: 705-719.

29. Kenny AJ, Cato I, Desprez M, Fader G, Schu¨ttenhelm RTE et al. (2003) An overview of seabed-mapping technologies in the context of marine-habitat classification. ICES J Mar Sci 60: 411-418.

30. Stanic S, Briggs KB, Fleischer P, Sawyer WB, Ray RI (1989) High-frequency acoustic backscattering from a coarse shell ocean bottom. J. Acoust. Soc. Am. 85:125-136.

2

Morphology of the Niger Delta: Local Facies Belts Orientation versus Depobelts and Growth Fault Orientations

Durogbitan A Abimbola*

Petrfac IES, Research and Development, Manchester, UK

Abstract

This work presents for the first time, evidence and the relationship between the ancient coastal belt and the modern coastal belt in terms of local facies belts orientation versus depobelt and fault orientations within the Niger Delta. This observation suggests that the ancient coastal belt of the delta is more lobate /arcuate than the modern Niger delta. The delta became broadly convex to the sea during the late Miocene.

Deltas are influenced by a variety of fluvial and marine processes and these processes controlled their morphology. The most widely used classification scheme today is that of Galloway who subdivided delta according to their dominant processes i.e. rivers, waves and tide. Fluvially dominated deltas tend to display lobate-elongate morphology, e.g. the Balize delta- Mississippi. In contrast wave-dominated deltas tend to be more lobate and have smooth, arcuate to cuspate margins, e.g. the Nile delta. Tide dominated deltas tended to be estuarine to irregular in geometry. Although the Niger delta has been characterized to be mixed influenced delta showing the combination of the effects of river, wave and tidal processes, the modern Niger delta has been described to be more wave dominated and more lobate, and the classification falls within the wave dominated section on Galloway classifications. This implies that during Miocene, the delta was more fluvially dominated in relation to sediment supply and basinal processes (wave and tide). This also implies that the modern Niger delta cannot be used as a direct analogue for the ancient delta.

Keywords: Niger delta; Depobelts; Facies belts

Introduction

The morphology of the Niger Delta can be subdivided into an early stage, spanning the Paleocene to early Eocene, and a later stage of delta development beginning in Miocene time [1]. Early coastlines were concave to the sea and depositional patterns strongly influenced by basement topography [1]. Delta progradation has been assumed to occur along two major axes [2]. The first paralleled the Niger River, where sediments supply exceeded subsidence rate. The second which is smaller than the first became active basinward of the Cross River during the Eocene to Oligocene. Late stages of deposition said to have began in the early to middle Miocene, as these separate eastern and western depocentres merged. In the late Miocene the delta prograded and shorelines became broadly convex into the sea. Rapid sediment loading by this delta progradation mobilized underlying unstable shales (Akata Formation). These shales rose into diapiric walls, deforming overlying strata. The resulting complex deformation structures caused local uplift, which has resulted in major erosion events into the leading progradational edge of the Niger Delta [3] (Figure 1).

The Cenozoic Niger Delta complex has been described as an arcuate-lobate shape by many authors (e.g [3-5]). Structural analysis has identified a number of faults within the study area. The growth faults are listric in cross section and concave to the basin in plan-view. The Niger delta has been subdivided into seven mega units called depobelts [3]. Which are self-entities with respect to stratigraphy, structure-building and hydrocarbon distribution. Each depobelt is parallel to the modern coastal belt in which has been distinguished by their ages and are defined by sedimentary faults. These faults are also observed to be parallel to these regional depobelts and the present day facies belts (Figures 2 and 3).

Sesimic and Sequence Stratigraphy

This section summarised a seismic and sequence stratigraphic framework (studied area) for a growth-faulted delta deposits of the Agbada Formation (Middle Miocene) within the Ewan and Oloye Fields, northwestern Niger delta, (Figure 4). The application of sequence stratigraphic methods to Niger Delta sequences is complex

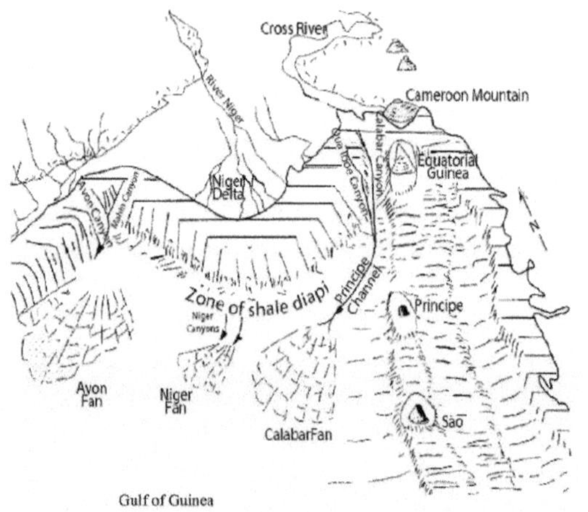

Figure 1: Physiographic sketch of the deep marine sediments in the Gulf of Guinea offshore Niger Delta [2].

***Corresponding author:** Durogbitan A Abimbola, Principal Geoscientist, Petrfac IES, Research and Development, Manchester, UK
E-mail: adewoledurogbitan@yahoo.co.uk

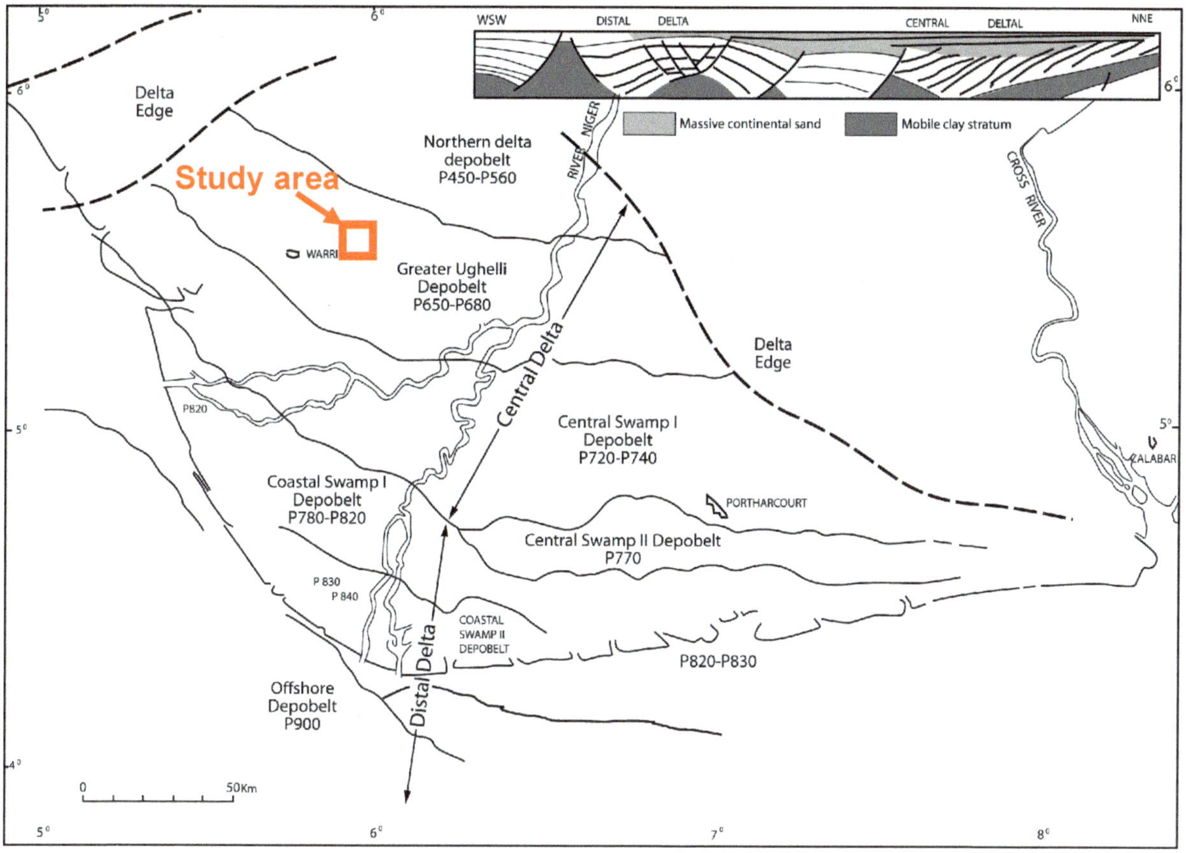

Figure 2: Age of deltaic sequences in depobelt and relationship to the broad changes in tectonic style [3].

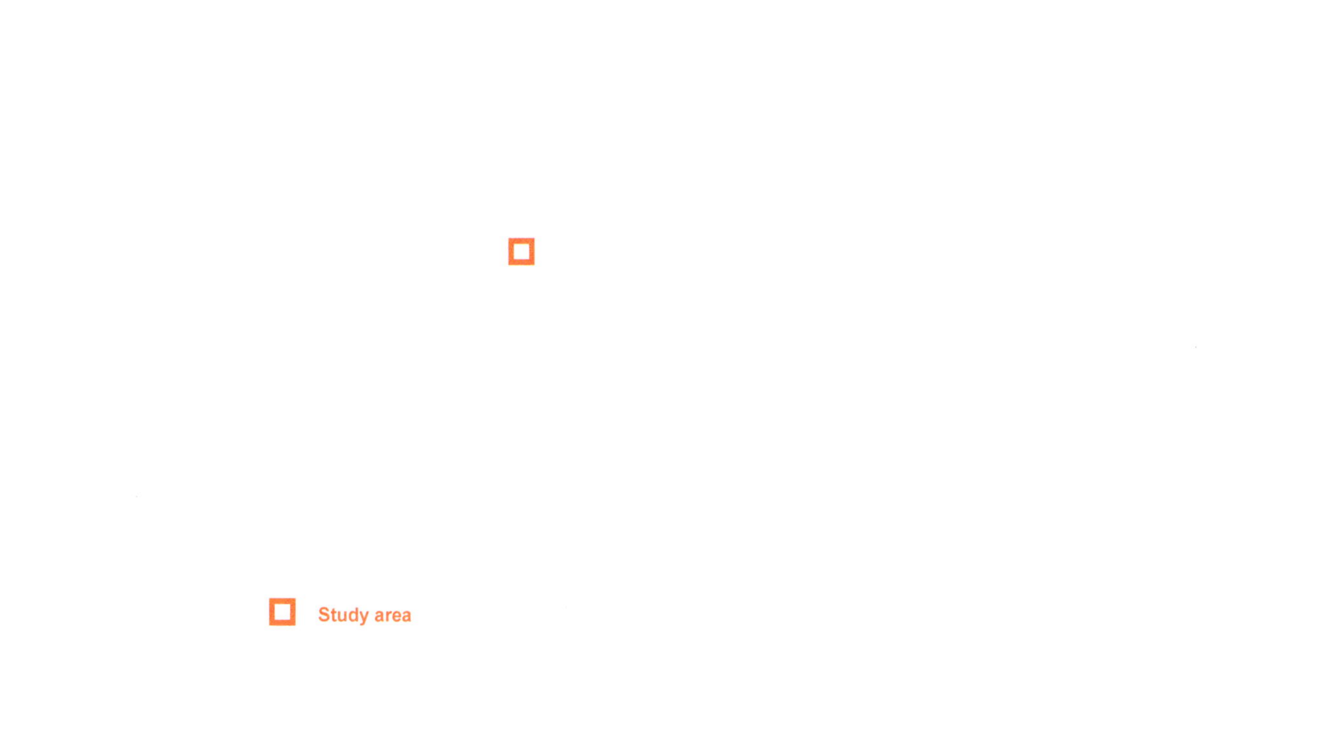

Figure 3: Growth faults and known hydrocarbon accumulation [14].

because the key surfaces, strata geometry or stacking patterns are difficult to identify. These complexes are due to changes in relative sea level which are strongly influenced by the interplay between shoreline advance and retreat, concomitant isostatic loading and rebound of the continental shelf growth faulting and basin physiography.

Detailed sequence stratigraphic analyses showed that the middle Miocene-early Pliocene strata in the Ewan and Oloye fields of the northwestern Niger delta consist of six depositional sequences [6] (Figures 5-8). In general, sequences are widespread. They are made up of progradational HST, TST are thin but widespread, and LST are restricted to within valley fills. Outside the LST valleys, transgressive surfaces and sequence boundaries are coincident on interfluves. This

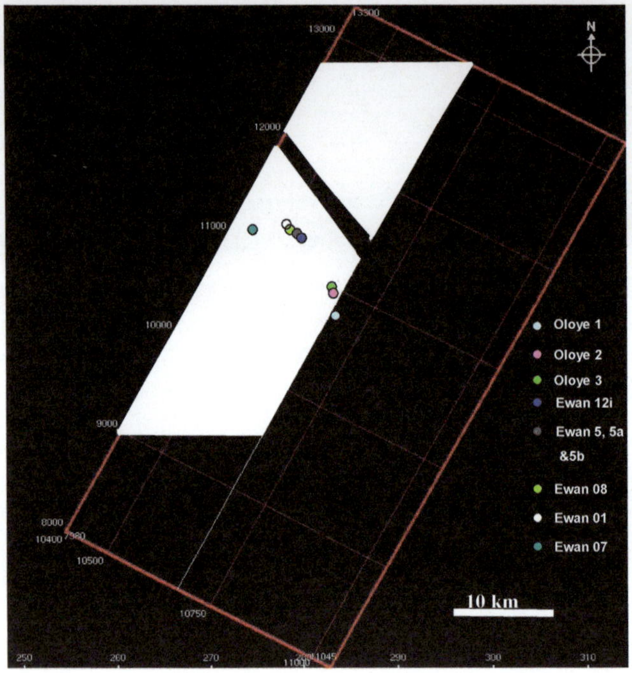

Figure 4: Base map of seismic survey showing position of seismic cross sections and wells. Area under study is enclosed within the white shaded portion.

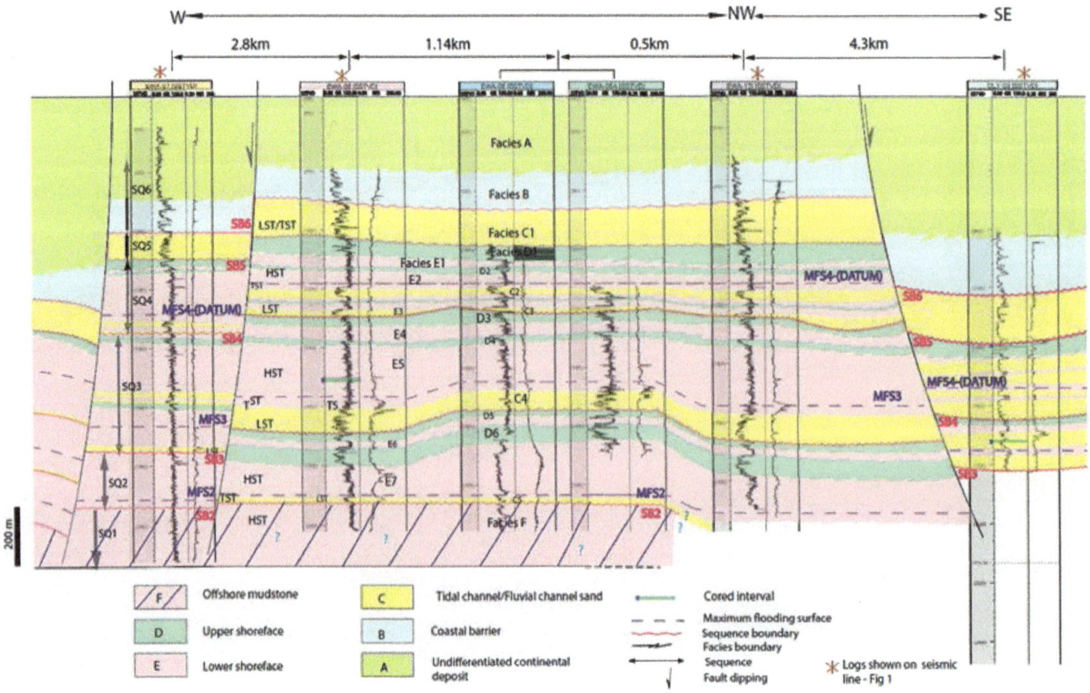

Figure 5: Correlation panel of Ewan and Oloye logs illustrating sequences component, parasequence setd and facies within the middle Miocene Agabda Formation [6].

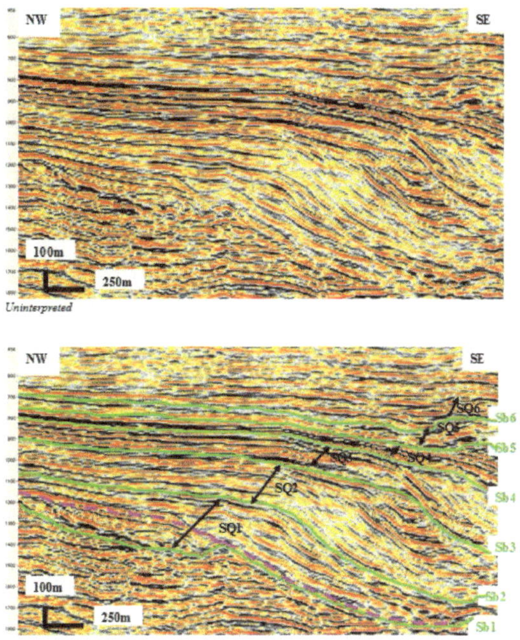

Figure 6: Seismic section showing identified sequence boundaries and transgressive surface (Crossline 12291- upper patches). Transgressive surfaces are denoted by light purple dashed line. Generally shown progradation stacking pattern, most of the incisions occurred at the shoreline break (s). See position in Figure 4.

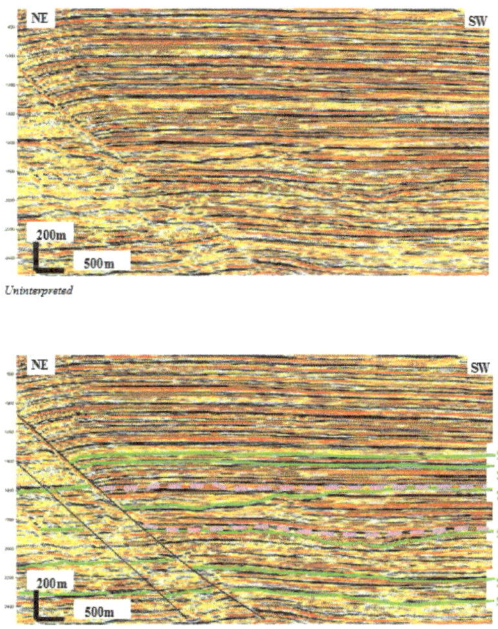

Figure 7: Seismic section showing interpreted sequence boundaries and the transgressive surfaces (inline 10620 - lower patches). Transgressive surfaces are denoted by light purple dashed line. The reflecting pattern at the lower patches is more aggradational compared with the northern part. See position in Figure 4.

suggests that during LST times most of the area was periodically subaerially exposed. During lowstand period, shorelines and facies belts prograde basinward (south) and there are development of canyons which serve as a conduit for high amplitude deposit (deep water sand) on the slope and within the basin floor [6,7]. This mechanism is interpreted to have led to the deposition of considerable amount of sandstones basinward (E, F, G;[6]).

Individual sequences have been mapped across the study area and biostratigraphic data has allowed a correlation with regional biozonal ages for the Niger delta complex [6]. However, there is strong local growth fault control on sequence variability. Further local controls on sequence geometry are shelf instability and slope collapse structures which have a strong geomorphological impact on the area [6]. The slope scars act as a locus for lowstand incision and affect overall sequence thickness. The incisions are narrower and deeper in lower sequences whereas those at the higher sequences are broader and less deeply incised. The additional accommodation space above slope scars allows local development of thick HST with seismically resolvable clinoforms. The occurrence of the several incisions in the middle Miocene succession is interpreted as evidence of significant relative sea level fluctuations, and the presence of type-1 sequence boundaries may be the stratigraphic signature of major drops in relative sea level during Miocene and Pliocene.

By and large, sequences that developed within the study area are seen to be controlled locally by episodic shelf instability and fault growth, with relative sea level providing major controls on sequence development and deep channel incisions [6].

Discussion

Structural and seismic facies analysis of the study area shows that

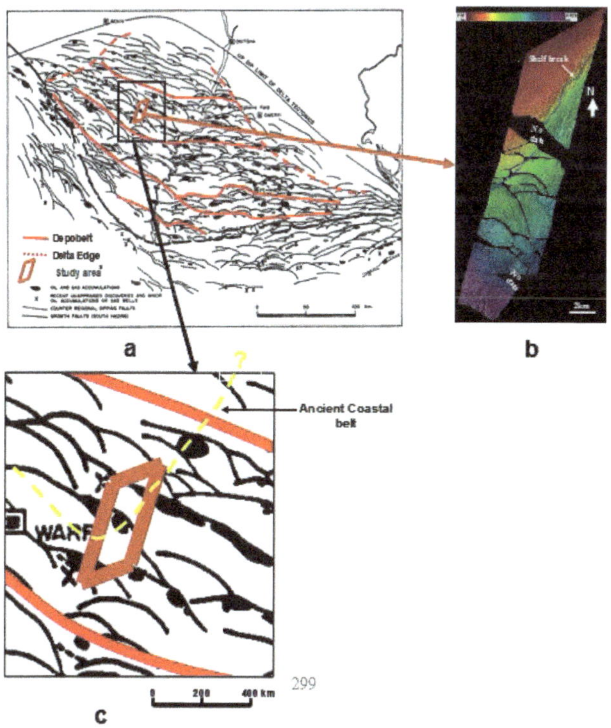

Figure 8: (a) Schematic relationship between the regional depobelt and fault orientations (Modified from [3,14]). (b) Depth structure map of one of the interpreted key surfaces from the study area- Ewan and Oloye Fields (sequence boundary 4- Horizon 1. Local facies belts are observed to be oblique to the modern regional depobelts and the interpreted growth faults, compared with the present day coastal belt which is parallel to the depobelts. (c) Enlarged section of the study area (from 8a), illustrating relationship between the local facies belt, depobelt and growth fault.

the local facies belts have a different orientation in relation to the growth faults, and the depobelts mapped using the high resolution 3D seismic integrated with available well data, clearly do not mirror the present day orientation. From this detailed study, local facies belts are observed to be oblique to the modern regional depobelts and the interpreted growth faults, compared with the present day coastal belt which is parallel to the depobelts.

This work presents for the first time, evidence and the relationship between the ancient coastal belt and the modern coastal belt in terms of local facies belts orientation versus depobelt and fault orientations within the Niger Delta. This observation suggests that the ancient coastal belt of the delta is more lobate /arcuate than the modern Niger delta. The delta became broadly convex to the sea during the late Miocene.

Deltas are influenced by a variety of fluvial and marine processes and these processes controlled their morphology [8-10]. The most widely used classification scheme today is that of Galloway [8], who subdivided delta according to their dominant processes i.e. rivers, waves and tide (Figure 9). Fluvially dominated deltas tend to display lobate-elongate morphology, e.g. the Balize delta- Mississippi. In contrast wave-dominated deltas tend to be more lobate and have smooth, arcuate to cuspate margins, e.g. the Nile delta. Tide dominated deltas tended to be estuarine to irregular in geometry [8] (Figure 10). Although the Niger delta has been characterized to be mixed influenced delta showing the combination of the effects of river, wave and tidal processes [3,11,12]. The modern Niger delta has been described to be more wave dominated and more lobate [1], and the classification falls within the wave dominated section on Galloway classifications. This implies that during Miocene, the delta was more fluvially dominated in relation to sediment supply and basinal processes (wave and tide) (Figure 11). This also implies that the modern Niger delta cannot be used as a direct analogue for the ancient delta [13,14] (Figure 12).

Conclusion

In summary, seismic, sequence stratigraphic and depositional models developed for the Ewan and Oloye fields, enable detailed facies

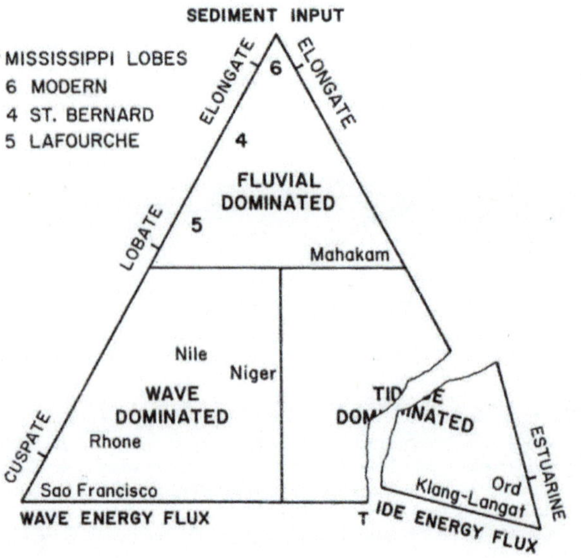

Figure 9: Classification of deltas in terms of river, wave and tide influence [8,10].

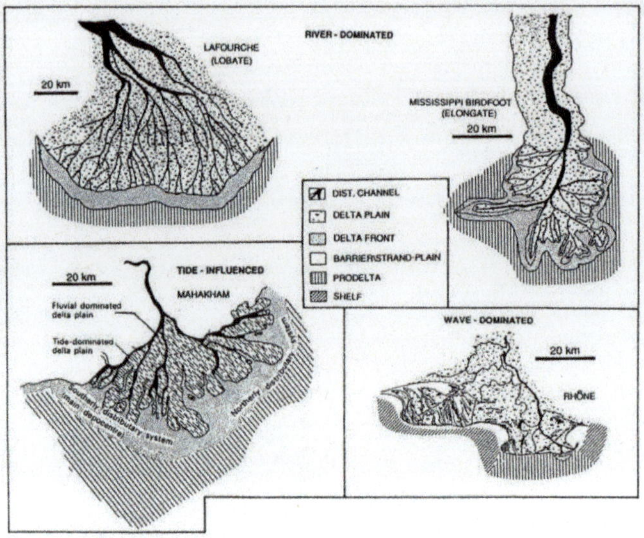

Figure 10: Representative modern examples of river-dominated, wave-dominated and tide influenced deltas [9,10,13].

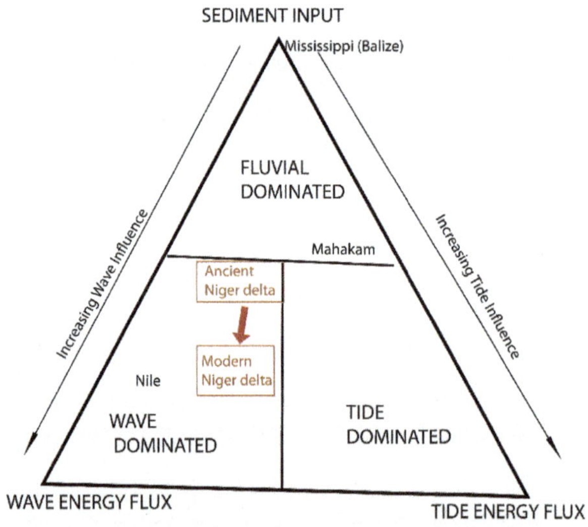

Figure 11: Schematic tripartite diagram showing relationship of Ancient Niger delta (Miocene) to present day Niger delta as observed from the study area [8,10].

correlation on the field scale, and offer predictive models that can be used on a regional scale

The implication of this interpretation is that the modern Niger delta cannot be used as a direct analogue for the ancient delta (from the point view of sedimentary process). The data from the study area suggests that the ancient delta was more lobate with more complex facies distribution, and the delta was more fluvially influenced, wave dominated in relation to sediment supply and basinal processes fluvial input. It can be inferred and interpreted from the study that the ancient delta is characterized by lowstand shelf edge while the present day modern delta are of high stand shelf edge, because there was a believed that relative sea level over 20,000 yrs ago was at high stand. This defined major difference between the present day delta and the ancient delta in term of shelf edge location.

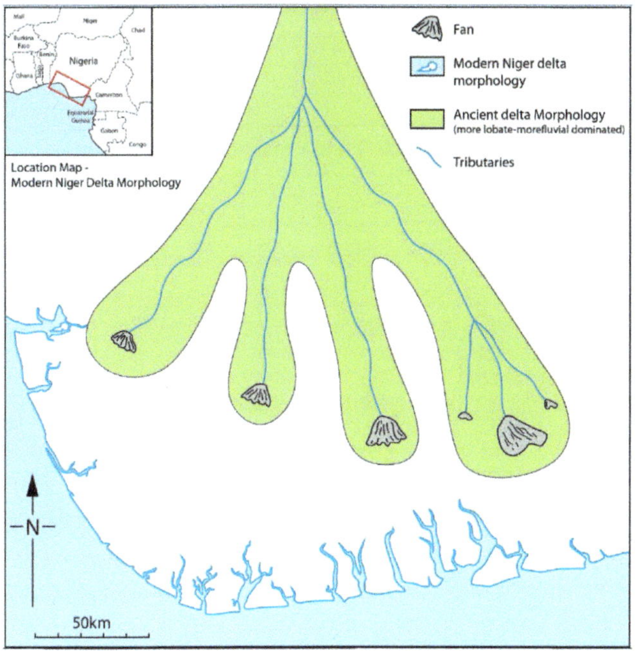

Figure 12: Conceptual model and Schematic morphology of the ancient Niger Delta.

References

1. Doust HE, Omatsola (1989) Niger Delta: American Association of Petroleum Geologists Memoir 48: 210-238.
2. Reijers TJA, Petters SW, Nwajide CS (1997) The Niger delta basin in RC Selley, African basins-sedimentary basins of the world, Amsterdam, Elsevier Science pp: 151-172.
3. Doust HE, Omatsola (1990) Niger delta in Edwards JD and Santogrossi PA divergent/passive margin basin AAPG 48: 201-238.
4. Weber KJ, Daukoru EM (1975) Petroleum Geology of the Niger Delta. Proc. Ninth World Petrol. Congress 2: 209-221.
5. Evamy BD, Haremboure J, Kimberling P, Knapp WA, Molloy FA, et al. (1978) Hydrocarbon habitat of tertiary Niger delta. AAPG Bull 62: 1-39.
6. Durogbitan AA (2010) Seismic, sequence stratigraphic and structural analysis of Ewan and Oloye fields, Northwestern Niger Delta. Earth Science Department, University of Manchester pp: 337.
7. Durogbitan AA, Gawthorpe RL, Redfern J (2008) Seismic straigraphy and development of deep fluvial incisions in the middle miocene deposits, Ewan and Oloye Fields, Northwestern Niger Delta. Proceeding of NAPE Annual Convention and Exhibition.
8. Galloway WE (1975) Process framework for describing the morphologic and stratigraphic evolution of deltaic depositional systems. In Broussard ML, Deltas, models for exploration: Houston, TX, Houston Geological Society pp: 87-98.
9. Bhattacharya JP, Walker GR (1992) Deltas. In: Walker GR and James PN. Facies models: response to sea level change. Geological Assoc of Canada pp: 157-177.
10. Walker RG (1992) Facies models and modern stratigraphic concepts. In Walker RG and James NP, Facies Models: Response to seal level changes. Geological Assoc of Canada pp: 1-14.
11. Allen JRL (1965) Late quaternary Niger delta and adjacent areas- sedimentary environments and lithofacies: AAPG Bull 49: 547-600.
12. Allen JRL (1970) Sediments of the modern Niger delta: a summary and review. In: Morgan JP, Deltaic sedimentation, modern and ancient. SEPM Spec Publ 15: 138-151.
13. Fishers WL, Brown LF, Scott AJ, McGowen JH (1969) Delta system in the exploration for oil and gas, a research colloquium, Austin, TX, Texas Bureau of Economic Geology p: 204.
14. Whiteman AJ (1982) Nigeria: Its petroleum geology, resources and potential. London, Graham & Totman 1: 166.

3

Diversity of Phytoplankton in Different Stations from Muthupettai, South East Coast of India

Babu A*, Varadharajan D, Vengadesh Perumal N, Thilagavathi B, Manikandarajan T, Sampathkumar P and Balasubramanian T

Faculty of Marine Sciences, Annamalai University, Parangipettai -608 502, Tamil Nadu, India

Abstract

The distributions of phytoplankton and its community structure in relation to physico-chemical parameters in aquaculture discharge waters and adjacent waters of Muthupettai lagoon were studied. Physico-chemical parameters and nutrients concentration were fluctuated depends on the seasonal chances, whereas the total of 101 phytoplankton species were recorded, Among them, 76 species of diatoms (Bacillariophyceae), 17 species of dinoflagellates (Dinophyceae), 5 species of blue greens (Cyanophyceae), 2 species of green algae (Chlorophyceae) and 1 species of silicoflagellates (Chrysophyceae) were recorded, where, *Nitzschia seriata, Coscinodiscus centralis, Thalassiothrix frauenfeldii,* and *Ceratium furca* were the most abundant in phytoplankton community. The percentage composition were recorded as Diatom>Dinoflagellates>Blue greens>Greens>Silicoflagellate respectively. The study revealed that there are no effects due to the aquaculture discharge waters for the mangrove ecosystem. This study also revealed seasonal environmental variables and phytoplankton distributions in Muthupettai lagoon.

Keywords: Phytoplankton; Muthupettai; Lagoon; Nutrients; Aquaculture

Introduction

The global demand for seafood continues to rise despite most wild fisheries being at their maximum level of sustainable exploitation [1]. Predictions expected that by 2030, more than 50% of fisheries production will need to come from aquaculture due to human population growth, continuing demand for seafood and static or declining fish harvests [2]. In tropical and subtropical coastal areas worldwide, no economical activity has evolved as quickly as shrimp farming in the last 15 years. However, such an enormous development has been accompanied by strong controversies on the environmental, economic and social impacts of shrimp farming [3]. Phytoplankton can be use as good bio-indicator of water quality including pollution, since they reflect even the suitable changes taking place in their environment by changing their species composition, biomass, community structure and productivity. More over the abundance of phytoplankton can be taking as the best mean for quantitative assessment of potential fisheries of an area [4]. Thus, pond discharges are potential sources of pollution in receiving waters. There is a need for a more comprehensive approach to assessing the influence of shrimp farm discharge on the surrounding environment, with ability to predict changes in the environment following further changes in farm management practices. Therefore, in the present study has been undertaken to study the phytoplankton density and distribution in relation to physico-chemical parameters at Muthupettai, for a period of one year from April 2009-March 2010.

Materials and Methods

The present study was carried out the all the shrimp farm of Muthupettai, Station-1, situated in the outlet of the pond. Station- 2, situated in the farm discharges mixing in the back water area (junction of Kandakruchnar and Koraiar). Station- 3, situated in the mouth of the Muthupettai lagoon (Lat 10° 20' 58N: Long 79° 32' 07E), (Lat 10° 19' 47N: Long 79° 33' 07 E), (Lat 10° 18' 44 N; Long 79° 31' 32E) is bounded by a part of the Bay of Bengal and the northeast and Palk Strait on the Southwest and which embraces as vast swamp area (Figure 1). Three different points where selected at equidistance downstream from the farm, where the discharge water from a shrimp farm gradually mixes with the river water and flows into mouth of the Muthupettai lagoon. The shrimp farm was a pump fed type and stocked with the black tiger shrimp *Penaeus monodon*. All the shrimp farmers used CP dry pellets feed. A total of 75 shrimp farms, one side are surrounded by the mangrove fields and the other side had wasteland and cultivable lands with paddy fields. The farm has a total area of 150 acres with 18 excavated ponds including two reservoirs. The water spread area of the farm ranged from 0.7 to 0.9 ha. The depth of the pond water body was approximately 1.5m. The selected study areas are given bellow. Surface

Figure 1: Map showing the study area.

***Corresponding author:** Babu A, Faculty of Marine Sciences, Centre of Advanced Study in Marine Biology, Annamalai University, Parangipettai-608 502, Tamil Nadu, India, E-mail: babumarine@gmail.com

water samples were collected at monthly interval from the stations 1 and 2 for a period of one year from April, 2009 to March, 2010, for the estimation of various physico-chemical parameters. Rainfall data were obtained from the meteorological unit (Govt. of India.) located at Thiruvarur. Temperature was measured using a standard centigrade thermometer. Salinity was estimated with the help of a refractometer (ERMA, Hand Refractometer, and Japan) and pH was measured using an ELICO Grip pH meter. Dissolved oxygen was estimated by the modified Winkler's method and is expressed as ml/l. For the analysis of nutrients, surface water samples were collected in clean polyethylene bottles, kept immediately in an icebox and transported to the laboratory [5]. The collected water samples were filtered by using a Millipore filtering system and analysed for dissolved inorganic nitrate, nitrite, phosphate and reactive silicate adopting the standard procedures described by Strickland and Parsons [5]. Phytoplankton samples were collected of monthly intervals from the surface water by towing a plankton net (mouth diameter 0.35 m) made up of bolting silk cloth (N.0.3 mess size- 48 μm) for half an hour. The collected samples were preserved in 5% neutralized formalin and used for qualitative analysis for the quantitative analysis of phytoplankton the settling method described by was adapted and the numerical plankton analysis was carried out using light microscope [6]. For the quantitative analysis of phytoplankton, 500 ml of water was filtered through a bagnet of same mesh size and the numerical plankton analysis was carried out using a binocular microscope. Phytoplanktons were identified by using the standard works done [7-12]. Biodiversity indices were calculated following the standard formulae: Biodiversity: H'=-S Pi log Pi; I=1; richness: D=1-C; C=SPi^2; Pi=ni/N and evenness: J'=H'/log^{2}S [13-15]. A calendar year was divided into four seasons viz., Monsoon (October to December), Postmonsoon (January to March), summer (April to June) and Premonsoon (July to September) based on the northeast monsoon which is prevalent in the study area. Correlation coefficients (r) were calculated for phytoplankton density and physico-chemical parameters and the Analysis of Variance (F) tests were made for hydrological parameters in relation to stations and seasons.

Results and Discussion

Total rainfall (mm)

Total rainfall of 1672 mm was recorded from April 2009 to March 2010. It varied from 13 mm to 561 mm and rain did not occur in the months of June 2009, February and March 2010 (Figure 2). Rainfall is the most important cyclic phenomenon in tropical countries as it brings important changes in the hydrographical characteristics of the marine and estuarine environments. In the present study, the peak values of rainfall were recorded during the monsoon month of October 2009. The rainfall in India is largely influenced by two monsoons viz., southwest monsoon on the west coast, northern and northeastern India and by the northeast monsoon on the southeast coast [16]. On the other hand tidal rhythm, water current and evaporation in summer produced only little variation in those parameters or more or less stable in the absence of rainfall. Maruthanayagam C and Subramanian P have also reported the bulk of rainfall during northeast monsoon season along the southeast coast of India [17].

Water temperature (°C)

The surface water temperature fluctuated from 22 to 31. The maximum temperature was recorded during summer and minimum temperature during monsoon at all the stations (Figure 3) Generally, surface water temperature is influenced by the intensity of solar radiation, evaporation, freshwater influx and cooling and mix up with ebb and flow from adjoining neritic waters. The water temperature during October was low because of strong land sea breeze and precipitation and the recorded high value during summer could be attributed to high solar radiation [18,19]. The observed spatial variation in temperature could be due to the viable intensity of prevailing streams and the resulting mixing of water [20]. Statistically analysis showed a positive correlation (r=0.712 at station 1 and r=0.812 at station 2) between atmospheric and surface water temperature for both stations.

Salinity (psu)

The salinity acts as a limiting factor in the distribution of living organisms, and its variation caused by dilution and evaporation is most likely to influence the fauna in the coastal ecosystem [21]. The salinity ranged from 1 to 34 psu. The highest salinity was recorded during summer season (April and May, 2009) and least salinity recorded during monsoon (October 2009) at all the stations (Figure 4 and Table 3). Generally, changes in the salinity in the brackish water habitats such as estuaries, backwaters and mangroves are due to the influx of freshwater from land run off, caused by monsoon or by tidal variations. This is presently evidenced by the negative correlation (r=-0.696 at station 1 and r=-0.746 at station 2) obtained between salinity and rainfall. Further, salinity also influenced by the higher temperature

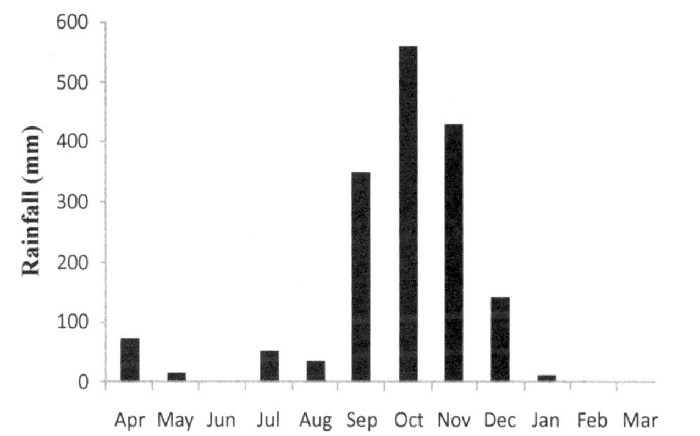

Figure 2: Showing rainfall changes in during the year of April 2009- March 2010.

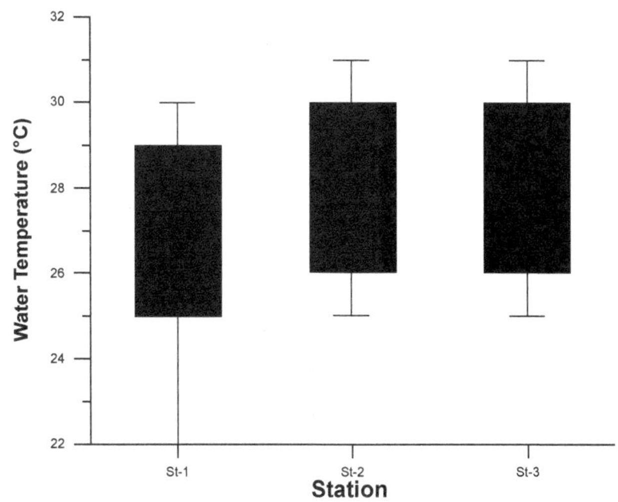

Figure 3: Water temperature at different sampling stations at Muthupettai lagoon.

as is evident from the obtained significant positive correlation with temperature. The salinity was found to be high during summer season and low during the monsoon season at all the stations. The recorded higher values (34.0%) could be attributed to the low amount of rainfall, higher rate of evaporation and also due to neritic water dominance, as reported by earlier workers in other areas [22-24]. During the monsoon season (4.0%), the rainfall and the freshwater inflow from the land in turn moderately reduced the salinity, as reported by Mitra et al. [25] in the Godavari estuary in the Bay of Bengal and coastal waters of Kalpakkam [25-27].

Hydrogen ion concentration (pH)

Hydrogen ion concentration (pH) Hydrogen ion concentration (pH) in surface waters varied from 7.6 to 8.2. The maximum pH was recorded during summer and post monsoon and minimum pH during monsoon season at all stations (Figure 5 and Table 3). Generally, fluctuations in pH values during different seasons of the year is attributed to factors like removal of CO_2 by photosynthesis through bicarbonate degradation, dilution of seawater by freshwater influx, low primary productivity, reduction of salinity and temperature and decomposition of organic materials as stated by Subramanian and Mahadevan and Vijayakumar et al. [28,29]. The recorded high summer pH might be due to the influence of seawater penetration and high biological activity and due to the occurrence of high photosynthetic activity [30,31]. The statistical analysis also revealed that salinity show highly significant negative correlation with rainfall.

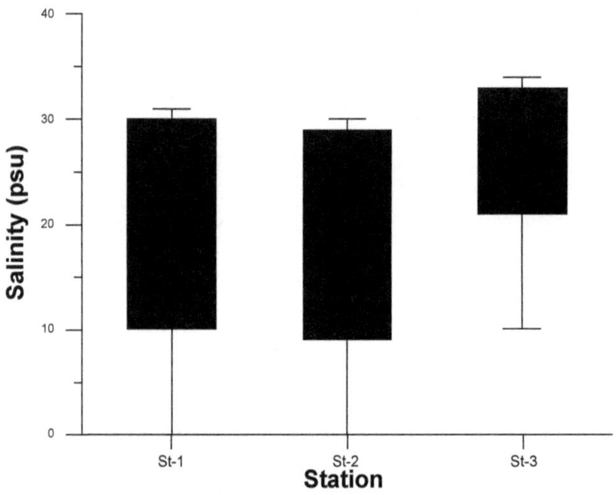

Figure 4: Salinity at different sampling station in Muthupettai lagoon.

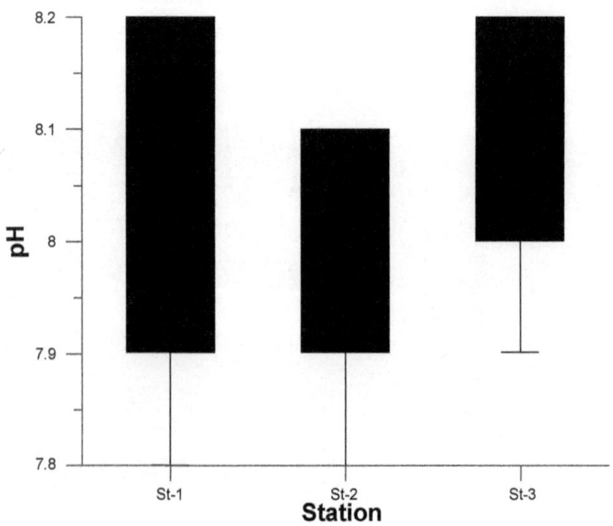

Figure 5: pH at different sampling station in Muthupettai lagoon.

Dissolved Oxygen

The variation in dissolved oxygen concentration varied from 2.21 to 6.65. The maximum values were recorded during monsoon season and minimum values were recorded during summer season at station 1 (Figure 6 and Table 3). It is well known that the temperature and salinity affect the dissolution of oxygen [32]. In the present investigation, higher values of dissolved oxygen were recorded during monsoon months at all the stations. Season-wise observation of dissolved oxygen showed an inverse trend against temperature and salinity. The observed high monsoonal values might be due to the cumulative effect of higher wind velocity coupled with heavy rainfall and the resultant freshwater mixing [33]. Segar and Hariharan [34] have mainly attributed seasonal variation of dissolved oxygen to freshwater flow and terrigenous impact of sediments. Further, significant inverse relationship between rainfall and nutrients indicated that freshwater flow constituted the main source of the nutrients in the estuaries and lagoons.

Micronutrients

Nutrients are considered as one of the most important parameters in the estuarine and lagoon environment influencing growth, reproduction and metabolic activities of living being. Distribution of nutrients is mainly based on the season, tidal conditions and freshwater flow from land source.

Ammonia

The ammonia concentration fluctuated from 0.962 to 8.098, 0.162 to 3.174, and 0.051 to 0.187 at stations 1, 2 and 3 respectively. The maximum values were recorded during summer (stations 1) and minimum values during post monsoon season (station 3). Higher concentration of ammonia (8.098 μM) was observed during the summer season in station 1 due to aquaculture pond discharge water mix the station. Lower concentration of ammonia (0.051 μM) was observed during the post monsoon season in station 3 (Figure 7). In station 2 and 3 higher concentration could be partially due to the death and subsequent decomposition of phytoplankton and also due to the excretion of ammonia by planktonic organisms [34].

Dissolved inorganic nitrate

The dissolved inorganic nitrate concentration ranges from 1.42 to 6.151, 1.557 to 4.243, and 2.571 to 5.231 at stations 1, 2 and 3 respectively. The maximum values were recorded at stations 1 during post monsoon and minimum values were recorded at station 1 during monsoon season. It could be mainly due to the organic materials received from the pond discharge (Figure 8). In station 3 the maximum nitrates values recorded at monsoon season the increased nitrates level was due to fresh water inflow, mangrove leaves (litter fall) decomposition and terrestrial run-off during the monsoon season [28]. Another possible way of nitrates entry is through oxidation of ammonia form of nitrogen to nitrite formation [25]. The recorded low values (1.292 μM) during non-monsoon period may be due to its utilization by phytoplankton as evidenced by high photosynthetic activity and also due to the neritic water dominance, which contained only negligible amount of nitrate [19,28].

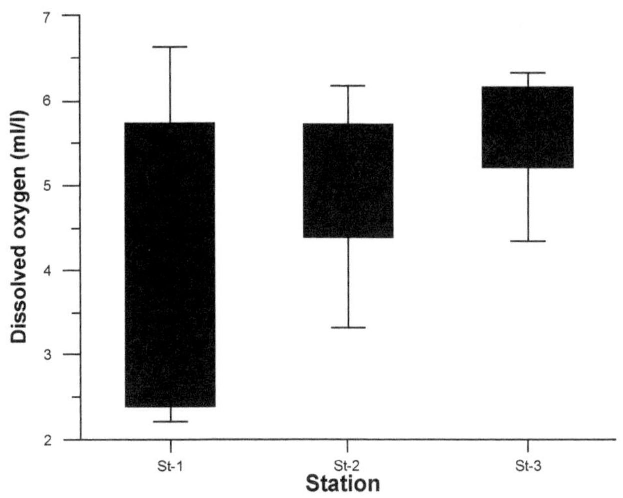

Figure 6: DO at different sampling station in Muthupettai lagoon.

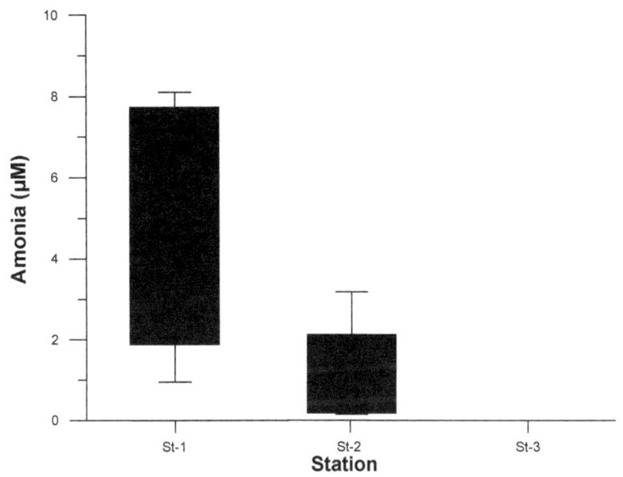

Figure 7: Ammonia at different sampling station in muttupettai lagoon.

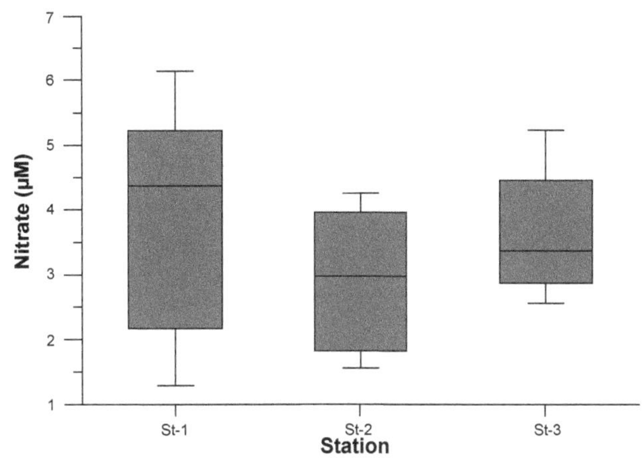

Figure 8: Nitrate at different sampling station in Muthupettai lagoon.

Dissolved inorganic nitrite

The dissolved inorganic nitrite concentration values ranged from 0.389 to 3.331 at station 1, 0.162 to 2.574 at station 2, and 0.301 to 0.861 at station 3. The maximum value was recorded at station 1 during post monsoon and minimum value at station 2 during monsoon. The recorded higher nitrite values during post monsoon season (3.331 µM) could be due to the increased pond discharge, oxidation of ammonia and reduction of nitrate and by recycling of nitrogen and also due to bacterial decomposition of planktonic detritus present in the environment [17]. Further, the denitrification and air-sea interaction exchange of chemicals are also responsible for this increased value (Figure 9 and Table 3) [30]. The recorded low nitrite value (0.162 µM) during summer and monsoon seasons may be due to high freshwater inflow and low salinity [31,32].

Dissolved inorganic phosphate

The dissolved Inorganic phosphate concentration ranges varied from 1.161 to 1.684, 0.159 to 0.921, and 0.157 to 0.687 at stations 1, 2, and 3 respectively. The maximum values were recorded during post monsoon at station 1 and minimum values at station 3 during summer season. The recorded high concentration of inorganic phosphates (1.684 µM) during post monsoon season might possibly be due to pond discharge, in station 2 and 3 the high values recorded at monsoon season due to monsoonal fluctuation, which in turn increased the level of phosphate (Figure 10) [35]. Further, regeneration and release of total phosphorus from bottom mud into the water column by turbulence and mixing also attributed to the higher monsoonal values [36]. The recorded low phosphates value (0.157 µM) during summer could be attributed to the limited flow of freshwater, high salinity and utilization of phosphate by phytoplankton [37]. The variation may also be due to the processes like adsorption and desorption of phosphates and buffering action of sediment under varying environmental conditions [23]. Moreover, the weatherings of rocks soluble alkali metal phosphates, the bulk of which are carried into the estuaries are also responsible for the recorded higher values [22]. The addition of super phosphates applied in the agricultural fields as fertilizers and alkyl phosphates used in households as detergents can be other sources of inorganic phosphates during the season [18].

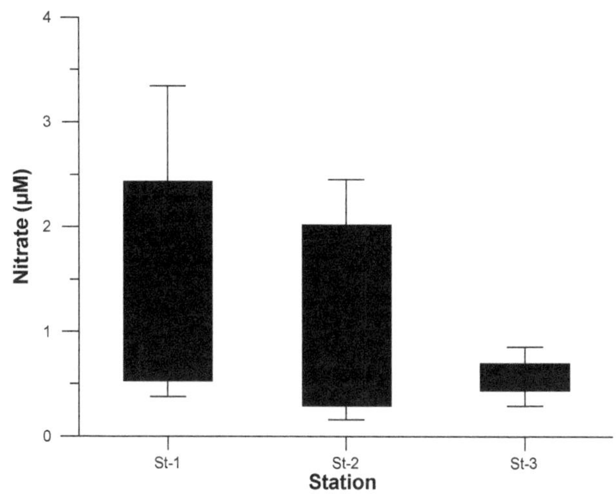

Figure 9: Nitrite at different sampling station in Muthupettai lagoon.

Dissolved silicate

The dissolved silicate concentration ranged from 25.914 to 54.662, 24.147 to 50.857 and 25.515 to 65.705 at stations 1, 2, 3 respectively. The maximum values were recorded during monsoon season at station 3, and minimum values were recorded during summer at station 2. The silicate content was higher than that of the other nutrients (NO_3, NO_2 and PO_4) and the recorded high monsoon values (65.705 µM) may be due to heavy inflow of monsoonal freshwater derived from land drainage carrying silicate leached out from rocks. Further, due to the turbulent nature of water, the silicate from the bottom sediment might have been exchanged with overlying water in this estuarine and lagoon environment (Figure 11) [18]. Besides this, the dissolution of particulate silicon carried by the river, the removal of silicates by adsorption and co-precipitation of soluble silicate silicon with humic compounds and iron [21]. The observed low summer values (24.147 µM) could be attributed to uptake of silicates by phytoplankton for their biological activity [35,36].

In the present study a totally 101 species of phytoplankton were recorded during the study period. Among them, 76 species of diatoms (Bacillariophyceae), 17 species of dinoflagellates (Dinophyceae), 5 species of blue greens (Cyanophyceae), 2 species of green algae (Chlorophyceae) and 1 species of silicoflagellates (Chrysophyceae) were found. The maximum phytoplankton species was recorded in station III and minimum species was recorded in station I than there other station. Among the station I were recorded phytoplankton maximum contribution of Bacillariophyceae (44) followed by Dinophyceae (9), Cyanophyceae (2) and Chlorophyceae (1), as well as station II phytoplankton species was recorded maximum in Bacillariophyceae (61) followed by Dinophyceae (11), Cyanophyceae (3), Chlorophyceae (1) and Chrysophyceae (1) and the station III maximum species was recorded in Bacillariophyceae (70) following them Dinophyceae (15), Cyanophyceae (4), Chlorophyceae (2) and Chrysophyceae (1) (Table 1). Among various species, *Nitzschia seriata, Coscinodiscus centralis, Thalassiothrix frauenfeldii,* and *Ceratium furca* were the most abundant forms. In general, the distribution and abundance

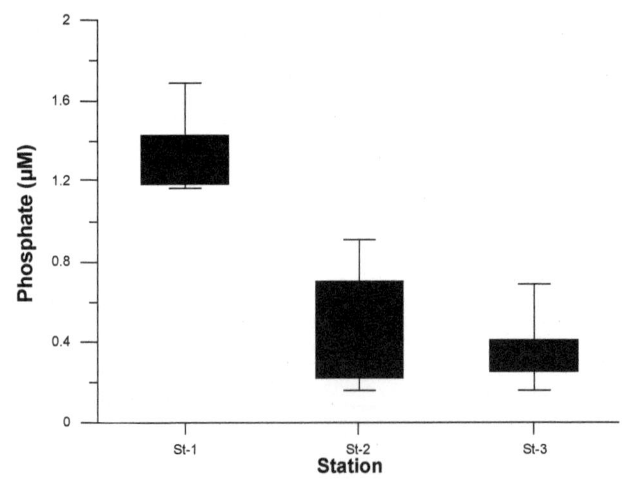

Figure 10: Phosphate at different sampling station in muttupettai lagoon.

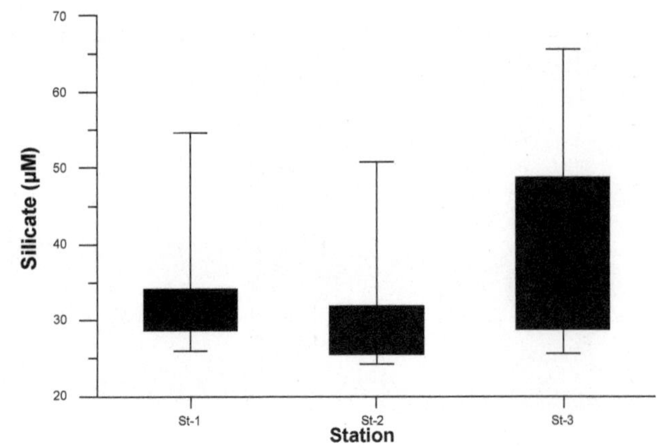

Figure 11: Silicate at different sampling station in muttupettai lagoon.

S.No	Bacillariophyceae (Diatoms)	St-1	St-2	St-3
1	*Asterionella glacialis*	*	*	+
2	*Bacillaria paradox*	*	+	+
3	*Bacteriastrum comosum*	*	+	+
4	*Bacteriastrum delicatilum*	+	+	+
5	*Bacteriastrum hylinium*	+	+	+
6	*Bellerochea malleus*	*	+	+
7	*Biddulphia biddulpia*	+	*	+
8	*Biddulphia heteroceros*	+	+	*
9	*Biddulphia obtuse*	*	*	+
10	*Biddulphia pulchella*	+	+	+
11	*Biddulphia reticulata*	*	+	+
12	*Biddulphia* sp.	+	+	+
13	*Chaetoceros affinis*	+	+	+
14	*Chaetoceros brevis*	+	+	+
15	*Chaetoceros coarctatus*	*	+	+
16	*Chaetoceros currivisetes*	+	*	+
17	*Chaetoceros danicus*	*	+	+
18	*Chaetoceros deblis*	+	*	*
19	*Chaetoceros decipiens*	*	+	+
20	*Chaetoceros diversus*	+	*	+
21	*Chaetoceros indicus*	*	+	+
22	*Chaetoceros messanensis*	+	*	+
23	*Chaetoceros peruvian*	+	*	+
24	*Climocodium fraunfeldium*	*	*	+
25	*Coscinodiscus centralis*	+	+	+
26	*Coscinodiscus concinnus*	*	+	+
27	*Coscinodiscus ecentricus*	*	+	+
28	*Coscinodiscus gigas*	+	+	+
29	*Coscinodiscus granii*	*	+	+
30	*Coscinodiscus radiatus*	+	*	+
31	*Coscinodiscus thori*	+	+	+
32	*Cyclotella* sp.	+	+	+
33	*Cyclotella striata*	*	+	+
34	*Diatoma anceps*	+	+	+
35	*Ditylum brightwelli*	*	+	+
36	*Eucambia groenlandica*	*	+	+
37	*Eucambia zoodiocus*	+	+	+
38	*Fragillaria* sp.	+	+	+
39	*Guinordia* sp.	+	+	+
40	*Gyrosigma* sp.	+	+	+
41	*Hemiaulus sinensis*	*	+	+
42	*Hemidiscus hardmanianus*	*	*	+
43	*Lauderia* sp.	+	+	+
44	*Leptocylindrus danicus*	+	+	+
45	*Lithodesmium undulatum*	*	+	+
46	*Navicula henneydii*	*	+	+

47	*Navicula* sp.	+	+	+
48	*Navicula venhoffeni*	*	+	+
49	*Nitzschia closterium*	+	+	+
50	*Nitzschia longissima*	*	+	+
51	*Nitzschia pungens*	+	+	+
52	*Nitzschia seriata*	+	+	+
53	*Nitzschia* sp.	+	+	+
54	*Odentella mobiliensis*	+	+	+
55	*Odentella sinensis*	*	+	+
56	*Planktoniella sol*	+	+	+
57	*Pleurosigma angulatum*	+	+	+
58	*Pleurosigma depressum*	+	+	+
59	*Pleurosigma directum*	*	*	+
60	*Pleurosigma elongatum*	+	+	+
61	*Pleurosigma normanii*	*	*	+
62	*Rhizosolenia alata*	+	*	*
63	*Rhizosolenia cylindrus*	+	+	+
64	*Rhizosolenia imbricata*	*	+	+
65	*Rhizosolenia robusta*	+	+	+
66	*Rhizosolenia setigera*	*	+	+
67	*Rhizosolenia styliformis*	+	+	+
68	*Skeletonema costatum*	+	+	+
69	*Stephanophysis palmariana*	*	+	*
70	*Streptothaeca indicus*	*	+	*
71	*Thalassionema nitzschioides*	+	+	+
72	*Thalassiosira* sp.	+	+	+
73	*Thalassiothrix fraunfeldii*	*	*	+
74	*Triceratium favus*	*	+	+
75	*Triceratium reticulatum*	+	+	+
76	*Triceratium robertsianum*	+	+	*
	Dinophyceae			
1	*Ceratium extensum*	+	+	+
2	*Ceratium furca*	+	+	+
3	*Ceratium fusus*	*	*	+
4	*Ceratium inflatum*	+	+	+
5	*Ceratium lineatum*	*	+	+
6	*Ceratium macroceros*	*	+	+
7	*Ceratium trichoceros*	*	+	+
8	*Ceratium tripos*	*	*	+
9	*Ceratocorys horrida*	+	*	*
10	*Dinophysis caudata*	*	+	+
11	*Noctiluca* sp	+	+	+
12	*Ornithocercus steinii*	+	*	*
13	*Prorocentrum micans*	*	*	+
14	*Protoperidinium depressum*	+	+	+
15	*Protoperidinium oceanicum*	+	+	+
16	*Protoperidinium* sp.	+	+	+
17	*Pyrophacus stenii*	*	+	+
	Cyanophyceae (Blue-greens)			
1	*Anabena* sp	+	+	+
2	*Trichodesmium erythraea*	*	+	+
3	*Spirulina major*	*	*	+
4	*Lyngbya* sp.	+	+	*
5	*Oscillatoria limosa*	*	*	+
	Chlorophyceae (Greens)			
1	*Volvox* sp.	+	+	+
2	*Pediastrum simplex*	*	*	+
	Chrysophyceae (Silicoflagellate)			
1	*Dichtyocha fibula*	*	+	+
	Total	**56**	**76**	**91**

(+ present &* absent)

Table 1: Study period recorded phytoplankton species from Muthupettai region.

of phytoplankton in tropical waters, varied remarkably due to the seasonal environmental fluctuations, and these variations are well pronounced in the sheltered system of estuarine waters. The percentage contribution of each group of phytoplankton was in the following order: Diatom>Dinoflagellates>Blue greens>Greens>Silicoflagellate.

The phytoplankton species composition was comparatively more in station 3 than in stations 1 and 2. Generally, diatoms were found to be dominant in Lagoon mouth waters, which could well thrive in widely changing hydrographical conditions [22,37-40]. Depending on several factors in coastal lagoons, there are dominant groups: diatoms, dinoflagellates, cyanobacteria, chlorophytes, phytoflagellates, silicoflagellates and euglenophytes. Among the most important variables for the dominance of a given group, there are sources of water supply (a river or the sea), salinity and dynamic aspects of a lagoonal system. Among chlorophytes, several genuses have been registered during the winter, when fresh water supply occurs.

Presently observed high population density and species diversity during summer and pre monsoon might be due to the predominance of diatoms viz: *Nitzschia seriata, Thalassiothrix frauenfeldii, Odontella sinensis, Bacteriastrum comosum, Chaetoceros affinis,* and *Coscinodiscus centralis*. The phytoplankton abundance during summer season could be attributed to the increased salinity, pH, high temperature and high intensity of light penetration during the season [31].

The abundance of phytoplankton was lowest during monsoon months, when the water column was remarkably stratified to a large extent because of heavy rainfall, high turbidity caused by run-off, reduced salinity, decreased temperature and pH, overcast sky and cool conditions. However, during this season, freshwater algal forms like *Anabaena* sp., *Oscillatoria* sp., *Chlorella* sp., *Nostoc* sp., *Lynbya* sp., *Spirogyra* sp. *Volvox* sp. and *Spirulina* sp. were noticed. The phytoplankton counts were high during southwest monsoon season as reported in some of the studies in Bay of Bengal [38]. Similar observations were made earlier [41-43]. This is being supported by who stated that phytoplankton and their growth depend on several environmental factors, which are variable in different seasons and regions [44]. This kind of cyclic change in the species composition of phytoplankton was a characteristic feature of the Pichavaram coastal mangroves [45].

The ranges of phytoplankton population density (cells/l) were: 5906-763284 (st- 1), 9064-185070 (st. 2) and 20949-104072 (st. 3) (Figure 12). The maximum density was recorded during summer and minimum during monsoon. The observed high density during the summer could be attributed to more stable hydrographical conditions prevailed during that period. Station 1 showed comparatively high population density due to high nutrient concentrations due to aquaculture pond discharge and optimal salinity [46]. The salinity showed positive correlation with zooplankton density (r=0.801 st. 1, r=0.649 st. 2 and r= 0.802 st. 3) [42-48]. Phytoplankton composition and abundance in supply water is modified in shrimp ponds. In some culturing systems, where salinity decreases because of the mixing with fresh water from rivers, there are ponds where diatoms, cyanobacteria, chlorophytes and dinoflagellates dominate, depending on several environmental factors (e.g. light, salinity, and temperature and nutrient levels). The occurrence of some species can be temporal or can last longer. Sometimes there are blooms of short periods, but a very high abundance of one or a few species that can alter shrimp growth due to oxygen depletion at nights [47].

The ranges of species diversity, richness and evenness at stations

1, 2 and 3 were: 0.49-3.82, 0.15-0.97 and 0.16-93 respectively (Figures 13-15). The least values of biodiversity indices were recorded during summer and post monsoon season, but were higher during premonsoon season at all the stations could be correlated with the recorded lower and higher salinity values, as reported by Ramakrishnan et al. [39]. This is further evidenced by the positive correlation obtained between salinity and species diversity, richness and evenness [42,45,48].

CCA biplot of phytoplankton community structure and environmental variables scored for March 2009 to April 2010 (Figures 16-18). Active variables selected by the analysis (P<0.01) are underlined. A positive correlation is expressed by relative long vectors pointing approximately in the same direction, and a negative correlation is indicated by arrows pointing in opposite directions. The higher the correlation is the longer the arrow in the diagram. The variables year is also represented (Table 2).

In the present study, farm discharge mixing area had high polluted, also this area minimum density of phytoplankton was recorded (stations 1, 2 and 3), when compared with open sea. Probably, because high organic load at stations 1 and 2. A major problem experienced with shrimp farms is the discharge of pond waters. Approximately 80% of nitrogen added to ponds as shrimp feed is not retained as shrimp biomass [49,50]. Remaining nitrogen acts to fuel plankton and microbial production within ponds, often resulting in negative effects on pond water and sediment quality such as anoxia, nutrient toxicity, and blooms of undesirable algal species [1,51]. Pollution of the receiving water includes nutritional enrichment, organic matter and settling of solids [51,52].

The most microalgae are obligate photoautotrophs, and their growth is strictly dependent of photosynthetic activity; in the case of dinoflagellates, the nutritional strategies are diversed, and show varying degrees of mixotrophy and heterotrophy through a combination of phototrophy and phagotrophy in response to rapid changes in the environmental conditions such as light availability, inorganic and organic nutrient concentrations and food particle abundance. In most shrimp farms from the Gulf of California, coastal waters are used for supplying shrimp ponds; in some cases water is pumped directly from the coast and in other cases indirectly through coastal lagoons. In the Gulf of California, the most abundant and diverse groups of the phytoplankton are diatoms (415 species) and dinoflagellates (270 species). Based on this data, it would suggest that the influence of discharge is contained the phytoplankton density. Although, the coastal zone is also a dynamic area with many cyclic and random processes owing to a variety of resources and habitats. Studies of the abundance,

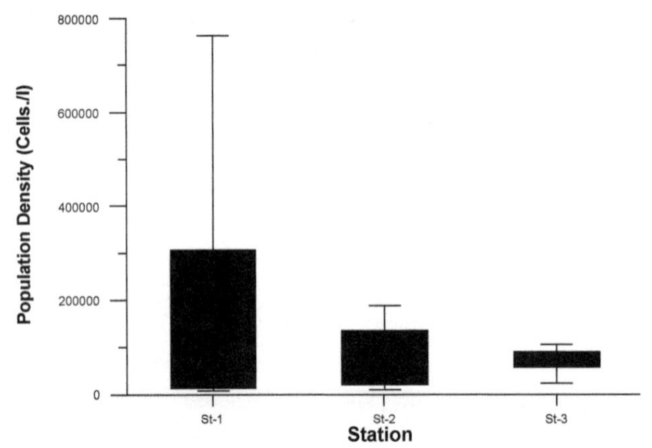

Figure 12: Phytoplankton population density recorded from stations 1, 2 & 3.

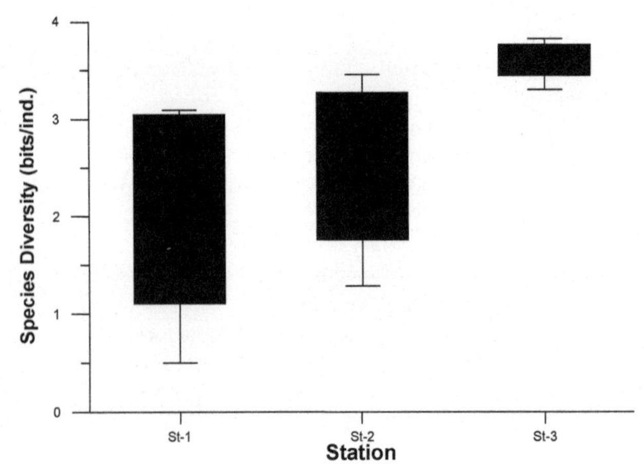

Figure 13: Phytoplankton species diversity recorded from stations 1, 2 and 3.

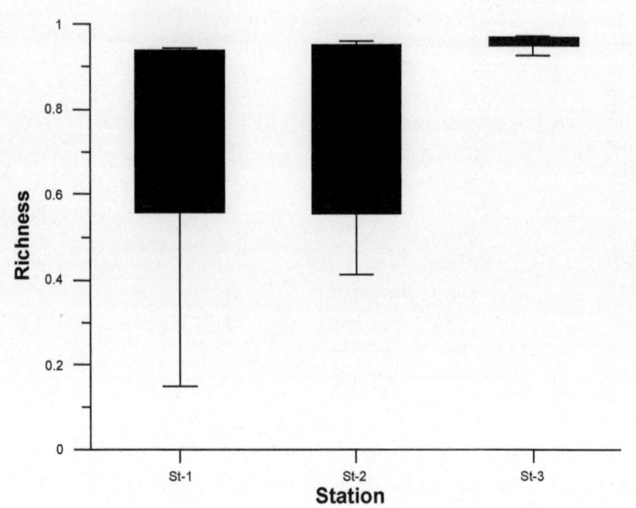

Figure 14: Phytoplankton species richness recorded from stations 1, 2 and 3.

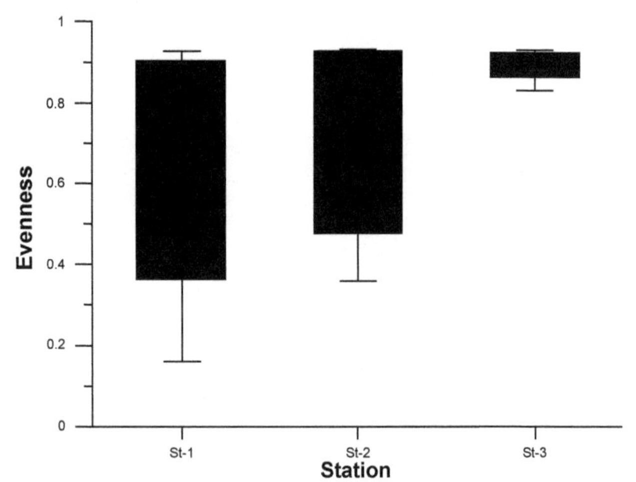

Figure 15: Seasonal variations of evenness recorded from stations 1, 2 and 3.

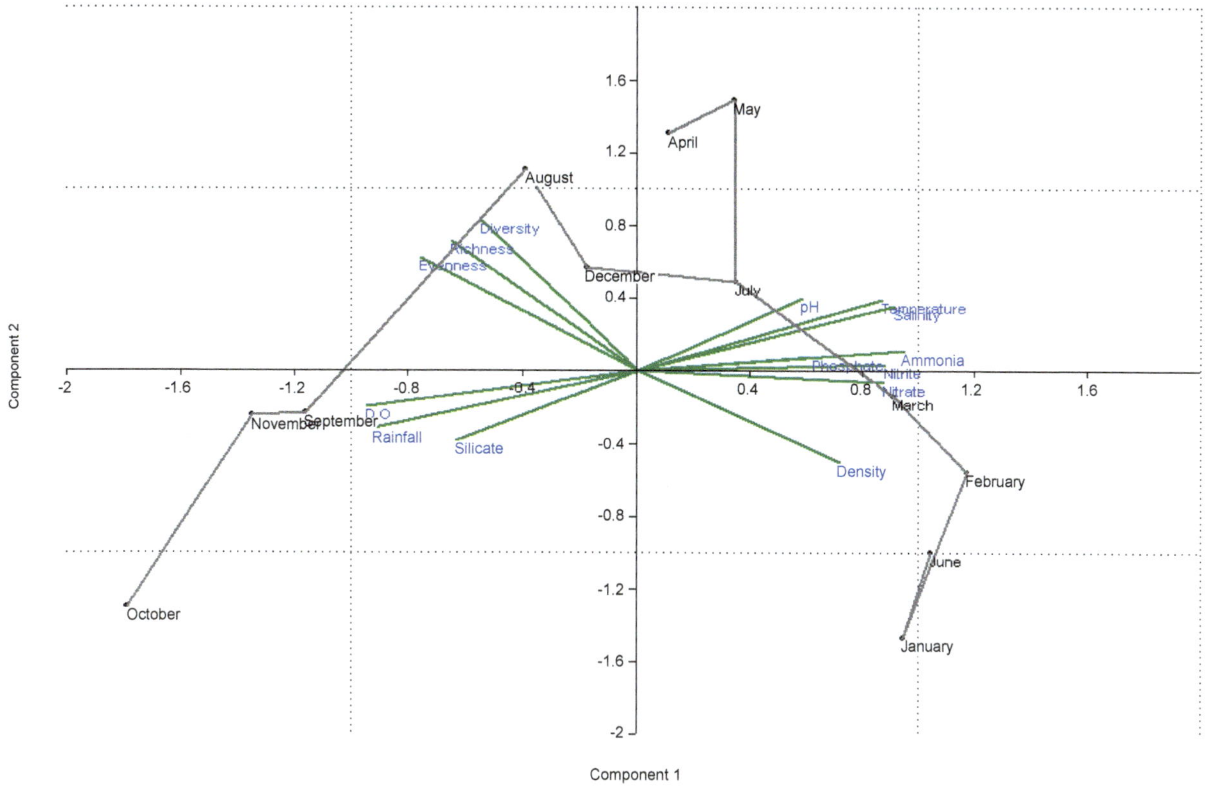

Figure 16: Correspondence analysis ordination diagram of phytoplankton community structure and environmental parameters at station 1.

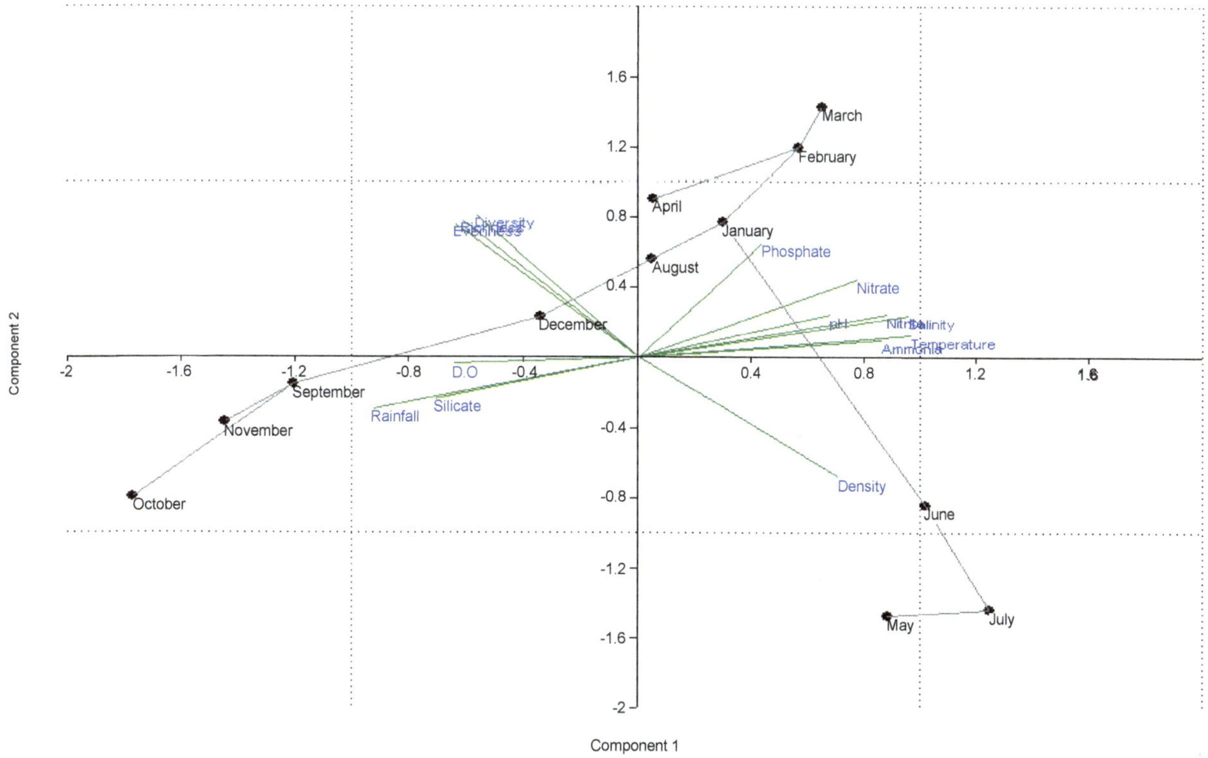

Figure 17: Correspondence analysis ordination diagram of phytoplankton community structure and environmental parameters at station 2.

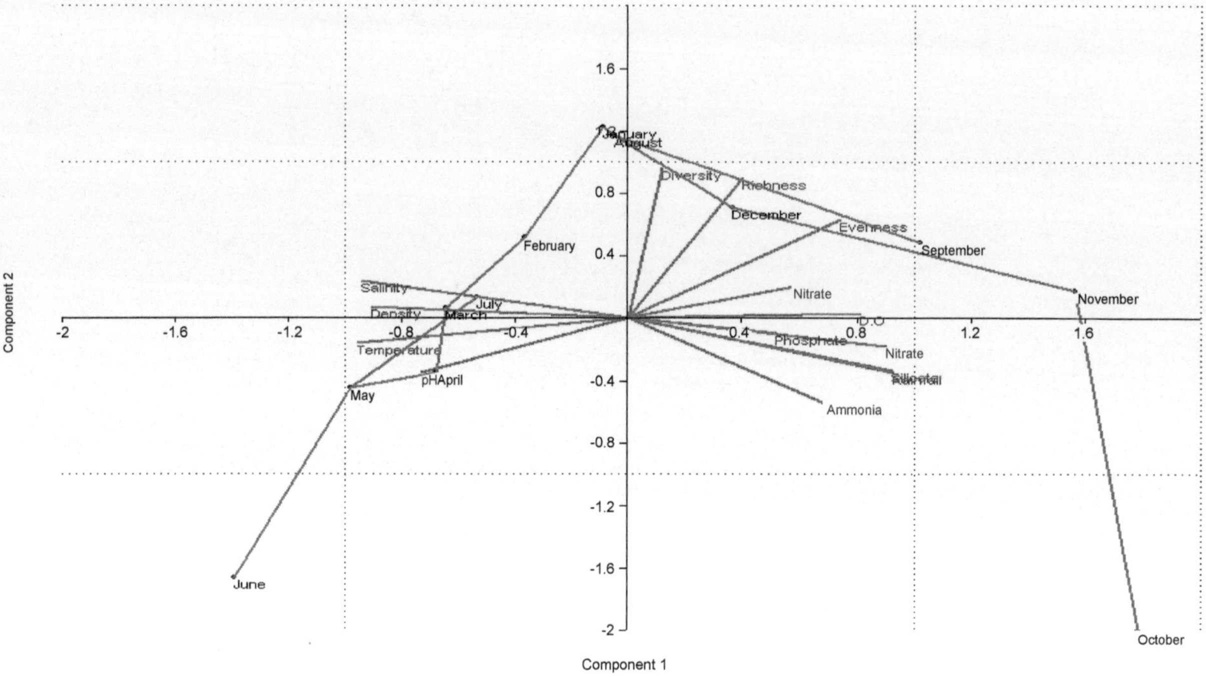

Figure 18: Correspondence analysis ordination diagram of phytoplankton community structure and environmental parameters at station 3.

	Engine value			% of Variance		
Month	**Station 1**	**Station 2**	**Station 3**	**Station 1**	**Station 2**	**Station 3**
April	9.36415	8.95609	9.04365	62.428	59.707	60.291
May	2.80704	3.29397	2.94883	18.714	21.96	19.659
June	1.04594	0.838987	1.15584	6.9729	5.5932	7.7056
July	0.933509	0.671245	0.84853	6.2234	4.475	5.6569
August	0.383715	0.551213	0.564902	2.5581	3.6748	3.766
September	0.187896	0.440726	0.236751	1.2526	2.9382	1.5783
October	0.098986	0.145756	0.106567	0.65991	0.97171	0.71045
November	0.085436	0.048737	0.05505	0.56957	0.32491	0.367
December	0.058135	0.035112	0.023633	0.38757	0.23408	0.15755
January	0.024714	0.011814	0.013279	0.16476	0.078758	0.088523
February	0.01048	0.006352	0.002967	0.069868	0.042346	0.019783
March	1.94E-30	7.23E-31	2.68E-29	1.30E-29	4.82E-30	1.79E-28

Table 2: Analyses of correlation; r=correlation coefficient & % of Variance; p=significance level (alpha level).

Parameters	**Station**	**April**	**May**	**June**	**July**	**Aug**	**Sept**	**Oct**	**Nov**	**Dec**	**Jan**	**Feb**	**Mar**
Rainfall		74.00	15.00	0.00	53.00	34.00	351.00	561.00	430.00	142.00	12.00	0.00	0.00
W.T (°C)	I	28.00	29.00	29.00	30.00	28.00	25.00	22.00	24.00	27.00	28.00	28.00	29.00
	II	29.00	30.00	30.00	31.00	30.00	26.00	25.00	25.00	28.00	29.00	30.00	30.00
	III	29.00	30.00	31.00	29.00	28.00	27.00	26.00	25.00	27.00	28.00	29.00	30.00
D.O. (ml/l)	I	3.38	2.35	2.21	3.74	4.21	5.74	6.65	5.74	4.62	3.27	2.38	3.02
	II	4.38	5.21	3.32	5.02	5.62	5.72	6.17	5.74	5.62	5.02	4.22	5.62
	III	5.21	5.62	4.34	5.71	5.62	6.17	6.33	6.21	5.62	5.21	4.62	5.21
Salinity (‰)	I	29.00	30.00	30.00	31.00	28.00	10.00	0.00	5.00	23.00	29.00	30.00	31.00
	II	27.00	28.00	29.00	30.00	26.00	9.00	0.00	4.00	21.00	26.00	29.00	29.00
	III	34.00	34.00	33.00	30.00	28.00	21.00	10.00	20.00	24.00	29.00	32.00	33.00
pH	I	8.20	8.20	8.10	8.00	7.90	7.90	7.80	7.90	7.80	7.90	8.00	8.20
	II	8.00	8.00	8.10	8.00	8.10	8.00	7.90	7.80	7.90	8.00	8.00	8.10
	III	8.20	8.20	8.20	8.00	8.00	8.00	8.00	7.90	7.90	8.10	8.00	8.20
NO2 (μM)	I	1.43	2.03	2.14	2.21	1.41	0.53	0.45	0.39	2.74	2.43	3.33	2.39
	II	1.12	1.82	2.02	2.01	1.02	0.24	0.16	0.29	1.80	2.02	2.33	2.45
	III	0.57	0.51	0.30	0.44	0.51	0.86	0.70	0.65	0.40	0.69	0.61	0.68

NO3 (µM)	I	5.23	4.66	4.94	3.72	2.17	2.15	2.72	1.29	4.11	5.34	6.15	5.09
	II	3.21	2.18	3.51	3.85	2.74	1.82	1.56	1.76	2.06	3.95	4.05	4.24
	III	3.30	2.58	2.88	3.51	3.25	4.80	5.23	4.47	3.33	3.40	3.64	2.57
NH3 (µM)	I	6.06	7.72	8.10	6.09	3.07	1.07	0.96	1.87	6.10	7.03	8.04	7.05
	II	1.13	2.12	1.24	3.17	1.12	0.17	0.16	0.18	1.05	2.14	2.11	2.85
	III	0.05	0.07	0.10	0.09	0.07	0.13	0.19	0.10	0.10	0.07	0.09	0.10
IP (µM)	I	1.40	1.36	1.16	1.25	1.30	1.18	1.16	1.26	1.34	1.43	1.68	1.58
	II	0.74	0.56	0.25	0.22	0.35	0.16	0.18	0.26	0.35	0.41	0.70	0.90
	III	0.41	0.30	0.25	0.20	0.28	0.16	0.45	0.69	0.40	0.35	0.30	0.32
SiO3 (µM)	I	32.93	33.16	29.90	25.91	26.81	33.99	54.66	35.82	28.63	29.01	30.97	32.93
	II	30.11	30.25	27.15	25.38	24.15	36.84	50.86	31.78	24.67	28.15	26.87	29.95
	III	32.54	31.30	28.68	25.52	35.58	48.73	65.71	58.67	36.63	29.01	28.54	29.15
Density (Cells/l)	I	26354	38612	763284	165957	19163	8319	5906	11372	14052	307817	430552	268144
	II	24841	185070	133232	189074	23125	20412	9064	14277	36298	51639	34502	34796
	III	83578	104072	89458	103576	72884	60480	20949	25531	56794	61348	67214	79284
Diversity	I	3.09	3.05	0.95	2.04	3.09	2.25	2.06	2.52	2.61	0.49	1.10	1.26
	II	3.33	1.35	1.76	1.29	3.11	3.46	2.78	3.22	3.12	3.25	3.27	3.21
	III	3.44	3.51	3.36	3.60	3.82	3.76	3.30	3.63	3.72	3.82	3.67	3.53
Richness	I	0.94	0.94	0.41	0.79	0.94	0.86	0.86	0.90	0.91	0.15	0.56	0.57
	II	0.95	0.41	0.56	0.41	0.93	0.96	0.93	0.95	0.94	0.95	0.95	0.95
	III	0.95	0.95	0.93	0.96	0.97	0.97	0.94	0.97	0.97	0.97	0.96	0.96
Evenness	I	0.90	0.91	0.29	0.61	0.93	0.85	0.89	0.89	0.90	0.16	0.36	0.41
	II	0.90	0.36	0.48	0.36	0.87	0.93	0.93	0.91	0.89	0.91	0.93	0.93
	III	0.86	0.85	0.83	0.87	0.92	0.92	0.90	0.93	0.92	0.92	0.90	0.89

Table 3: Physico-chemical parameters and density, diversity, richness and evenness of phytoplankton species in different stations during April 2009–March 2010.

distribution and composition of phytoplankton communities are, therefore, a fundamental contribution to our understanding of the structure and function of marine ecosystems.

References

1. David JWM (1997) The role of microorganisms in aquaculture ponds. Aquacult 151: 333-349.
2. Simon JFS, Matthew RPB (1998) Nutrient budgets in intensive shrimp ponds: implications for sustainability. Aquacult 164: 117-133.
3. Federico Páez-Osuna (2001) The environmental impact of shrimp aquaculture: causes, effects, and mitigating alternatives. Environ Manag 28: 131-140.
4. Gouda R, Panigrahy RC (1996) Ecology of phytoplankton in coastal waters off Gopalpur, east coast of India. Indian J Mar Sci 25: 81-84.
5. Strickland JDH, Parsons TR (1972) A Practical Hand Book Of Seawater Analysis. (2nd Edn) Bull Krishivigan Res Board, Canada.
6. Anonymous (1979) Settling without the inverted microscope.
7. Subrahmanyan R (1946) A systematic account of the marine plankton diatoms of the Madras coast. Proceedings of the Indian Academy of Sciences - Section B 24: 85-197.
8. Prescott GW (1978) How to know the freshwater algae. Brown company publishers Lowa 239.
9. Steidenger KA, Williams J (1976) Dinoflagellates. Mem. Hourglass Cruises 2: 1-251.
10. Taylor FJR (1976) Dinoflagellates from the International Indian Ocean Expedition : a report on material collected by the R.V. "Anton Bruun" 1963-1964. Bibil Botan 132: 1-226.
11. Mackay MA, Norton RS, Borowitzka LJ (1986) Marine blue-green algae have a unique osmoregulatory system 73: 301-307.
12. Santhanam R, Ramanathan N, Venkataramanujam K, Jegatheesan G (1987) Phytoplankton of the Indian seas. An aspects of Marine Botany. Daya Pub house, Delhi, India.
13. Shannon CE (1948) A Mathematical Theory of Communication. The Bell System Technical Journal 27: 379-423.
14. Gleason HA (1922) On the relation between species and area. Ecol 3: 158-162.
15. Pielou EC (1966) The measurement of diversity in different types of biological collections. J Theo Biol 13: 144.
16. Perumal P (1993) The information of meterological phenomena on the ecosystem of a tropical region, southeast coast of India- A case study. Ciencias marinas 19: 343-351.
17. Maruthanayagam C, Subramanian P (2000) Hydrological and zooplankton biomass variation in Palk Bay and Gulf of Mannar along the east coasts of India. J Mar Biol Ass India 41: 7-18.
18. Das J, Das SN, Sahoo RK (1997) Semidiurnal variation of some physico-chemical parameters in the Mahanadi estuary, east coast of India. Indian J Mar Sci 26: 323-326.
19. Govindasamy C, Kannan L, Jayapaul A (2000) Seasonal variation in physico-chemical properties and primary production in the coastal water biotopes of Coromandel coast, India. J Environ Biol 21: 1-7.
20. Reddi KR, Jayaraju N, Suriyakumar I, Sreenivas K (1993) Tidal fluctuation in relation to certain physio-chemical parameters in Swarnamukhi river estuary, east coast of India. Indian J Mar Sci 22: 232-234.
21. Mohan PC, Sreenivas N (1998) Diel variations in zooplankton populations in mangrove ecosystem at Gaderu canal, southeast coast of India. Indian J Mar Sci 27: 486-488.
22. Gowda G, Gupta TRC, Rajesh KM, Gowda H, Lingadhal C, et al. (2001) Seasonal distribution of phytoplankton in Nethravathi estuary, Mangalore. J Mar Biol Ass India 43: 31-40.
23. Rajasegar M (2003) Physico-chemical characteristics of the Vellar estuary in relation to shrimp farming. J Environ Biol 24: 95-101.
24. Sai Sastry AG, Chandramohan R (1990) Physicochemical characteristics of Vasishta Godavari estuary, east coast of India: pre pollution status. Indian J Mar Sci 19: 42-46.
25. Mitra A, Patra KC, Panigrahy RC (1990) Seasonal variations of some hydrographical parameters in a tidal creek opening in to the Bay of Bengal. Mahasagar 23: 55-62.
26. Satpathy KK (1996) Seasonal distribution of nutrients in the coastal waters of Kalpakkam, east coast of India, Indian J Mar Sci 25: 221-224.
27. Karuppasamy PK, Perumal P (2000) Biodiversity of zooplankton at Pichavaram mangroves, South India. Advan Biosci 19: 23-32.
28. Subramanian B, Mahadevan A (1999) Seasonal and diurnal variations of hydro

biological characters of coastal waters of Chennai (Madras) Bay of Bengal. Indian J Mar Sci 28: 429-433.

29. Vijayakumar S, Rajan K, Mridula Mendon M, Hariharan R (2000) Seasonal distribution and behavior of nutrients with reference to tidal rhythm in the Mulki estuary, Southwest coast of India. J Mar Boil Ass India 42: 21-23.

30. Choudhury SB, Panigrahy RC (1991) Seasonal distribution and behavior of nutrients in the creek and coastal waters of Gopalpur, east coast of India. Mahasagar. Bull Nat Ins Oceanog 24: 81-88.

31. Mani P, Krishnamurthy K (1989) Variation of phytoplankton in a tropical estuary (Vellar estuary, Bay of Bengal, India). Inter Review Hydrobiol 74: 109-115.

32. Murugan A, Ayyakannu K (1991) Ecology of Uppanar backwaters, Cuddalore: 2. nitrients. Mahasagar 24: 103-108.

33. Rajashree G, Panigrahy RC (1995) Seasonal distribution and behaviour of nitrate and phosphate in Rushikulya estuary, east coast of India. Indian J Mar Sci 24: 233-235.

34. Segar K, Hariharan V (1989) Seasonal distribution of nitrate, nitrite, ammonia and plankton in effluent discharge area of Mangalore, west coast of India. Indian J Mar Sci 18: 170- 173.

35. Nair PVR, Gopinathan CP, Balachandran VK, Mathew KJ, Regunathan A, et al. (1984) Ecology of mud banks: Phytoplankton productivity I Alleppey mud bank. Bull Centre Mar Fish Res Ins 31: 28-34.

36. Chandran R, Ramamoorthi K (1984) Hydrobiological studies in the gradient zone of the Vellar estuary. I Physico-chemical parameters. Mahasagar Bull Nat Inst Oceanog 17: 69-78.

37. Senthilkumar S, Santhanam P, Perumal P (2002) Diversity of phytoplankton in Vellar estuary, Southeast coast of India. The 5th Indian Fisheries Forum Proceedings, (*Eds*: S. Ayyappan, J.K. Jena and M. Mohan Joseph). Published by AFSIB, Mangalore and AeA, Bhubanewar, India 245-248.

38. Sujatha M, Panda D, Panigrahy RC (1993) Physico – chemical characteristics of the Bahuda estuary (Orissa), east coast of India. Indian J Mar Sci 22: 75-77.

39. Ramakrishnan R, Perumal P, Santhanam P (1999) Spatio-temporal variations of hydrographical features in the Pichavaram mangroves and Mohi aqua farm, Southeast coast of India. Intl Sem Appl Hydrogeochem Annamalai University, India 197-203.

40. Perumal NV, Rajkumar M, Perumal P, Rajasekar KT (2009) Plankton diversity in the Kaduviyar estuary, Nagapattinam, South east coast of India. J Environ Biol 30: 1035-1046.

41. Marichamy R, Gopinathan CP, Sivameetan P (1985) Studies on primary and secondary production in relation to hydrography in the inshore waters of Tuticorin. J Mar Biol Ass India 27: 129-137

42. Patterson Edward JK, Ayyakkannu K (1991) Studies on the ecology of plankton community of Kollidam estuary, southeast coast of India. I Phytoplankton Mahasagar Bull Nat Ins Oceanog 24: 89-97.

43. Rajasegar M, Srinivasan M, Rajaram R (2000) Phytoplankton diversity associated with the shrimp farm development in Vellar estuary, south India. Seaweed Res Utilin 22: 125-131.

44. Ei-Gindy AAH, Dorghan MM (1992) Interrelation of phytoplankton, chlorophyll and Physio-chemical factors in Arabian Gulf and Gulf of Oman during summer. Indian J Mar Sci 21: 257-261.

45. Mani P (1992) Natural phytoplankton communities in Pitchavaram mangroves. Indian J Mar Sci 23: 278-280.

46. Rajasekar KT, Rajkumar M, Jun Sun, Prabu VA, Perumal P (2005) Seasonal variations of phytoplankton diversity in the Coleroon coastal waters, southeast coast of India. Acta Oceanologica Sinica 29: 97-108.

47. Boyd CE, Tucker CS (1998) Waste Management. Pond Aquacult Water Qual Manage 541-575.

48. Kannan L, Vasantha V (1992) Micro phytoplankton of the Pichavaram mangroves, southeast coast of India. Species composition and population density. Hydrobiol 247: 77-86.

49. Briggs MRP, Funge-Smith SJ (1994) A nutrient budget of some intensive marine shrimp ponds in Thailand. Aquacult Fish Manage 25: 789-811.

50. Christopher Jackson, Nigel Preston, Peter J Thompson, Michele Burford (2003) Nitrogen budget and effluent nitrogen components at an intensive shrimp farm. Aquacult 218: 397-411.

51. Anonymous (1978) Fish farms have to watch their waste. Fish Farming Inter 5: 19.

52. Burford MA, Glibert PM (1999) Short-term nitrogen uptake and regeneration in early Cairns Jr., J., Mount, D.I., 1990. Aquatic Toxicol Environ Sci Technol 24: 154-161.

4

Environmental Influences on the Abundance of Dominant Fishes in a Very Shallow Tropical Coastal Lagoon in Northwestern Yucatan Peninsula, Mexico

José Luis Bonilla-Gómez[1*], Maribel Badillo[2], Korynthia López[3], Alfredo Gallardo[2], Carmen Galindo[2], Daniel Arceo[2] and Xavier Chiappa-Carrara[2]

[1]*Department of Marine Sciences and Limnology, National Autonomous University of Mexico, Mexico D. F., Mexico*
[2]*UNAM-Sisal, Faculty of Science, National University of Mexico, Sisal, Yucatan, Mexico*
[3]*UNAM-Sisal, Faculty of Chemistry, National Autonomous University of Mexico, Sisal, Yucatan, Mexico*

Abstract

Spatial-seasonal patterns of six abundant fishes were analyzed in La Carbonera, a lagoon on the northwestern coast of the Yucatan peninsula, Mexico; between April 2009 and May 2010, via non-parametric multivariate analysis. The null hypothesis tested was that no differences existed in the overlapping of the abundance of fish species among the lagoon defined by types of habitats (channel, seagrass, mangrove and hypersaline), and by three climatic seasons (dry, rainy, and north winds). Significant differences were found in the environmental variables between climatic season and the types of habitat, except for the temperature and pH. Result of the multidimensional scaling suggests a significant difference in habitats with characteristics of channel and presence of mangrove. The redundancy analysis showed that depth and salinity were the best related to species abundance. Oxygen and temperature made an additional significant contribution to the canonical model, it was strongly related to the north winds season, due to the fact that the strong winds, which can last a week, increase the oxygen content of the water but decrease the temperature. Relative abundances were strongly associated with depth (dry; r=0.76), salinity (rainy; r=0.83) and oxygen (north winds; =-0.80). Results suggest that the different habitats are used as feeding zones; and that during the dry season, there was a overlapping in abundance of species. Thus, information proposed protection strategies in this season to contribute to fish biodiversity conservation in a valuable area of Yucatan Peninsula.

Keywords: Spatial-seasonal patterns; Multidimensional scaling; Redundancy analysis; La Carbonera lagoon

Introduction

The Yucatan peninsula has numerous coastal lagoons, where their structure and function are being modified by wastewater discharge, groundwater pumping, and land use changes [1]. They are generally standing water with the calm hydrodynamics that are essential for many juvenile and adult fish [2]. Knowledge of the response of these fishes to changes in environmental conditions will contribute to our understanding of the potential effects of anthropogenic impacts on estuarine fish species [3].

The habitat patterns; such as mangrove and seagrass can be viewed as a reference of an assemblage of the ichthyofauna for tropical karstic lagoons [4]. In Yucatan, information is also available about fish assemblages in tropical coastal systems influenced by freshwater inputs [5,6] and fish assemblages in a coastal hypersaline lagoon [7]. There are many studies of fish assemblages in temperate systems [8], but little information exists about tropical karstic estuarine systems, particularly along the northwestern coast of the Yucatan peninsula. Several species, such as: *Lagodon rhomboides* (pinfish), *Sphoeroides testudineus* (checkered puffer), *Mugil curema* (white mullet), *Chriodorus atherinoides* (hardhead halfbeak), *Eucinostomus argenteus* (silver mojarra) and *Eucinostomus gula* (Jenny mojarra) have been already described as "abundant" in the coastal lagoons of Yucatan, contributing with more than 50% considering and analysis of the entire fish assemblage caught with seine net and their importance value index [5,9]. However, few studies analyze the population dynamics of wetland fishes for possible correlations with hydrological variables [10].

In terms of the climatic seasons, the Yucatan peninsula has three marked seasons annually: dry, rainy and north winds [9,11,12]. In small karstic coastal lagoons of the Yucatan, which are a typical system along this coast, the dry season like the hottest reduced and altered the habitat available for fish species [13]. La Carbonera, a karstic coastal lagoon, is a transitional environment between a karstic land and a marine habitat with high environmental variability over short time scales, and this lagoonal system is important for local fisheries and as bird habitat (e.g., flamingos *Phoenicopterus ruber*). The present study investigates extreme shallowness of the system along with the susceptibility of environmental changes, the presence of an overlapping between the abundance of six abundant population of fish among the La Carbonera lagoon defined by four habitat types (channel, seagrass, mangrove and hypersaline), and about the climatic season basis (dry, rainy, and north winds).

Materials and Methods

Study area

La Carbonera coastal lagoon lies on the northwestern Yucatan

***Corresponding author:** José Luis Bonilla-Gómez, Department of Marine Sciences and Limnology, National Autonomous University of Mexico, Circuito Exterior S/N, Ciudad Universitaria, Mexico DF, 04510, Mexico
E-mail: jose.bonilla@comunidad.unam.mx

peninsula, Mexico (21°13'-21°14' N; 89°52'-89°54' W; Figure 1). It is connected to the sea by a channel formed in 1988 by Hurricane Gilbert, and is fringed with *Rhizophora mangle* stands and a sand barrier. The lagoon is very shallow (average depth 0.5 m) and covers an area of ~16.5 km². It is a very low karstic marine-biogenic littoral [14]. Freshwater input is restricted to underground seeps and rainfall. There is a *cenote* (sinkhole) in the southern part of the lagoon, in a patch of mangroves. There are three climatic seasons in the zone: dry (D; March-June), rainy (R; July-October) and north winds (NW; November-February) [11]. The system presents three different types of vegetation: mangroves, seagrass, and spring with bushes patch called colloquially "*petenes*"; and the same karstic substratum throughout [15] with three types of sediments: sandy, rocky bed and muddy. Hence, four types of habitat were identified *in situ* in the nine stations according to [16], which are based on substrate, salinity, vegetation: hyperhaline (H), seagrass (S), channel (C), and mangrove (M) (Table 1).

Sampling and environmental data

Fourteen monthly sampling including fish and physical variables were performed between April 2009 and May 2010 during the three climate seasons (D, R, and NW) in nine sampling stations distributed in different types of habitats (H, S, C, and M). We recorded water temperature (°C), salinity, pH, dissolved oxygen (mg/L) and total dissolved solids (TDS, mg/L) at each sampling station (measured at 30 cm depth) with a Yellow Springs Instrument (YSI), model 556 MPS. Wind speed (km/h) and direction was determined with an anemometer, Sper Scientific 850070. Depth (cm) was recorded at each sampling station.

Juveniles of *L. rhomboides*, *S. testudineus*, *M. curema*, *C. atherinoides*, *E. argenteus*, and *E. gula* were caught monthly with a beach seine net (40 m long×1 m depth×1.27 cm stretch mesh size; Mexican government permit No. DOPA/04031/310510.1940). The sampling area was 400 m², with one sample per station; the net was cast once in each site for each month. All sites were sampled within a 6 h period during high tide on the same day. Specimens were euthanized in ice slurry, preserved in formaldehyde (4%) and transported to the

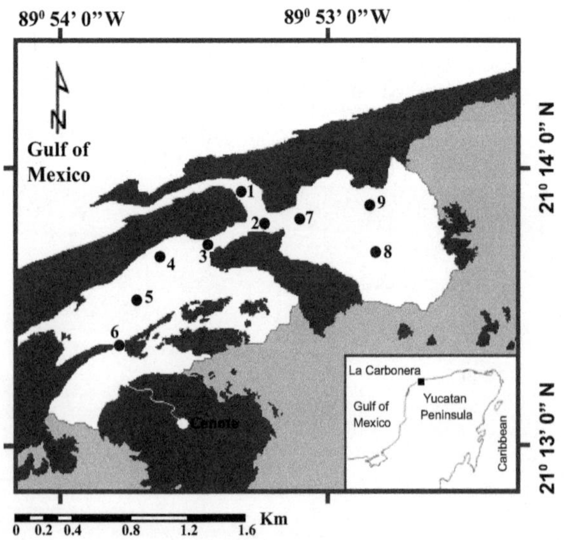

Figure 1: La Carbonera coastal lagoon lies on the northwestern Yucatan peninsula.

laboratory, where they were identified and weighed (± 0.01 g precision). All the six species were considered euryhaline marine.

Data Analysis

The environmental variable data collected in the sampling stations were grouped per type of habitats (C, S, H, and M; Table 1), and the monthly data were performed per climate seasons (D, R, and NW). A factorial two-way variance analysis (ANOVA) assessed spatial and seasonal differences in the environmental data and the interaction between habitat and season, having previously verified the assumptions of normality (Kolmogorov Smirnov) and homogeneity of variance (Bartlett and Cochran) [17]. Analyses used STATISTICA 7.0 (StatSoft Inc., 2004).

Multivariate statistical analyses of spatial-seasonal fish data used the PRIMER v.6 and Permanova add program [18]. Similarities between samples were computed by a Bray-Curtis dissimilarity matrix using fish species abundances with a Log (x+1) transformed data; they were graphically represented by cluster analysis (using group-average linking based on Bray-Curtis similarities), which was used to define groups of site preferences and ordination plots by species (using total abundances). A similarity matrix was used to generate a non-metric multidimensional scaling (MDS) ordination [19]. MDS constructs a 2-dimensional ordination in a manner that best represents relationships among samples in a similarity matrix [20,21]. In ordination plots, the relative distance between points reflects the similarity of species abundance in those samples. MDS was verified by taking into account the value of stress: low values (<0.1) suggest good separations among the groups, while higher values (>0.2) indicate that the results should be interpreted with caution [21]. An analysis of similarity (ANOSIM) was performed between the fish of each habitat and climatic seasons [22], with a non parametric test that uses a Bray-Curtis similarity matrix with abundance data. This procedure consists of a statistical test (R), which is analogous to an ANOVA. The null hypothesis tested was that no differences existed between the fish abundances in the different type of habitats.

Detrended Correspondence Analysis (DCA) on species data assessed the length of the gradients [23]; these were 1.12 for the first axis and <2 for the rest of the axes, suggesting a linear response model [24]. Based thereon, Redundancy Analysis (RDA) examined the relationships between the environmental variables of the lagoon and between the species abundance and sampling stations, using the program CANOCO for Windows 4.5 [25]. RDA computations were performed on log-transformation [$y'=\log(a*y+b)$] and centered by species (no centering for samples was performed). Scaling focused on inter-sample distances and there was no post-transformation of species scores. The RDA used the correlation matrix option, and scaling was conducted on a correlation triplot ordination. The significance of each environmental variable was tested through 499 unrestricted Monte Carlo permutations.

Results

Environmental conditions

Mean temperatures (Figure 2) were significantly different by season (Table 2), and were highest in the rainy months; particularly registered in August (34.2°C) and October (32.6°C), and were lowest when north winds prevailed, with a minimum of 20.3°C registered in February. No spatial (per habitat) variation was observed ($p>0.05$). Mean salinity was 35.1 (range 3.8-52.7) in the sampling period; seasonal

Station	Habitat	Substrate	Depth (cm)	Temp. (°C)	Salinity	O_2 (mg/L)	TSD (mg/L)	pH
1	Channel	Rocky, Sandy	56.6 ± 24.7	28.8+4.4	37.7 ± 4.6	7.5 ± 1.7	30.8 ± 9.9	8.2 ± 0.2
2	Mangrove (60%)	Sandy	39.9 ± 24.5	27.4 ± 3.1	36.9 ± 5.2	6.1 ± 1.5	30.7 ± 9.4	8.1 ± 0.2
3	Channel	Muddy	72.6 ± 33.0	27.3 ± 3.8	33.2 ± 3.5	6.4 ± 1.6	25.1 ± 13.5	8.2 ± 0.3
4	Seagrass (80%)	Sandy, Muddy	27.7 ± 12.0	28.8 ± 3.8	30.3 ± 6.4	7.6 ± 1.6	24.1 ± 11.5	8.2 ± 0.3
5	Hyperhaline	Muddy	35.1 ± 13.4	29.0 ± 4.0	31.3 ± 14.2	7.6 ± 2.3	24.4 ± 12.5	8.2 ± 0.3
6	Channel	Sandy	57.1 ± 17.0	29.8 ± 4.2	28.8 ± 4.6	7.9 ± 1.1	24.0 ± 13.7	8.2 ± 0.3
7	Channel	Sandy	63.5 ± 41.2	27.9 ± 2.7	34.8 ± 5.0	5.9 ± 1.8	29.9 ± 11.2	8.0 ± 0.4
8	Hyperhaline	Sandy	41.5 ± 10.5	28.8 ± 3.4	38.1 ± 8.9	7.3 ± 1.6	32.3 ± 12.1	8.2 ± 0.2
9	Mangrove (70%)	Muddy	32.6 ± 12.0	30.3 ± 4.9	34.1 ± 6.2	7.5 ± 1.9	34.6 ± 8.9	8.2 ± 0.2

Table 1: Characteristics of the different sampling stations, type of habitats (based on substrate, salinity, vegetation observed *in situ*) and mean values (± standard deviation) of the environmental variables present in each sampling stations. % refers to the percentage cover in each habitat.

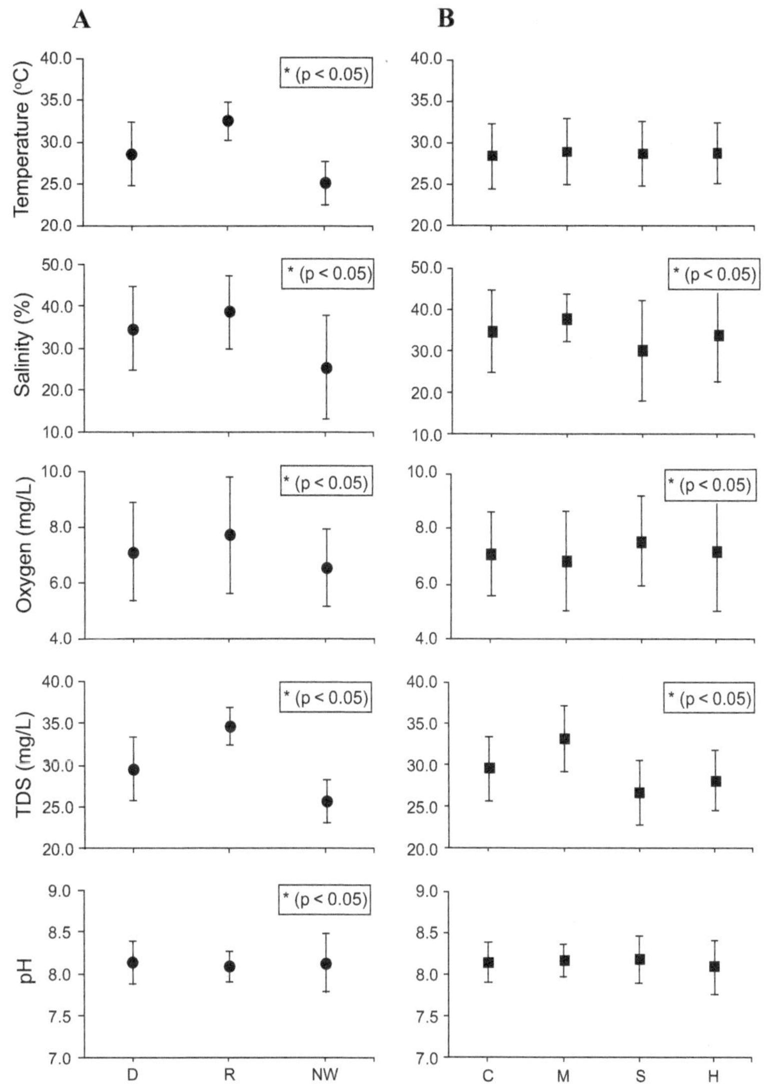

Figure 2: Figure representing mean temperatures were significantly different by season and were highest in the rainy months.

and spatial variation were significant different ($p<0.05$; Table 2) and were higher in M (Figure 2). Dissolved oxygen varied significantly with season and with the habitats, especially in S, with highest and lowest mean values occurring in August and September (Figure 2). Test supported Development (TSD) and pH both indicated significant variations with the season values ranging from 27 to 35 mg/L and 8.02 to 8.19 respectively. TDS showed spatial differences, being higher in M ($p<0.05$), but the pH was no different between habitats (Figure 2). Depth showed seasonal and spatial differences, and were higher in C (Figure 3) being shallow in the dry season and, spatially, being shallowest in stations 1, 4 and 5 and deepest in stations 3 and 7 (Table 1). Wind velocity varied with season variations (Figure 3).

Spatial and Seasonal Variations in Fish Abundance

A total of 8370 specimens were captured and each species occurred in all sampling stations, except that *C. atherinoides* was never caught in station 4 with a sea grass habitat type (Table 3, Figure 4). Also, *C. atherinoides*, *M. curema* and *E. argenteus* were not found in some months, whereas *L. rhomboides*, *S. testudineus* and *E. gula* occurred in all months. Most species were most abundant in the dry season, but *M. curema* and *E. gula* were more abundant in the rainy season (Figure 4). Cluster analysis showed four groups of habitat (sampling stations) preferences with a similarity level between groups (Bray-Curtis similarity) of 22 to 82% (Figure 5). Group I; which included the habitat S, and was distinct from other groups, showing 22% similarity. *Sphoeroides testudineus* occurred in the group IV, *L. rhomboides* and *C. atherinoides* in group I, III and IV; *M. curema* in groups I, II and IV; *E. gula* in groups III and IV; and while *E. argenteus* in group IV.

Variables/Effects	*S. S.*	*D. F.*	*M. S.*	*F*	*p*
Temperature					
Season	786.53	2	393.27	55.79	*0.0000
Habitat	8.52	3	2.84	0.40	0.7513
Season-Habitat	13.09	6	2.18	0.31	0.9309
Salinity					
Season	3711.0	2	1855.5	30.261	*0.0000
Habitat	823.8	3	274.6	4.478	*0.0052
Season-Habitat	624.8	6	104.1	1.698	0.1279
Dissolved oxygen					
Season	18.281	2	9.140	2.943	0.0568
Habitat	4.387	3	1.462	0.471	0.7032
Season-Habitat	4.261	6	0.710	0.229	0.9665
TDS					
Season	1124.68	2	562.34	4.5829	*0.0125
Habitat	640.65	3	213.55	1.7404	0.1637
Season-Habitat	445.35	6	74.22	0.6049	0.7258
pH					
Season	0.129	2	0.065	0.94	0.3948
Habitat	0.093	3	0.031	0.45	0.7175
Season-Habitat	0.238	6	0.040	0.57	0.7503
Depth					
Season	481.7	2	240.9	0.4413	0.6443
Habitat	28173.1	3	9391.0	17.2063	*0.0000
Season-Habitat	5458.6	6	909.8	1.6669	0.1357
Wind velocity					
Season	1565.84	2	782.92	5.3511	*0.0060
Habitat	29.54	3	9.85	0.0673	0.9772
Season-Habitat	579.72	6	96.62	0.6604	0.6817

Table 2: Results of the factorial two-way ANOVA to test spatial and seasonal significant differences (*; $p<0.05$) in the environmental variables registered.

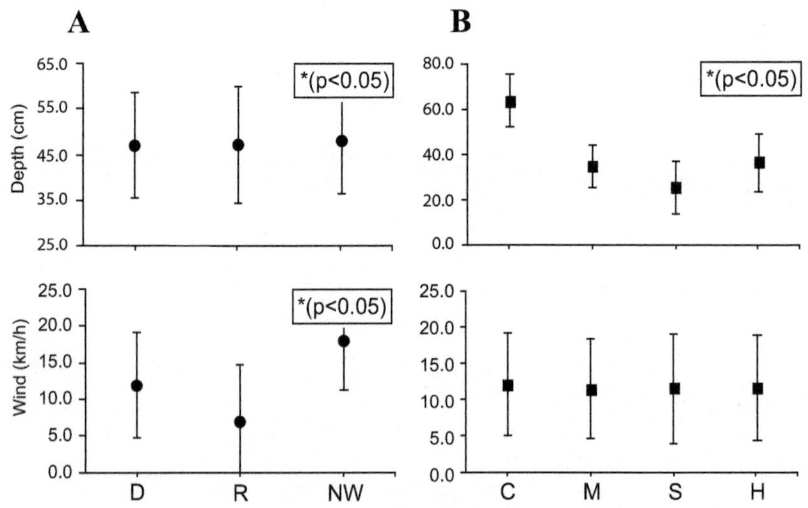

Figure 3: Depth showed seasonal and spatial differences.

The fish populations were most frequent in deeper sites, high salinities in the lagoon and preference by Channel and mangrove habitats. This pattern of species preferences was similar in the two-dimensional MDS ordination plot with a stress of 0.08 and 0.03, which is a good representation of the diagram (Figure 6a). Also an overlapping was showed between the dry season in the species abundance with the others seasons (Figure 6b). The one-way ANOSIM with 999 permutations indicated significant differences in the composition of fish abundance between the different habitats (r=354, p<0.05).

Effect of Environmental Variables on Spatial-seasonal Variation of Fishes

RDA yielded a first canonical ordination axis explaining 23.4% of the variance of species data and 45% of the species-environment relationship (Table 4). This axis had a very high species-environment correlation. Joint examination of canonical coefficients and inter-set correlations helped to identify this first axis reflecting a salinity and depth gradient (Table 4). The second canonical axis explained less variation (15.2% of species data and 29.3% of species-environment relation) but had also a high species-environment correlation. This axis can be clearly identified with a negative gradient of oxygen and temperature. Through a triplot, a comprehensive graphical representation of RDA ordination examined the number of samples, species and environmental variables reflected the salinity and depth arrow and the first canonical ordination axis a high correlation between them (Figure 7). The pH and TDS are also correlated with the first axis. The arrows representing wind, temperature and oxygen forms an acute angle with the second ordination axis, reflecting the high correlation between them. Projections of centroids of the nominal variable used to represent climatic seasons show clearly that this variable has a very low discriminatory value due to the position of the centroids of its classes near the origin of the axes. However, during the north winds season, a group of samples was clearly shown. During the dry season, the species were highly correlated among them, and with depth, salinity, and TDS. Meanwhile, *M. curema, S. testudineus, L. rhomboides*, and *E. argenteus*

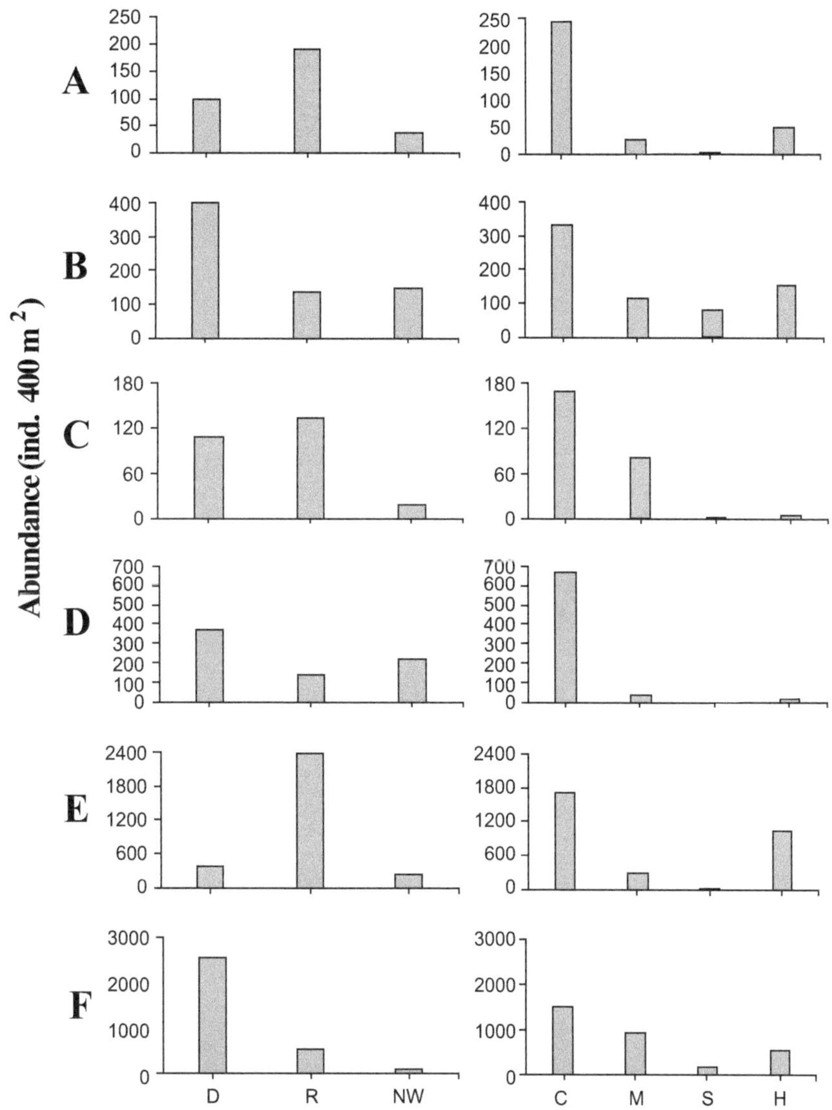

Figure 4: The frequency of occurrence (%) of the fish species are showed per habitat.

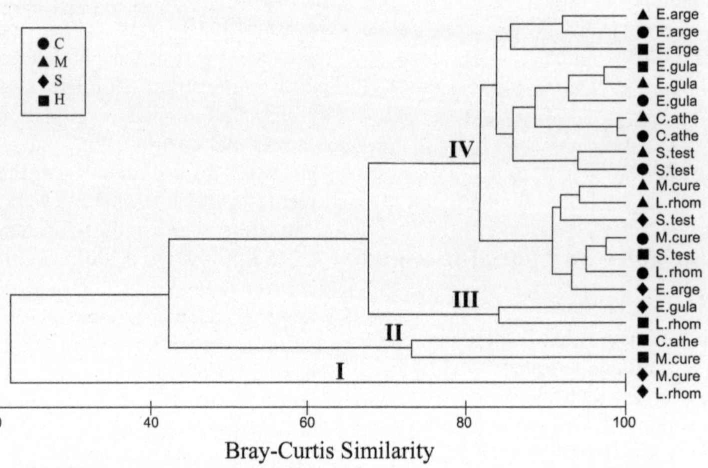

Figure 5: The figure showing the cluster analysis of all the groups of habitat.

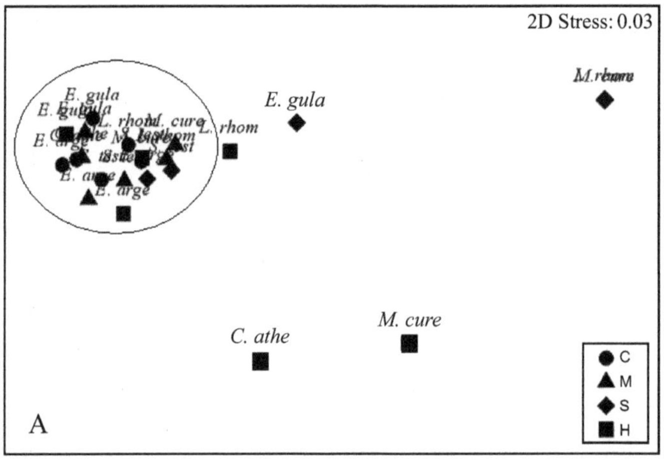

Figure 6a: Figure showing the pattern of species preferences was similar in the two-dimensional MDS ordination plot with a stress.

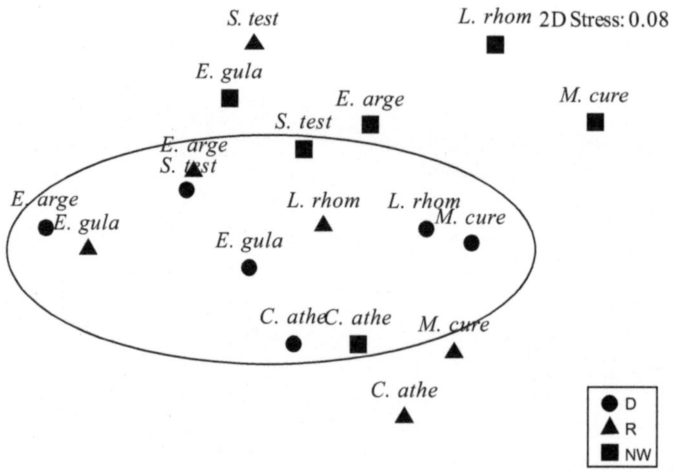

Figure 6b: Figure showing overlapping was showed between the dry season in the species abundance with the others seasons.

Species	Family	N	%D	%B	Frequency of occurrence (%)						
					Habitat				Climatic season		
					Channel	Mangrove	Seagrass	Hypersaline	Dry	Rainy	North
E. argenteus	Gerreidae	3079	36.68	7.39	5.32	1.87	0.73	2.79	86.45	11.33	2.22
E. gula	Gerreidae	3094	36.97	7.66	7.19	7.88	30.55	8.70	15.04	77.22	7.74
C. atherinoides	Hemiramphidae	699	8.35	10.88	3.68	5.60	0.73	0.28	42.90	22.71	34.38
M. curema	Mugilidae	252	3.01	3.15	14.60	2.42	0.00	1.00	45.23	47.72	7.05
L. rhomboides	Sparidae	569	6.80	8.80	37.00	18.80	8.73	57.30	64.11	29.16	6.73
S. testudineus	Tetraodontidae	686	8.20	63.13	32.21	63.44	59.27	29.93	63.67	13.27	23.06

Table 3: Total of individuals (N), relative density (%D 400 m^2) and the relative biomass (%B 400 m^2) found during the period of study (April 2009-May 2010). The frequency of occurrence (%) of the fish species are showed per habitat (spatial variation) and per climatic seasons (temporal variation), both analysis with independient data. (NW=Norths winds).

Summary of the ordination axes	1	2	3	4
Eigenvalues	0.234	0.152	0.089	0.033
Species-environment correlations	0.789	0.800	0.653	0.616
Cumulative percentage variance				
of species data	23.4	38.6	47.5	50.8
of species-environment relation	45.0	74.3	91.3	97.6
Canonical coefficients for standardized variables and inter-set correlations of environmental variables with species axes	Canonical coefficient		Inter-set correlation	
	Axis 1	**Axis 2**	**Axis 1**	**Axis 2**
Depth	0.437	0.434	0.345	0.347
Oxygen	-0.319	-0.845	-0.252	-0.676
Temperature	-0.068	-0.718	-0.054	-0.574
Salinity	0.458	-0.401	0.361	-0.321
pH	-0.274	-0.193	-0.216	-0.154
TDS	0.186	-0.265	0.147	-0.212
Wind	-0.056	0.653	-0.044	0.522
Dry season	0.720	-0.216	0.569	-0.173
Rainy season	-0.191	-0.447	-0.151	-0.358
North winds season	-0.529	0.663	-0.418	0.530

Table 4: Summary of redundancy analysis.

were highly correlated mainly to salinity and were associated to the dry season (included as dummy variable), in relation with the first canonical axis (45%). However, *E. gula* showed a highly correlation with the temperature and an inverse relation with the wind velocity, and *C. atherinoides* showed a high relation with depth and an inverse relation with the pH (Figure 7).

In the triplots of species, samples, and environmental variables (Figure 8), the distance between sampling stations (focused in samples) in the plot does not reflect the Euclidean distances between them due to scaling used, basically focused on species correlations. The projections of species, environmental variables and sampling stations on canonical axes is a way to understand main environmental variables associated to each group of climatic season. RDA results indicated that in the dry season (Figure 8a), axis 1 explained about 45.3% and axis 2 explained 38.1% of the relationship between species and environment. In the rainy season, axis 1 explained 57.5% and axis 2 explained 20.1% of the species and environment relation (Figure 8b). When north winds prevailed, axis 1 explained 82.3% and axis 2 explained 13.7% (Figure 8c). Depth showed the vector of length best correlated in the dry season with axis 2, r=0.76 and an inflation factor of 1.37 (Table 5), followed by salinity, which was best correlated with axis 1, r=-0.63 and an inflation factor of 1.67. In the dry season, abundance of *C. atherinoides* was highly correlated with depth; *E. gula* and *E. argenteus* were highly correlated with salinity and temperature and abundant at stations 8 and 9, and were inversely correlated with *S. testudineus* and station 2. *M. curema* and *L. rhomboides* were inversely correlated with pH and preferred station 7. No preferences or correlations were observed for station 3, the deepest station (Figure 8a).

In the rainy season, the salinity had the vector length best correlated with axis 1, r=0.83 and an inflation factor of 1.63 (Table 5); temperature was the second best correlated with axis 1, r=0.63 (Figure 8b). *Lagodon rhomboides*, *E. gula* and *S. testudineus* were highly correlated with depth and inversely with temperature and oxygen, and correlated with station 2. Abundance of *M. curema* and *E. argenteus* was highly correlated with stations 2 and 4, and inversely correlated to TDS and pH. *Chriodurus atherinoides* was correlated with station 8. There were no preferences or direct correlation to environmental variables for stations 1, 6, and 7.

During the season of north winds, fish abundance was inversely correlated to oxygen, it showed a best correlation with axis 1 (r=-0.80 with inflation factor of 2.19) (Table 5), followed by the depth vector length with a best correlated with axis 1 (r=0.70) (Figure 8c). Projections of abundance of all species on canonical axes and environmental factors showed correlated with lagoon depth and salinity, and inversely correlated to oxygen and temperature, including their prolongations in the direction opposite to the arrowhead. The species showed correlation with stations 1 and 3 and inversely with station 2, 8, and 9. *E. gula and E. argenteus* are not correlated with environmental variables, but show a clear high correlation with the second axis. Station 7 did not show direct correlations with the environmental variables.

Discussion

Spatial and temporal differences of environmental variables

In the lagoon system of La Carbonera, high salinity and temperature were found in the dry season. One exception corresponds to October 2009, which was an atypical month since high temperatures and low

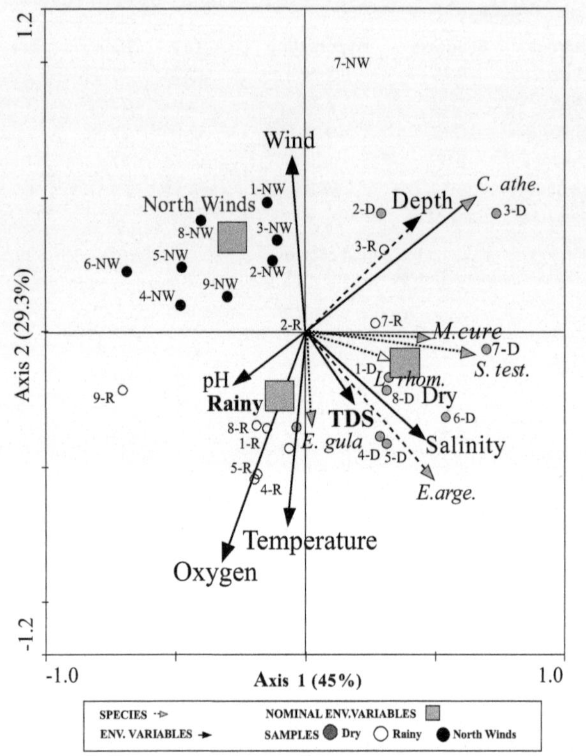

Figure 7: Figure representing the comprehensive graphical representation of RDA ordination.

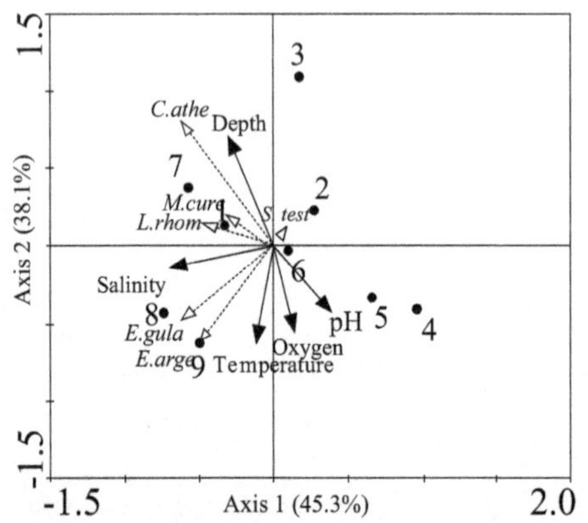

Figure 8A: Graph representing RDA results indicated that in the dry season.

precipitation led to the salinity increase during the rainy season instead the usual dilution of the rainfall that is normally observed. A salinity peak (~50) in October 2009 was reported by [26], which is different with the lowest salinity (~15) found in December 2009 at the end of the rainy season. Both salinity and temperature gradually decrease until the lowest values were registered in the north winds season, following the general pattern found also by [16] in another hypersaline coastal system on the northeastern coast of the Yucatan peninsula. The presence of northerly winds, usually colder and faster, is important in the spatial distribution of the environmental features of this shallow system. This is especially true when considering the dissolved oxygen values, which increase due to turbulence and vertical mixing of the water column (Figure 2). Rainfall affects salinity due to dilution and low air temperatures accompanied with low radiation which are responsible for the drop in water temperature values [26]. In general, the inner stations presented higher temperature, pH and oxygen that gradually decrease towards to the mouth of the lagoon; however, the behavior of salinity and TDS were opposite. A complex environmental variability is created over short time scales, which regulate the abundance and distribution of ichthyofauna.

Only dissolved oxygen showed differences between stations, associated with differences in depth: their variation had been influenced by the wind stress and could be influenced with tidal effects, but were not studied. Dissolved oxygen concentrations are usually high in lagoons as the result of primary production and the constant influx and mixing of fresh and salt water, although naturally high levels of organic matter can reduce oxygen concentrations during the period of low water intake [27].

Relationship between fish species distribution and environmental variation

The MDS showed that fish species had different spatial preferences in all measured seasons; were the most frequent habitats occurred mainly with characteristic of channel (particularly stations 1 and 7)

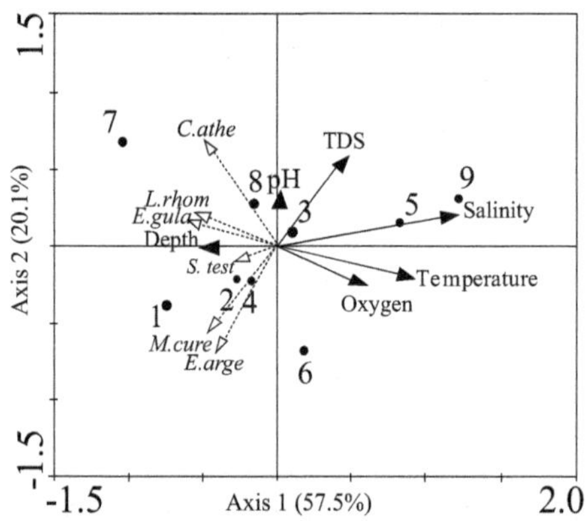

Figure 8B: Graph representing the statistics of the species and environment relation in the rainy season.

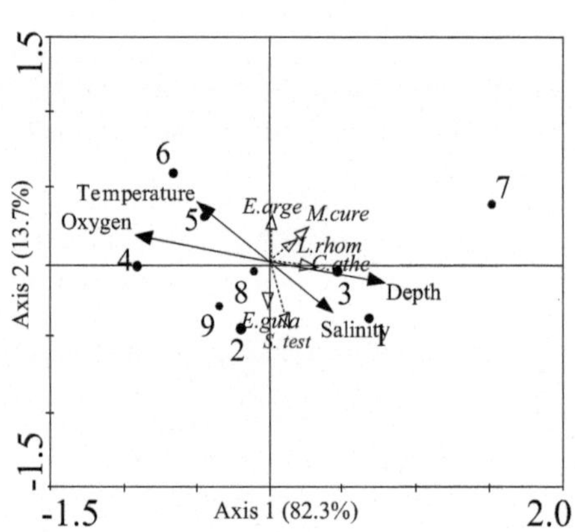

Figure 8C: Figure representing the axis 1 statistics.

Season	Variables	Inflation factor	Canonical coeff.-axis 1	Canonical coeff.-axis 2	Correlation (r) species-axis 1	Correlation (r) species-axis 2
Dry	Depth	1.37	-0.341	0.867	-0.27	0.76
	Oxygen	9.97	0.155	-0.689	0.13	-0.60
	Temperature	6.75	-0.134	-0.775	-0.11	-0.68
	Salinity	1.67	-0.786	-0.173	-0.63	-0.15
	pH	3.67	0.433	-0.531	0.35	-0.47
	TDS*	-	-	-	-	-
	Wind*	-	-	-	-	-
Rainy	Depth	2.35	-0.376	-0.011	-0.34	-0.01
	Oxygen	2.99	0.449	-0.329	0.41	-0.33
	Temperature	3.72	0.696	-0.266	0.63	-0.26
	Salinity	1.63	0.911	0.258	0.83	0.26
	pH	1.60	0.022	0.449	0.02	0.45
	TDS	1.94	0.362	0.747	0.33	0.74
	Wind*	-	-	-	-	-
North winds	Depth	1.64	0.711	-0.265	0.70	-0.22
	Oxygen	2.19	-0.821	0.339	-0.80	0.28
	Temperature	1.50	-0.433	-0.748	-0.42	0.61
	Salinity	2.18	0.390	-0.643	0.38	-0.53
	pH*	-	-	-	-	-
	TDS*	-	-	-	-	-
	Wind*	-	-	-	-	-

*Variables not included due to collinearity

Table 5: Results of the inflation factor, canonical coefficients and the best correlation (r) using redundancy analysis (RDA). The table show the environmental variables per climatic season.

and secondary with mangrove, and sites slightly deeper and with high salinities in the lagoon (Table 1, Figures 2 and 3). In these sampling stations, the distribution patterns of fish species overlap and coincide with the maximum observed abundance. Slightly deeper sites have been recognized as preferred habitats of some fish species [28].

No statistical differences in the spatial distribution of the environmental variables were observed throughout the system. However, three different zones were found; two inner zones of La Carbonera lagoon (corresponding to station 5 and 8 in the present work) with a high variability in temperature (between <29°C and <31°C) and salinity (between <38 and <40) and another zone with lowest temperature (<27°C) and salinity (<4) beside to the freshwater seep "*cenote*" of the lagoon (Figure 1) [26].

It must be kept in mind that the study area is a transitional zone between the sea and a series of small freshwater seeps; where populations of marine, estuarine and freshwater fish species are found. Although it is well known that underground rivers exist in the Yucatan peninsula, their contribution to this ecosystem has not been quantified.

In other hand, it was pointed out [27] that fluctuations of the environmental variables are related to the distance between one sampling sites to another along coastal lagoon systems Although statistical spatial patterns were not found in this study, it must be kept in mind that such variables produces that fish move to optimize the supply of resources in the lagoon. The distribution of fish species may be prompted by spatial gradients and seasonal environmental changes [29,30]; and also by physiological or morphological changes in juveniles that increase the susceptibility to environmental gradients. The number, density and biomass of the fish species found by [31] in a tropical estuary exhibit spatial and seasonal change due to fluctuations in salinity, and that salinity influenced also fish migration in the same spatial scale. A strong relationship has been reported between the distribution of the salinity and the distribution of the ichthyoplankton in a karstic coastal lagoon system in eastern Yucatan peninsula [32].

Seasonal differences in temperature and salinity influence the distribution patterns of fish communities in coastal systems [3,33]. Despite the environmental variability that characterizes La Carbonera, this lagoon system provides a favorable habitat and nursery area for juvenile fish due to its shallow, sandy bottom and the presence of mangroves. In particular, euryhaline species, such as *S. testudineus*, were found along the lagoon, regardless of depth, and were important contribution to the overall fish biomass in all months. As expected, environmental variables varied significantly along the year, influencing both abundance (Table 3) and the temporal distribution (Table 5). During the north winds season, for instance the group of data concerning of fish abundance, data is clearly separated from other seasons (Figure 7). There is an overlapping of the vectors during the dry season that suggests a strong correlation between the hydrological variables, especially salinity and depth of some sites (Figure 7 and 8a). It was reported that temperature and depth are important in influencing spatial distribution of *E. argenteus* abundance. Salinity can be one of the most important influences on the use of coastal system by organisms [34]. In a hypersaline coastal lagoon of the Yucatan coast, very high salinity (>80) decreases the number of fish species [7].

Data show the existence of a dependence of water depth and *L. rhomboides* abundance that can be the results of depth partitioning of both marsh and flat habitat-types during high tide. The same pattern, but considering salinity was described in a tropical estuarine system in the Gulf of Mexico, focusing on the abundance of *L. rhomboides* and *M. curema* and other fish species [35]. According to [10] fish population dynamics in areas that experience frequent dry periods are more likely to be dominated by hydrological changes. Species such as *M. curema* and *S. testudineus* that feed mainly on detritus [6,36] were more abundant during the dry season, which can be the result of increasing of organic matter across the lagoon [37].

The use of environmental variables as indicators of resident fish species preferences is based primarily on the tenet that fish species effectively integrate and reflect possible healthy conditions in the La Carbonera lagoon system, according the proposal of Whitfield and Harrison [33] considering other estuarine systems. It is important to consider that biological interactions, such as predator-prey relationships, food availability and often the reproductive biology of a species, influence the composition of fishes in a coastal lagoon [6]. Although not measured here, such variables may determine fish movements to optimize their performance and can be responsible for variations in the fish distribution in La Carbonera.

Multivariate analysis: implications on fish data

The ever-changing environmental conditions of shallow coastal lagoons are difficult to relate to the spatial and seasonal distribution of fish. In this study, combination of clustering and ordination analysis was the most effective way to check the adequacy and mutual consistency of fish species distribution [21]. Although similarity indices have commonly been used to evaluate differences or similarities in fish assemblages, these measures are mainly descriptive coefficients, rather than estimators of statistical parameters [35,38]. A constrained ordination produces as many canonical axes as there are explanatory variables, but each of these axes are linear combinations (a multiple regression model) of all explanatory variables. Examination of the canonical coefficients (i.e., the regression coefficients of the models) of the explanatory variables on each axis allows identification the set of variables that are significant in explaining the first and second axis. In this study, the explanatory variables were not always dimensionally homogeneous; thus canonical ordinations were obtained with standardized explanatory variables. In RDA, this does not affected the choice between running the analysis on a covariance or a correlation matrix, since this latter choice relates to the response of the environmental variables [24]. Hence, RDA was applied to the data matrix of fish species abundance in order to evaluate the changes between sites (sampling stations), environmental variables, and climatic seasons (dry, rainy, and north winds). The clustering showed that the six species followed spatial patterns of occurrence associated to water depth; with the exception of *C. atherinoides* which was not found at station 4 (seagrass habitat), and had an overlap, especially between the stations 1 and 7 (Figure 6). Main differences reported by [6] among fish assemblages in a long karstic coastal system (northwestern Yucatan peninsula) was the species composition in space and time owing to the behavior and biology of each species at a given time and place. The variables analyzed played an important role in regulating the annual patterns of fish species abundance in the La Carbonera coastal lagoon. Environmental variables did not show a spatial gradient as in other coastal systems of Yucatan [6,39], but rather a clear seasonal pattern influenced by climatic conditions.

In conclusion, we suggest that hydrological changes are an important factor shaping spatial and seasonal patterns; especially in habitats with characteristics of channel and mainly during the dry season, in the local fish populations across La Carbonera coastal lagoon. Furthermore, results did not differ significantly when analyzing the same abundance data set and abiotic variables with different multivariate techniques; rather it was shown that the combination of those techniques enhanced the interpretation of fish distribution [40]. Thus, patterns of distribution and habitat preferences of the abundant fish populations can be attributed to a synergy between abiotic variables which suggest that the different habitats are used as feeding zones; and that during the dry season mainly showed an overlapping in abundance of species. As such, investigating the role of those interactions for future research are necessary to propose protection strategies to contribute to fish biodiversity conservation in a valuable area.

Acknowledgements

This work was supported by PAPIIT (IN207609, IN213012) and CONACYT (103229) research grants. We thank Joel Loera Pérez, María Camarena, Andrea Camacho, and Juani Tzeek Tuz for technical support. We also want to express our gratitude to Ann Grant and Chelsea Combest-Friedman whose comments improved the manuscript on earlier versions for constructive comments. J. L. Bonilla-Gómez is grateful to the Consejo Nacional de Ciencia y Tecnología de México (CONACyT) for the scholarship awarded to pursue doctoral studies (CVU/ Becario: 178371/240833) in the Posgrado en Ciencias del Mar y Limnología at the Universidad Nacional Autónoma de México (UNAM).

References

1. Herrera-Silveira JA, Morales-Ojeda SM (2010) Subtropical karstic coastal lagoon assessment, Southeast Mexico: The Yucatan Peninsula case. In: Coastal lagoons: Critical habitats of environmental change. Kennish MJ & Paerl HW, eds., CRC Press, New York 307-333.
2. Day WJ, Hall CAS, Kemp WM, Yáñez-Arancibia A (1989) Estuarine Ecology. John Wiley and Sons, Toronto.
3. Selleslagh J, Amara R (2008) Environmental factors structuring fish composition and assemblages in a small macrotidal estuary (eastern English Channel). Estuar Coast Shelf Sci 79: 507-517.
4. Medina-Gómez I, Herrera-Silveira JA (2003) Spatial characterization of water quality in a karstic coastal lagoon without anthropogenic disturbance: a multivariate approach. Estuar Coast Shelf Sci 58: 455-465.
5. Arceo-Carranza D, Vega-Cendejas ME (2009) Spatial and temporal characterization of fish assemblages in a tropical coastal system influenced by freshwater inputs: northwestern Yucatan peninsula. Rev Biol Trop 57: 89-103.
6. Arceo-Carranza D, Vega-Cendejas ME, Montero-Muñoz JL, Hernández de Santillana MJ (2010) Influence of habitat type on diel fish associations in a tropical coastal lagoon. Rev Mex Biodivers 81: 823-837.
7. Vega-Cendejas ME, Hernández de Santillana M (2004) Fish community structure and dynamics in a coastal hypersaline lagoon: Rio Lagartos, Yucatan, Mexico. Estuar Coast Shelf Sci 60: 285-299.
8. Martino EJ, Able KW (2003) Fish assemblages across the marine to low salinity transition zone of a temperate estuary. Estuar Coast Shelf Sci 56: 969-987.
9. Vega-Cendejas ME, de Santillana MH, Arceo D (2012) Length-weight relationships for selected fish species from a coastal lagoon influenced by freshwater seeps: Yucatan peninsula, Mexico. J Appl Icthyol 28: 140-142.
10. Ruetz CR, Trexler JC, Jordan F, Loftus WF, Perry SA (2005) Population dynamics of wetland fishes: spatio-temporal patterns synchronized by hydrological disturbance? J Anim Ecol 74: 322-332.
11. Bonilla-Gómez JL, López-Rocha J, Badillo M, Tzeek-Tuz J, Chiappa-Carrara X (2011) Growth and mortality of *Lagodon rhomboides* (Pisces: Sparidae) in a tropical coastal lagoon in northwestern Yucatan, Mexico. Rev Mar Cost 3: 99-109.
12. Bonilla-Gómez JL, Tzeek-tuz J, Badillo M, Chiappa-Carrara X (2012) Growth and mortality of *Sphoeroides testudineus* (Tetraodontiformes: Tetraodontidae) in a tropical coastal lagoon in northwestern Yucatán Peninsula. Rev Mex Biodivers 83: 1237-1240.
13. Tapia FU, Herrera-Silveira JA, Aguire-Macedo ML (2008) Water quality variability and eutrofic tends in karstic tropical coastal lagoons of the Yucatan Peninsula. Estuar Coast Shelf Sci 76: 418-430.
14. Batllori-Sampedro E, González-Piedra JI, Días-Sosan J, Febles-Patrón JL (2006) Hydrological characterization of the Yucatan´s northwest coastal region, Mexico. Invest Geog 59: 74-92.
15. Lankford RR (1977) Coastal lagoons of Mexico: their origin and classification. In: Estuarine Processes. Wiley M, (ed.). Academic Press, New York 182-215.
16. Peralta-Meixueiro MA, Vega-Cendejas ME (2011) Spatial and temporal structure of fish assemblages in a hyperhaline coastal system: Ría Lagartos, Mexico. Neotropical Ichthyol 9: 673-682.
17. Zar JH (1999) Biostatistical analysis. Prentice Hall, Upper Saddle River, New Jersey.
18. Clarke KR (2001) PRIMER V5: User Manual/ Tutorial PRIMER-E. Plymouth, UK.

19. Kruskal JB, Wish M (1978) Multidimensional Scaling. Sage Publications, Beverley Hills.

20. Field JG, Clarke KR, Warwick RM (1982) A practical strategy for analyzing multispecies distribution patterns. Mar Ecol Prog Ser 8: 37-52.

21. Clarke KR, Warwick RM (2001) Change in marine communities: an approach to statistical analysis and interpretation. Natural Environmental Research Council, Plymouth, UK.

22. Clarke KR, Green RH (1988) Statistical Design and Analysis for a 'Biological Effects' Study. Mar Ecol Prog Ser 46: 280.

23. Ter Braak CJF, Prentice IC (2004) A theory of gradient analysis. Advances in Ecological Research 34: 235-282.

24. Lepš J, Šmilauer P (2003) Multivariate analysis of ecological data using CANOCO™. Cambridge University press, UK.

25. Ter Braak CJF, Šmilauer P (2002) CANOCO Reference Manual and CanoDraw for Windows User's Guide: Software for Canonical Community Ordination (Version 4.5). Microcomputer power, Ithaca, New York, USA.

26. Jeronimo G, Gómez-Valdés J, Badillo M, López-Aguiar K, Galindo C, et al. (2012) VARIACIONES ESTACIONALES DE TEMPERATURA Y SALINIDAD EN LA REGION COSTERA DE LA CORRIENTE DE CALIFORNLA 8: 167-178.

27. Moyle PB, Cech JJ (2000). Fishes: An introduction to ichthyology. (5thedn), Pearson Prentice Hall.

28. Lawror LR (1980) Overlap, similarity and competition coefficients. Ecology 61: 245-251.

29. Koenig WD (1999) Spatial autocorrelation of ecological phenomena. Trends Ecol Evol 14: 22-26.

30. Cocheret de la Morinière E, Pollux BJA, Nagelkerken I, van der Velde G (2003) Diet shifts of Caribbean grunts (Haemulidae) and snappers (Lutjanidae) and the relation with nursery to coral reef migrations. Estuar Coast Shelf Sci 57: 1079-1089.

31. Barletta M, Barletta-Bergan A, Saint-Paul U, Hubold G (2005) The role of salinity in structuring the fish assemblages in a tropical estuary. J Fish Biol 66: 45-72.

32. Chiappa-Carrara X, San Vicente-Añorve L, Monreal-Gómez A, Salas de León D (2003) Ichthyoplankton distribution as an indicator of hydrodynamic conditions of a lagoon system in the Mexican Caribbean. J Plankton Res 25: 687-696.

33. Harrison TD, Whitfield AK (2006) Temperature and salinity as primary determinants influencing the biogeography of fishes in South African estuaries. Estuar Coast Shelf Sci 66: 335-345.

34. Marshall S, Elliott M (1998) Environmental influences on the fish assemblages of the Humber Estuary, U.K. Estuar Coast Shelf Sci 46: 175-184.

35. Castillo-Rivera M, Zavala-Hurtado JA, Zarate R (2002) Exploration of spatial and temporal patterns of fish diversity and composition in a tropical estuarine system of Mexico. Rev Fish Biol Fisher 12: 167-177.

36. Franco L, Bashirullah KMB (1992). Alimentación de la lisa (*Mugil curema*) del golfo de Cariaco-Estado Suche, Venezuela. Zootecnia Tropical 10: 219-238.

37. Elliott M, de Jonge VN (2002) The management of nutrients and potential eutrophication in estuaries and other restricted water bodies. Hydrobiologia 475-476: 513-524.

38. Krebs CJ (1998) Ecological Methodology. Addison-Wesley Educational publishers, New York.

39. Rueda M (2001) Spatial distribution of fish species in a tropical estuarine lagoon: a geostatistical appraisal. Mar Ecol Prog Ser 222: 217-226.

40. González-Sansón G, Aguilar C, Hernández I, Cabrera Y (2009) Effects of depth and bottom communities on the distribution of highly territorial reef fish in the northwestern region of Cuba. J Appl Ichthyol 25: 652-660.

5

Impact of Human Activity on Marine and Coastal Environment in the Gulf of Tadjourah

Awaleh MO[1]*, Hoch Farhan Bouraleh[1], Soubaneh YD[2], Badran M[3], Bahga HO[1] and Samaleh AI[1]

[1]*Chemistry laboratory, Institute of Earth Sciences, Centre for studies and Research in Djibouti (CERD), Airport Road, PO Box 486, Djibouti - city, Republic of Djibouti*

[2]*Department of Biology, Chemistry and Geography, University of Quebec at Rimouski, 300, Allée des Ursulines, Rimouski, QC, G5L 3A1, Canada*

[3]*Regional Organization for the Conservation of Environment of the Red Sea and Gulf of Aden "PERSGA", PO Box: 53662 Jeddah 21583 Saudi Arabia*

Abstract

Extensive research with the aim of establishing seawater quality monitoring is considered essential of any integrated coastal management program. The present study reports for the first time the republic of Djibouti costal water quality. The state of the sixteen stations' seawater quality was assessed on the basis of determination of temporal and spatial variability of inorganic nutrients with physicochemical variables. The samples were collected seasonally from different areas such as harbor and important touristic area in the Gulf of Tadjourah for three years (2008, 2009 and 2012). The seawater temperatures, pH or Chlorophyll a of sampling sites were evaluated and compared to those of the Gulf of Tadjourah or Red Sea waters. Relatively high concentrations of nutrients (for some sites) and very low chlorophyll a concentrations (0.006 to 0.06 $\mu g.l^{-1}$) were observed at sampling sites. The seawater concentrations of trace metals in ten stations across the Doraleh coast, where is located the main port of Djibouti, were also investigated in 2012 and the values were compared to the normal range of concentrations for seawater. The levels of microbial concentrations were also determined for the main beaches of Djibouti-city and showed relatively higher concentrations for stations beaches close to sewage outfall.

Keywords: Djibouti; Gulf of Tadjourah; Seawater; Inorganic nutriments; Heavy metals; Coliform

Introduction

Seawater and beach quality monitoring assessment is considered essential parts of any integrated coastal management program [1]. Extensive research with the aim of establishing guidelines and standards for recreational water quality has been conducted all over the world. In this context, social, cultural, environmental, and economic factors should be taken into consideration because of the great variation from one area to another [2]. The republic of Djibouti (Horn of Africa), with a coastal length of about 372 Km lies between 10-13°N and 41-44°E. About 58% of the population of the Republic of Djibouti, which has been estimated at about 800000 inhabitants, lives in the capital city of Djibouti [3]. In Djibouti-city, untreated waste waters were directly discharged into the coastline. This situation is of great concern for the coastal environment as well as for the health of the people. Djibouti coastline is experiencing a rapid increase of coastal utilization for several purposes such as recreation and refreshment, fishing or industrial development. This makes coastal water quality and proper environmental management highly pressing. At present very little information is available on the state of the water quality of Djibouti coastline. Realizing the importance of this information for tourism development and consequently on the national economy, the Government of Djibouti in collaboration with the Regional Organization for Conservation of the Environment of the Red Sea and Gulf of Aden (PERSGA) promoted the initiation of a sustainable monitoring program of the coastal and marine environment, aiming at establishing a national database that serves the Republic of Djibouti and feeds in a regional database of the Red Sea and Gulf of Aden. A database will help significantly in the scientifically based management of the coastal region. The aims of this work are to investigate the spatial and temporal pattern of the water quality along the Djibouti coasts and thus provide a basic profile of seawater characteristics on the coasts of Djibouti. The area under investigation includes industrial effluent sites, recreational resorts, urban agglomeration, fishing ports, shipping and industrial activities area as well as coastal site away from development and a reference point five Kilometers offshore.

Material and Methods

Reagents and standards

Seawater certified reference materials (MOOS-2, NASS-5, and CASS-4), were provided by National Research Council (NRC), Canada. Standard stock solutions of each metal (Cd, Co, Cr, Cu, Fe, Mn, Ni, Pb and Zn) were purchased from Sigma-Aldrich (Germany). Hydrochloric and nitric acid were purchase from Sigma-Aldrich (Germany). Reagents and standard solutions used during the analyses were of supra pure grade.

Sampling site selection and identification

The sampling campaigns were conducted at 26 selected sampling sites for the investigations of seawater in Djibouti shoreline area (Table 1 and Figure 1). Sixteen sites were positioned along the Djibouti coasts and were selected on the basis of human activity, such as recreation, on the coast. Seawater for inorganic nutriments (nitrate, nitrite, phosphate, and silicate in μM) analyses was collected from these locations during summer (August) and winter (December) in three years (2008, 2009 and 2012). Ten sampling sites were located in Doraleh Port (Figure 1) and were selected due to specific activities take place.

***Corresponding author:** Mohamed Osman Awaleh, Laboratoire de Chimie, Institut des Sciences de la Terre, Centre d'Etudes et de Recherches de Djibouti (CERD), Route de l'aéroport, B.P. 486, Djibouti – ville, République de Djibouti
E-mail: awaleh@gmail.com

Sites #	Sampling site	GPS Coordinates	
1	Off shore	011°40'11,0"	043°10'43,75"
2	Abattoir	011°34'57,34"	043° 09' 32,78"
3	Siesta Beach	011°35'56,02"	043°09'13,62"
4	Peltier Hospital	011°37'03,61"	043°09'24,13"
5	Kempesky Hotel	011°37'23,76"	043°08'49,01"
6	Escale – Venus	011°36'04,33"	043°08'34,85"
7	Fishing Port	011°35'59,77"	043°07'49,53"
8	Doraleh Port	011°34'51,75"	043°29'31,85"
9	Petit Ambado Beach	011°35'33,90"	043°03'27,11"
10	Khor Ambado	011°35'37,11"	043°01'20,21"
11	Arta Beach	011°35'04,19"	042°49'13,59"
12	Goubet	011°32'08,48"	042°31'53,93"
13	Tadjourah	011°46'42,07"	042°54' 08,63"
14	Sable Blanc Beach	011°46'41,08"	042°55'30,47"
15	Maskali Island	011°42'39,04"	043°08'57,38"
16	Moucha Island	011°43'04,00"	043°11'19,90"
17	Doraleh coastline	011°35'43,4"	043°04'19,9"
18	Doraleh coastline	011°35'43,5"	043°04'08,0"
19	Doraleh coastline	011°35'43,7"	043°04'06,8"
20	Doraleh coastline	011°35'45,7"	043°04'00,1"
21	Doraleh coastline	011°35'43,4"	043°03'56,6"
22	Doraleh coastline	011°35'43,3"	043°03'52,6"
23	Doraleh coastline	011°35'42,0"	043°03'50,2"
24	Doraleh coastline	011°35'41,6"	043°03'49,1"
25	Doraleh coastline	011°35'42,06"	043°03'43,03"
26	Doraleh coastline	011°35'43,37"	043°03'43,04"

Table 1: Sampling sites identification information.

Sample collection

All containers used to collect and store seawater were soaked in 10% HCl for 24 hours and washed with deionized distilled water prior to using to prevent contamination. All containers used to collect and process the samples were rinsed at least three times with water samples prior to their collection. Water samples for the determination of nutrients and chlorophyll a were taken in polyethylene containers and kept on an ice box during the transportation. Upon return to the laboratory, the samples were filtered through Nylon Millipore membrane filters (0.45 µm). The Nylon filters were used for the determination of chlorophyll a and the filtrate for nutrients analysis. For heavy metal analysis, water samples were acidified to pH2 using concentrated HNO_3 immediately after collection, and stored at 4°C until analysis.

Physicochemical analysis

For the sixteen sampling sites # 1-16 (Figure 1) inorganic nutriments analysis were carried out. While for samples from Doraleh area (sites # 17-27), dissolved oxygen (DO), suspended solid (SS), Nitrite, Ammonium, Phosphate, Silicate, Chlorophyll a and Heavy metals (Cd, Co, Cr, Cu, Fe, Mn, Ni, Pb, Zn) were analysed. All analyses were made in triplicate. The method detection limit for Cd, Co, Cr, Cu, Fe, Mn, Pb, Zn, and Ni, were 0.1, 0.2, 0.2, 0.2, 0.2, 0.05, 1.5, 0.2 and 0.3 µg/L respectively. Two certified reference materials, CASS-4 (near shore seawater) and NASS-5 (seawater), were used as a control for the analytical methods of heavy metals in seawater (e.g. Cd, Co, Cr, Cu, Fe, Mn, Pb, Zn and Ni). On the other hand, seawater certified reference material for nutrients (MOOS-2) was used as a control for the analytical methods of nutrients. Temperature, electrical conductivity, salinity, dissolved oxygen and pH were measured in situ with a C862

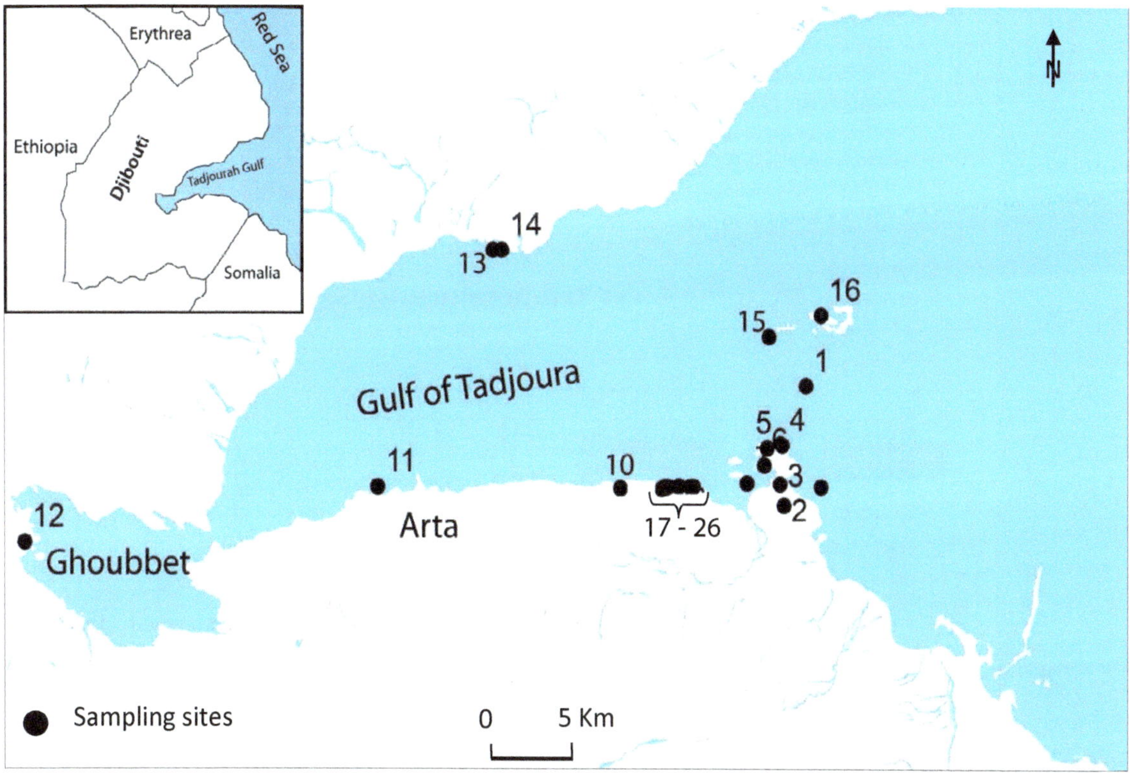

Figure 1: Djibouti National Monitoring program sampling sites (windows : location of the Republic of Djibouti).

multi-parameter meter and associated electrodes. Calibration of each electrode was checked every day prior to the start of sampling. Nitrite, Ammonium, Phosphate, Silicate, Chlorophyll a analyses were done according to protocol described by Strickland and Parsons [4]. Analyses of heavy metals were conducted using an Ultima 2 (Horiba Jobin Yvon) Inductively Coupled Plasma Atomic Emission Spectrometer (ICP-AES). Calibration was performed using NASS-5, to minimize matrix and other associated interference effects, while CASS-4 was used to check the precision and accuracy of the analysis.

Statistical analysis of data

The computer program R2.15.1 [5] was used to conduct statistical analysis. Data obtained from seawater samples from Djibouti recreational area (sites # 1-16) were also analyzed using the Pearson correlation coefficient (a measure of linear association) and paired sample t test.

Determination of bacteria concentration

Seawater samples for Fecal coliform (FC) and fecal streptococci (FS) analyses were collected with sterile 500 mL bottles, which was then kept in the dark and on ice during its transfer to the laboratory, and analyzed within 4 h of sample collection. An appropriate volume of a seawater sample (100 mL) was filtered through a 0.45-mm membrane filter. The membrane filters were aseptically removed and placed in the center of a pre labeled Deoxycholate Agar culture plate for the analysis of Total Fecal coliform and Fecal coliform, which were then sealed and incubated for 24 hours at 44.5 ± 2°C and at 37.1 ± 1°C, respectively. Three to five colonies from each plate were picked and biochemical tests were performed to confirm the identity [6]. Fecal streptococci (FS) were analyzed similarly to fecal coliforms, however, the membrane filters were transferred to Baird Parker medium which was then sealed and incubated at 37.1 ± 1C for 24 hours. API 20E was used to confirm FC identity, gram stain, catalase, and bile esculin tests were used to confirm FS. From each sample lot representative colonies of *S. aureus* were chosen and subjected to the coagulase test. For Salmonella and Shigella analyses, 1 L of seawater sample was filtered through a 0.45-mm membrane filter. The membrane filter was placed at 37°C in Hecktoen enrichment medium for 24 hours. Subcultures were made to xylose lysine dextrose agar (XLD) and Salmonella Shigella agar (SSA) plates (Oxoid).

Results and Discussions

Recreation area

The recreational areas in Djibouti-city such as Siesta Beach (site#3), Petit Ambado Beach (site #9), Khor Ambado Beach (site #10) as well as Escale-Venus (site #6), were selected for Djibouti National Monitoring Program initiated by PERSGA (Table 1). Two island beaches, Moucha (site #16) and Maskali (site #15), having hotels for the tourist visiting the area as well as Sable Blanc beach (site #14) and Arta beach (site #11) were also selected for this monitoring program because of touristic activity. Wastewater from the Djibouti slaughterhouse are discharged without any treatment to the seawater, at Abattoir site (site #2), which has been selected for monitoring. Furthermore, the Djibouti fishing port (site #7), which hosts some seafood industrial activities, was also selected as monitoring point. Tadjourah (site #13) has been chosen for the monitoring program as a coastal city and capital of Tadjourah District. The Goubet site (site #12), located at the embouchure of the Gulf of Tadjourah, was selected because it hosts a motel for tourist activity. The Djibouti general hospital discharges its wastewaters to the seawater. Therefore, this point of discharge, Peltier hospital (site #4), was chosen as one of the monitoring site. The coast of the largest hotel in Djibouti-city, kempeski (site #5), has also been sampled in this monitoring program. Finally, a point five kilometres offshore was selected as reference site (site #1). Seawater temperature, pH and concentration of the inorganic nutriments recorded at Djibouti monitoring sampling points for four sampling periods (April 2008, December 2008, April 2009 and December 2009) are shown in Figures 2-7. The seawater temperatures in April ranged between 30°C and 33°C and in December between 27°C and 28°C. The results indicated that upper layer temperatures of sea water are usually more important in April (summer) than in December (winter). This is in accordance with the temperatures that were found to fluctuate between 26°C and 27°C for March and 27°C to 30°C for July to September in the upper layer of the sea [7]. In the Gulf of Tadjourah, the thermocline was found around 60 m in March (end of winter). At this period, the upper layer temperature was 26.5°C and the temperature decreased to about 22.5°C between 60 and 110 m [7]. The water temperature variations of the Gulf of Tadjourah are influenced by the mixing deep water circulation from the Gulf of Aden. The vertical distribution of temperature from surface

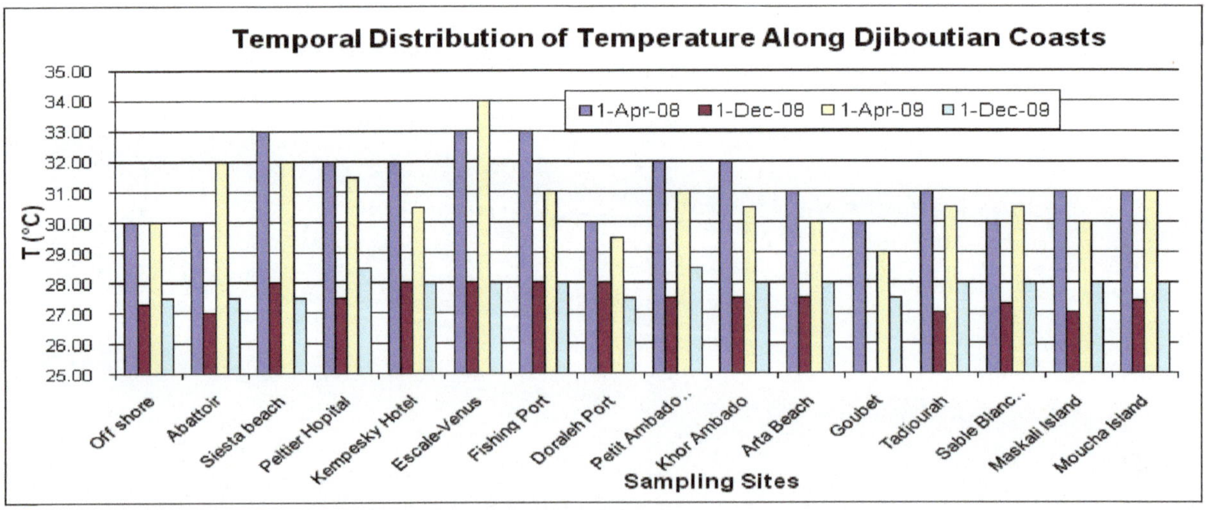

Figure 2: Seawater surface temperature at Djibouti monitoring sites during April and December 2008, 2009.

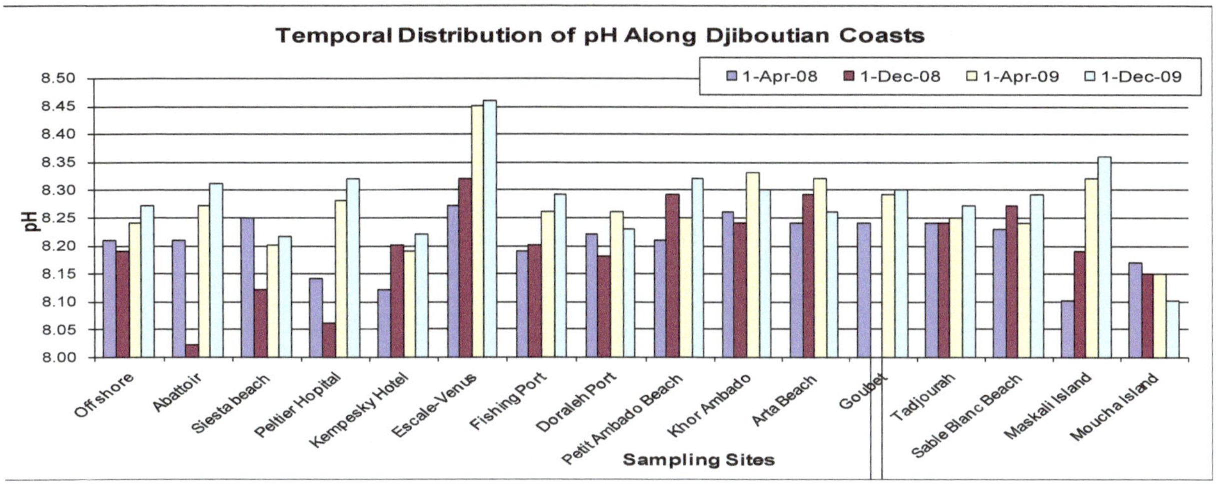

Figure 3: Surface seawater pH at Djibouti monitoring sites during April and December 2008, 2009.

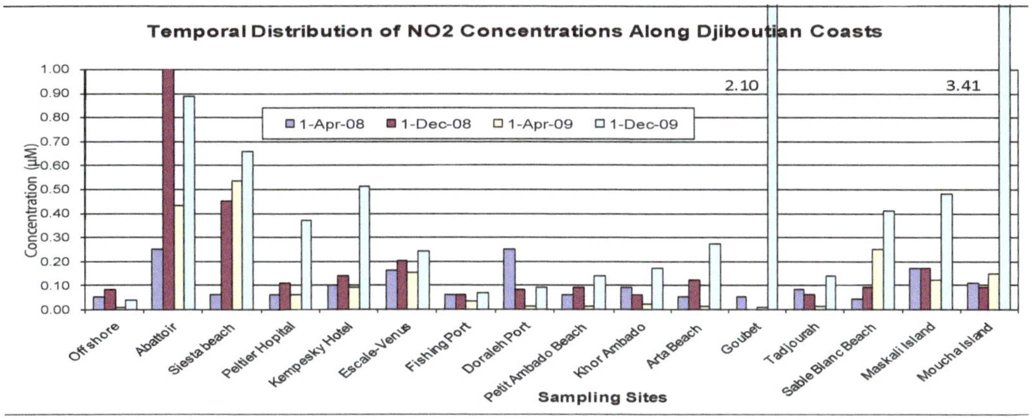

Figure 4: Surface seawater nitrite concentration at Djibouti monitoring sites during April and December 2008, 2009.

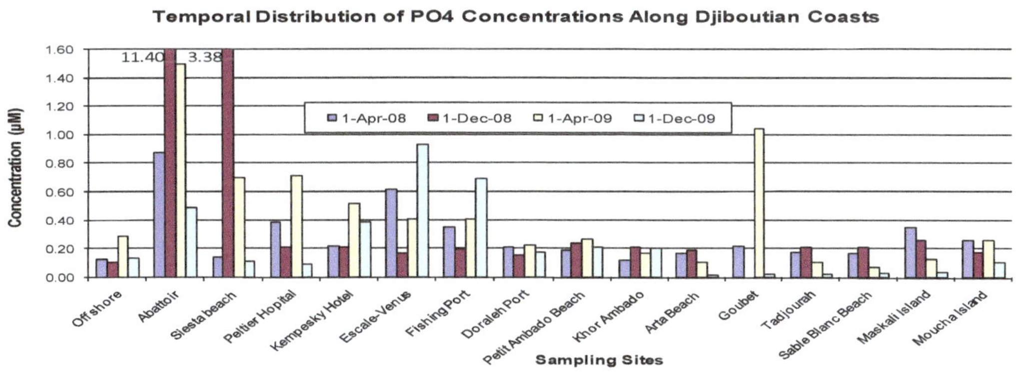

Figure 5: Surface seawater phosphate concentration at Djibouti monitoring sites during April and December 2008, 2009.

to 400 m depth in the Gulf of Aden shows that the thermocline is weaker in winter (from 16 to 25.5°C) than summer (from 16.5 to 29°C) [8]. Moreover, the deep water of the Gulf of Aden shows a steady decrease of temperature as low as 10°C from 1000 to 2000 m depth. This is a result of its direct connection with the Indian Ocean [9]. Seawater temperature is a conservative characteristic of a marine ecosystem that along with salinity are amongst the most important physical variables determining basic features of a water body. In addition of setting tolerance limits to most living organisms, these two variables combined determine the density of a water body, which in turn plays a significant role in controlling water transport both vertically and horizontally. The shore of the Gulf of Aden is considered as one of the hottest climates in

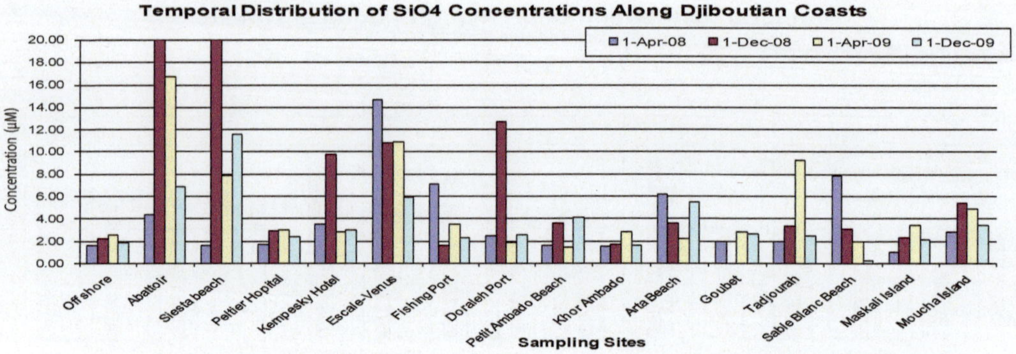

Figure 6: Surface seawater silicate concentration at Djibouti monitoring sites during April and December 2008, 2009.

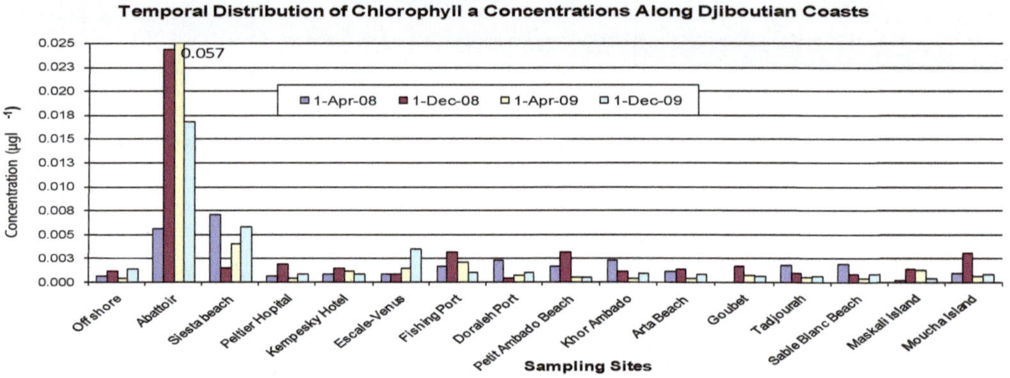

Figure 7: Surface seawater chl a concentration at Djibouti monitoring sites during April and December 2008, 2009.

the world [9]. The highest surface water temperatures recorded in April in the Djibouti costal water were positively correlated to the high air temperatures found (30°C to 40°C) at this season (April to September). The mixing process seems to have little influence on the temperature of the coastal water compared to temperature variations observed in the Gulf of Tadjourah or the Gulf of Aden. The surface seawater pH ranged between 8.0 and 8.5. All pH measurement were found to be in the normal seawater pH range, 7.5-8.5 [10] and typical of waters of the Red Sea and Gulf of Aden [11]. These pH values suggest the absence of upwelling waters in the sampled costal sites during the sampling periods. The pH values are lower in the upwelling waters body in the equatorial regions and are proportional to the temperature [12]. Chlorophyll a (Chl a) concentrations were generally low in all sampling sites, rarely exceeding 0.06 $\mu g.l^{-1}$. These concentrations are to be considered indicative as the amount of seawater filtered for the spectrophotometric analysis of Chl a was relatively small. However, the spatial distribution of Chl a concentrations indicated clearly that Abattoir site had the highest value in three sampling events (Dec. 2008, Apr. 2009, and Dec. 2009) as compared to the other sites. Low Chl *a* concentration was reported in the Red Sea from 1998 to 2009 with concentrations as low as 0.08 $\mu g.l^{-1}$ [13]. Strong co-variations between Chl a and the physical parameters (e.g. wind) showed that physical processes have effect on the changes in the phytoplankton blooms in the Red Sea [13] (Table 2). Relatively high concentrations of nutrients were recorded, such as a nitrite concentrations of 2.10 and 3.41 µM at Goubet and Moucha Island respectively in December 2009; phosphate concentration of 3.38 µM at Siesta Beach in December 2008; and silicate concentration of 52.58 µM at Siesta Beach in December 2008. The Gulf of Aden is known for upwelling that occurs in late summer and results in increased nutrient concentrations in the surface water. High nutrient concentrations have been recorded in August and September with nitrite concentrations reaching up to 20 µM during upwelling of Somalia and southern of Yemen [14]. With all sampling events in the present study not coinciding with upwelling season, the elevated nutrient concentrations need further investigation before solid judgments on the cause could be made. Wastewater discharge into the Djibouti shoreline influences the quality of the marine environment, which is extremely important to Djibouti as a renewable source of food, a source of desalinated water for drinking and a place for recreational activities. On the other hand, microbial contamination of marine waters worldwide is estimated to cause millions of gastrointestinal and acute respiratory infections [15] and numerous skin infections [16] every year. Runoff, floods, and sewage discharges release large numbers of microorganisms, including fecal indicators, into seawater. This was clearly demonstrated by the high fecal coliform (FC) and fecal streptococci (FS) contents of Djibouti-city Beaches (Table 3). The highest concentration of fecal indicators was found in Siesta beach (site #3) (Table 3). This beach is the most frequented one in Djibouti-city and was far about 1 Km from sewage outfall. It is noteworthy that during the sampling process there are no bathers. Abattoir (site #2) and Peltier (site # 4) locations, adjacent to sewage outfalls, have shown also high level of FC and FS contents (Table 3). On the other hand, Escale-Venus location (site # 6), where restaurants discharged directly their wastewaters into the sea, is much less polluted than locations 2-4 (Table 3). This could be due to the fact that the sites 2-4 were located near sewage outfalls that discharged mainly septic wastewater with long retention times compared to site 6 where restaurant discharge directly their wastewater into the sea. Generally, domestic wastewater in

Site #	pH	EC (mS/cm)	T (°C)	NO_2 (µM)	PO_4 (µM)	SiO_2 (µM)	Chl *a* (µg/l)	DO ($mg.l^{-1}$)	Cd (µg/L)	Co (µg/L)	Cr (µg/L)	Cu (µg/L)	Fe (µg/L)	Mn (µg/L)	Pb (µg/L)	Zn (µg/L)	Ni (µg/L)
17	8.41 ± 0.03	50.41 ± 0.05	28	0.343 ± 0.01	1.383 ± 0.02	8.53 ± 0.3	0.020 ± 0.01	7.1 ± 0.1	< 0.1	< 0.2	0.88 ± 0.003	< 0.2	0.56 ± 0.002	04.39 ± 0.004	< 1.5	< 0.2	< 0.3
18	8.42 ± 0.03	50.43 ± 0.04	28	0.167 ± 0.01	0.510 ± 0.01	4.4 ± 0.2	0.012 ± 0.01	7.5 ± 0.2	< 0.1	< 0.2	0.76 ± 0.001	< 0.2	0.34 ± 0.001	04.91 ± 0.004	< 1.5	< 0.2	< 0.3
19	8.48 ± 0.05	50.93 ± 0.06	28	0.114 ± 0.02	0.874 ± 0.01	5.25 ± 0.1	0.013 ± 0.01	7.8 ± 0.3	< 0.1	< 0.2	0.55 ± 0.003	< 0.2	0.55 ± 0.004	05.56 ± 0.002	< 1.5	< 0.2	< 0.3
20	8.51 ± 0.06	50.12 ± 0.03	29	0.202 ± 0.01	0.874 ± 0.01	4.4 ± 0.1	0.010 ± 0.03	7.9 ± 0.2	< 0.1	< 0.2	0.57 ± 0.004	< 0.2	0.43 ± 0.002	03.82 ± 0.004	< 1.5	< 0.2	< 0.3
21	8.48 ± 0.06	50.15 ± 0.04	28	0.158 ± 0.01	0.620 ± 0.03	4.94 ± 0.1	0.010 ± 0.01	8.2 ± 0.1	< 0.1	< 0.2	0.84 ± 0.002	< 0.2	0.36 ± 0.001	03.97 ± 0.005	< 1.5	< 0.2	< 0.3
22	8.43 ± 0.05	50.01 ± 0.04	29	0.097 ± 0.03	0.764 ± 0.02	8.13 ± 0.3	0.013 ± 0.01	7.9 ± 0.1	< 0.1	< 0.2	0.42 ± 0.001	< 0.2	0.74 ± 0.004	12.08 ± 0.001	< 1.5	14.4	< 0.3
23	8.50 ± 0.03	50.06 ± 0.07	29	0.132 ± 0.01	0.655 ± 0.01	8.67 ± 0.3	0.006 ± 0.02	8.1 ± 0.2	< 0.1	< 0.2	0.86 ± 0.003	< 0.2	0.32 ± 0.005	06.45 ± 0.004	< 1.5	12.5	< 0.3
24	8.52 ± 0.04	50.03 ± 0.06	29	0.035 ± 0.03	0.692 ± 0.01	18.1 ± 0.5	0.007 ± 0.01	8.0 ± 0.1	< 0.1	< 0.2	0.74 ± 0.001	< 0.2	0.54 ± 0.006	09.78 ± 0.002	< 1.5	6.3	< 0.3
25	8.49 ± 0.06	50.03 ± 0.04	29	0.079 ± 0.01	0.692 ± 0.03	9.57 ± 0.4	0.01 ± 0.04	7.9 ± 0.3	< 0.1	< 0.2	0.86 ± 0.003	< 0.2	0.41 ± 0.002	06.12 ± 0.005	< 1.5	< 0.2	< 0.3
26	8.48 ± 0.05	50.12 ± 0.03	29	0.062 ± 0.01	0.656 ± 0.02	13.4 ± 0.4	0.01 ± 0.03	7.6 ± 0.1	< 0.1	< 0.2	0.79 ± 0.002	< 0.2	0.64 ± 0.001	05.52 ± 0.004	< 1.5	< 0.2	< 0.3

Table 2: Physical and chemical parameters of the Doraleh coastline seawaters.

Site #	Sampling locations	TC (UFC/100 ml)	FC (UFC/100 ml)	FS (UFC/100 ml)	SAL (UFC/100 ml)	SHI (UFC/100 ml)
2	Abattoir	45000	40000	2750	0	0
3	Siesta Beach	75000	71500	2500	0	2000
4	Peltier Hospital	55000	25000	600	0	9000
5	Kempesky Beach	60000	55000	650	0	0
6	Escale – Venus	100	24	17	0	0
7	Fishing Port	140	49	32	0	0
10	Khor Ambado Beach	0	0	0	0	0
11	Arta Beach	0	0	0	0	0

Table 3: Levels of Total Coliform (TC), Fecal Coliform (FC), Fecal Streptococci (FS), Salmonella (SAL), and Shigella (SHI).

Djibouti is of high strength and is often septic because of low flows, long retention times and high temperatures [17,18]. Kempesky beach (site # 5) seawater has shown high level of FC and FS concentrations (Table 3) even though this location was far from any sewage outfall. However, this beach is located less than 1 Km from Djibouti Port. Therefore, these results could be interpreted by the fact that the current direction was mainly from the Djibouti Port toward Kempesky beach (site # 5) most of the monitoring period. This finding suggests that any future monitoring program should take into consideration a daily record of current direction and, if possible, current speed. Khor Ambado (site # 10) and Arta (site # 11) beaches were found to be clean (Table 2). These results could be explained by the fact that there are no sewage outfalls near these beaches. During the 2012 monitoring, E. coli were isolated from all sampling points except Khor Ambado (site # 10) and Arta (site # 11) beaches. Djibouti-city beaches were mainly evaluated using the European Community (EU) standards for FC (EU mandatory: 2000 cfu/100 ml and EU guidelines: 100 cfu/100 ml) and FS (EU guidelines: 100 cfu/100 ml) [19]. It can be observed from Table 3 that, in general, most of the Djibouti-city beaches located near sewage outfall failed to comply with the European Community Bathing Water Directive. Furthermore, the presence of *Staphylococcus aureus* (*S. aureus*) in marine coastal environments may be considered as a risk indicator for skin, eye and ear diseases [20]. It should be noted that *S. aureus* were identified in seawater from Abattoir (site 2), Siesta Beach (site #3), Peltier (site #4), and Venus-Escale (site #6).

Doraleh port area

Doraleh port is one of the two main ports of Djibouti where almost all commodities as well as oil is delivered. Thus, the Doraleh port was selected as one of the monitoring point for the Djibouti Sustainable National Monitoring Program. Furthermore, in 2012, an elaborated campaign was conducted to collect water samples from ten (10) stations at the Doraleh littoral area for nutriment and heavy metals analyses (Figure 1). The locations of these stations are summarized in Table 1. In the Red Sea the horizontal and vertical circulation, water temperature and salinity mainly determine the concentration of dissolved oxygen [21]. Indeed, the dissolved oxygen in water is usually depending on its temperature and salinity. It is also depending on a considerable degree on the quantity of organic matter present in the aquatic environment. The dissolved oxygen levels in Doraleh coastal waters ranged 7.1-8.2 mg/L, with an average mean level of dissolved oxygen 7.8 mg/L (Table 2). These values were in the range of the concentration of dissolved oxygen in the surface water of the Red Sea during winter [21]. Nitrite concentrations ranged from 0.04 to 0.34 μM (Table 2). These values are within the reported nitrite concentrations in Red Sea, Aqabe Gulf (Jordanian) and Jeddah Red Sea [22]. Phosphate concentrations at Doraleh coastal area ranged from about 0.50 to 1.38 μM (Table 2). These values are slightly higher than the phosphate concentration reported for Jeddah Red Sea [22]. However, phosphate concentrations at Doraleh coastal are in the same range than to other sites in Djibouti Monitoring Program (Figure 5). Silicate Concentrations in the present campaign, ranged from 4.40 to 18.15 μM. These values were in the range of Djibouti coastline silicate concentration [23]. It has been reported that diatoms always numerically dominated the phytoplankton community when concentrations of silica were in excess of 2.2 μM [24]. On the other hand, the diatoms depend also on N/PO_4 [25] and SiO/N ratios [26]. Dominance by diatoms ceased or became more variable when concentrations of Si were less than this value. Previous study in Doraleh area reported that the silicate ranged from 2.55 to 4.11 μM in summer and 1.38-1.87 μM in winter [23]. On the basis of this study, one can speculate that diatoms dominate the summer phytoplankton community in Doraleh coastline. Chlorophyll a concentrations ranged from 0.006 to 0.02 $\mu g.l^{-1}$ for the Doraleh coastal area. These values are well below the reported value for chl a in Aqaba Gulf and Suez Gulf, 0.1-1.0 μg/L and 0.5-4.0 μg/L respectively [26]. However, Chl a concentrations in Doraleh shoreline are in the same order of magnitude that the Gulf of Tadjourah seawaters (Figure 7). In addition, similar Chl a concentrations were obtained when 5 or 10 liters of seawater were filtered for the spectrophotometric analysis of Chl a. In other terms, Tadjourah Gulf seawaters featured low Chl a concentrations in summer as well in winter. In order to further investigate the natural cycle that is responsible for this low Chl a concentration, three stations were set up in 2014 into the Gulf of Tadjourah to monitor at different depths the temperature, salinity, dissolved oxygen, and Chl a on a daily basis. Heavy metal concentrations at the 10 sampling stations within Doraleh coastline are shown in Table 2. The copper, lead, cadmium, nickel and cobalt were below the detection limits for all station. Zinc was revealed in three stations (sites # 22-24) while iron, chromium and manganese could be detected at all stations (Table 2). The chromium concentrations of the Doraleh coastal waters ranged from 0.42 to 0.88 $\mu g.l^{-1}$ (Table 2). These values are within the normal range for seawater [26]. On the other hand, Tring Xuan Gi [26] reported that in general the zinc concentration in seawater (over the world) range from 0.65 μg/L (Pacific Ocean) to 42 $\mu g.l^{-1}$ (South China sea) passing through 7.92 $\mu g.l^{-1}$ (Atlantic Ocean) and 27 $\mu g.l^{-1}$ (South African Coast). For Doraleh coastal water, the zinc concentrations for stations 17-22 and 25-26 are below the detection limits (0.2 $\mu g.l^{-1}$), while the zinc concentrations of stations 6-8 range 6.32-14.4 $\mu g.l^{-1}$. These values were within the normal range for seawater [26,27]. Iron is present in very low concentrations in the oceans (0.003-0.11 $\mu g.l^{-1}$) [28-30], despite its enhanced abundance in the earth's crust (5.6%) [31], and is a vital constituent of plant life. The element plays an important role in plant metabolism where it is essential for photosynthetic and respiratory electron transport, nitrate reduction, chlorophyll synthesis and detoxification of reactive oxygen species [32]. The iron concentrations in Doraleh coastal waters range 0.32-0.74 $\mu g.l^{-1}$ (Table 2). These values were greater than typical iron concentration in ocean. However, samples taken from a point 1 kilometer offshore in Doraleh coastline area showed iron concentrations below 0.1 $\mu g.l^{-}$

[1], which was within the normal range for seawater [28-30]. In other terms, the iron concentration in seawater sampled in the vicinity of the Doraleh Port were slightly high (0.32-0.74 $\mu g.l^{-1}$), while those from a point 1 kilometer offshore were in the normal range for seawater (0.1 $\mu g.l^{-1}$). Therefore, the slightly high iron concentrations of Doraleh coastline would probably originated from the Doraleh Port activity.

Statistical analysis

Hierarchical cluster analysis (HCA) is a commonly used method to classify observations or variables, in order to define more or less homogeneous groups and emphasize their genetic relations [33]. A classification scheme using Euclidean distance for similarity measurement, together with Ward's method for linkage [34] produces the most distinctive groups where each member within the group is more similar to its fellow members than to any member outside the group [35]. In this study, HCA was applied to the Djibouti recreation area data. Twenty four variables (T-Summer-2008, pH-Summer-2008, NO_2^--Summer-2008, PO_4^{3}-Summer-2008, SiO_2-Summer-2008, Chl a-Summer-2008, T-Winter-2008, pH-Winter-2008, NO_2^--Winter-2008, PO_4^{3}-Winter-2008, SiO_2-Winter-2008, Chl a-Winter-2008, T-Summer-2009, pH-Summer-2009, NO_2^--Summer-2009, PO_4^{3}-Summer-2009, SiO_2-Summer-2009, Chl a-Summer-2009, T-Winter-2009, pH-Winter-2009, NO_2^--Winter-2009, PO_4^{3}-Winter-2009, SiO_2-Winter-2009, Chl a-Winter-2009) were considered to classify all the water samples from this study. In HCA the variables are log transformed and normalized so that each variable has equal weight. The HCA results were presented in a dendrogram (Figure 8). The hierarchical analysis allowed us to distinguish four groups of water: Cluster 1 (samples # 1, 4, 5, 7-16), Cluster 2 (sample # 3), Cluster 3 (sample 6), and Cluster 4 (sample # 2). Principal components analysis (PCA) allows one to define eigenvectors of a variance –covariance or a correlation matrix from a data set corresponding to a raw matrix of n rows of observations by p columns of variables [33]. PCA was performed on a data set of twenty four variables (T-Summer-2008, pH-Summer-2008, NO_2^--Summer-2008, PO_4^{3-}-Summer-2008, SiO_2-Summer-2008, Chl a-Summer-2008, T-Winter-2008, pH-Winter-2008, NO_2^--Winter-2008, PO_4^{3-}-Winter-2008, SiO_2-Winter-2008, Chl a-Winter-2008, T-Summer-2009, pH-Summer-2009, NO_2^--Summer-2009, PO_4^{3}-Summer-2009, SiO_2-Summer-2009, Chl a-Summer-2009, T-Winter-2009, pH-Winter-2009, NO_2^--Winter-2009, PO_4^{3-}-Winter-2009, SiO_2-Winter-2009, Chl a-Winter-2009) and sixteen sea water samples. Principal components analysis of water chemical variables produced three components (Table 4), accounting for 72.31% of the total variance of the dataset. Table 4 presents the principal component loadings, as well as their respective explained variance. Loadings, that represent the importance of the variables for the components are in bold for values greater than 0.7 (Table 4). Most of the variance is contained in the PC1 (43.25%) which is associated with the variables PO_4-Summer 2008, Chl a-Summer 2008, pH-Winter 2008, NO_2-Winter 2008, PO_4-Winter 2008, SiO_2-Winter 2008, NO_2-Summer 2009, Chl a-Summer 2009, PO4-Winter 2009, SiO_2-Winter2009, and Chl a-Winter with squared regression coefficients of 0.783, 0.893, -0.845, 0.866, 0.970, 0.924, 0.766, 0.928, 0.789, 0.869, and 0.986 respectively. PC2 (19.03%) is mainly related to SiO_2-Summer 2008, T-Summer 2009, pH-Summer 2009, pH-Winter 2009 with squared regression coefficients of 0.829, 0.746, 0.889, and 0.873 respectively. The water sample observations of the two principal component axes are plotted in Figure 9. In score plot PC1 vs. PC2 (Figure 9) the water samples are divided into four distinct clusters mainly according to their geotectonic units: Cluster 1 (samples # 1, 4, 5, 7-16), Cluster 2 (sample # 3), Cluster 3 (sample 6), and Cluster 4 (sample # 2). The principal component analyses of the water samples show the same groupings that were apparent in the hierarchical clusters analyses (Figures 8 and 9). It is of interest to note that the Cluster 1 (sample # 2) corresponds to the seawater sample from the shoreline that receives wastewater from the Djibouti slaughterhouse (Figure 1). The Cluster 2 (sample # 3) and Cluster 3 (sample # 6) were related to the seawater samples from Djibouti-city shoreline that receives untreated domestic wastewaters (Figure 1). On the other hand, the Cluster 4 (samples # 1, 4-5, 7-16) include seawater samples that were not subject to any wastewater discharge. Therefore, the spatial difference in the water quality parameters in the Gulf of Tadjourah is mainly related to

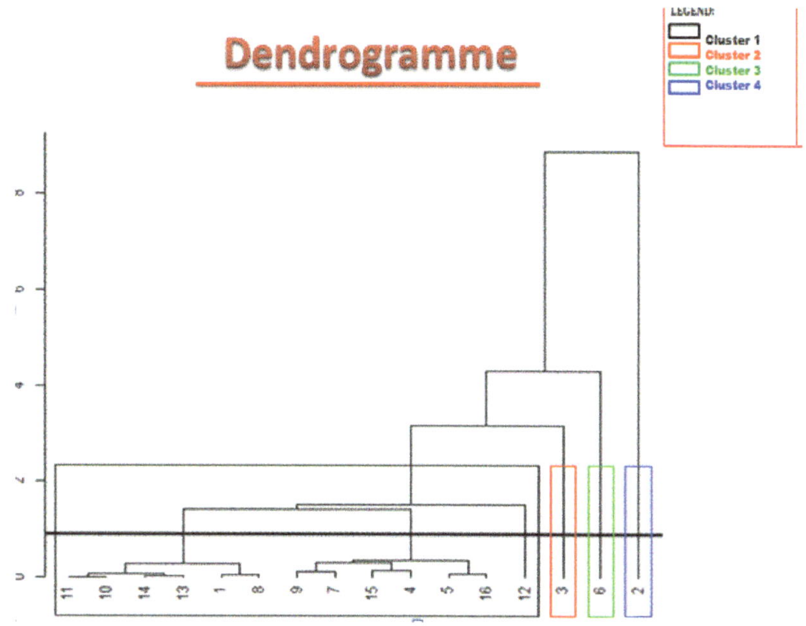

Figure 8: Four clusters in the dendrogram for the water sample in the Lake Abhe area. The black line is "phenon line".

parameters	PC1	PC2	PC2
PO_4-Summer 2008	0.783	0.396	-0.432
SiO_2-Summer 2008	0.133	0.829	0.011
Chloa-Summer 2008	0.893	-0.146	-0.325
pH-Winter 2008	-0.845	0.298	0.274
NO_2-Winter 2008	0.866	-0.214	-0.001
PO_4-Winter 2008	0.97	-0.198	-0.055
SiO_2-Winter 2008	0.924	-0.1518	0.289
T-Summer 2009	0.505	0.746	0.156
pH-Summer 2009	0.111	0.889	-0.189
NO_2-Summer 2009	0.766	-0.045	0.385
Chl *a*-Summer 2009	0.928	-0.133	-0.282
pH-Winter 2009	0.065	0.873	-0.095
PO_4-Winter 2009	0.789	-0.137	0.017
SiO_2-Winter2009	0.869	0.261	-0.031
Chlo *a*-Winter 2009	0.986	-0.032	-0.049
Variance explained	10.38	4.567	2.406
% Total variance	43.25	19.03	10.03
Cumulative (%)	43.25	62.28	72.31

Table 4: Results from the principal component analysis. Vectors, eigenvalues and cumulative variance (principal vectors are in bold).

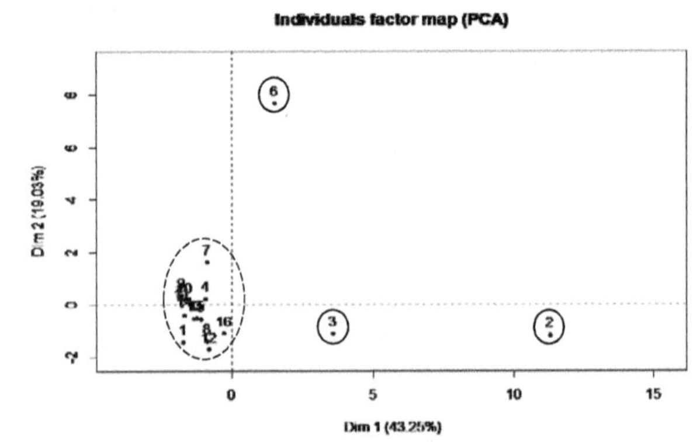

Figure 9: Principal Component's analysis results. PC1 versus PC2 plot for samples.

the impact of human activity.

Paired t tests were used to detect significant variations among parameters in the different four cluster locations (Cluster 1, Cluster 2, Cluster 3, and Cluster 4). Pearson correlation was used to detect linear correlation between the four cluster locations. Table 5 summarizes the paired t test results and the Pearson correlations of pH, NO_2, PO_4, SiO_2, and Chl a in seawater samples. The results indicated no significant difference in the parameters measured at all locations (Table 5). These results may be explained by the fact that the available data is not sufficient to be much more interpreted by means of statistical analyses. Therefore, the Djibouti seawater should be at least monitored fortnightly at each sampling site so as to have sufficient data for future statistical analyses.

Conclusions and Perspectives

The present study documents the first results on inorganic nutriments (NH_4, PO_4, SiO_2, NO_2), microbiology (TC, FC, FS, Salmonella, Shigella), and heavy metals (Cd, Co, Cr, Cu, Fe, Mn, Ni, Pb, Zn) concentrations in Djibouti marine and coastal environment. The Gulf of Tadjourah seawater features very low chlorophyll a concentrations. In order to get deeper understanding of this particular low chlorophyll a, three stations monitoring daily the temperature, salinity, dissolved oxygen and chlorophyll a have been set up in the Gulf of Tadjourah in the end of 2014 and the results of this monitoring will be further reported. The results of the present study demonstrate that higher microbial concentrations in seawater were associated with wastewater discharge. Beaches located near wastewater outfalls failed to comply with the European Community Bathing Water Directive, while beaches far from the point source were in good compliance. This study highlights the importance of wastewater treatment before discharging in the sea for reducing the risk of anthropogenic pollution in the Djibouti-city beaches. A more effective management practice would be treating wastewater and reusing the treated water as Djibouti suffers from scarcity of fresh water resources. Preventing wastewater discharge to the sea will also reduce the risks associated with microbial contamination on seafood and on bather health directly. The heavy metals concentrations in Doraleh coastline seawater were within the normal range for seawater except for iron, which is slightly high and only for the sampling points located in the vicinity of the Doraleh port. Thus, establishing a regular monitoring of the aquatic system should

	pH		NO_2		PO_4		SiO_2		Chlo_a	
	Pearson	*t* test	Pearson	*t* test	Pearson	*t* test	Pearson	*t* test	Pearson	*t* test
2&1	1.955	-0.809	1.106	3.1032	-0.5874	1.2912	0.6522	1.5167	-0.8255	2.2476
2&3	2.5045	-0.6018	0.9647	1.5007	17.0388	1.3677	6.4749	1.2561	-0.724	1.8327
2&4	2.3366	-0.7955	0.6763	2.7425	-0.4070	1.2099	1.0627	1.4928	-0.3554	2.2467
2&5	-0.1739	-0.3882	0.5818	2.3863	-0.8610	1.2174	5.223	1.4833	0.5719	2.266
2&6	1.0874	-3.1346	1.4445	2.4699	-1.8471	1.1041	0.0427	0.8855	-0.0508	2.1917
2&7	1.09130	-0.8656	0.6765	2.9666	-1.4794	1.1711	-1.0810	1.3177	0.4279	2.1925
2&8	2.5537	-0.7712	-0.6413	2.4038	-1.2395	1.2843	5.5160	1.5880	-1.1728	2.1936
2&9	-0.2483	-0.9758	1.0967	3.0494	0.4421	1.2783	0.5539	1.4865	-0.5237	2.1688
2&10	1.6808	-1.2638	0.3110	2.8446	0.8568	1.3009	-0.0547	1.5163	-2.2547	2.1728
2&11	-0.2656	-1.0589	1.0304	3.0515	1.1141	1.3284	-0.7738	1.2964	-1.6460	2.2294
2&12	-0.3954	-1.0921	1.0118	-0.0559	0.1096	1.2023	0.2146	1.4806	-0.3283	2.2499
2&13	1.0386	-0.9188	0.4041	2.9227	1.2575	1.3275	0.0175	1.3399	-1.4730	2.2207
2&14	-0.2483	-0.9258	0.2632	2.2667	1.4723	1.3317	-0.1985	1.3841	-2.0677	2.2051
2&15	0.6687	-0.9259	0.4948	2.2252	0.4916	1.2997	0.3545	1.5081	1.5496	2.3203
2&16	1.1092	-1.1574	0.4086	-0.3072	-0.2208	1.2825	2.0712	1.4066	-0.2214	2.2075

3&1	0.7707	-1.2741	-0.4382	2.8466	-0.4593	1.1412	0.4372	1.4206	-0.1786	2.9674
3&4	0.8178	-0.0714	1.1667	2.7545	-0.3184	0.9105	0.9098	1.3924	-1.6106	2.4676
3&5	-0.9371	0.3163	1.0863	2.0506	-0.7121	0.9339	5.5192	1.3695	-5.5980	2.5978
3&6	0.2042	-2.6769	1.0846	2.0092	-2.0280	0.6462	-0.1726	0.6598	0.4453	2.4903
3&7	0.2496	-1.1675	-0.0864	2.8534	-1.4718	0.81068	-1.3249	1.1874	-2.468	1.5923
3&8	1.1222	-1.2185	-2.3254	1.8203	-1.1259	1.1173	7.3789	1.4939	2.3748	3.9834
3&9	-0.7999	-1.6026	0.4823	2.8548	0.5743	1.0976	0.9350	1.4025	-1.0418	1.8737
3&10	0.5839	-3.1726	0.2971	2.6955	0.9109	1.1625	-0.2659	1.4188	0.9305	3.2474
3&11	-0.9156	-2.0409	0.9138	2.8722	1.0644	1.2410	-0.5243	1.1788	-0.2958	2.8608
3&12	-1.3431	-2.2227	1.1897	-0.7070	0.2151	0.8683	0.1956	1.3811	-1.3598	2.6188
3&13	0.3764	-2.0233	0.1784	2.7556	1.1840	1.2375	-0.2017	1.1877	0.6271	3.2585
3&14	-0.6461	-1.7256	2.2401	3.1084	1.3354	1.2493	-0.3219	1.2545	1.1644	3.5004
3&15	-0.0835	-0.6877	0.8575	1.7060	0.4202	1.1506	0.2425	1.4085	-3.4503	2.5257
3&16	0.2401	-1.2652	1.0650	-0.6844	-0.1907	1.1108	1.5665	1.2825	-1.8938	1.9153
6&1	2.71352	2.7718	0.4855	6.9581	-0.1911	2.1157	-0.1861	4.6299	1.1069	1.5238
6&4	3.4427	5.1288	3.5183	0.6731	-0.5397	0.7393	-0.5718	4.1710	-0.3806	0.9169
6&5	1.2877	3.2062	3.0752	-0.2644	0.3294	1.2294	0.1539	2.4844	-0.8621	0.8370
6&7	5.8070	2.9505	1.6278	8.9982	3.3224	1.5303	1.6350	5.8274	-1.5164	-0.3486
6&8	1.4968	2.7113	-0.2107	1.4077	0.0942	2.0977	0.0574	1.8116	-0.2596	0.6605
6&9	0.8129	2.1885	4.1832	10.6950	-0.9293	1.7735	-1.6423	3.3251	-1.1898	0.1470
6&10	2.9300	2.2194	1.9808	5.3452	-0.1715	2.1456	-0.0376	4.7230	-0.5864	0.5206
6&11	0.7199	2.0203	8.2651	2.0161	-1.9225	2.1116	0.0799	3.1477	-0.5546	0.9484
6&12	1.0231	1.9901	7.0161	-1.1332	-1.0560	0.3011	-1.2502	4.2146	-1.1864	0.8574
6&13	2.0545	2.2796	1.9922	7.5128	-1.8382	2.0468	0.0222	2.5781	-1.0535	0.8555
6&14	0.5906	2.1761	1.1036	-0.1236	-1.4717	2.1183	3.3334	10.3198	-0.5030	0.8371
6&15	5.6191	4.7730	2.8277	-0.7135	-0.7637	1.6596	-0.6415	4.0584	-0.5474	1.0524
6&16	-1.4398	2.8771	2.3070	-0.9319	-0.6899	1.8203	-0.1392	3.2901	-0.6746	0.3106

Table 5: Paired t test and the Pearson correlation results for pH, nitrite, phosphate, silicate, and Chlorophyl *a* in seawater samples.

be helpful in setting standards for regulating discharge in the sea. Noteworthy is that Djibouti is blessed with blooming development that needs to be rationally managed to avoid adverse impacts and secure sustainability and profitability for the investors and for the country.

Acknowledgments

This research was support by the Centre d'Etudes et de Recherche de Djibouti (CERD). We are very thankful to Djibouti Ministry of Housing, Urban Planning and Environment for coordinating with PERSGA to support Djibouti National Monitoring Program. We highly appreciated the collaboration of the Djibouti Navy during this monitoring.

References

1. Elmanama AA, Afifi S, Bahr S (2006) Seasonal and spatial variation in the monitoring parameters of Gaza Beach during 2002-2003. Environ Res 101: 25-33.
2. Andreadakis AD (1997) Wastewater Treatment and Disposal for the preservation of Bathing and Coastal Water Quality in Touristic Areas. Mar Chem 58: 389-395.
3. Amina SC (2007) Atlas of Africa, Djibouti Atlas, Les editions Young Africa 36-37.
4. Strickland JDH, Parsons TR (1972) In: A Pratical Hand Book of Sea Water Analysis (2nd edn.) Fisheries Research Board of Canada Bulletin Ottawa, Canada, Fish Res Bd Can Bull 167.
5. R Development Core Team (2012) Statistical Analysis: an Introduction using R/R basics. The R Foundation for Statistical Computing.
6. American Public Health Association, Standard Methods for the Examination of Water and Wastewater (1998) (20th edn.) APHA Washington DC.
7. Myklevoll S (1982) A Pelagic Fish Survey in the EEZ of the Republic of Djibouti. Reports on Surveys with the R/V "Dr Fridtjof Nansen", Institute of Marine Research Bergen.
8. Levitus S, Boyer TP (1994) World Ocean Atlas. Temperature NOAA ATLAS NESDIS 4: 117.
9. Morcos SA (1970) Physical and chemical oceanography of the Red Sea. Oceanogr Mar Biol Ann Rev 8: 73-202.
10. Harvey HW (1955) The chemistry and Fertility of seawaters, C.U.P. London.
11. PERSGA (2014) State of the Marine Environment Report for the Red Sea and Gulf of Aden. Jeddah.
12. Millero FJ (1996) Chemical Oceanography. CRC Press, New York 70-73.
13. Elawad AES (2012) Study of inter annual variability of chlorophyll in the Red Sea. University of Bergen 37-42.
14. Najah MK (2005) The Red Sea and Gulf of Aden. GIWA Regional Assessment Sub-regions.
15. Shuval H (2003) Estimating the global burden of thalassogenic diseases: human infectious diseases caused by wastewater pollution of the marine environment. J of Water Health 1: 53-64.
16. Yau V, Wade TJ, Wilde CKD, Colford JM (2009) Skin related symptoms following exposure to recreational water: a systematic review and meta-analysis. Water Qual Expo Health 1: 79-103.
17. Awaleh MO, Ahmed MM, Soubaneh YD, Bouh SM, Hoch FB, et al. (2013) Wastewater reclamation using discarded reverse osmosis membranes for reuse in irrigation in Djibouti, an arid country. Water Sci and Technol 67 : 1362-1369.
18. Moussa MA (2009) Study of the performance of purifying stations biopack. University Cergy Pantoise, France.
19. Cabelli V, Dufour P, McCabe L (1982) Swimming-associated gastroenteritis and water quality. J Epidemiol 115: 606-616.
20. Gabutti G, DeDonno A, Bagordo F, Montagna MT (2000) Comparative Survival of Faecal and Human Contaminants and Use of *Staphylococcus aureus* as an Effective Indicator of Human Pollution. Marine Pollut Bull 40: 697-700.
21. Douabul A, Haddad AM (1999) In: Douabul A, Rouphael TS, Marchant R (eds) The Red Sea and Yemen's Red Sea Environments. Protection of Marine Ecosystems of the Red Sea Coast of Yemen: Hassell & Assoc., AMSAT and

UNOPS 1-16.

22. PERSGA (2006) State of the Marine Environment Report for the Red Sea and Gulf of Aden Jeddah.
23. Awaleh MO (2009) Djibouti National Monitoring program of the marine and Coastal Environment-First Technical Report. Centre d'Etudes et de Recherche de Djibouti (CERD).
24. Egge JK, Aknes DL (1992) Silicate as regulating nutriment in phytoplankton competition. Marine Ecol-Prog Ser 83: 281-289.
25. Hillebrand H, Sommer U (1997) Response of epilithic microphy- tobenthos of the Western Baltic Sea to in situ experiments with nutrient enrichment. Mar Ecol Prog Ser 160: 35-46.
26. Brzezinsky MA (1985) The Si:C:N ratio of marine diatoms: Interspecific variability and the effect of some environmental variables. J Phycol 21: 347-357.
27. Trinh XG (1999) Analysis and Estimation of Trace Metals in Seawater and Sediment in the South China Sea, Area IV: Vietnamese Waters. Southeast Asian Fisheries Development Center. Proceedings of the SEAFDEC Seminar on Fishery Resources in the South China Sea.
28. Landing WM, Bruland KW (1987) The contrasting biogeochemistry of iron and manganese in the Pacific Ocean. Geochim Cosmochim Ac 51: 29-43.
29. John H, Martin R, Gordon M, Fitzwater SE (1990) Iron in Antarctic Waters-Nature. 345: 156-158.
30. Bruland KW, Orians KJ, Cowen JP (1994) Reactive trace metals in the stratified central North Pacific. Geochim Cosmochim Ac 58: 3171-3182.
31. Taylor SR (1964) Abundance of chemical elements in the continental crust. A new table, Geochim Cosmochim Ac 28: 1273-1285.
32. Sunda WG, Huntsman SA (1995) Iron uptake and growth limitation in oceanic and coastal phytoplankton. Mar Chem 50: 189-206.
33. Davis JC (1986) Statistics and Data Analysis in Geology. John Wiley & Sons, New York.
34. Ward JH (1963) Hierarchical grouping to optimize an objective function. Journal of the American Statistical Association 58: 236-244.
35. Guler C, Thyne GD, McCray JE, Turner AK (2002) Evaluation of graphical and multivariate statistical methods for classification of water chemistry data. Hydrogeology Journal 10: 455-474

6

Climatic Consequences of Long-term Global Salination of Ocean

Gaspar Banfalvi*

Department of Biotechnology and Microbiology, University of Debrecen, Hungary

Abstract

The long-term salination of two major osmolyte systems, the ocean and the inner environment of vertebrates has been compared. The average osmolality of today's sea (1.09 Osm) is more than three times higher than that of the blood of land vertebrates (~0.3 Osm). Of the two major strategies for ionic adaptation, in the first pattern the osmolarity of organisms (unicells, invertebrates, primitive vertebrates) equals that of the surrounding water, but the qualitative composition of body or cells fluid differ from those of the environment. In the second pattern of ionic adaptation advanced invertebrates and vertebrates maintain their energy consuming osmotic content and ionic composition of their extracellular environment. In spite of geological changes, terrestrial vertebrates maintained their physiological ionic concentration that corresponded to the sea at the time of their emergence in the Devonian era, rather than the osmolality of the ocean maintained its constancy. Paradoxically, the recent global melting of ice and snow with its temporary dilution effect is opposed by the long-term salination of the ocean. To resolve the contradiction between the salination process and its short term oscillations supported by Raoult's law of dilute solutions, glacial periods favoring salination and interglacial periods of dilutions are: a) characterized as salinity fluctuations over geological ages, b) representing a dynamic osmolyte system against a general geochemical balance, c) directed towards a general salination process of oceans interrupted by glacial and interglacial oscillations. The gradually increasing salinity of ocean poses a long-term threat to the biodiversity and global life. It is assumed that unless mitigative measures of global proportion will be implemented, biodiversity of life on Earth will be endangered by the salinity of oceans and the shrinkage of the fresh water resources.

Keywords: Osmolytes; Salination; Glacial concentration; Interglacial dilution; Biodiversity

Introduction

Climate change is a global phenomenon affecting both terrestrial and aquatic life. Global warming and the melting of polar and Greenland ice have a diluting effect on oceans. The sea level rise at low sea temperature is hardly affected by the thermal expansion of oceans, but seriously impacted by the huge amount of water released by the melting snow and ice of the arctic ice caps and the Greenland ice sheet.

Another threat of global importance that has not been dealt with is known as salinity oscillation discussed here in terms of global vapor pressure changes governed by Rault's law of dilute solutions. Thermosteric changes have received considerable attention, but salinity-driven halosteric patterns of long-term sea level change have not been throughly investigated [1]. Only regional halosteric anomalies have been considered as important drivers of the sea level variability [2-7]. Medium-term, multidecadal changes have been ignored because the halosteric fluctuations could not be detected or were close to zero [8-11], as well as historical salinity measurements [12,13] were not sufficient to come to any conclusion and primarily thermosteric effects have been taken into consideration in climatic changes [14,15].

This review focuses primarily on the second aspect of global threat related to the long-term halosteric changes of ocean starting with the evolution of dilute solutions, followed by sea level rises and falls, ocean as a dilute solution, global water circulation, salination and dilution processes, long-term salination versus temporary dilutions and future aspects of salination affecting life on Earth.

Evolution of Dilute Solutions

The thermodynamic law of dilute solutions established by François-Marie Raoult is summarized in Figure 1A showing the equilibrium of various phase-transitions of water around the freezing point. The freezing point of distilled water at the sea level is 0°C. By multiplying the osmolarity of seawater (1.09 Osm) with the molar freezing point depression (-1.86°C) the freezing point for saline (0.3 Osm × -1.86°C) is -0.56°C and for seawater (-1.86 × 1.09) -2°C. The vapor pressure of dilute solutions including the ocean is a fundamental property of ice-vapor-liquid equilibrium system of Earth known as biosphere. This system implies that at all temperatures the vapor pressure of dilute solutions (e.g. saline solution, seawater) is lower than that of the pure solvent (distilled water) and indicates that the freezing point of a solution depends strongly on its vapor pressure. In dilute solutions, the vapor pressure will be higher at each temperature than that of the pure solvent, and the freezing-point will be lower than 0°C. Similarly the boiling point elevation and osmolality of a dilute solution is dependent on its concentration, the higher the concentration of the solution, the higher its boiling point elevation and osmotic concentration will be. The graph also shows that below a certain vapor pressure (0.6 kPa) water is not in liquid form but exists as ice. Important to mention that the higher the temperature, the larger the difference in vapor pressure will be between between distilled water and solutions. For more information on the thermodynamics of water see [16].

Major evolutionary events on Earth including the evolutionary aspects of ocean life and the progress of increasing salt content of ocean are summarized in Figure 1B. Prebiotic life could have started 4.2 billion years ago. The "RNA World" could have lasted for an estimated

***Corresponding author:** Gaspar Banfalvi, University of Debrecen, Department of Biotechnology and Microbiology, Life Sciences Building 1.1021, Egyetem Square, Debrecen 4010, Hungary, E-mail: bgaspar@unideb.hu

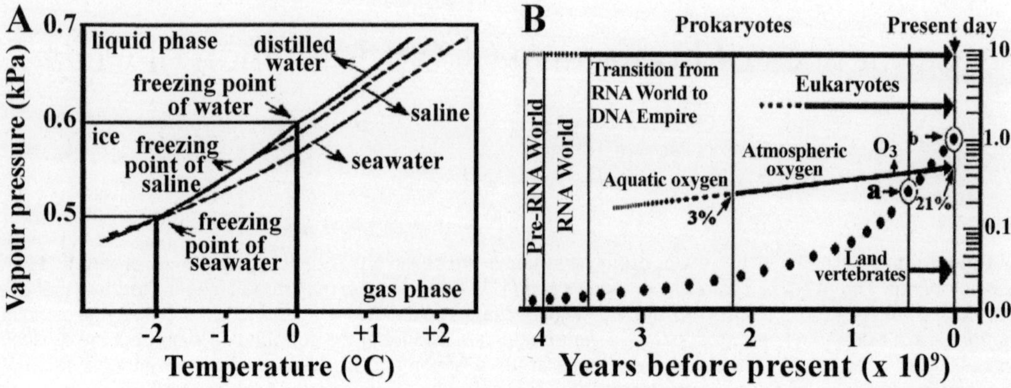

Figure 1: Long-term reduction of vapor pressure as a consequence of increasing salinity of sea. A) Reduction of vapor pressure based on Raoult's law of dilute solutions. Lowered vapor pressure of a saline (0.9% NaCl) and seawater versus distilled water. B) Evolutionary events of ocean life and long-term salination. Modified with permission [15].

400 million years, and replaced by DNA-based cells. An estimated 3.6 billion years ago photosynthetic stromatolytes and cyanobacteria started to hydrolyze water leading to the appearance of atmospheric oxygen ~2.2 billion years ago. Eukaryotic cells came into being ~2-1.5 billion, plants appeared between 1.2-1.5 billion, land vertebrates some 450 million years ago. The hypothetical salination of ocean is indicated by the dotted curve. Unfortunately the initial salt concentration after the World Ocean cooled down to the temperature tolerable for primitive life is unknown, but was probably much lower than the recent salinity value. Attention is called to a and b salt concentrations (circled black dots) in Figure 1: a) Devonian salt concentration of ocean is believed to have corresponded to the concentration of blood electrolytes (0.3 Osm) at the time when land vertebrates left the sea [17-20]. b) This dot in Figure 2 corresponds to present day's sea osmolality (1.09 Osm) after 4.55 billion years of global history.

Sea Level Rise above its Present Level

The global temperature increase and recent sea level rise raise the question: How far the recent sea level increase could go taking into account the available fresh water reservoirs? These reservoires include the arctic ice, glaciers and the permanent snow. To answer the question, fundamental data are given in Table 1.

These data include:

r1 = Average radius of Earth, the distance from Earth's center to its solid surface without sea (6,368 km),

r2 = The radius of Earth at sea level (6,371 km) [21,22],

r3 = Earth radius after a hypothetical 100 m sea level rise (6,371.1 km)

V1 = Volume of Earth + sea = 1.08321×10^{12} km^3

Volume of sea 1.332×10^9 km^3 [23,24].

Volume of Earth + hypothetical sea volume after 100 m sea level rise: 1.08326×10^{12} km^3

Volume of hypothetical sea level rise: 5×10^9 km^3.

Volume of fresh water reserve (ice, galciers, permanent snow) ~2.4 $\times 10^9$ km^3 [25,26].

Data of fresh water supply are given in Table 1. These data have been used to estimate the potential sea level rise as a consequence of global warming and melting.

The potential sea level rise is demonstrated in Figure 2. From the data of Table 1 one can estimate that the available fresh water reservoir would not be sufficient to supply a 100 m sea level rise. The volume of the available fresh water reserve (2.4×10^9 km^3) would cause less than 60 m elevation of sea water. Taking into account that the continental shelf narrows upwards increasingly more water will be necessary as the sea level rises. Based on above estimations and to answer the question how far the recent sea level increase could go, the maximum of sea level rise would be closer to 50 than to 60 m. One can also expect that the thaw of fresh water reserve will never be complete, which further reduces the possible sea level rise. One can also predict from the available small buffer capacity of the fresh water reservoirs, that future interglacial oscillations of sea level rises will take place with higher frequencies but decreasing amplitudes, in long-term not be able to compensate the progression of salination process of the ocean.

A further danger concerns the constancy of water volume on Earth, challenged by those who suspected that Earth's air gradually leaks back into space and Earth could become dry like Venus [27]. The quote of Benjamin Franklin that "a small leak can sink a great ship" could be perplexing at global proportion by the observation that in the highest reaches the atmosphere is shrinking [28-30]. Air leakage on Earth would mean fresh water loss in the form of water vapor. Earlier changes leading to the recent climate of Mars reflect what Earth could also become, such as the Martian polar ice caps, characteristic seasonal changes and weather patterns with very little atmosphere. The interaction of the ionosphere of Mars with solar wind gradually stripping away atoms from the upper atmosphere [31] could be another similarity between these sister planets.

Ocean as a Dilute Solution

By viewing the ocean as the oldest dilute osmolyte system, the laws of dilute solutions should apply to the ocean.

Arguments that supply this notion

In dilute aqueous solutions the molar excess of solvent over solute is more than one hundred. The molar concentration of water is 55.5 M (1 M = 18 g/l, 55.5 M = 1000 g/l) consequently the border between a dilute and concentrated solutions can be drawn at 0.555 Osm (3.34×10^{23} particles/kg). The osmotic concentration calculated from the salinity and from the freezing-point depression of seawater (-2°C) corresponds to 1.09 molar non-dissociable solute concentration. In dilute solutions there are practically no interactions between the molecules of the

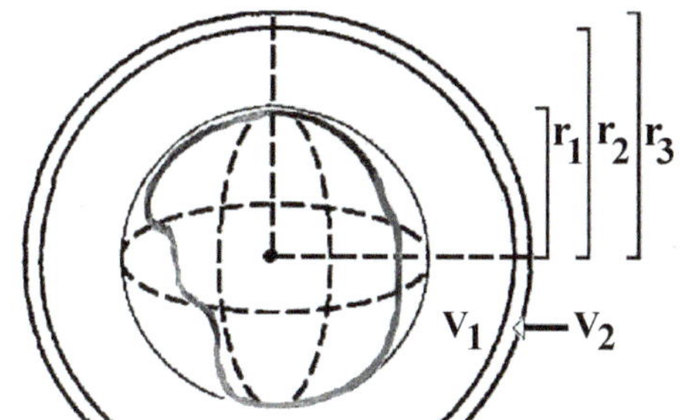

Figure 2: Demonstration of potential sea level rise. r1: radius of the Earth. Earth is not round, shown by the "potato shaped" gray line. r2: radius at sea level; r3: radius after sea level rise. V1: volume of sea before global melting. V2: volume of sea after global melting.

Earth	Mean Radius	Volume	References
	m	m3	
Earth radius	6,367,450	1.0814×10^{21}	[7, 8, 12]
Sea volume		1.33×10^{18}	[9]
Sea volume	6,363,767.8	1.332×10^{18}	[10]
Hypothetical radius			
after 100 m sea level rise	6,367,550	1.08145×10^{21}	Estimated
after 50 m sea level rise	6,367,500	1.08142×10^{21}	Estimated
Volume needed for 100m sea level rise		$\sim 5 \times 10^{16}$	Estimated
50 m sea level rise Estimated		$\sim 2.0 \times 10^{16}$	Estimated
Fresh water reserv (ice, galciers, permanent snow)		$\sim 2.4 \times 10^{16}$	[11]

Equatorial radius at sea level: 6378.1 km [7, 12]. Polar radius at sea level: 6356.8 km [8]. Average radius at sea level approximated by the averages of equatorial and polar radii: 6367.45 km. Average ocean depth: 3682.2 m corresponding to average volume of 1.332 x 109 km3 [10].

Table 1: Highest potential sea level rise.

solute. Although, the osmolality of the present day ocean (1.09 Osm) is almost two times higher than in ideal dilute solutions (<0.555 Osm) the solute-solvent interactions are still negligible.

The solutes are non-volatile. The sea contains primarily non-volatile salts. One notable exception is iodine, but present in very low concentration (<0.001 ppm = 10^{-7}%). Due to its volatility, the iodine content of rain inside the continents diminishes with the distance from the coast.

The solvent evaporates. More than 80% of the evaporated water close to distilled water quality comes from the sea surface, less from the surface of the land and from fresh water reservoirs, lakes and rivers (Figure 3C). As far as evaporation is concerned beside the concentration, the surface temperature is an important factor.

More water molecules break away from the surface at higher temperature, resulting generally in higher vapor pressure over the sea and more precipitate formation. The so called Southern Oscillation causes variations in the temperature of the surface with warming accompanied by high air surface pressure of the Tropical Eastern Pacific (El Niño) and cooling with low pressure (La Niña) in the Tropical Western Pacific. The Tropical Western Pacific area (Manus, Papua New Guinea, Naru, Northern Australia) with its warmest sea-surface temperature is the "warm pool" of the Pacific ocean causing a high-altitude cirrus cloud system affecting both the solar energy absorption and the suspected escape of heat energy into the space. The Western Pacific system demonstrates that the biosphere is a semi-closed system the permeability of which is vulnerable.

Evaporation takes place in this semi-closed aquatic biome, which consists of marine and freshwater. In this system the salinity of the fresh water carried to the ocean constantly increases, while the total mass of water plus water vapor of the atmosphere remains relatively constant. This constancy has been challenged by the assumption that the atmosphere could be leaking back to the space [27,29,30]. Leakage of the thermosphere questions the Earth as a closed system as the equilibrium of the water cycle would be shifted, with gradual loss of upper atmosphere including fresh water as water vapor. Leakage in the water cycle would also contribute to the long-term salination process.

Balance between evaporation and precipitation in a closed system is reached when the numbers of water molecules leaving the surface of the sea are equal to the number of molecules returning to it. Evaporation and precipitation near the sea level are closer to equilibrium than at higher altitudes particularly at distant locations above the continents. Although, global balance could never be reached, local saturation occurs frequently, with the vertical profile of temperature determining the form of precipitation (rain, snow, sleet, etc.). The lack of saturation, when much less water returns to the surface than evaporated is generating and spreading deserts at an alarming rate. Cutting and burning plants in forests to create fields, poor conservation, overtaxed

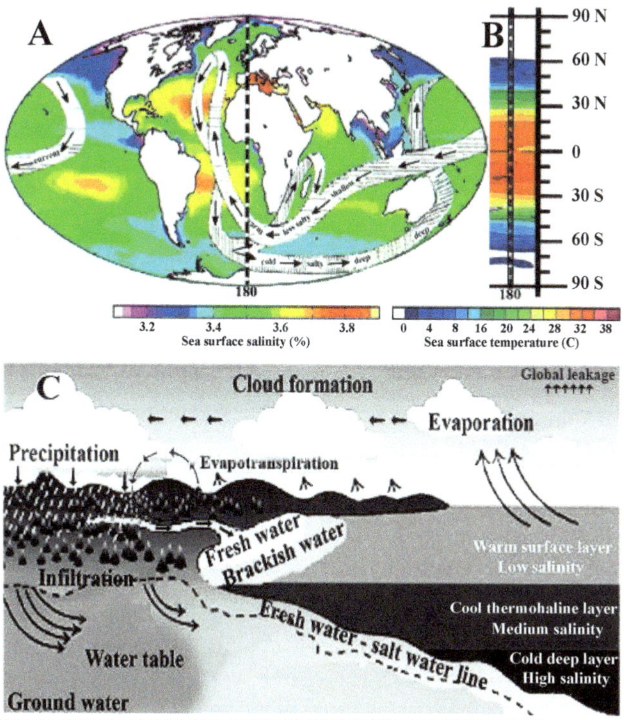

Figure 3: Sea surface salinity, temperature and evaporation. A. False-color composite of global distribution of sea surface salinity and representation of thermohaline circulation. Sea surface salinity generally ranges between 3.3 and 3.7% in the open ocean [41]. The Great Ocean Conveyer shows the recent global thermohaline circulation. Combination of the annual mean sea surface salinity data obtained from the World Ocean Atlas http://www.nodc.noaa.gov/OC5/WOA09/pr_woa09.html) and thermohaline circulation (http://www.killerinourmidst.com/THC.html). B. Sea surface temperature during the first week of February 2011 at the 180th meridian (Date Line) passing through Russia, Fiji and Antarctica.
(http://en.wikipedia.org/wiki/Sea_surface_temperature#mediaviewer/File:Weeklysst.gif).
C. Water cycle with warm (low density), cool (medium) and cold (higher density) stratification of sea water. Arrows in the upper right corner refer to the leakage from the upper atmosphere to the space.

water supply, soaring population and recently global warming are blamed without mentioning the global effect of salinity increase and as a consequence the global reduction of vapor pressure in air.

The extent of cloud formation on Earth is influenced by Raoult's law of dilute solutions. The higher the osmolality of the sea, the less the vapor pressure, evaporation and precipitation will be at global scale. To the contrary, the more dilute and higher the temperature of the sea, the more water molecules will evaporate increasing the vapor tension in air resulting in more cloud and precipitate formation. That the dilution effect at global scale is not uniform is seen at three major areas including the intensive melting in the two polar regions and near the Greenland Ice sheet.

That Charles' law also prevails is evident from the fact that increasing temperature expands the volume of the ocean.

Global Water Circulation

The three physical variables in the ocean are temperature, salinity and density. The physical properties of seawater include both 'thermodynamic properties' like density and freezing point, as well as 'transport properties' like the electrical conductivity and viscosity. Density is an important property in ocean science because small spatial changes in density result in spatial variations in pressure at a given depth, which in turn drive the ocean circulation [32].

The Arctic ice contains at least 15% sea ice and its fresh water stream is carried by ocean currents. The driving mechanism of global circulation is known as thermohaline circulation that depends on the complex interplay of sea water temperature and salinity across the globe. The seawater is circulated by ocean currents of warm or cold water generated and driven by solar warming, volcanic heat, hydrothermal circulation of hot Earth's crust, arctic cooling, Earth's rotation, gravity, gravitation of the moon, winds and temperature, all influencing the climates of the continents. Nevertheless, it is the salinity at the ocean surface that is the fundamental factor in the large-scale ocean climate, affected to a large extent by air-sea boundary processes such as evaporation and precipitation, and to a lesser extent by the horizontal and vertical advection of adjacent water. Consequently, the changes in surface-layer salinity over a certain period allow the estimation of the corresponding changes in evaporation and precipitation [33].

The Great Ocean Conveyor belt drives the global thermohaline circulation including the Gulf Stream – North Atlantic – Norway Current in the Atlantic Ocean, the Humboldt Current off South America, and the West Australian Current (Figure 3). Reduced circulation is causing severe meteorological activities near the circulating currents. As a consequence of local dilution in the north Atlantic region the storm intensity and frequency could be mentioned that have been doubled in the past 30 years. Due to the increased evaporation from diluted and warmer seawater, cloud formation, storm frequency and severity have been intensified. These events are not restricted to the North Atlantic region; more hurricanes hit the US South mainland, especially Texas than the average had been per decade in the previous 40 years.

The sea surface temperature is measured near the surface layer confined to the top portion of the ocean from 1 mm to 20 m below the surface. The top 2-3 m layer has a heat capacity resembling the entire atmosphere above [34]. The sea surface temperature map created by satellite temperature measurements shows the regional variations across the globe in the vicinity of the 180th meridian. The sea surface temperature is high near the equator (27-30°C) and below the freezing point (- 2°C) within the polar region. Due to the horizontal movement of the conveyor belt, the eastern tropical regions are cooler than the western tropical regions subjected to more vertical upwelling and sinking. As temperature and salinity are related inversely, cold water is denser and warm water has lower density (Figure 3) (lower right corner).

Global Aspects of Salination

Detailed spatial salinity distributions were missing due to the lack of observational data. Satellite-based, ship-based and Argo-float measurements started at the beginning of the 21st century. It was the Argo Project that allowed to investigate medium-term, multidecadal salinity changes and continues to provide an immense amount of surface salinity data of the World Ocean [35,36]. Estimates of global and basin zonal-mean near-surface salinity changes date back to the 1950 - 2000 period showing moderately increasing values of local freshening and salination areas of the oceans. There is a general agreement regarding the dilution of the Pacific ocean between the 0 and 60 N latitudes, salinity increase in the Northern latitudes of the Atlantic and Indian oceans [33,37,38].

As far as the long-term constancy of salinity of the oceans is concerned we do not have good records. The salinity of the ancient oceans could not be measured directly. However, the glacial and interglacial ages reflect dynamic changes known as concentration and dilution periods with lowering and rising sea levels. Both experimental and theoretical studies have dealt mostly with the long-term mean of salt concentration. The idea of a general geochemical balance provided a model to make such constancy plausible [39,40]. However, by viewing the oceans as a dynamic osmolyte system rather than a steady-state one, long-term salinity changes cannot be disregarded.

There is a general consensus that extracellular and intracellular salt concentrations of living organisms were identical and vertebrates remained in osmotic equilibrium with the sea before the migration to land in the early Devonian Era some 450 million years ago [17-20]. This view needs to be reconciled with the fact that the average osmotic concentration of seawater is now 1.09 Osm (35 ppt). Incomplete mixing is indicated by some variation in salinity; in open seas, but this stratification ranges only between 33 and 37 ppt [41-45]. Incomplete mixing does not explain why the salinity of the sea became more than three times higher than the isotonic concentration of the blood of land vertebrates (~0.3 Osm) (Table 2). What distinguished the long-term changes at the origin of tetrapods of all land-living vertebrates were the many selective factors associated with the shift in environment from sea to fresh water and onto land. These changes included locomotion, skeletal support, feeding, respiration, sensory apparatus and reproduction [30]. All changed in the long run, but in spite of the diversification the salinity of their body fluids remained constant. The uniform osmolality of blood in land vertebrates (300 ± 15 mOsm) with the notable exception of amphibians (160-240 mOsm) is a reflection of osmolality of seawater at the time of their evolutionary emergence from the marine environment (Table 2). Although, water vapor of the ocean greatly impacts the humidity and precipitation formation not only over the sea, but also over the land, but in this respect not only long-term, but also short-term humidity and precipitation data are missing.

Convergent evolution, an extremely important topic within biodiversity, refers to the independent development of similar features in the descendants of different ancestral groups including land vertebrates. In spite of their convergent evolution land vertebrates maintained their uniform osmolality. It has been reasoned that at higher than 0.4 Osm the function of informational macromolecules

Osmotic values of vertebrates mOsm	
Terrestrial vertebrates [42]	
Mammals	295
Birds	317
Reptiles	286
Lizards	307
Snakes	300
Turtles	287
Amphibians	160-240
Marine vertebrates	
Fishes	60-80
Teleosts (bonny fish)	
African lungfish (P. aethiopicus) [43]	
Preaestivation	234 (urea: 4.2 mM)
Aestivation (13 mo)	650 (urea: 203 mM)
Cartlilagenous	
Shark (Bonnethead)	1094 [44]
Seawater	1094 [42]

Table 2: Characteristic data of the two major osmolyte systems
1 Osm = 6×10^{23} dissolved particles/kg water.

(DNA, RNA, protein) would be threatened. Because of the strong concentration dependence of salt effect it would be difficult to design a protein that could maintain its optimal functional abilities over a wide range of salt concentration [46]. This explanation is in conformity with the existence of many isoenzymes that compensate small microenvironmental changes in different tissues and organs. Isozymes consist of slightly different amino acid sequences, but catalyze the same biochemical reaction.

Marine organisms are highly exposed to salinity changes in a process known as osmosis. As the dissolved salt generally does not easily cross biological membranes, it is the water that flows in or out to create equilibrium. Poikilosmotic conformers are marine organisms that maintain the salinity of the sea in their ionic makeup. These osmoconformers did not develop mechanisms to control osmosis and their cells have the same or similar salt content as the sea. Osmotic regulators (homeosmotic organisms) developed a variety of mechanisms to maintain a relatively constant bodily osmotic pressure independent of the osmotic pressure of the external environment.

Beside marine osmotic regulators, land vertebrates also encountered changes in salinity. The stability of the inner environment of land vertebrates versus the salination taking place in the sea dwelling animals indicates that salination in sea was followed mainly by osmoconformers rather than prevented by osmoregulation [42]. Major processes that could have caused the salination of ocean will be detailed.

As already mentioned global loss of fresh water reservoirs is likely to lead to diminishing oscillations in sea level rises and falls. The chronology of the Paleozoic sea-level changes seems to support this notion, with a gradual sea level rise through the Cambrian, reaching an ~200 m zenith in the Late Ordovician [47]. Sea-level data also reflect that even at the highest sea level rise the dilution effect was less than 10% taking into account the average radius of the Earth and the average depth of oceans across the planet (3682 m) [23]. These estimations also show that salinity oscillations could not have caused the >3-fold salinity increase of ocean (1.09 Osm) relative to the osmotic concentration of land vertebrates (0.3 Osm). The current Phanerozoic eon [48,49] dating back to 500 million years is crucial from the point of view of terrestrial evolution, constituting the development of terrestrial plants, multicellular animals, emergence of terrestrial animals and the development of modern faunas. Within the Phanerozoic the recent wet, warm interglacial period with an average 3.4% salinity is more in favor of specietion than the earlier somewhat saltier sea containing less dissolved oxygen. It is expected that the recent interglacial period is likely to be extended in the future [50,51]. By the end of the latest ice age the volume of the ice was ~5×10^7 km^3. An ~ 4×10^7 km^3 loss of all polar and glacial ice has been predicted, corresponding to 80% loss of our fresh water reservoir [52]. However, others reported that the recent fresh water reserve (ice, glaciers, permanent snow) is only 2.4×10^7 km^3 [25]. An 80% loss (1.9×10^7 km^3) of this fresh water would cause not more than an estimated 50 m sea level rise by ignoring the thermal expansion of water. The attenuation of oscillations of wet and dry climatic periods provides convincing evidence of eustatic changes due to salinity variations and thermal expansions over geologic ages and contradicts the general geochemical balance and long-term constancy of ocean salinity.

Salination Processes

Ice formation

In Polar Regions and in mountains snow and ice formation cause solvent deficiency and salination of oceans. During the advances of ice ages the sea level decreased significantly which in turn increased its salinity. Based on Charles' law the cooling effect of ice ages lowered the temperature of sea, decreased its volume and increased salt concentration. Lowered water temperature increased the solubility of oxygen, which in turn contributed to the diversification of aqueous life.

Continental drift and outpouring lava

The constant reshaping of seafloor occurs in the oceanic lithosphere (Earth's crust plus upper solid mantle) where the oceanic plates are colliding, moving apart and moving one under or above the other causing an immense pollution and small but constant increase in the salt content of the sea. The upper part of the crust is formed from episodic extrusions and intrusions of basaltic melt. The lower crust is formed by rapid hydrothermal alterations of the mantle [53]. The volcanic gases (CO_2, H_2S, HCl, HCN, SO_2) and the soluble salts of the new crustal material of the erupting lava are dissolved pollutants of the ocean.

Weathering and denudation

These processes carry away the surface waste of land, especially if not protected by forests. Endogenic processes (volcanoes, earthquakes, plate tectonic uplifts) also expose the continental crust to exogenic denudation processes of weathering and erosion. Mass wasting is accelerated by man. Most of the salt of the eroded surface of the soil and sand especially that of spreading deserts is deposited in the ocean as sedimentary rocks, the salt content of which increases the salinity of the sea. From osmotic point of view only the number of dissolved particles matters irrespective of the quality of the salinating material.

Hydrological cycle

This cycle also known as water cycle circulates from the land to the sky and back to the sea. It consists of evaporation from sea, evapotranspiration from soil and vegetation and precipitation. The flow back returns the fresh water bodies to sea carrying diluted dissolved salt (<0.1%). The "salt clock" method that served to determine the age of Earth was based on the continuous salinity intake of ocean by the water cycle [39]. Although, the "salt clock" did not work, but led to the recognition that salination is a long lasting process [54-56]. The salination effect of hydrological cycle is strengthened by the chemical pollution of rivers that has already been noticed e.g. in Amazonia due to the mining on industrial scale, forest burning and uncontrolled use of pesticides [57].

Sea-level regression

After the Cretaceous high from 304 to 119 million years ago [58], the sea level was lower in the final stage of the Cretaceous by more than at any other time in the Mesozoic Era, but there is no direct evidence for the cause of this regression. According to the currently accepted explanation, the mid-ocean ridges could have sunk under their own weight [59]. Such a severe regression could have resulted in a virtual loss of sea. The higher salt content of the ridges could have increased the salinity of sea. Other disturbances responsible for sea to sea variations in chemical composition, even if small, may have caused significant global climatic changes, the consequences of which may not all be known.

Seawater reverse osmosis

One would expect that desalination of ocean to obtain fresh water without removing salt would increase the salinity of sea. Reverse osmosis is a desalination membrane process based on semi-permeable membranes that has been commercially used since the 1970s [60]. The volume of desalinated water is 10^9-times less than the volume of the ocean and does not represent a real threat to affect ocean salinity. In fact the most important waste management programs include the use of nuclear energy, solar energy, tidal power, wave power of ocean surface waves and wind power could serve not only the generation of electricity, but also the removal of salt from sea by desalination. Desalination of seawater by reverse osmosis in huge quantities would have an opposite effect and could pump fresh water into fresh water reservoirs.

River pollution and dumping of contaminated wastes

The danger of river pollution and ocean dumping has been recognized and banned worldwide, but its prohibition has not been implemented. As already referred to it, irrespective of the origin of waste, salination increases osmosis, lowers the rate of evaporation and precipitation formation.

Local pollution: Convincing evidence was provided that tropical waters became saltier over the past 40 years, while polar seas that injected immense amount of fresh water into the ocean, lowered its salinity [61,62]. In spite of local chemical pollutions the mean ocean salinity has not been changed. For this illusorical constancy the mass action law gives a reasonable explanation. The mass of the ocean exceeds about 270-times the mass of the atmosphere (1.38×10^{21} kg/5.1×10^{18} kg = 270). It will probably take more time to trace the global metahaline salinity increase than air pollution, which became measurable within one century.

Themohaline circulation

Another threat of large and rapid salinity changes may be the disruption of the thermohaline circulation, a possibility to be considered by the sea surface salinity (SSS) programs. These programs are being executed by the European Space Administration (ESA), by the Aquarius/SAC-D mission of NASA and by the Space Agency of Argentina (Comisión Nacional de Actividades Espaciales, CONAE). Refining data of local changes in the open ocean by these measurements are expected to clarify several phenomena of the water cycle and ocean circulation, representing the two major components of the climate system of Earth. Estimating global ocean heat content and global steric sea surface level with temperature/salinity data from the Argo network revealed increasing values during the years from 2005 to 2012 [63].

Dilution Processes

Salt extraction

Beside the several processes of salination we have also to consider global dilution processes. Rising sea beds cause obstruction, limited flow, loss of contact of inner shallow seas to the ocean. Evaporate deposits could have temporarily decreased ocean salinity [48,64,65]. Others thought that salt extraction did not decrease significantly the salinity of sea [66].

Interglacial periods

These sea level rises are part of sea level oscillations and provide only temporary dilutions followed by temporary by glacial salination periods. The high sea level rises of the the Paleozoic sea-levels (~ 460-470 million years ago) were followed by other smaller rises. During the latest glacial, the buildup of ice over the continents and polar regions lowered the global sea level by an estimated 130 m compared to its present level [67] increasing its salinity [68-70]. The predicted 80% loss of all polar and glacial ice being under way could be extended by the greenhouse effect that is expected to lengthen the recent interglacial period [50,61,62]. Wet and dry climatic periods provide convincing evidence of eustatic and euhaline changes due to thermal expansion and salinity variations over geologic time periods contradicting the general geochemical balance and constancy of ocean salinity.

Effect of thermal expansion on volume

Water expands at higher temperature but contracts as it is warmed up from 0 to 4°C. The average deep sea temperature is 2°C, the average sea surface temperature 17°C. Assuming a 2°C increase (from 2 to 4°C) in global sea temperature the volumetric contraction would still be negligible if any and similarly the expansion would also be within a few %. By eliminating the negligible changes of thermal expansion and contraction of seawater, the global melting remains to be restricted to the availability of the fresh water reserve with its modest (2%) dilution effect on World Ocean.

Considering the many glacial advances and retreats that occurred in the past, it is assumed that the concentration of the ocean was not simply fluctuating during these wet and dry climatic periods. The glacial-interglacial cycles provided evidence that the ocean concentration was changing over geologic periods, and Halley's "salt clock" method led to the recognition of a long-lasting salination process [54,55,71]. The diminishing amplitudes of interglacial sea level rises over geological time support the notion of long-term salination and gradual shrinkage of fresh water reservoirs. The acceptance of salinity changes implies that a long-lasting geochemical balance that would provide constancy to the ocean does not exist.

Long-term Salination vs Temporary Fluctuations

The dilution and salination of ocean can be summarized as

1. Assuming a gradual change, at least a 1 million year period would be necessary to measure the long-term salination effect. We have been able to trace only local salinity changes and the short-term dilution effect of global warming [72]. The global surface air temperature increased by approximately 0.85°C since 1880 [73]. This temperature elevation is attributed predominantly to the increasing atmospheric

concentration of greenhouse gases, which in turn have a cooling effect in the upper atmosphere known as "greenhouse cooling" [74].

2. Water with less salt warms up faster and generates higher vapor pressure. Even a small decrease in salinity may significantly contribute to cloud and precipitation formation. Local slow dilution of ocean generates increasing amount of precipitation, fast dilution causes storm intensification.

3. The recent slight decrease in ocean salinity [48] is likely to mask the small global effect of increasing chemical pollution. Nevertheless, local pollution is already taking its toll (e.g. Aral Sea, estuaries), and exerts its devastating effect on life in these areas.

4. Salinity imbalances with climatic consequences are noticed at locations where:

a) Upwelling returns deep water to surface (e.g. Indian, Pacific Ocean), explaining the high frequency of floods in Eastern Australia.

b) Radiation emitted by the sun is heating up the diluted surface water, generating saturation vapor pressure, tropical and subtropical storms (e.g. North Atlantic tropical cyclones). An estimated >90% of the excess energy of global warming is absorbed by the ocean. The rest of heat energy is melting sea and land ice, heating the land surface and atmosphere [75-77].

c) The input of fresh water by glacier melting and runoff into the North Atlantic could weaken the conveyor belt.

d) The estimated fresh water reservoir is about 2.4×10^6 Km^3 [25] limiting the potential sea level rise to 50 m. Earlier much higher sea level rises (>200 m) indicate that a significant portion of the fresh water reservoir could have been consumed during the salination process. The reduction of the total amount of sea water could be the consequence of water leakage.

Assuming the continuing tendency of the natural long-term salination process alone or in combination with global atmospheric leakage and antropogenic pollution, the osmolality increase of sea is likely to speed up in the future. By extrapolating the recent speed the salination could be as high as ~5 Osm in 500 million years unless the pollution would reach a saturation level, slowed down and follow a sigmoidal curve with an extended period of life on Earth beyond 500 million years. Although, this conjecture for which the only evidence is the diminishing fresh water supply remains speculative, but the salination process cannot be ignored and mitigative measures are needed to slow down the process

Global temperature and salinity rise follow their own pattern in accord to the changing hydrological parameters. The absorption and radiation of heat by the atmosphere known as the natural greenhouse-effect has been contributed by man that has increased the average global surface temperature between 1884 and 2015 by 0.87°C [78]. The 6th annual Sustainable Innovation Forum (SIF15) at Stade de France with the participation of more than 190 countries was the largest event held during the annual Conference of Parties (COP21). The most important mitigative measure of SIF15 against the "dangerous anthropogenic interference" was the decision to keep the global mean surface temperature increase below 2°C relative to the pre-industrial times. Similar global decision would be necessary to prohibit dumping of wastes into water bodies and keep the fresh water of rivers and lakes under a determined osmolarity minimum to prevent the anthropogenic salinity increase of oceans.

Conclusion

To summarize the euhaline and eustatic changes of ocean, it is predicted that salination is becoming a serious concern of global life. Inversely correlated intermittent fluctuations of ocean volume and fresh water reservoires are restricted to glacial and interglacial periods. The long-term salination process leads to the conclusion that the glacial-interglacial cycles will take place at increasingly higher salt concentrations of oceans resulting in gradually shortened and consequently more frequently occurring ice ages with gradually diminishing fresh water reservoirs. The widening gap between the increasing concentration of oceans, containing less dissolved oxygen *versus* the diminishing fresh water, the declining rate of evaporation and the spread of deserts are warning signals. The global survival strategy to extend life on Earth needs to turn to the preservation of fresh water and slow down the salination process.

Competing Interest

The author declares to have no competing interest.

Acknowledgement

This work was supported by a grant of the Hungarian National Science and Research Foundation to GB (OTKA T42762 grant. The critical reading of the manuscript by Prof Henry Paulus is gratefully acknowledged.

References

1. Paul J, Durack PJ, Susan E, Wijffels SE, Peter J, et al. (2014) Long-term sea-level change revisited: the role of salinity. Environ Res Letts 9: 114017.
2. Pattullo J W, Munk R, Revelle R, Strong E (1955) The seasonal oscillation in sea level. J Mar Res 14: 88-156.
3. Tabata S, Thomas B, Ramsen D (1986) Annual and interannual variability of steric sea level along line P in the Northeast Pacific Ocean. J Phys Oceanogr 16: 1378-1398.
4. Maes C (1998) Estimating the influence of salinity on sea level anomaly in the ocean Geophy. Res Lett 25: 3551-3554.
5. Sato OT, Polito PS, Liu WT (2000) Importance of salinity measurements in the heat storage estimation from TOPEX/POSEIDON. Geophys Res Lett pp: 27549-27551.
6. Wunsch C, Ponte RM, Heimbach P (2007) Decadal trends in sea level patterns: 1993–2004. J Clim 20: 5889-5911.
7. Suzuki T, Ishii M (2011) Long-term regional sea level changes due to variations in water mass density during the period 1981-2007. Geophys Res Lett 38: L21604.
8. Antonov JI, Levitus S, Boyer TP (2005) Thermosteric sea level rise, 1955–2003 Geophys Res Lett 32: L12602.
9. Munk W (2003) Ocean freshening, sea level rising. Science 300: 2041-2043.
10. Ishii M, Kimoto M, Sakamoto K, Iwasaki SI (2006) Steric sea level changes estimated from historical ocean subsurface temperature and salinity analyses. J Oceanogr 62: 155-170.
11. Gregory JM, Banks HT, Stott PA, Lowe JA, Palmer MD (2004) Simulated and observed decadal variability in ocean heat content. Geophys Res Lett 31: L15312.
12. Boyer T P, Antonov JI, Baranova OK, Coleman C, Garcia HE, et al. (2013) World Ocean Database 2013 edS Levitus and A Mishonov 209 (Technical ed). National Oceanographoc Data Center, Silver Spring, MD.
13. Durack PJ, Wijffels SE, Boyer TP (2013) Long-term salinity changes and implications for the global water cycle Ocean Circulation and Climate, A 21st Century Perspective 2nd edn. Eds. G Siedler, SM Griffies, J Gould, JA Church Oxford: International Geophysics, Academic, Elsevier, vol 103 chapter 28: 727-757.
14. Church JA, Clark PU, Casenave JM, Gregory JM, Jevrejeva S, et al. (2013) The Physical Science Basis (eds TF Stocker, D Qin, G-K Plattner, M Tignor, SK Allen, J Boschung, A Nauels, Y Xia, V Bex and P M Midgley. IPCC, Univ. Press, Cambridge, UK pp: 1137-1216.

15. Banfalvi G (2009) DNA Empire. In: Banfalvi G (ed) Apoptotic chromatin changes. Springer, Dordrecht p: 10.

16. Iribarne JV, Godson WL (1973) Atmospheric Thermodynamics. Geophysics and Astrophysics Monographs, D. Reidel Publ, Dordrecht Holland p: 222.

17. Gradstein FM, Ogg JG, Smith AG (2004) Construction and summary of a geological time scale. In. A geologic time scale. Gradstein FM, Ogg JG, Smith AG (eds) Cambridge University Press, Cambridge pp: 455-462.

18. McFarlan WN, Heiser JB (1979) Life in water: Its influence on basic vertebrate functions. In Vertebrate Life, Macmillan, New York pp: 221-284.

19. Smith HW (1943) Lectures on the Kidney', Porter Lectures IX, University of Kansas Press, Lawrence.

20. Smith HW (1953) From fish to filosopher', Little, Brown and Co, Boston.

21. Selected Astronomical Constants (2011) The Astronomical Almanac. Retrieved 2011-02-25 World Geodetic System (WGS-84) (1984) Available online from National Geospatial-Intelligence Agency.

22. Cazenave A (1995) Geoid, topography and distribution of landforms. In Ahrens TJ (ed). Global earth physics a handbook of physical constants. Washington, DC: American Geophysical Union.

23. Charette MA, Smith W (2010) The volume of Earth's Ocean. Oceanography 23: 112-114.

24. Charette MA, Smith WHF (2010) The volume of Earth's ocean. Oceanography 23: 104-106.

25. Shiklomanov IA (1995) World fresh water resources. In: Gleick PH (ed) Water in crisis: A guide to the world's fresh water resources. Oxford University Press, New York p: 13-24.

26. World Geodetic System (1984).

27. Catling DC, Zahnle KI (2009) The planetary air leak. Scientific American 300: 36-43.

28. Lastovicka J, Akmaev RA, Beig G, Bremer J, Emmert JT (2006) Global change in the upper atmosphere. Science 314: 1253-1254.

29. Emmert JT, Picone JM, Meier RR (2008) Thermospheric global average density trends, 1967-2007, derived from orbits of 5000 near-Earth objects. Geophys Res Lett 35: L05101.

30. Englert CR, Harlander JM, Emmert JT, Babcock DD, Roesler FL (2010) Initial ground-based thermospheric wind measurements using Doppler asymmetric spatial heterodyne spectroscopy (DASH). Opt Express 18: 27416-27430.

31. The Solar Wind at Mars.

32. Pawlowicz R (2013) Key Physical Variables in the Ocean: Temperature, Salinity, and Density. Nature Education Knowledge 4: 13.

33. Hosoda S, Suga T, Shikama N, Mizunol K (2009) Global surface layer salinity change detected by Argo and its implication for hydrological cycle Intensification. J Oceanography 65: 579-586.

34. Soloviev A, Lukas R (2006) The near-surface layer of the ocean: structure, dynamics and applications. Athmospheric and Oceanographic Sciences Library, Springer, Dordrecht, The Netherlands pp: 1-79.

35. Argo Science Team (2001) Report of the Argo Science Team Second Meeting. Argo Science Team Second Meeting, Southampton Oceanogr. Cent. Southampton, UK.

36. Argo Steering Team (2007) Eighth meeting of the International Argo Steering Team. Argo Steering Team Eighth Meeting, Paris, France.

37. Boyer TP, Levitus S, Antonov JI, Locarnini RA, Garcia HE (2005) Linear trends in salinity for the World Ocean, 1955-1998. Geophys Res Letts 32: L01604.

38. Durack PJ, Wijffels SE (2010) Fifty-year trends in global ocean salinities and their relationship to broad-scale warming. J Climate 23: 4342-4362.

39. Rubey WW (1951) Geologic history of sea water. Geol Soc Am Bull 62:1111-1147.

40. Railsback LB, Anderson TF, Ackerly SC, Cisne JL (1989) Paleoceanographic modeling of temperature-salinity profiles from stable isotopic data. Paleoceanography 4: 585-591.

41. Worthington LV (1981) The water masses of the World Ocean: Some results of a finescale census. In Evolution of Physical Oceanography. Warren B, Wunsch C (eds) MIT Press, Cambridge, MA pp: 43-69.

42. Banfalvi G (1991) Evolution of osmolyte systems. Biochem Education 19: 136-139.

43. Delaney RG, Lahiri S, Hamilton R, Fishman AP (1977) Acid-base balance and plasma composition in the aestivating lungfish (Protopterus). Am J Physiol 232: R10-R17.

44. Harms C, Ross T, Segars A (2002) Plasma biochemistry reference values of wild bonnethead sharks, Sphyrna tiburo. Vet Clin Pathol 31: 1-5.

45. Carrol RL (2001) The origin and early radiation of terrestrial vertebrates. J Paleont 75: 1202-1213.

46. Yancey PH, Clark ME, Hand SC, Bowlus RD, Somero GN (1982) Living with water stress: evolution of osmolyte systems. Science 217: 1214-1222.

47. Haq BU, Schutter SR (2008) A chronology of Paleozoic sea-level changes. Science 322: 64-68.

48. Hay WW, Migdisov A, Balukhovsky AN, Wold CN, Flögel S, et al. (2006) Evaporites and the salinity of the ocean during the Phanerozoic: Implications for climate, ocean circulation and life. Palaeogeography, Palaeoclimatology, Palaeoecology 240: 3-46.

49. Holser WT (1984) Gradual and abrupt shifts in ocean chemistry during Phanerozoic time. In: H.D. Holland and AF Trendall (Editors), Patterns of Change in Earth Evolution. Springer-Verlag, Berlin pp: 123-144.

50. Berger A, Loutre MF (2002) An exceptionally long interglacial ahead? Science 297: 1287-1288.

51. Augustin L, Barbante C, Barnes PR, Barnola JM, Bigler M, et al. (2004) Eight glacial cycles from an Antarctic ice core. Nature 429: 623-628.

52. Berger A, Loutre MF (1996) Modelling the climate response to astronomical and CO2 forcings. Ext Geophys Climate Environ C. R. Acad. $ci. Paris 323: 1-16.

53. Lewis BTR (1983) The process of formation of ocean crust. Science 220: 151-157.

54. Joly J (1899) An estimate the geological age of the Earth. Trans Royal Soc Dublin 2: 23-66.

55. Burchfield JD (1998) The age of the Earth and the invention of geological time. Geological Society, London, Special Publications 143: 137-143.

56. Dalrymple GB (2001) The age of the Earth in the twentieth century: a problem (mostly) solved. Special Publications, Geological Society of London 190: 205-221.

57. Dolbeck J, Mergler D, Larribe F, Roulet M, Lebel J, et al. (2001) Mercury contamination and fish diet of an Amazonian population (Brazil). Sci Total Environ 271: 87-97.

58. Miller KG, Kominz MA, Browning JV, Wright JD, Mountain GS, et al. (2005) The Phanerozoic record of global sea-level change. Science 310: 1293-1298.

59. Liangquan L, Keller G (1998) Abrupt deep-sea warming at the end of the Cretaceous. Geology 26: 995-998.

60. Loeb S, Van Hessen F, Shahaf D (1976) Production of energy from concentrated brines by pressure-retarded osmosis: II. Experimental results and projected energy costs. J Membrane Sci 1: 249-269.

61. Dickson B, Yashayaev I, Meincke J, Turell B, Dye S, et al. (2002) Rapid freshening of deep North Atlantic Ocean over the past four decades. Nature 416: 832-836.

62. Curry R, Mauritzen C (2005) Dilution of the Northern North Atlantic Ocean in recent decades. Science 308: 1772-1774.

63. Von Schuckmann K, Sallée JB, Chambers D, Le Traon PY, Cabanes C, et al. (2014) Consistency of the current global ocean observing systems from an Argo perspective. Ocean Sci 10: 547-557.

64. Southam JR, Hay WW (1981) Global sedimentary mass balance and sea level changes. In: The Sea, Wiley-Interscience, New York 7: 1617-1684.

65. Zharkov MA (1981) History of Paleozoic salt accumulation, Springer-Verlag, Berlin.

66. Holland HD, Lazar B, McCaffrey M (1986) Evolution of the atmosphare and oceans. Nature 320: 27-33.

67. Lambeck K, Chappel J (2001) Sea level change through the last glacial cycle. Science 292: 679-686.

68. Shackleton NJ (1987) Oxygen isotopes, ice volume and sea level. Quat Sci Rev 6: 183-190.

69. Duplessy JC, Bard E, Labeyrie L, Duprat J, Moyes J (1993) Oxygen isotope records and salinity changes in the northeastern Atlantic Ocean during the last 18,000 years. Paleoceanography 8: 341-350.

70. Zachos JC, Stott LD, Lohmann KC (1994) Evolution of early Cenozoic marine temperatures. Paleoceanography 9: 353-387.

71. Dalrymple GB (1991) The Age of the Earth. Stanford California: Stanford University Press.

72. Antonov JI, Locarnini RA, Boyer TP, Mishonov AV, Garcia HE (2006) World Ocean Atlas 2005: Salinity S Levitus S (ed) (2006). NOAA Atlas NESDIS 62, US Government Printing Office, Washington, DC p: 182.

73. Intergovernmental Panel on Climate Change (IPCC) (2013) The Physical Science Basis.

74. Cicerone RJ (1990) Greenhouse cooling up high. Nature 344: 104-105.

75. Hansen J, Sato M, Kharecha P, von Schuckmann K (2011) Earth's energy imbalance and implications. Atmos Chem Phys 11: 13421-13449.

76. Church JA, White NJ, Konikow LF, Domingues CM, Cogley JG, et al. (2011) Revisiting the Earth's sea-level and energy budgets from 1961 to 2008. Geophys Res Lett 38: L18601.

77. Cazenave A, Llovel W (2010) Contemporary sea level rise, Ann Rev Marine Sci 2: 145-173.

78. http://climate.nasa.gov/vital-signs/global-temperature.

Nephthea sp.: Correlation Between Natural Products Production and Pressure from Local Environmental Stressors

Hedi Indra Januar[1,2], Boedi Hendrarto[2], Ekowati Chasanah[1] and Anthony D. Wright[3*]

[1]*Indonesia Research Center for Marine and Fisheries Product Processing and Biotechnology, Jl. KS Tubun Petamburan VI, Jakarta 10260, Indonesia*
[2]*Magister of Environmental Science, Diponegoro University at Indonesia, Jl. Imam Bardjo, SH No. 5 Semarang, Indonesia*
[3]*University of Hawaii, College of Pharmacy, 34 Rainbow Drive, Hilo, Hawaii 96720, USA*

Abstract

The presented study looked in a preliminary and speculative way for a correlation between production by a single soft coral species, *Nephthea* sp., of two secondary metabolites and local environmental stresses, pollution, around the Seribu Islands, Indonesia. The study region was selected because of the significant land-based run-off from the heavily populated Indonesian capital, Jakarta. A *Nephthea* sp., was chosen based on the well known bioactive cembranoid diterpene content of animals of the genus that is often involved in protecting them from predation. For the study, levels of the recently reported 3,4-epoxy-nephthenol acetate (**1**) and the previously known 15-hydroxy-cembrane (**2**) were monitored. Results showed levels of **1** and **2** to decrease as pollutants increased. A contour map of levels of **1** and **2** showed a likely different pattern of their occurrence, with **2** being found more in polluted, higher stress, areas and **1** likely only to occur in areas with low environmental stresses. It is postulated that low levels of anthropogenic stress, pollution, in the reef environment result in no inhibition of biodiversity and number of fishes that prey on *Nephthea* sp., likely causing enhanced secondary metabolite production in these species in order to protect them from predation. This may mean levels and/or presence or absence of **1** might be indicators of pollution entering a coral reef area, with the resultant threat to a reef's potential biopotency. Results also show where anthropogenic pollution increases in a reef system the overall level of biodiversity drops, as a consequence so does chemical diversity and with it the chance of discovering new and biological active secondary metabolites.

Keywords: *Nephthea* sp.; Cembranoids; Diterpenes; Secondary Metabolites; Environmental Stressors; Pollution; Seribu Islands; Indonesia

Introduction

In biotechnology and natural products research terms, coral reef communities are seen as one of the best sources of potentially new chemical entities that may have useful biological properties. Increasing anthropogenic pressures, pollution, and global warming are causing considerable damage to coral reef structures through direct and indirect means which leads to the conclusion that coral reef biodiversity is generally under threat [1]. The loss, reduction or unexpected appearance of any species to a natural system can change the overall dynamic of that system dramatically, in turn causing more and more change in that system and typically not in a positive way [2,3]. In other words, when populations are depleted genetic variation is reduced which in turn reduces an organisms' ability to adapt to new environmental changes and stresses [4]. And, due to the interdependencies among species, the demise of one can lead to the decrease or demise of others. Eventually this leads to either significant reduction or loss of biodiversity. Another way to look at this kind of change, particularly significant loss of species, is in terms of the likelihood that natural chemical diversity, or better said, natural products diversity, and everything this implies, that currently exists in reef communities is also threatened. The most obvious examples of this type of change would be through the loss of habitat, appearance of an invasive species [5] or rapid population growth of a predatory species [3] that may stop or enhance the establishment of certain sessile organisms like algae, soft corals and sponges, while allowing others to grow and thrive, while others die. These balanced predator prey relationships are key to the overall health and diversity of almost all living systems and dramatic external changes in these can be catastrophic [5]. Some species require challenges to stimulate the production of certain chemical defense substances [6,7], and if these challenges are not present the substances are simply not produced, this is known as an inducible or plastic response compared to a constant response where substances are produced continuously [8,9]. If the induced substances are not produced chemical diversity decreases and at some point the production of certain substance may stop altogether in some species. These negative responses may also occur as a result of the change of local conditions [10,11]. Thus, loss of predatory fish through overfishing or loss of other competitive organism in some other way, for example, by an environment becoming polluted with excessive amount of nutrients [12] will result in not just the reduction and or loss of biodiversity but also in the reduction and or loss of chemical diversity. Based on this premise, that local environmental stress, pollution, may affect chemical diversity, the research undertaken and reported in this current paper documents in a general and speculative way a possible relationship between natural product diversity in coral reef organisms and local environmental stresses placed on a given environment.

Materials and Methods

Study site and organism selection

The Seribu Islands is one of Indonesia's small island areas located in DKI (Special Capitol District) Jakarta province. Because of the islands' location close to the Indonesian capital, Jakarta, the marine

***Corresponding author:** Anthony D. Wright, University of Hawaii, College of Pharmacy, 34 Rainbow Drive, Hilo,Hawaii 96720,USA, E-mail: adwright@hawaii.edu

habitat around these islands has local environmental stressors from land-based industrial and heavily populated domestic run-off that first enters Jakarta Bay and then flows out around them. In 2004, metal contaminants from Jakarta Bay were already considered to be above normal levels for marine organisms. At this time, the so called red tide phenomenon also occurred several times in Jakarta Bay as the concentration of phosphate and nitrate ions were high in that area [13]. In the Seribu Islands themselves there were also local domestic run-off and fish aquaculture conditions that could possibly trigger or be responsible for the high levels of phosphate and nitrate ions measured in the surrounding waters. In simple visual terms the amount of observable (July 2009) solid waste of all kinds flowing from Jakarta Bay to the Seribu Islands can only be described as significant. Based on these observations the Seribu Islands were chosen as the site for the current study. The organisms selected for investigation were *Nephthea* sp., based on their common occurrence on local coral reefs around these islands and their apparent successful deterrence of predators [14,15]. This successful defense strategy of soft corals is usually correlated with a defensive chemical anti-predatory system based on bioactive natural products known as terpenes, specifically cembranoid diterpenes [7,16]. The cembranoids focused on in this study were the recently reported 3,4-epoxy-nephthenol acetate (**1**) and the known 15-hydroxy-cembrane (**2**) [17].

Animal Material

All *Nephthea* sp., samples were collected on the basis of their morphological similarities to the sample used in the recently reported investigation of *Nephthea* sp. [17]. All samples were indentified on the basis of their overall morphological and spicule profiles. The samples' profiles were compared with the morphology and spicules profile database of soft corals [18]. From the profiles of sample spicules, all collected samples were identified as belonging to a single species within the genus *Nephthea*. Full taxonomic assignment of this likely new species will be the subject of a future discrete taxonomic investigation.

Samples were collected from twelve locations ranging from southern to northern Seribu Islands, Indonesia, during the period April 19-22, 2010. Precise locations were recorded using a GPS; Garmin GPSMAP 60CSXA. A plot of each site was made on maps from Holtrof [19], and are shown in Figure 1(a). At each site, at 5m depth, a 30m transect was laid based on established line methods [20] to estimate live coral cover. From each 10m section of the 30m line transect, a sample of the soft coral *Nephthea* sp., was taken giving a total of three from each transect. The fresh coral sample (ca. 5g) was immediately placed

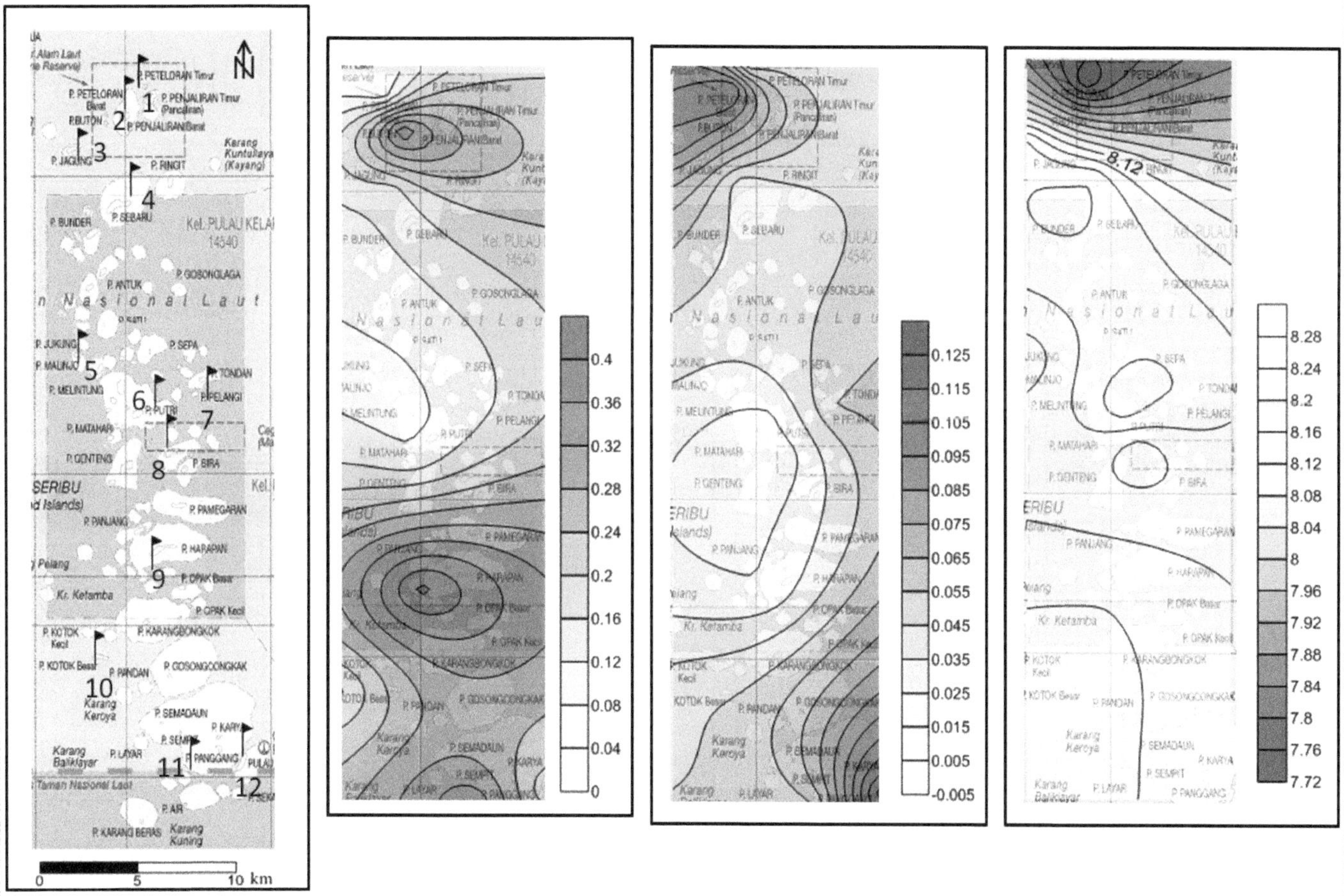

(a) (b) (c) (d)

Figure 1: (a) Map of sampling sites in the Seribu Islands, and contour maps of; (b) active phosphate in ppm (c) inorganic nitrogen in ppm (d) pH

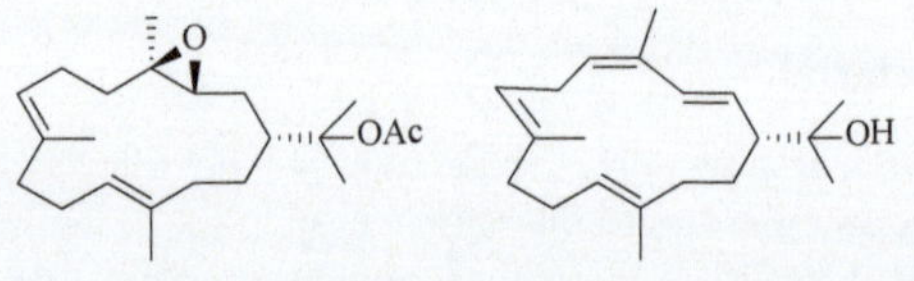

1 2

Figure 2: Cembranoid diterpenes quantified from *Nephthea* sp., collected from Seribu Islands; 3,4-epoxy-nephthenol-acetate (**1**) and 15-hydroxy-cembrane (**2**).

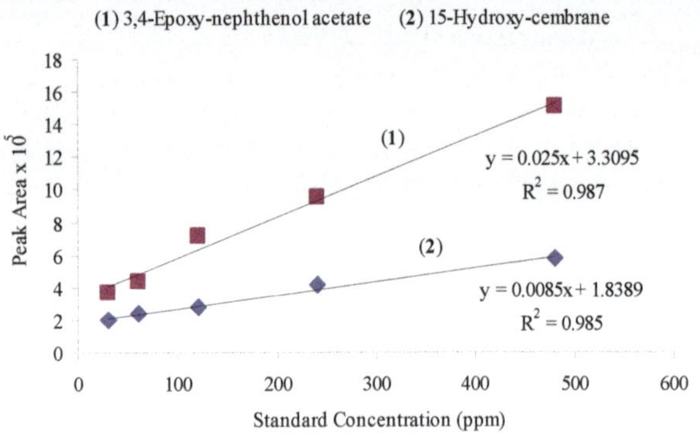

Figure 3: Regression curves of 3,4-epoxy-nephthenol acetate (**1**) and 15-hydroxy-cembrane (**2**) standards).

in 10mL of HPLC grade methanol and maintained at 0°C in a cool box containing ice water. On return to the laboratory, samples were exhaustively extracted with methanol to yield the extract that was used for further study.

Environmental Water Analysis

At each sampling site, water samples within one meter of *Nephthea* sp., (5m depth) were taken. Since three animal samples were collected from each site, three water samples were also collected at corresponding locations. Analysis of PO_4^-, NO_2^-, NO_3^-, NH_3, and pH were undertaken immediately on board above the sampling site using a portable colorimeter Hach DR-890 and a pH-meter. Dissolved inorganic nitrogen (DIN) concentration was determined as the sum of NO_2^-, NO_3^-, and NH_3 according to an established method [21].

Determination of Cembranoid Diterpenes

The two cembranoid diterpene standards were first isolated and characterized as 3,4-epoxy-nephthenol acetate (**1**) and 15-hydroxy-cembrane (**2**) (Figure 2) in a previous study [17]. For each compound, a dilution series ranging from 480–30ppm was prepared and a standard curve of concentration constructed (Figure 3). Quantification of cembranoid diterpenes in collected samples was achieved by injection of 20 µL of methanol extract *Nephthea* sp. (10mg/mL) into a Shimadzu 2010A HPLC system equipped with a Shimadzu ODS RP-C_{18} column (150×2.0 mm) and mobile phase gradient mixtures of water containing increasing amounts of acetonitrile (ACN). Determination of the concentration of each compound was achieved by comparison

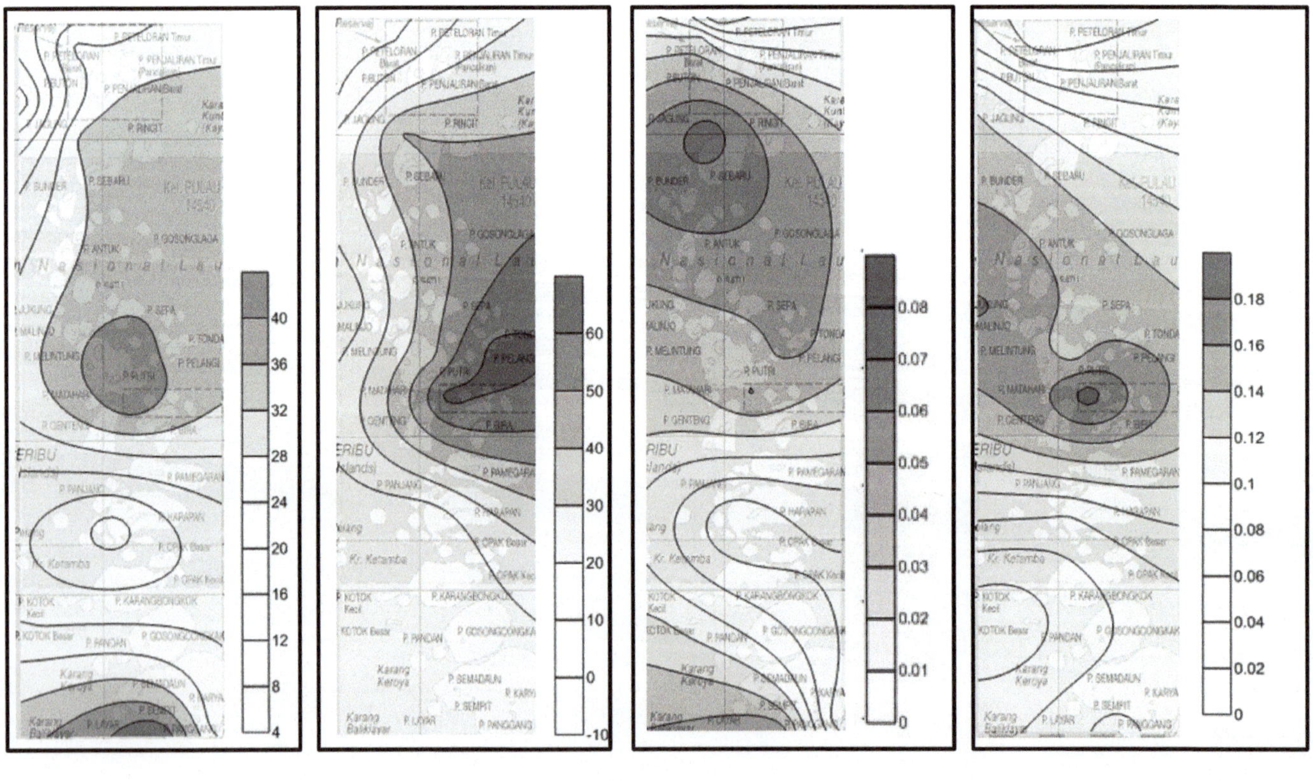

(a) (b) (c) (d)

Figure 4: (a) Contour maps of % live coral cover in Seribu Islands waters (b) % Cytotoxicity of *Nephthea* sp., extracts (c) % Yield of 15-hydroxy-cembrane (**2**) (d) % Yield of 3,4-epoxy-nephthenol acetate (**1**)

of values obtained from regression plot standard concentration curves (Figure 3) and peak area of each correlated compound in resultant HPLC chromatograms.

Cytotoxic Testing

Screening of samples for their cytotoxic was conducted employing a single human cancer cell line, T47D (breast cancer), based on the methods of Zachary [22]. Each sample was analysed in triplicate at 30mg/L, based on limit inhibition potency from Fajarningsih *et al.*, [23].

Contour Maps

Each variable measured; phosphate concentration, inorganic nitrogen concentration, pH (Figure 1 b-d), live coral cover, extract cytotoxicity towards T47D cells, and concentrations of 3,4-epoxy-nephthenol acetate (**1**) and 15-hydroxy-cembrane (**2**) (Figure 4 a-d), was plotted to previously obtained GPS location coordinates. Contour maps of each variable were generated using Surfer Golden Software Version 9.9.78Results

Table 1 shows data for active phosphate concentration (PO_4^{3-}), dissolved inorganic nitrogen (DIN) concentration, and pH in sea water samples collected from twelve different locations during the time of this study; resultant contour maps are shown in Figure 1.

Figure 2 shows the chemical structures of the two cembranoid diterpenes (**1**-**2**) whose concentration in *Nephthea* sp., samples were analyzed in this study.

Table 2 shows percentage of live coral cover on reefs from where samples were collected, cytotoxicity of *Nephthea* sp., extracts towards T47D cells, and yields of **1** and **2**. Quantification of both cembranoid diterpenes was based on standard regression curves shown in Figure 3. Contour maps generated from these data are shown in Figure 4.

Discussion

It was evident at several study sites that levels of phosphate and dissolved inorganic nitrogen were above threshold values (0.1μM and 1μM, respectively) for a minimally impaired coral reef [24] (Figure 1). High levels of phosphate were probably caused by effluent from domestic run-off, as the majority occurred in southern areas where there are local communities living on the Seribu Islands. In northern regions of the islands, all parameters analyzed showed high concentrations of phosphate and inorganic nitrogen. The reason for high values in these areas is unknown, but it seems likely that the levels observed may come from non-point sources, such as movement of ions in local ocean currents. The northern part of Seribu Islands is an open sea area, which has oil drilling rig activities and is also a busy pathway for boat traffic to and from Jakarta province, which may have an influence on the variables being measured.

Secondary metabolite investigation of collected samples found them in most cases to contain both investigated compounds (**1**-**2**).

In a general sense, the results in Table 2 and Figure 4 (a) show the impact of anthropogenic stressors, pollutants, on the Seribu Islands coral reef environment. The contour map of live coral cover shows the maximum live cover in these areas was only around 40%. The biopotency, as measured by cytotoxicity, and not to be confused with the presence or absence of **1** and **2**, of extracts from *Nephthea* sp., loosely correlates with the observed coral coverage and vice versa, in all but one case, Site 10. In the southern region, where Site 10 was located, the cytotoxicity levels appeared not to show the same correlation with live coral cover as their central and northern counterparts.

Although speculative, the occurrence of cytotoxic compounds in marine invertebrates is usually correlated with an organisms attempt to win and or maintain living space in a given benthic area against their competitor(s) [7,25-27]. The high levels of nutrient ions detected in the southern part of the Seribu Islands, that are considered less than optimal to support a wide variety of life on the coral reefs, might be a cause of the lack of biopotency vs coral cover observed there and that the high coral cover was observed from only a few species that appear to be able tolerate such conditions, e.g., *Sarcophyton* sp., (Soft corals) and *Montipora* sp., (Hard corals) together with organisms from other groups, e.g., zoanthids, echinoderms, and algae, but the actual species diversity is low. The abiotic part of the environment in these regions was predominantly covered with rubble and sand.

Visual observation at sampling sites in the southern part of the Seribu Islands showed that these areas were dominated by soft corals

Site	Location	Phospate (ppm)		Dissolved Inorganic Nitrogen (ppm)		pH	
		Value	±SD	Value	±SD	Value	±SD
1	East Peteloran Island	0.043*	0.006	0.029	0.001	7.77	0.06
2	West Penjaliran Island	0.017*	0.006	0.097	0.016	7.73	0.06
3	Jagung Island	0.357	0.031	0.082	0.000	8.20	0.00
4	Sebaru Besar Island	0.103	0.015	0.020	0.003	8.23	0.06
5	Melinjo Island	0.033*	0.012	0.018	0.002	8.13	0.06
6	Putri Island	0.037*	0.012	0.004*	0.002	8.27	0.06
7	Pelangi Island	0.083	0.006	0.035	0.003	8.23	0.06
8	Kayu Angin Bira Island	0.047*	0.015	0.000	0.000	8.13	0.06
9	Kelapa Island	0.413	0.006	0.006	0.002	8.23	0.06
10	Kotok Island	0.110	0.000	0.033	0.010	8.27	0.06
11	Panggang Island	0.357	0.006	0.043	0.007	8.23	0.06
12	Pramuka Island	0.170	0.000	0.128	0.006	8.20	0.00

*Value below stated instrument detection limit (0.05 ppm for Phosphate and 0.005 ppm for nitrogen)

Table 1: Concentration of active phosphate (ppm) and inorganic nitrogen (ppm), and pH at sampling sites in the Seribu Islands.

Site	Location	Weight (%) 3,4-Epoxy-nephthenol acetate (1) in wet weight of *Nephthea* sp.		Weight (%) 15-Hydroxy-cembrane(2) in wet weight of *Nephthea* sp.		Inhibition effect of 30 ppm *Nephthea* sp. extract towards T47D cells (%)		Live Coral Cover (%)
		Value	±SD	Value	±SD	Value	±SD	Value
1	E. Peteloran Island	0.00	0.00	0.00	0.00	0.00	0.00	20
2	W. Penjaliran Island	0.01	0.00	0.04	0.01	0.00	0.00	25
3	Jagung Island	0.11	0.02	0.06	0.01	0.00	0.00	5
4	Sebaru Besar Island	0.16	0.02	0.09	0.01	40.52	0.78	30
5	Melinjo Island	0.13	0.02	0.06	0.00	0.00	0.00	25
6	Putri Island	0.14	0.01	0.05	0.01	37.71	1.56	35
7	Pelangi Island	0.19	0.01	0.06	0.00	61.92	0.62	30
8	Kayu A. Bira Island	0.17	0.04	0.06	0.01	63.10	1.98	35
9	Kelapa Island	0.00	0.00	0.00	0.00	0.00	0.00	10
10	Kotok Island	0.05	0.01	0.04	0.01	8.84	1.08	20
11	Panggang Island	0.07	0.00	0.07	0.00	10.13	1.51	40
12	Pramuka Island	0.05	0.00	0.00	0.00	0.00	0.00	20

Table 2: Yield of 3,4 epoxy-nephthenol acetate (**1**) and 15-hydroxy-cembrane (**2**) in wet weight of *Nephthea* sp., inhibition of 30 ppm *Nephthea* sp., extract to T47D cell lines, and live coral cover in each sampling location.

belonging to the genus *Sarcophyton*. *Sarcophyton* sp., are common on inshore reefs, and so it is not surprising that they could accommodate a wide range of nutrient conditions without showing adverse effects [28]. This proposed effect has also been discussed for some reefs on the Great Barrier Reef - Australia, where higher nutrient concentrations in inshore waters may contribute to the dominance and possible expansion by octocorals in that region [29]. The results of this latter study may also correlate in some way with those found for the soft coral *Sarcophyton ehrenbergi* that showed there was a general increase in the "physiological-change indicator", which is defined as the ratio of stress metabolites (terpenes) to energy storage metabolites (lipids), when local stress was increased by relocation next to a hard coral species but was variable when the nutrients nitrogen and phosphorus were considered [11]. Lower overall diversity may also mean lower competition for living space and so *Nephthea* sp., as a response might produce lower amounts of cytotoxic (antipredatory) compounds [8,9]. High cytotoxic activity of extracts observed in the central part of the Seribu Islands that has lower levels of nutrient ions compared to other sites combined with high coral cover supports the tentative correlation between high diversity of reef species and competition for living space [30]; in the particular case being studied the soft coral would produce higher levels of cytotoxic compounds in order to survive, win and, or maintain a space on the reef [31,32].

Generally, as seen in contour maps Figures 4(c) and 4(d), the cembranoid diterpenes **2** and **1** from *Nephthea* sp., occurred in a similar distribution pattern as that of live coral cover Figure 4(a). As there is a positive relationship between live coral cover and total number of fish species [33], it is probable that local coral fishes are in abundance in those areas with live coral cover. As a logical consequence, and based on the observation that many cembranoid diterpene compounds in soft corals are known to be toxic, i.e., serve as local coral fish feeding deterrents, as antifouling agents, and as agents assisting with successful competition for space with others corals [14,16], as well as being implicated in spawning [14], it is likely that *Nephthea* sp., growing in areas of high live coral cover will produce larger amounts of feeding deterrent cembranoid diterpenes than those from sites with less cover, so as to effectively deter fish feeding [9,31]. In contrast, in areas with higher levels of pollutants marine organisms would have difficulty surviving; that is, conditions might cause a reduction of hard coral cover and hence the number of reef fish species [34]. Under these circumstances, as the local number of predatory fishes decrease it is probable that *Nephthea* sp., would no longer require prior high levels of their antipredatory substances, like cembranoid diterpenes.

The finding of different concentration patterns of 3,4-epoxy-nephthenol acetate (**1**) and 15-hydroxy-cembrane (**2**) throughout the Islands [see Figures 4(c) and (d)] showed both of these compounds to likely serve as anti-predatory systems, and that each of them may address a different type(s) of predator(s) [8]. The known 15-hydroxy-cembrane (**2**) occurred more generally at every site that had live coral cover, while 3,4-epoxy-nephthenol acetate (**1**) occurred only in specific areas of the Seribu Islands, those with lower levels of environmental stressors. From these observations it might be likely that in the latter case the predator(s) or effector(s) that triggered this compounds production is (are) only found in what are considered "clean" environments. As production of **1** seems to be significantly compromised through pollution and other human activities, this study shows that one source of marine biodiversity may be being rapidly compromised or lost as a result of such occurrences [35].

In summary, the results of this study showed in a very preliminary and speculative way that *Nephthea* sp., sampled from the Seribu Islands produced lower amounts of cembranoid diterpenes as nutrification entered their habitats. This finding may be the result of anthropogenic stressors and its negative correlation on the total number of soft coral fish predators. In other words, when cembranoid diterpenes serve as an anti-predatory system, the production of those natural products might be conditional on predator challenge. As lower amounts of bioactive compounds were found in areas considered polluted, it might reasonably be concluded that anthropogenic pollution is a major threat to reef biopotency and hence, natural product discovery.

Acknowledgement

Taxonomic assignments were made by Beginner Subhan, Agricultural University, Department of Marine Science and Technology, Jl. Raya Darmaga

Kampus IPB Darmaga, Bogor 16680, Indonesia. We thank, Satibi, Sahiran, and Hamdan at Kepulauan Seribu National Park for their assistance with sample collection. Our thanks also go to S. Iswani, A. Pratitis, and G. Patantis for assisting with water quality and cytotoxic analyses.

References

1. Rajendran N, Ravichandran S, Raghunathan S, Vijayalakshmi L, Swaminathan V, et al. (2001) In: Coral reefs of India: State-of-the-Art Report. Balasubramanian T, Ajmal Khan S Eds. ENVIS Publication Series, India.
2. Goreau T (1992) Coral Reef Protection in Western Jamaica. In: Protecting Jamaica's Coral. Reefs: Water Quality Issues, pg: 39-65.
3. Engelhardt U, Hartcher M, Taylor N, Cruise J, Engelhardt D, et al. (2001) Crown-of-thorns starfish (Acanthaster planci) in the central Great Barrier Reef region. Results of fine-scale surveys conducted in 1999-2000.CRC Reef research centre technical report No. 32.
4. Lacy RC (1997) Importance of genetic variation to the viability of mammalian populations. J Mammal 78: 320-335.
5. Meinesz A, de Vaugelas J, Hesse B, Mari X (1993) Spread of the introduced tropical green alga Caulerpa taxifolia in northern Mediterranean waters. J App Phycol 5: 141-147.
6. Cetrulo GL, Hay ME (2000) Activated chemical defenses in tropical versus temperate seaweeds. Mar Ecol Prog Ser 22: 243-253.
7. Pawlik JR (1993) Marine invertebrate chemical defenses. Chem Rev 93: 1911-1922.
8. Wright JT, de Nys R, Steinberg PD (2000) Geographic variation in halogenated furanones from the red alga Delisea pulchra and associated herbivores and epiphytes. Mar Ecol Prog Ser 22: 227-241.
9. Hoover CA, Slattery M, Targett NM, Marsh AG (2008) Transcriptome and metabolite responses to predation in a south pacific soft coral. Biol Bull 214: 319-328.
10. Leone PA, Bowden BF, Carroll AR, Coll JC (1995) Chemical consequences of relocation of the soft coral Lobophytum compactum and its placement in contact with the red alga Plocamium hamatum. Mar Biol 122: 675-679.
11. Fleury B, Coll JC, Tentori E, Duquesne S, Figueiredo L (2000) Effect of nutrient enrichment on the complementary (secondary) metabolite composition of the soft coral Sarcophyton ehrenbergi (Cnidaria: Octocorallia: Alcyonacea) of the Great Barrier Reef. Mar Biol 136: 63-68.
12. Camargo JA, Alonso Á (2006) Ecological and toxicological effects of inorganic nitrogen pollution in aquatic ecosystems: A global assessment. Environ Int 32: 831-849.
13. Lestari, Edward (2004) The effect of heavy metal contamination to sea water quality and fisheries resources (Study Case mass death of fishes in Jakarta bay). Makara Sains 8: 52-58.
14. Sammarco P, Coll JC (1992) Chemical adaptations in the Octocorallia: Evolutionary considerations. Mar Ecol Prog Ser 88: 93-104.
15. Fabricius KE (1995) Slow population turnover in the soft coral genera Sinularia and Sarcophyton on mid- and outer- shelf reefs of the Great Barrier Reef. Mar Ecol Prog Ser 126: 145-152.
16. Coll JC (1992) The chemistry and chemical ecology of octocorals (Coelenterata, Anthozoa, Octocorallia). Chem Rev 92: 613-631.
17. Januar HI, Chasanah E, Motti CA, Tapiolas DM, Liptrot CH, et al. (2010) Cytotoxic cembranes from Indonesia specimens of the soft coral Nephthea sp. Mar Drugs 8: 2142-2152.
18. Manuputty AEW, Pemanfaatan P, Kelautan DI (2002) Karang lunak (soft coral) perairan Indonesia. buku 1. Laut Jawa & Selat Sunda. Lembaga Ilmu Pengetahuan Indonesia, Pusat Penelitian Oseanografi, Proyek Pemanfaatan dan Diseminasi Iptek Kelautan, Jakarta.
19. http://catalogue.nla.gov.au/Record/980812.
20. Holtrof GW (2004) Jakarta Maps - Street atlas and index with CD-ROM. PT. Java Books Jakarta, Indonesia.
21. English S, Wilkinson C, Baker V (1994) Survey Manual for Tropical Marine resources. Australian Institute for Marine Science (AIMS), Townsville, Queensland, Australia.
22. Kroger R, Holland MM, Moore MT, Cooper CM (2009) Seasonal patterns of nitrogen and phosphorus losses in agricultural drainage ditches in northern Mississipi. In: Agricultural Runoff: Coastal Engineering and Flooding. Hudspeth CA, dan Reeve T.E Eds. Nova Science Publishers Inc. pg: 279-289.
23. Zachary I (2003) Determination of cell number. In: Cell proliferation and apoptosis. Hughes D, Mehmet H Eds. Bios Scientific Publishers Ltd. Pg: 13-35.
24. Fajarningsih ND, Januar HI, Nursid M, Wikanta T (2006) Antitumor potency of sponge Crella papilata from Seribu Islands Marine National Park. Indonesian J Mar Fish Postharvest Biotechnol 1: 35-41.
25. Jameson SC, Kelty RA (2004) A Review of indicators of land-based pollution stress on coral reefs: A background paper for joint EPA/NOAA/USGA/DOI workshop on assessing pollution stress on coral reefs, August 31-September 2, Honolulu, Hawaii, USA.
26. Puyana M, Fenical W, Pawlik JR (2003) Are there activated chemical defenses in sponges of the genus Aplysina from the Caribbean? Mar Ecol Prog Ser 246: 127-135.
27. Kelly SR, Jensen PR, Henkel TP, Fenical W, Pawlik JR (2003) Effects of Caribbean sponge extracts on bacterial attachment. Aqua Micro Biol 31: 175-182.
28. Sotka E, Forbey J, Horn M, Poore A, Raubenheimer D, et al. (2009) The emerging role of pharmacology in understanding consumer-prey interactions in marine and freshwater systems. Integr Comp Biol 49: 291-313.
29. Koop K, Booth D, Broadbent A, Brodie J, Bucher D, et al. (2001) ENCORE: The effect of nutrient enrichment on coral reefs. Synthesis of results and conclusions. Mar Pollut Bull 42: 91-120.
30. Fleury BG, Coll JC, Sammarco PW, Tentori E, Duquesne S (2004) Complementary (secondary) metabolites in an octocoral competing with a scleractinian coral: Effects of varying nutrient regimes. J Exp Mar Biol Ecol 303: 115-131.
31. Hay ME, Fenical W (1996) Chemical ecology and marine biodiversity: Insights and products from the sea. Oceanogr 9: 10-20.
32. Rossi S, Snyder MJ (2001) Competition for space among sessile marine invertebrates: Changes in HSP70 expression in two Pacific Cnidarians. Biol Bull 201: 385-393.
33. Thakur NL, Thakur AN, Muller WEG (2005) Marine natural products in drug discovery. Nat Prod Rad 4: 471-477.
34. Bell D, Galzin R (1984) Influence of live coral cover on coral reef fish communities. Marine Ecol 15: 265-274.
35. Jones GP, McCormick MI, Srinivasan M, Eangle JV (2004) Coral decline threatens fish biodiversity in marine reserves. Proc Nat Acad Sci 101: 8251-8253.
36. Battershill C (2007) Southern ocean biodiversity and discovery: As we see more, we may be seeing less, In Proceedings of 12th International Marine Natural Products Symposium, Queenstown, New Zealand, February 4-9, Conference Abstracts 14.

8

Floating Desalination and Water Pumping Plants as Harmful Algal Bloom Control Technologies

Kevin C Owen[1]*, Jenna L Owen[2] and Daniel P Owen[3]

[1]Höegh LNG AS, 15469 Osprey Glen Drive, Lithia, FL, USA
[2]Tindale-Oliver and Associates, Tampa, FL, USA
[3]Department of Biological Sciences, Florida State University, Tallahassee, FL, USA

Abstract

Harmful Algal Blooms (HABs) have significant ecological and economic effects on the marine environment and use. In recent years, researchers have been increasingly developing and testing methods to treat and control HABs. General categories or strategies proposed as HAB control technologies include mechanical, biological, chemical, genetic, and environmental controls.

The authors of this paper suggest using floating desalination plants to treat or control red tides. HAB producing dinoflagellates have been shown to be sensitive to physical and chemical changes in the environment, such as changes in temperature and salinity. The typical response of dinoflagellates is to form cysts that then settle out of the water column.

The discharges from a floating desalination and water pumping plant can rapidly change the temperature and salinity in the water column. These changes could be expected to induce encystment in the dinoflagellate species that form cyst and could cause mortality in those species unable to form temporary cysts. Preventing population growth, inducing encystment, or causing mortality would effectively end a HAB.

Discharges from a desalination plant are temporary in nature and include hypersaline water, freshwater (hyposaline water), and heated warmer. By discharging the heated hyposaline water at a low depth in the water column, due to its lower density than seawater, the discharge would move upward towards the surface. Since the hypersaline water would be denser than seawater, by discharging it at the surface, the hypersaline water would sink. In environments where a stratified water column exists, pumping water could disrupt the stratification without the need for additional desalination. The discharges from a floating desalination plant would stress the red tide with surfacing warmer fresh water and sinking hypersaline water. The stresses caused by these disturbances can disrupt a HAB.

These temperature and salinity changes that could be created by a floating desalination plant would be achieved without the discharge of chemicals or other materials that could have other detrimental environmental impacts. A good aspect of this treatment is that, with continued mixing after discharge, the water would return to ambient temperature and salinity relatively quickly with minimal effect on the marine environment. Since the dinoflagellates have been shown to react quickly to environmental changes, the temperature and salinity of the discharges could be controlled to reduce adverse impacts on other marine organisms. Bench-scale and field tests should establish optimal temperature and salinity ranges for HAB treatment and control.

Keywords: Harmful Algal Blooms (HAB); HAB control; HAB mitigation

Introduction

Harmful Algal Blooms (HABs), commonly referred to as red tides, have significant ecological and economic effects on the marine environment and use. Over the lasts ever aldecades, there has been a dramatic increase in [1]:

- Areas affected by HABs;
- The number of HABs that have occurred;
- Economic losses attributed to HABs;
- The types of resources affected by HABs;
- The number of toxic species identified as causing HABs; and
- The number of toxins that have been identified in HAB causing species.

Many aspects of HAB ecology have been studied including the processes or mechanisms that lead to the termination of HAB. Potential HAB termination mechanisms include the dispersal of the HAB by physical advection and the breakdown of the oceanographic fronts that might have originally acted to initiate and perpetrate the HAB [2]. The authors of this paper suggest that the introduction of sudden and significant environmental perturbations could disrupt a HAB and that the introduction of such disruptive perturbations can be used as a mechanism of HAB control.

Researchers have been developing and testing methods to treat and control HABs. However, since HABs are complex oceanographic phenomena, challenges to developing effective HAB control challenge

***Corresponding author:** Kevin C. Owen, Höegh LNG AS, 15469 Osprey Glen Drive, Lithia, FL, USA, E-mail: kowen1972@yahoo.com

measures exist. These challenges include the diversity in the species that create HABs and of the impacts caused by the different types of HABs. As such, investigations into HABs generally require multidisciplinary studies and approaches [1].

Given the complex nature and diversity of HABs, the strategies, needs, and objectives of HAB controls are generally thought to be diverse. Researchers generally feel that HAB control strategies and objectives could vary [1]:

- Between locations;
- Among HAB types;
- To protect fisheries;
- To minimize economic losses;
- To minimize ecosystem impacts; and
- To protect public health.

HAB controls are considered to be the actions taken with the objective to suppress or destroy HABs. HAB control is generally considered to be one the most challenging and controversial aspects of HAB management (1). HAB control technologies can generally be grouped into the following categories or strategies [1]:

- Mechanical control;
- Biological control;
- Chemical control;
- Genetic control; and
- Environmental control.

Several of these technologies or strategies have been applied with varying success to control HAB species. For all of the technologies; there are logistical, regulatory, and environmental issues. Due to the complexity of HABs, the science of HAB control is still considered to be rudimentary and the development of this science is considered to be slow [1].

However, as pasteurization was found to control a number of disease-causative organisms in a number of liquids, the authors of this people suggest that certain control approaches may be suitable for the control of a wide range of HAB species across a wide-spread geographic range. The authors of this paper suggest that floating desalination and water pumping plants can be used to treat or control HABs.

General Strategy for HAB Control

Research suggests that the dynamics and impacts of HABs are frequently controlled by physiological responses of the causative dinoflagellate with local environmental conditions and by interactions between biological and physical processes that occur over a broad range of temporal and spatial scales [3]. HAB producing dinoflagellates have been shown to be sensitive to physical and chemical changes in the environment, such as changes in temperature and salinity. The typical response of dinoflagellates to environmental perturbations is to form either permanent or temporary cysts that eventually settle out of the water column. Temporary encystment is a survival mechanism by which dinoflagellates respond to sudden or significant environmental perturbations that may be of limited duration. Factors influencing temporary cyst formation include [4-6]:

- Salinity changes;
- Temperature changes;
- Light changes;
- Chemical changes;
- Nutrient depletion; and
- Dissolved oxygen changes.

Permanent cysts are generally the result of sexual reproduction in the dinoflagellates and often accompany the final phases of a HAB.

Given the characteristics of HABs and dinoflagellate life cycles, the primary objectives of HAB control strategies should be as follows:

- Disrupt the environmental conditions that support the persistence and propagation of HABs;
- Stop dinoflagellate cell division and therefore, population growth within the HAB;
- Induce the formation of temporary cyst formation for the dinoflagellates that have this life cycle stage; and
- Most drastically, cause cell mortality in the dinoflagellates that have created the HAB.

Disrupting the environmental conditions that support the HAB may be sufficient, in and of itself, to lead to the termination of the HAB. Without the proper environmental conditions, the cell division and population will decrease rather than continuing to grow. Due to this disruption, without population growth, a HAB would be effectively terminated.

Rather than terminating a HAB, a change in environmental conditions may induce the formation of temporary cysts. After a short encystment period, the dinoflagellates may excyst and, if the disruptions have passed, a HAB could be re-established. The ability of certain dinoflagellate species to form temporary cysts may allow a HAB to become re-established after the application of a potential control technology that is of limited effective duration, such as clay application [7]. A floating desalination and water pumping plant would allow for a HAB area to be continuously or repeatedly treated so that, as the dinoflagellate sex cyst from the temporary cysts, significant environmental perturbations can be reintroduced. Repeated treatments would ensure that the temporary cyst formation and excystment cycle is counter-balanced and terminate a HAB.

In practical terms there are only a limited number of environmental conditions supporting HABs that could be disrupted on a scale to potentially terminate a HAB. The primary conditions that can be disrupted on such a scale are most likely salinity, water temperature, and stratification of the water column. These three parameters can be significantly altered by a floating desalination plant. In areas where a stratified water column exists, a floating water pumping plant could be adequately efficient without the need for a desalination operation.

If the environmental disruption is significant enough, the mortality of the cells in the dinoflagellate population can actually be affected. An extreme form of HAB control would be to cause a die-off of the dinoflagellate cells, which could be achieved through the use of a floating desalination plant. The major challenge of using a floating desalination plant would be to terminate a HAB without causing mortality in other marine organisms. By controlling the salinity and temperature of the discharges from a floating desalination plant, the mortality of other marine organisms should be avoidable while still terminating a HAB.

Floating Desalination and Water Pumping Plant for HAB Control

A floating desalination and water pumping plant is suggested as a potential technology for the control and suppression of a HAB. This control and suppression is possible because the discharges from a floating desalination plant or a water pumping plant could create sudden, large scale environmental perturbations. These environmental perturbations would include:

- Hyper saline discharges;
- Hypo saline - freshwater discharges;
- Heated water discharges; and
- Disrupt water column stratification through advection currents.

The environmental perturbations could induce simultaneous formation of temporary cysts in the species that form temporary cysts. Repetitive discharges from a floating desalination plant could counteract the cyst cycle and prevent the re-establishment of a population of motile, free-swimming dinoflagellates. For both species that form temporary cysts and species that lack this life-cycle stage, the hyper saline, hypo saline, and heated discharges could create conditions that would lead to the ultimate death of the cell, permanently ending a HAB.

A floating desalination plant can cause significant, temporary changes in salinity, water temperature, and water stratification. A floating desalination plant could be designed so that the water intake is near the top, middle, or bottom of the water column. Once the water intake begins, a current would be created, which would begin to disrupt any water stratification that may be present.

After the desalination is completed, there will be a hyper saline discharge and a hypo saline discharge. These discharges will both by heat due to the desalination process. The salinity, temperature, and volume of these discharges can be controlled during the desalination process. Since the hyper saline discharge will have greater density than the ambient conditions, the discharge of the hyper saline water at or near the sea surface will allow the effluent to sink and mix in the water column. If the hyper saline water is cooled before discharge, such as running the hyper saline discharge line inside the line for the cooler intake water, the density difference can be increased, which would enhance the sinking and mixing of the effluent.

The heated, hypo saline discharge will be less dense than the ambient water. This less dense effluent can be discharged at or near the sea bottom and, due to the density difference would rise and mix through the water column. The water intake, hyper saline discharge, and hypo saline discharge will create a complex, dynamic environmental perturbation with rapidly changing water temperature and salinity and a disruption of stratification through currents created by the intake, the discharges, and advection. The creation of such a complex and dynamic environment will destabilize the conditions supporting the HAB and, given the dinoflagellate ecology, is predicted to induce the termination of a HAB.

If a HAB exists in a strongly stratified area, HAB termination could be achieved more simply though a floating water pumping plant without the need for a desalination process. In a strongly stratified water column, the water intake and discharge structures can be placed above and below the thermocline or pyncnocline. Similar to a floating desalination plant, a water pumping plant can create currents through the intake and discharge of the water on either side of the stratified boundary. Currents could also be created through advection due to the discharge of warmer water into a colder water strata and the discharge of colder water into a warmer water strata. Significant environmental perturbations can be caused by significant changes in salinity and water temperature and the disruption of water stratification, leading to the termination of a HAB.

Salinity limitations

All dinoflagellate species have a limited range of salinity tolerance. Both the upper and lower limits of this salinity tolerance range can be surpassed by the hyper saline and hypo saline discharges from a floating desalination plant. In strongly stratified waters, the discharges could also be sufficient to surpass these tolerance limits. For example: *K. brevis* [8]:

- Could not be acclimated to salinities above 45 practical salinity units (psu);
- Could not be acclimated to salinitiesbelow25 psu; and
- Exhibited poor growth poor at salinity above 36psucombined with water temperature above23°C.

Hyper saline discharges from desalination plants can achieve water salinities greater than the maximum limit while the hypo saline discharges can produce salinities less than the lower limit. Although a water pumping plant will not be able to achieve the extremes of a desalination plant, significant variations in salinity could be obtained.

A concern with discharges from a desalination plant is that the salinities may be at levels that could cause mortality in other marine organisms. However, the salinity of the discharges can be controlled to levels that disrupt the HAB without causing mortality in other marine species.

Water temperature limits

A desalination plant utilizing a distillation process driven by heating the water would result in water discharges that could be significantly above the temperature limits tolerated by the species forming a HAB. For example, in laboratory experiments, all attempts to increase the water temperature to over 30 °C resulted in a rapid death in *K. brevis* cells in the experiments [8]. Additionally, *K. brevis* exhibited poor growth at a salinity greater than 36psu combined with water temperature above 23°C [8]. Therefore, heated discharges, either by themselves or in combination with altered salinity levels could disrupt HABs at a combination of temperatures and salinities that may not cause mortality in other marine organisms. By pumping water in from one side of a thermocline and discharging the water into the opposite side of the thermocline, a water pumping plant may be able to alter the water temperature enough without the need of desalination.

As with salinity, a concern with discharges from a desalination plant is that the water temperature may be at levels that could cause mortality in other marine organisms. However, like salinity, the temperature of the discharges can be controlled to levels that disrupt the HAB without causing mortality in other marine species.

Disrupt stratification

For several species of dinoflagellates, stratification of the water column has been shown to contribute to the formation and propagation of HABs. Research has demonstrated that, in many instances [3,9].

- The development and propagation of certain dinoflagellate

population requires a stable and stratified water column;

- The existence of pyncnoclines are a necessary pre condition for the development to certain dinoflagellate populations; and
- Dense *Ceratiumfurca* populations developed near pyncnoclines in Japanese bays.

As discussed above, the water intake, hyper saline discharge and hypo saline discharges can create currents and cause advection and mixing that could disrupt water stratification. If the water column is strongly stratified, the intakes and discharges through a floating water pumping plant may adequately breakdown the water stratification without the added need for desalination.

Floating Desalination Plant – Environmental Impacts

In addition to controlling or terminating a HAB, there are a number of environmental impacts that could be expected from the use of a floating desalination plant. These environmental impacts could include [10,11].

- Increased salinity;
- Decreased salinity;
- Increased temperature;
- Modified water stratification;
- Potential for impingement and/or entrainment; and
- Greenhouse gas and air emissions.

The environmental impacts resulting from the use of a floating desalination plant would be temporary with only a short duration. The mixing of the hyper saline and hypo saline discharges with the marine waters would restore the ambient salinity and temperature conditions. With enough time, water column stratification can be expected to be re-established. In addition, the desalination process could be controlled to limit the salinity and temperature of the discharges to ranges that could disrupt a HAB while limiting the impact on the rest of the marine environment. Unlike other potential mechanisms being evaluated for HAB treatment, such as clay or chemical treatment or the use of biological controls, there would be no foreign chemicals, solids, or biological agents discharged into the environment.

Needed Research

There are several areas of research needed to advance the concept of using a floating desalination or water pumping plant as a HAB control technology. The needed research primarily involves optimization studies and field tests to establish:

- Minimum salinity limits for effective hyper saline discharge;
- Maximum salinity limits for effective hypo saline discharge; and
- Minimum temperatures for effective heated water discharge.

Salinity and temperature tolerances vary between dinoflagellate species and mixing conditions vary with water bodies and marine conditions. Therefore, the optimization studies and field tests may be needed for numerous HAB species under a variety of conditions.

Conclusions

Sudden and significant environmental perturbations disrupt HABs. A floating desalination plant could be an effective means of HAB control by inducing rapid changes in water temperature, salinity, and stratification. In areas with a stratified water column, a floating water pumping station could disrupt a HAB by temporarily breaking down the water stratification. As a feasible means of HAB control, a floating desalination and water pumping plant would be portable, reusable, and economical.

The expected environmental impacts would be limited in duration and areal extent. Optimization and field studies are needed to maximize the ability to disrupt a HAB while minimizing the adverse impacts on other marine organisms and the marine environment.

References

1. Anderson DM (2009) Approaches to monitoring, control, and management of harmful algal blooms. Ocean Coast Manag 52: 342-347.
2. Vargo GA (2009) A brief summary of the physiology and ecology of *Karenia brevis* Davis (G. Hansen and Moestrup comb nov) red tides on the West Florida Shelf and of hypotheses posed for their initiation, growth, maintenance, and termination. Harmful Algae 8: 573-584.
3. Donaghay PL, Osborn TR (1997) toward a theory of biological-physical control of harmful algal bloom dynamics and impacts. Lim & Ocean 42: 1283-1296.
4. Corliss JO, Esser SC (1974) Comments on the role of the cyst in the life cycle and survival of free - living protozoa. Trans Amer Micros Soc 93: 578-593.
5. Sarjeant W (1974) Fossil and living dinoflagellates. Academic Press, New York.
6. Owen KC, Norris DR (1984) Cysts of the thecate dinoflagellate *Fragilidium*, Balech ex Loeblich. Litoralia 1: 59-64.
7. Owen KC, Owen JL, Owen DP (2013) A potential mechanism for the regrowth of harmful algal blooms following clay treatment. Intl J Integr Bio 14: 10-16.
8. Magana HA, Villareal TA (2006) The effect of environmental factors on the growth rate of *Karenia brevis* (Davis) G. Hansen and Moestrup. Harmful Algae 5: 192-198.
9. Baek SH, Shimode S, Kikuchi T (2007) Reproductive ecology of the dominant dinoflagellate, *Ceratium fusus*, in Coastal Area of Sagami Bay, Japan. J Ocean 63: 35-45.
10. Menachem E, Phillip WA (2011) The Future of Seawater Desalination: Energy, Technology, and the Environment. Science 333: 712-717.
11. Dawoud MA, Al Mulla MM (2012) Environmental Impacts of Seawater Desalination: Arabian Gulf Case Study. Intl J Env Sustainability 1: 22-37.

Optimization of the Energy Efficiency Operational Indicator for M/V NSU JUSTICE 250,000 DWT by Grey Relational Analysis Method in Vietnam

Tien Anh Tran*

School of Energy and Power Engineering, Wuhan University of Technology, Wuhan 430063, P.R. China
Faculty of Marine Engineering, Vietnam Maritime University, Haiphong, Vietnam

Abstract

"The International Convention for the Prevention of Pollution from Ships (MARPOL 73/78), Annex VI" with Chapter IV: "Regulations on Energy Efficiency for Ships" has made effectively then enforcing all ships must follow regulations in aims with the energy efficiency in operating ships and reducing the environmental pollution. In this article, the calculation of the "Energy Efficiency Operational Indicator (EEOI)" has been conducted under the guidelines for using of the "ship energy efficiency operational indicator" (EEOI) of the "International Maritime Organization" (IMO), MEPC.1/Circ.684 adopted on 17th August 2009. The Grey Relational Analysis (GRA) method has been studied and applied in this research in order to optimize the energy efficiency for M/V NSU JUSTICE 250,000 DWT in the shipping transportation company, Vietnam. These results collected will be identified that the energy efficiency of ships achieves then the decreasing fuel consumption of engine and optimizing the routes of a certain ship. Besides that, this research is fundamental to these next studies about the energy efficiency of ships.

Keywords: Energy efficiency of ships; Grey relational analysis; MARPOL 73/78; EEOI; Bulk carriers

Introduction

The increase of carbon dioxide CO_2 emission into the environment has led to the big problem of the environmental pollution especially the greenhouse gas effects. The greenhouse gas emissions make the global warming potential, melting glaciers, or rising sea levels nowadays. This phenomenon has affected negatively to human and creature in the planet. So, they need to restrict and reduce the carbon dioxide CO_2 emission from actives and economical development at each nation.

The "International Maritime Organization" (IMO) with "Marine Environment Protection Committee" (MEPC 67) has proved the document of the third IMO GHG Study 2014. In where, it provides the updated emission estimates about greenhouse gas emissions from ship operational activities. This research also indicates the certain statistic about carbon dioxide CO_2 emissions from shipping transportation industries. In particular, the international shipping transportation (Figure 1) [1] "generated 796 million tonnes of carbon dioxide CO_2 in 2012 corresponding to 2.2% of the total carbon dioxide emissions from other industries. On the other hand, the global economic downturn, the international shipping transportation has counted with emitting 885 million tonnes of carbon dioxide CO_2 that was 2.8% of the total global carbon dioxide emissions for that year.

On the other hand, the "Marine Environment Protection Committee (MEPC)" has been considered to reduce the green house gas emissions for all ships at present. By method of revealing the energy efficiency measurement tool includes the operational measure and technical measure. In particular, the operational measure of energy efficiency is the "Energy Efficiency Operational Indicator" (EEOI) and the technical measure of energy efficiency is the "Energy Efficiency Design Index" (EEDI). Hence, the "Ship Energy Efficiency Management Plan" (SEEMP) is a tool in order to establish a mechanism in the field of development of the energy efficiency of ships nowadays.

The Grey Relation Analysis (GRA) method was formed by Ju-long Deng in 1982. In where, the internal message including architecture, operation mechanism, system characteristics and parameters is determined by a white system. On the other hand, a white system cannot get any information and characteristics about the system then it is called a black system. Grey relational analysis is adopted between the white system and black system. In a result, the minimal database requirement, simplicity of use and reasonably expected results are the advantages of the grey system over the traditional regression analysis. The Grey Relation Analysis (GRA) has been applied to various fields that relating to the energy efficiency. For instance, [2] the grey relational analysis model was used to examine the load of power system in Japan as well as

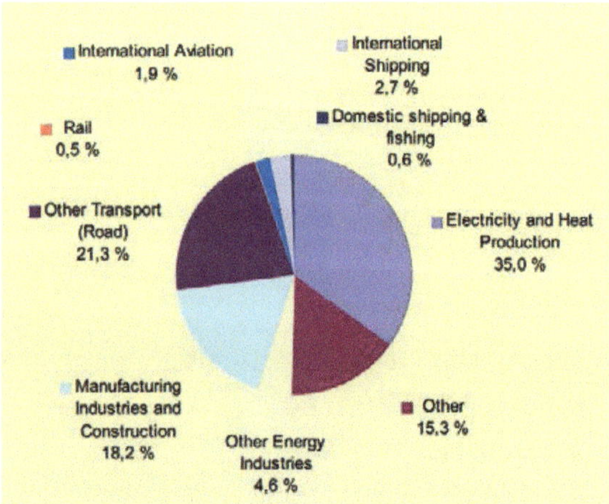

Figure 1: Carbon dioxide CO_2 emissions from shipping transportation with total global emissions [1].

***Corresponding author:** Tien Anh Tran, Faculty of Marine Engineering, Vietnam Maritime University, Haiphong, Vietnam, E-mail: trantienanhvimaru@gmail.com

was carried out by using the long-term load forecasting problem with a significance insteading of using the linear single regression model. Morita et al. [2] had proposed an interval prediction of the both upper and the lower bound values of load demand by using the grey dynamic model. Moreover, the interval prediction has been proposed by using grey dynamic model and the research results of Morita et al. had been compared with the actual cases from a linear single regression model. Besides that the relationship between energy consumption with relating carbon dioxide emissions [3]. Chang et al. had used grey relation analysis in order to analyze how energy-induced CO_2 emissions from 34 industries in Taiwan that caused by elements including: production, energy consumption, coal, oil, gas and electricity. On the other hand, the industrial production had the strongest relationship with CO_2 emissions and the economical development in Taiwan had made intensive industry development. Based on grey relational analysis method, the total energy consumption had a smaller and negative relational grade along with CO_2 gas emissions. Liang [4] also used this method to catch the preliminary schedules for the short-term hydro-electric. The aims of hydroelectric generation scheduling will indicate the optimal amount of generated power for the hydro in the system that used in the future. Shi et al. [5] has applied the Grey Relational Analysis in field of energy efficiency assessment for pump station. The core of Energy Efficiency Assessment (EEA) of energy consumers has been proposed in order to design an integrated assessment model. The grey relational analysis method has been used for applying to the assessment method along with using a motor driving pump system like as a case of study. Furthermore, the operating data of one pump station will be calculated and assessed by using the grey relational analysis method. In addition, the EEA methodology of energy system will be considered as a fresh idea for evaluating the energy efficiency.

In recent time, the methodological and theoretical in grey relational analysis have been improved through researches. Deng [6] has been proposed the information of synthesizing kenning mode and the theory of grey hazy set that consists of embryo, growing, mature and evidence subsets. Hsu and Chen were proposed for the combination between residual modification model and Artificial Neural Networks (ANNs) [7] in aims with enhancing the predictive reliability and accuracy of the original GM (1.1) model. Mao and Chirwa [8] used grey relational analysis for applying to survey motor vehicle fatality in USA and UK. Lin et al. [9] used grey relational analysis to determine a dynamic network in aims with analyzing the correlations and performances of the industrial productivity along with assumpting the energy consumption and combinational uses between fuel and carbon dioxide emissions. The aim of this research has indicated that the increase in electricity generation during the past 10 years was the reason for CO_2 gas emission in Taiwan.

The grey relational analysis can be provided a solution of a system that uses a model of the uncertain information [10]. Besides that the Grey theory system can be provided an optimal method to solve the multi-input, discrete data and uncertain information. On the other hand, the relationship between machining parameters and performance can be found in the Grey Relational Analysis (GRA) theory. In a result, the grey relational grade will utilize the discrete measurement method to measure the distance.

In general, the advantage of the Grey Relational Analysis (GRA) method is that it is designed to study uncertainty. Grey Relational Analysis theory is superior to other methods in field of theoretical analysis of systems with uncertain information and incomplete data experimental. Furthermore, it can be used if the experimental data collected is not supplemental or in case of the researchers does not sure that their experimental data collected is representative. On the other hand, it can be used in early effective factor assessment. However, having a limited number of researches referred in field of shipping transportation especially applying the Grey Relational Analysis (GRA) theory for energy efficiency of ships. Fong-Yuan Ma [11] has researched about the energy efficiency for bulk carrier that using grey relation method but in his research still had some drawbacks about analyzing the energy efficiency of ships. So, in this article, the author has conducted to research about energy efficiency operational indicator, EEOI by applying the Grey Relational Analysis (GRA) for M/V NSU JUSTICE 250,000 DWT of the shipping transportation company in Vietnam. This research included these sections in where section 1, Introduction; section 2, Energy Efficiency Operational Indicator; Section 3, Grey Relational Analysis; section 4, Case study; section 5, Results and discussion, and section 6, Conclusion.

Energy Efficiency Operational Indicator

"Energy Efficiency Operational Indicator", EEOI is operational measure that is established by International Maritime Organization – IMO following the "International Convention for the Prevention of Pollution from Ships (MARPOL 73/78)" in particular Annex VI for Prevention of Air Pollution from Ships. The Conference of Parties to the "International Convention for the Prevention of Pollution from Ships" was happened in 1973 and modified by the Protocol in 1978 (MARPOL 73/78). This conference held from 15th to 26th September 1997 that it has been debated the contents concerning about the sea marine environment protection in aims with responding the convention.

"International Maritime Organization" has adopted resolution A.963 (23) in order to reduce greenhouse gas emissions by meaning of "Marine Environment Protection Committee". MEPC is aims with identifying the mechanism to get the limitation of greenhouse has emission level for shipping. The establishment of carbon dioxide emission along with determing regulations and conventions of sea marine environment protection [1].

The EEOI index is presented for the energy efficiency of the ship operation over a consistent period which describes trading pattern of the vessel. On the other hand, IMO (International Maritime Organization) also indicated that in order to establish the EEOI (Energy Efficiency Operational Indicator) needs following steps below:

- "Define the period for which the EEOI is calculated*"
- "Define data sources for data collection";
- "Collect data";
- "Convert data to appropriate format" and
- "Calculate EEOI".

(*"Ballast voyages, as well as voyages which are not used for transport of cargo, such as voyage for docking service, should also be included. Voyages for the purpose of securing the safety of a ship or saving life at sea should be excluded").

Furthermore, the data recording method is used so that experimental data is collected and analyzed to manage the extraction of forced information. The experimental data must include: distance traveled, quantity of fuel used; kind of fuel used. All fuel used information may affect the amount of carbon dioxide emitted into the environment. For instance, the fuel information is provided on the bunker delivery notes that regulated under Regulation 18 of MARPOL 73/78 Annex

VI. On the other hand, the unit used for distance traveled and quantity of fuel should be expressed in nautical miles and metric tonnes. It is necessary to provide the significant information that is collected from experimental data of vessel along with regarding to the voyage characteristics like as fuel used amount, type of fuel, traveled distance, and amount of cargo carried. The distance traveled should be computed following the actual voyages and it is saved on Ship's Log-Book.

The calculation of EEOI index following Equation 1 below:

$$EEOI = \frac{\sum_j FC_j C_{Fj}}{m_{cargo} . D} \quad (1)$$

Rolling average of EEOI equation following Equation 2:

$$Average\text{EEOI} = \frac{\sum_i \sum_j (FC_{ij} . C_{Fj})}{\sum_i (m_{cargo,i} . D_i)} \quad c \quad (2)$$

Where:

j: is the fuel type;

i: is the voyage number;

FC_{ij} is the mass of consumed fuel j at voyage i;

C_{Fj}: is the fuel mass to CO_2 mass conversion factor for fuel j;

m_{cargo}: is the cargo carried (tonnes) or work done (number of TEU or Passengers) or gross tonnes for passenger ships; and

D: is the voyage distance (unit: nautical mile).

The value of EEOI concerns about navigation characteristics, for example: tonnes CO_2/(tonnes.nautical miles), tonnes CO_2/(TEU. nautical miles), tonnes CO_2/(person.nautical miles), etc.

The rolling average EEOI is used in the field of calculation of EEOI index in a suitable time period. For instance, six or ten voyages which are agreed as statistically relevant to the initial averaging period. The Equation (2) shows the rolling average EEOI formula with the certain elements.

On the another side, the fuel mass to CO_2 mass conversion factor C_F is a non-dimensional conversion factor between fuel consumption measured in gram and CO_2 emission also measured in gram based on carbon content (Table 1). The value of C_F is as follow:

It is possible to realize that the EEOI value is as smaller as better in the field of energy efficiency of ships at each certain voyage when considering operational condition on sea. On the other hand, the equations (1) and (2) can be established by adapting for a multiple numbers of voyages and an average EEOI value obtains during a period of time.

Energy Efficiency Operational Indicator itself is not mandatory. However, it is a monitoring tool for energy efficiency that implemented measures due to the International Maritime Organization reveals. It is a reason that to perform EEOI calculations would enable for the operators to measure the energy efficiency of ship and to analyze the effect of any modifies in operation such as voyage planning optimization, hull and propeller cleaning, trim optimization.

Grey Relational Analysis

Data preprocessing

The data processing is applied to give the data sequence into dimensional data sequence as well as transforming the original sequence to a comparable sequence. On the other hand, the original sequence and comparability sequence are represented as $x_0(k)$ and $x_i(k)$, i = 1,2,3, ..., m; k = 1,2,3, ..., n, respectively, where m is the total number of experiment to be considered, and n is the total number of observation data. Data preprocessing converts the original sequence to a comparable sequence. Several methodologies of preprocessing data can be used in Grey Relation Analysis, depending on the characteristics of the original sequence [12-14]. For the original sequence "the larger the better", the original sequence is normalized as follows [12]:

$$x_i(k) = \frac{x_i(k) - \min(x_i(k))}{\max(x_i(k)) - \min(x_i(k))} \quad (3)$$

For "the smaller the better" characteristics of the original sequence, the original sequence is normalized as follows [12]:

$$x_i(k) = \frac{\max(x_i(k)) - x_i(k)}{\max(x_i(k)) - \min(x_i(k))} \quad (4)$$

In case, if a defined target value, OB, exists, then the original sequence is normalized as follows [12]:

$$x_i(k) = 1 - \frac{|xi(k) - OB|}{\max\{\max(x_i(k) - OB, OB - \min(x_i(k))\}} \quad (5)$$

The original sequence is normalized by using an alternative simple method which is separated from the first value of the sequence i.e. $x_i(1)$ as follows:

$$x_i(k) = \frac{x_i(k)}{x_i(1)} \quad (6)$$

where, $x_i(k)$ us the original sequence, $x_i(k)$ is the sequence after the data preprocessing, $\max(x_i(k))$ is the largest value of $x_i(k)$ and $\min(x_i(k))$ is the smallest value of $x_i(k)$.

Grey relational analysis coefficients and grey relational analysis grades

After the data preprocessing, a grey relational analysis coefficient is calculated using the preprocessed sequences. The grey relational analysis coefficient can be calculated as [12]:

$$\gamma_i(k) = \gamma(x_o(k), x_i(k)) = \frac{\Delta_{\min} + \zeta\Delta_{\max}}{\Delta_{oi}(k) + \zeta\Delta_{\max}} \quad (7)$$

and, $0 < \gamma(x_o(k), x_i(k)) \le 1$

Type of Fuel	Reference	Carbon Content	C_F (t-CO_2/t-Fuel)
1.Diesel/Gas Oil	ISO 8217 Grades DMX through DMC	0.875	3.206000
2.Light Fuel Oil (LFO)	ISO 8217 Grades RMA through RMD	0.86	3.151040
3.Heavy Fuel Oil (HFO)	ISO 8217 Grades RME through RMK	0.85	3.114400
4.Liquified Petroleum Gas (LPG)	Propane Butane	0.819 0.827	3.000000 3.030000
5.Liquified Natural Gas (LNG)		0.75	2.750000

Table 1: Carbon dioxide conversion factor (C_F).

where $\Delta_{oi}(k)$ is the deviation sequence of the reference sequence $x_0(k)$ and comparability sequence $x_i(k)$;

$$\Delta_{oi}(k) = |x_o(k) - x_i(k)| \tag{8}$$

$$\Delta_{\max} = \max_{\forall j \in i} \max_{\forall k} |x_o(k) - x_j(k)| \tag{9}$$

$$\Delta \min = \min_{\forall j \in i} \min_{\forall k} |x_o(k) - x_j(k)| \tag{10}$$

with ζ is the distinguishing coefficient, $\zeta \in [0,1]$.

After computing the grey relational analysis coefficients, grey relational analysis grade is calculated using the following relationship [14,15].

$$\gamma = \gamma(x_o, x_i) = \sum_{k=1}^{n} \beta_k \gamma(x_o(k), x_i(k))$$

$$\text{Where, } \sum_{k=1}^{n} \beta_k = 1 \tag{11}$$

The grey relational analysis grade (x_o, x_i) represents the degree of correlation between the reference and comparability sequences. In case of grey relational analysis grade is equal to 1 then the reference and comparability sequences are identical. In a result, the comparability sequence is more important than other comparability sequence due to the grey relational analysis will be applied into this research by using grey relational analysis grade. On the other hand, the comparability sequence and the reference sequence will be overcome to compare the other grey relational analysis grades. The results of grey relational analysis will be compared to the absolute value of data including the sequence data along with the correlation of sequence data.

Case Study

M/V NSU JUSTICE 250,000 DWT

In Vietnam, the shipping transportation companies have a numerous number of ships with variety of sizes and types. There are a lot of shipping transportation companies where provide vessels and officers responding to develop the transportation business in domestic and international such as VINIC shipping transportation company associates with Vietnam Maritime University, Vietnam; VOSCO shipping transportation company, NISSHO shipping transportation company, etc. In this study, the target ship is used with a certain name M/V NSU JUSTICE 250,000 DWT of VINIC shipping transportation company. This is the biggest bulk carrier of company that operating on international routes. The experimental data has been collected from its operational activities. This vessel operates on the fixed routes between Japan - Australia - Brazil. Table 2 presents these specification parameters of M/V NSU JUSTICE 250,000 DWT.

In Figure 2 describes general of M/V NSU JUSTICE 250,000 DWT with design parameters show in Table 2 below.

M/V NSU JUSTICE 250,000 DWT is powered by two-stroke main engine, single acting, crosshead, exhaust gas turbo-charged marine diesel engine with engine type of MAN B&W 7S80MC-C (Mark 7). Its maximum continues rating (MCR): 21,910 kW at the shaft rotational speed of 74.5 rpm. Furthermore, this vessel is powered by three main diesel generators with engine type of 4-stroke, single acting, turbo-charged diesel engine (6N21AL-SV). The Maximum Continuous Rating (MCR) is 880 kW at the respective shaft rotational speed of 900 rpm. On the other hand, this vessel is fitted the Fixed Pith Propeller (FPP) diameter 9.60 m with 4 blades. Especially, the vessel is equipped with a data acquisition system for collecting the experimental data including the performance and navigation data. The performance and navigation data of this vessel with respect to the fuel consumption rates are considered in this study along with navigation conditions impact on ship.

Item	Category	Parameter
1	Vessel name	NSU JUSTICE
2	IMO Number	9441922
3	MMSI	373072000
4	Vessel type	Bulk carrier
5	Gross tonnage	132,868
6	Dead Weight Tonnage (DWT)	250,000
7	Flag	Panama
8	Draught	18 m
9	Speed Recorded (Max/Average)	13.3/10/4 knots
10	Length x Breadth (L x B)	329.95 m × 57 m
11	Year built	2012

Table 2: Specification parameters of M/V NSU JUSTICE 250,000 DWT.

Figure 2: M/V NSU JUSTICE 250,000 DWT.

Calculation of energy efficiency operational indicator

The EEOI calculation has an important position in enhancing the energy efficiency operation index of ships. In this case of study, the target ship is a bulk carrier with certain name M/V NSU JUSTICE 250,000 DWT. The author conducted to choose this type of ship because it is a popular ship that carries a numerous cargo in shipping transportation industry in the world and in particular, Vietnam. On the other hand, it has been chosen completely based on research situation and operational condition of shipping transportation companies in Vietnam.

Throughout the Figure 3 above, the Energy Efficiency Operational Indicator (EEOI) at each certain voyage is different. In this study, the recent voyages are used in calculation of EEOI index for M/V NSU JUSTICE 250,000 DWT that this vessel operating. These voyages almost carry bulk ore so there is nothing harmful good affects the carbon dioxide (CO_2) emission to the environment. In addition, it will not impact on the calculation of EEOI index following Equation 1. So, the calculation results above of EEOI index for M/V NSU JUSTICE 250,000 DWT completely depend on the fuel consumption level of main engine.

However, the name of voyage like as Voyage N°16A, N°16B, N°17A, N°17B, N°18A, N°18B, N°19A, N°19B, N°20A, and N°20B is separated from Voyage N°16, N°17, N°18, N°19, and N°20 due to its operational characteristics of routes, time on sea-going and at each port are individual. In fact, the voyage N°16 is total of voyage N°16A and voyage N°16B. In addition, these other voyages are similar like this. So, the value of the Energy Efficiency Operational Indicator (EEOI) at each certain voyage will be represented in Figure 4.

In Figure 4, the maximum EEOI index among these voyages is on Voyage N°16 with 9.83E-04, and the minimum EEOI index is on Voyage N°20 with 1.05E-05. Besides that, these left voyages have also the low values. i.g. EEOI of voyage N°17, N°18, N°19 in turns 1.08E-05, 1.13E-05, 1.14E-05. From these results calculated for M/V NSU JUSTICE 250,000 DWT will be fundamental background to carry out this research.

Optimization of energy efficiency of M/V NSU JUSTICE 250,000 DWT

In this study, the determination of grey relational analysis grades of experimental data for M/V NSU JUSTICE 250,000 DWT conducted in this section. This is the important working, which is analyzed from the data of voyages. The fundamental calculation is based completely from researching the theory before and referred into section 2.

In the experimental data was collected from the actual voyages of M/V NSU JUSTICE 250,000 DWT when this vessel carried a big amount of cargo traveled around routes between Japan-Australia-Brazil with certain name of voyages. The data of voyages has been taken from Noon-log Record on M/V NSU JUSTICE 250,000 DWT (Table 3). For instance, the Noon-log Record of a certain Voyage N°16 is showed in Figure 5 below.

In reality, there are five voyages that collected from the experimental data throughout actual travel with names Voyage N°16, N°17, N°18, N°19, N°20 due to n=5. On the other hand, the proof is based on the calculation of EEOI values for voyages as well as aims of this research

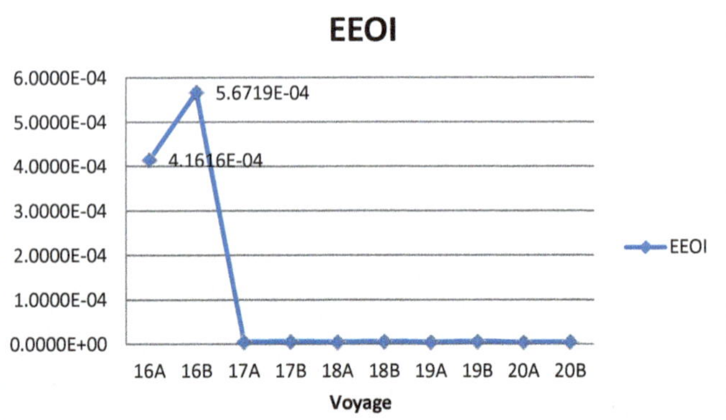

Figure 3: EEOI values for M/V NSU JUSTICE 250,000 DWT.

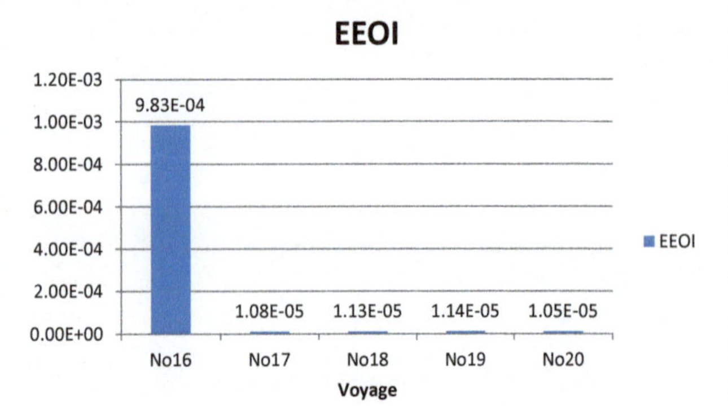

Figure 4: The total of EEOI value at each voyage No16, No17, No18, No19, No20.

Item	Voyage N°	EEOI
1	16A	4.1616E-04
	16B	5.6719E-04
2	17A	4.9504E-06
	17B	5.8613E-06
3	18A	5.0109E-06
	18B	6.2809E-06
4	19A	5.1751E-06
	19B	6.2269E-06
5	20A	5.0054E-06
	20B	5.509E-06

Table 3: EEOI value of M/V NSU JUSTICE 250,000 DWT.

that reduce the fuel consumption of engine on M/V NSU JUSTICE 250,000 DWT in order to increase the energy efficiency of ship in particular the Energy Efficiency Operational Indicator (EEOI). So, these parameters need to consider including the fuel consumption level (HFO + DO), distance traveled, and cargo carried among these studied voyages. From then, the indication of k is equal 4 (k=4);

In Table 4, the experimental data of M/V NSU JUSTICE 250,000 DWT has been taken out the Noon-log Record for Voyage N°16, N°17, N°18, N°19, and N°20. These parameters include the fuel consumption rate of engine corresponding to each certain voyage, types of fuel used both HFO (Heavy Fuel Oil) and DO (Diesel Oil), distance traveled, and amount of cargo carried. All of them were represented in Table 4 with these comparability sequences $x_i(k)$.

In where: Voyage N°16=N°16A+N°16B, voyage N°17=N°17A+N°17B, voyage N°18=N°18A+N°18B, voyage N°19=N°19A+N°19B, voyage N°20=N°20A+N°20B;

These comparability sequences have been taken from the actual experimental data of M/V NSU JUSTICE 250, DWT researched. These parameters considered in this study concern about the energy efficiency operation index including the fuel consumption rates of engine (HFO+DO), the distance traveled (unit: NM=Nautical Mile), and a weight of cargo carried (unit: T=Ton). Especially, the optimization of these parameters hase been conducted in this research in aims with increasing the energy efficiency of ships. So, these comparability sequences $x_i(k)$ have been studied concerning about these parameters above and described in Table 4 corresponding to each certain voyage.

In Table 5, the reference sequences $x_0(k)$ were indicated that under the certain parameters then applied the Grey Relational Analysis (GRA) method to optimize the experimental data of M/V NSU JUSTICE 250,000 DWT. The reference sequences $x_0(k)$ are the operational parameters that M/V NSU JUSTICE 250,000 DWT wants to achieve these values. This means that the reference sequences $x_0(k)$ are desirable values and they are established completely based on the Grey Relational Analysis (GRA) method associating with the operational experience on M/V NSU JUSTICE 250,000 DWT.

The reference sequence was designed to base on the actual experimental operation plus using the Grey Relational Analysis method. The reference sequence is corresponding to the desired sequence that the ship-owners and operators who want to gain these values. To optimize the energy efficiency operational indicator through the operational parameters then the fuel consumption rates of engine need to reduce along with ensuring the cargo carrying abilities of ships (distance traveled and amount of cargo carried). From the Grey Relational Analysis method, the fuel consumption level both HFO and DO (unit: Ton) is "the smaller the better" following the Equation (4)

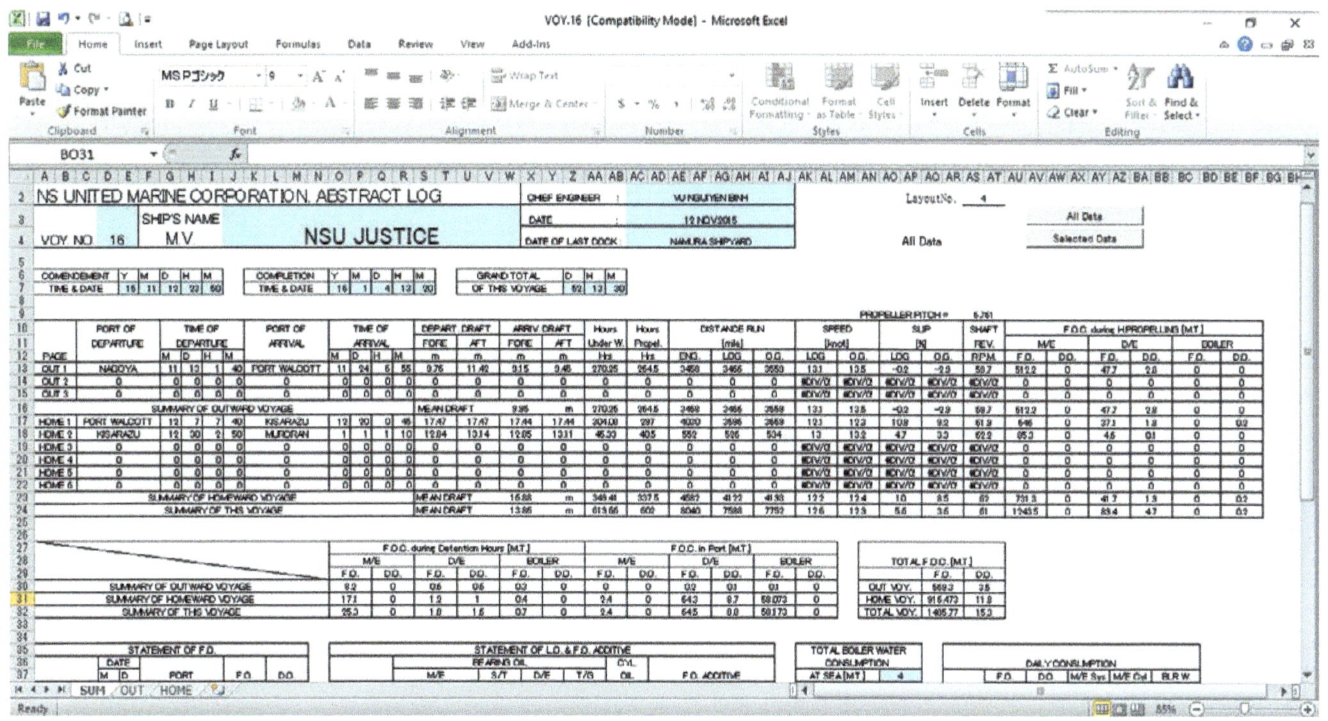

Figure 5: Noon-log Record of Voyage N°16, M/V NSU JUSTICE 250,000 DWT.

Item	Voyage	HFO (T)	DO (T)	Distance (NM)	Cargo (T)
1	N°16	1485.773	15.3	7588	247400
2	N°17	1663.1	13.3	7837	247500
3	N°18	5434.67	27.6	24469	246200
4	N°19	1680.941	17.1	7466	246876
5	N°20	1479.477	13.2	7156	246800

Table 4: These comparability sequences $x_i(k)$.

Item	HFO (T)	DO (T)	Distance (NM)	Cargo (T)
$x_0(k)$	1479.477	13.2	24469	247500

Table 5: The reference sequence $x_0(k)$.

above. In contrast, the distance traveled (unit: Knot or NM – Nautical Mile) and cargo carried (unit: Ton) are "the larger the better" applying the Equation (3).

The difference sequences Δ_{oi} between $x_0(k)$ and $x_i(k)$ have been computed based on Equation (8) in section 3.2 (Table 6). This calculation of the difference sequences Δ_{oi} of these parameters concerning about the energy efficiency operational index corresponding at each voyage separated. These voyages have been studied total of five voyages with the certain name: N° 16, N°17, N°18, N°19, and N°20 (n=5).

The grey relational analysis coefficient γ(k) has been calculated following the Equation (7) in section 3.2 above. These values have been recorded in Table 7 along with these values of grey relational analysis grade γ. The grey relational analysis coefficient of each voyage is a range of between 0 and 1 corresponding the each parameter needs to optimize in this study. In reality, there is a difference from these values collected and they will be compared and analyzed in section 5. On another side, the grey relational analysis grade will be indicated out at same Table 7 responding to each grey relational analysis coefficient γ(k). In where, there are four grey relational analysis coefficients $\gamma_1(k)$, $\gamma_2(k)$, $\gamma_3(k)$,

Δ_{oi}	Voyage	HFO (T)	DO (T)	Distance (NM)	Cargo (T)
Δ_{o1}	N°16	6.296	2.1	16881	100
Δ_{o2}	N°17	183.623	0.1	16632	0
Δ_{o3}	N°18	3955.193	14.4	0	1300
Δ_{o4}	N°19	201.464	3.9	17003	624
Δ_{o5}	N°20	0	0	17313	700

Table 6: The difference sequences Δ_{oi} between $x_0(k)$ and $x_i(k)$.

$\gamma_4(k)$ corresponding to each parameter that needs to optimize. The analysis of the grey relational analysis coefficient and grey relational analysis grade will be carried out in section 5 of this article. From then, the optimization of the experimental data of M/V NSU JUSTICE 250,000 DWT will be represented.

Results and Discussion

The energy efficiency of ships plays a significant role in the shipping transportation industry. Especially, the energy efficiency measures have been established through the International Maritime Organization (IMO) that regulated in MARPOL 73/78, Annex VI about the Prevention of Air Pollution from Ships. The Energy Efficiency Operational Indicator (EEOI) is an operational measure, which is a target objective in this research in aims with reduction of the Energy Efficiency Operational Indicator associating with the sea environmental protection. These operational parameters have been concentrated on this research including the fuel consumption rates of engine, types of fuel used, distance traveled of certain vessel, amount of cargo carried responding to each certain voyage. Especially, the novel of this research has applied to the Grey Relational Analysis method to solve the issue in field of the energy efficiency of ships. Since these previous researches have just referred to apply and optimize the process parameters at some different fields. However, the use of Grey Relational Analysis (GRA) method for the shipping transportation industries still limits. So, this research has also indicated out the innovative trend in applying the Grey Relational Analysis method in field of optimizing the experimental data of vessels. In particular, this article has been represented throughout these results below then based on the Grey Relational Analysis (GRA) method.

The combination of calculating the Energy Efficiency Operational Indicator (EEOI) along with the Grey Relational Analysis coefficients and the Grey Relational Analysis Grade is initial background to conduct the optimize the energy efficiency operational index for a certain vessel, M/V NSU JUSTICE 250,000 DWT of VINIC shipping transportation company in Vietnam. Especially, this vessel has been operated by the author and Officers of VINIC shipping transportation company in Vietnam that some of them who work as Lecturers at Vietnam Maritime University, Haiphong, Vietnam. So, the optimization of experimental data for M/V NSU JUSTICE 250,000 DWT has been carried out in this research beside the Grey Relational Analysis (GRA) method applied.

In Figure 6, the grey relational analysis coefficient $\gamma_1(k)$ is described for voyages including voyage N°16, N°17, N°18, N°19, and N°20 about using the heavy fuel oil (HFO) on ship. In general, the maximum values of grey relational analysis coefficients are voyage N°16 and N°20 with 0.9968 and 1.0000. In contrast, the lowest grey relational analysis coefficient is voyage N°18 with 0.3333. These left values for voyage N°17 and N°19 are in turns 0.9150 and 0.9075. In then, the grey relational analysis coefficient $_{l}(k)$ is represented for consuming the heavy fuel oil of engine on M/V NSU JUSTICE 250,000 DWT with the design value before then applied the Grey Relational Analysis method. The grey relational analysis coefficient is as higher as better. So, the voyage N°16, N°17, N°19, and N°20 have used effectively about the heavy fuel oil on ship reflecting to voyage N°18.

Similarly, the grey relational coefficient $\gamma_2(k)$ is also represented for all voyages in Figure 7. Throughout these results, the voyage N°17 and N°20 have the hight values corresponding to 0.9863 and 1.0000. In contrast, the similarity of grey relational analysis coefficient $_{l}(k)$ then the voyage N°18 reaches the minimal value with 0.3333. These left voyages include voyage N°16 and N°19 in turn 0.774 and 0.6486.

The grey relational analysis coefficient $\gamma_3(k)$ is represented in Figure 8 below corresponding to each different voyage.

The grey relational analysis coefficient $\gamma_3(k)$ also indicates that the highest value of it is voyage N°18 with 1.0000. These left voyages have these values similarly under 0.5000. In turns, the voyage N°16, N°17, N°19, and N°20 is 0.3389, 0.3423, 0.3374, and 0.3333.

The grey relational analysis coefficient $\gamma_4(k)$ for amount of cargo carried on M/V NSU JUSTICE 250,000 DWT also indicates in Figure 9. However, the highest value of grey relational analysis coefficient is voyage N°17 (1.0000). These left voyages are in turn 0.8667 for voyage N°16, 0.3333 for voyage N°18, 0.5102 for voyage N°19, and 0.4815 for voyage N°20.

Furthermore, the compound diagram between these gray relational analysis coefficients is represented in Figure 10 responding to each different voyage.

From the combination of the grey relational analysis coefficients each other also indicates that the distance traveled by ship has an important position in field of the energy efficiency of ships. Since it has been identified that voyage N°18 achieved low value of the Energy Efficiency Operational Indicator (EEOI index=1.13E-05) due to the grey coefficient analysis coefficient of this voyage gains the highest value ($\gamma_3(k)$=1.0000). Moreover, the travel route optimization method is a

$\gamma_i(k)$	Voyage N°16	Voyage N°17	Voyage N°18	Voyage N°19	Voyage N°20	γ
$\gamma_1(k)$	0.9968	0.9150	0.3333	0.9075	1.0000	0.8305
$\gamma_2(k)$	0.7741	0.9863	0.3333	0.6486	1.0000	0.7485
$\gamma_3(k)$	0.3389	0.3423	1.0000	0.3374	0.3333	0.4704
$\gamma_4(k)$	0.8667	1.0000	0.3333	0.5102	0.4815	0.6383

Table 7: The Grey Relational Coefficient $\gamma_i(k)$ and grey relational grade γ.

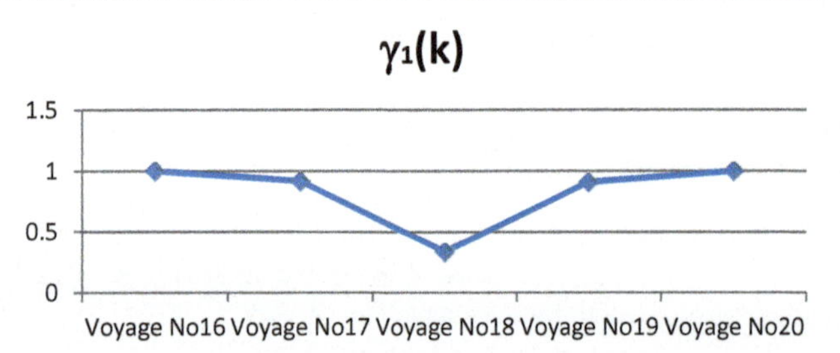

Figure 6: Grey relational coefficient $\gamma_1(k)$ for Heavy Fuel Oil (HFO).

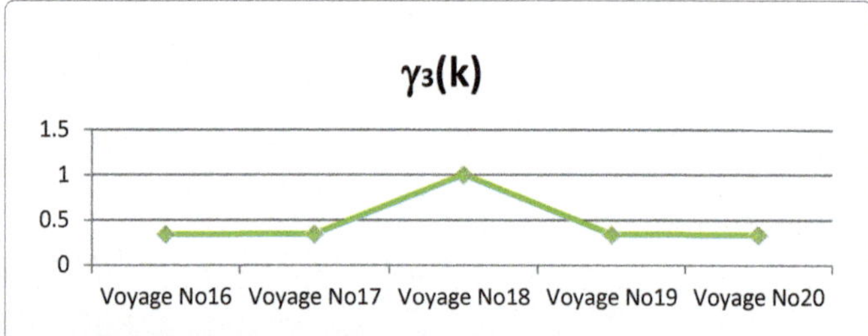

Figure 8: Grey relational coefficient $\gamma_3(k)$ for distance traveled.

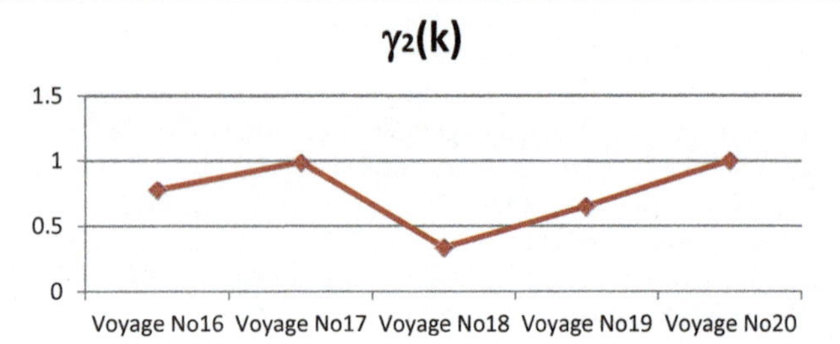

Figure 7: Grey relational coefficient $\gamma_2(k)$ for Diesel Oil (DO).

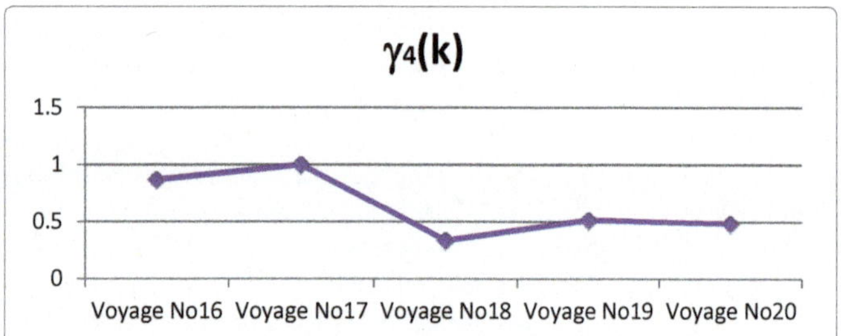

Figure 9: Grey relational coefficient $\gamma_4(k)$ for cargo carried.

starting point to carry out the energy efficiency of ships. This result has been performing throughout Figure 10. The different aspect of Figure 10 also indicates that the travel route optimization of ship will be the best method for energy efficiency management based on EEOI index associating with the grey relational analysis method in this research. In operational working, the significant selection of travel route will decide the energy efficiency of ship. The reduction of travel distance is a fundamental point in order to reduce the carbon dioxide emissions, fuel consumption of ship then enhancing the energy efficiency management of ship.

In Figure 11, the grey relational analysis grade γ at each different element of the energy efficiency operation index is identified. The response of experimental data is separate at each voyage. The grey relational analysis grade of the heavy fuel oil consumption is the highest with 0.8305 after that in turn the diesel oil fuel oil consumption, amount of cargo carrier, and distance traveled with these values are 0.7485, 0.6383, 0.4704.

In a result, the optimization of energy efficiency of ships through these operational parameters needs to consider under these external factors impact. Moreover, to gain the optimal operation in field of operating and managing the energy used on board must consider about the fuel consumption rates including the heavy fuel oil and diesel oil with the distance traveled of ship.

Conclusion

This article has referred to the use of energy efficiency of engine on ship by applying the Grey Relational Analysis method in research. The use of effective energy on ship will be reduced the fuel consumption rate of engine nowadays as well as protection of sea environment following the International Convention for Prevention of Pollution from Ships (MARPOL 73/78). Besides that, these results also indicated that the optimization of energy efficiency operation index through these operational parameters is able to do completely when considering about the distance traveled of ships. Moreover, the distance traveled can be conducted by Masters of their ships. This task can be ensured that the energy efficiency of ships increases along with remaining the ship's safety in case of operation. In this research, the author has applied effectively the Grey Relational Analysis method in optimizing the experimental data of M/V NSU JUSTICE 250,000 DWT in field of the energy efficiency of ships. This is a new trend in the energy efficiency of ships especially using the energy economically on ships as well as restricting the exhaust gas emissions in the environment. In addition, the mass of consumed fuel oil on ships and cargo carried depend on different type of ships including passenger ship, container ship, Ro-Ro, etc. So, the use of energy efficiency operational indicator (EEOI) and grey relational analysis methodology can be applied into other cases of ship type not only bulk carriers like as M/V NSU JUSTICE 250,000 DWT but also various type of ship in the field of energy efficiency management of ships nowadays.

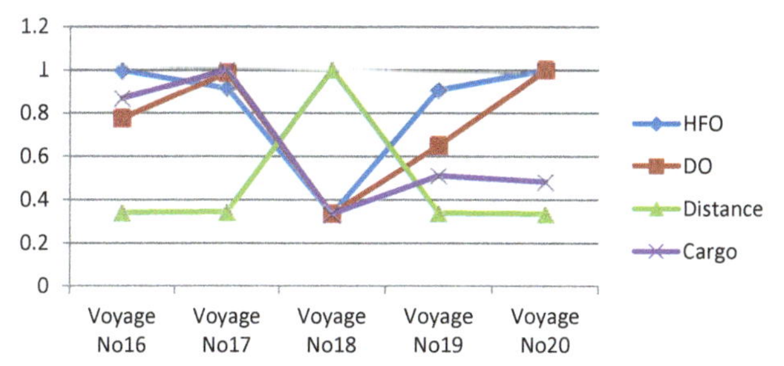

Figure 10: Grey relational analysis coefficients.

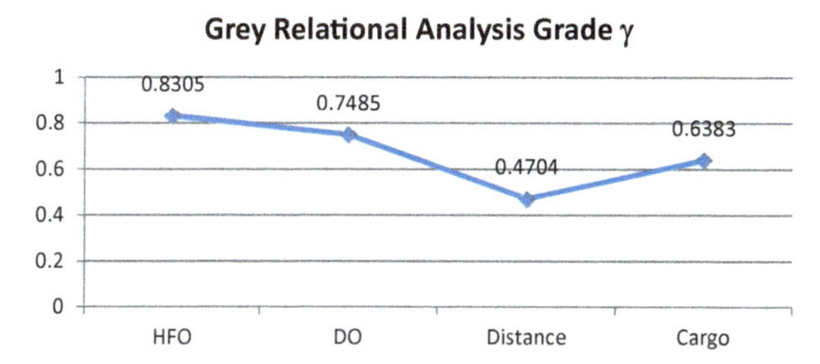

Figure 11: Grey relational analysis grades.

Acknowledgement

The author wants to give a kindly thank to Key Laboratory of Marine Power Engineering & Technology (Ministry of Transportation), School of Energy and Power Engineering, Wuhan University of Technology, 1178 Heping Avenue, Wuhan 430063, P.R. China to support in complement of this research.

References

1. International Maritime Organization (IMO), Second IMO GHG Study (2009). IHO, 4 Albert Embankment, London SE1 7 SR.
2. Morita H, Kase T, Tamura Y, Lwamoto S (1996) Interval Prediction of Annual Maximum Demand Using Grey Dynamic Model. Electrical Power & Energy Systems 18: 409-413.
3. Chang TC, Lin SJ (1999) Grey Relation Analysis of Carbon Dioxide Emissions from Industrial Production and Energy Uses in Taiwan. Journal of Environmental Management 56: 247-257.
4. Liang RH (1999) Application of Grey Relation Analysis to Hydroelectric Generation Scheduling. Electrical Power and Energy Systems 21: 357-364.
5. Shi MJ, Shi K, Xu GJ, Hu MY, Tong RM (2013) Applied Research of Grey Relational Analysis in the Energy Efficiency Assessment of Pump Station. Applied Mechanics and Materials 405-408: 3288-3292.
6. Deng JL (2001) Synthesizing Kenning Model and Grey Hazy Set. The Journal of Grey System 13: 254.
7. Hsu CC, Chen CY (2003) Applications of Improved Grey Prediction Model for Power Demand Forecasting. Energy Conversion and Management 44: 2241-2249.
8. Mao M, Chirwa EC (2006) Application of Grey Model GM (1,1) to Vehicle Fatality Risk Estimation. Technological Forecasting & Social Change 73: 588-605.
9. Lin Sue J, Lu IJ, Lewis C (2007) Grey Relation Performance Correlations Among Economics, Energy Use and Carbon Dioxide Emission in Taiwan. Energy Policy 35:1948-1955.
10. Deng JL (1990) A Course on Grey System Theory. HUST Press (in Chinese) Wuhan.
11. Fong-Yuan Ma (2014) Analysis of Energy Efficiency Operational Indicator of Bulk Carrier Operational Data Using Grey Relation Method. Journal of Oceanogrphy and Marine Science 5: 30-36.
12. Tzeng CJ, Lin YH, Yang YK, Jeng MC (2009) Optimization of turning operation with multiple performance characteristics using the Taguchi method and Grey Relational Analysis. Journal of Materials Processing Technology 209: 2753-2759.
13. Deng JL (1989) Introduction to Grey System Theory. Journal of Grey System 1: 1-24.
14. Yang YK, Shie JR, Huang CH (2007) Optimization of Dry Machining Parameters for High Purity Graphite in end Milling Process. Materials and Manufacturing Processes 21: 832-837.
15. Hu W, Yang C (2000) Grey Model of Direct Solar Radiation Intensity on the Horizontal Plane for Cooling Loads Calculation. Building and Environment 35: 587-593.

10

Ecosystem Approach based Methodology to Support Land use Planning: Egyptian Experience

Ghoneim SA[1]* and Ibrahim H[2]

[1]*Department of Environmental Planning, Cairo University, Egypt*
[2]*Regional Development Department, Cairo University, Egypt*

Abstract

This paper is focused on the Ecosystem Approach Framework for Planning and Management of Coastal Wetlands "EA-PMCW", as one of the novel Ecosystem Approach's customized frameworks. The EA-PMCW was developed by Ghoneim SA to support the land use planning and development decision in coastal wetlands as an example of the most sensitive and productive ecosystems on earth. However it translated the principles of the ecosystem approach into a step by step methodology considering the requirements of the planning process, it reflected the especial features of coastal wetlands and the challenges they are facing, including development pressures and increased impacts of climate change. The paper aims to test the applicability of the EA-PMCW within the context of the Egypt, as a rich representative case study. It starts with a brief description of the EA-PMCW and its main feature. Then it tests the applicability through exploring the potentials and challenges involved in applying different stages of the EA-PMW in Egypt. The results show high level of acceptability from the decision-makers and a promising level of applicability. Finally, some recommendations towards successful application could be extracted based on tackling the challenges and maximizes the potentials within the local context.

Keywords: Ecosystem approach; Coastal wetlands; Developing countries; Egypt

Introduction

The Ecosystem Approach Framework for Planning and Management of Coastal Wetlands 'EA-PMCW' was developed by Ghoneim [1], mainly based on developing the principles of the Ecosystem Approach 'EA' into a methodological framework. It aims to support land-use planning process and achieve sustainable development in coastal wetlands, as one of the most sensitive and productive natural ecosystems on earth. It should be mentioned here that the EA-PMCW was mainly learned from and built within four wider frameworks: 1. RAMSAR convention for wetlands [2-6]; 2. The Millennium Ecosystem Assessment [6]; 3. The principles and guidelines of the ICZM [7-9] and 4. The International Convention for Biodiversity. However, it is oriented to support the environmental land use planning process [9,10]. The EA-PMCW transferred the principles of the Ecosystem Approach 'EA' into a methodological framework, which integrates the distinguished features of the Watershed Approach and other suitable environmental planning approaches. It is also informed by the review of international planning and management experiences of EA applications and related frameworks and practical experiences, from the other EA customized frameworks. In order to test the applicability of this framework; the research used semi-structured interviews with the interested local experts. Moreover exploring the expected potentials and challenges could face the applicability of the EA-PMCW, a focus group workshop with the environmental managers and stakeholders helped to form some suitable recommendations for successful applications in Egypt.

The IUCN-CEM has been involved with the CBD since its drafting and through its further development, so it can be considered as one of the main supporters of EA and has sponsored and managed many EA-based projects all over the world (William 2000). IUCN–CEM has evaluated this history of EA applications according to five case studies (i.e. Niger/Nigeria, Mekong Delta, Congo, Panama, Indonesian Papua) representing different ecosystems with multidimensional issues. This evaluation highlighted some gaps to be covered in the EA to promote successful application, including [11]: The principles do not focus on how ecosystems, and the biodiversity within them, actually contribute to and affect livelihoods. There is some indirect consideration of economic benefits for local people, but issues related to livelihood security and the resilience of the ecosystem so that it can continue to produce fundamental goods and services, are not an explicit part of EA thinking; Institutions are hardly mentioned except for managers: the normal structure of various ministries with local responsibility for land-use planning and means they are not used to considering ecosystems as units of management; The EA needs to be seen as useful by several groups of national-level policy makers, who set the strategy; provincial/sub-regional agencies who adapt national to local conditions; and district and local-level officials who implement policy and who have been the main focus of these case studies. Several authors have critically analyzed the practical experience of applying EA to marine environmental planning and management. The main barriers and the key development issues in these applications in marine planning and management have been summarized, as follows [12]:

- Ineffective stakeholder participation in planning and management.
- The lack of capacity for decentralized and integrated management.
- The overriding influence of perverse incentives.
- Conflicting political priorities.

***Corresponding author:** Ghoneim SA, Faculty of Regional and Urban Planning, Department of Environmental Planning, Cairo University, Egypt
E-mail: h1_sadeldin@yahoo.com

Objectives and Main Features of the EA-PMCW

The main features of the EA-PMCW are summarized in Figure 1.

Dynamic

The proposed framework should be dynamic/ adaptive to reflect the continuous changes of the ecosystem.

Respecting the natural boundaries of the ecosystem

The unity of the ecosystem and flow of its natural functions should be maintained to achieve sustainability.

Multi-temporal (Long-term and Short-term)

To achieve suitability in the long term and to respond to short term challenges and needs, the framework should include different time scales of plans and ensure that the short-term managements plans are well fitting within the long-term plans.

Multi-spatial scales (local, regional, and global)

Applying the MEA's conceptual framework especially coastal wetlands play vital roles on different scales; the proposed framework should consider the development objectives and effects on these scales.

Multi-dimensional (Ecological, Social, Economic...)

However the environmental dimension should be at the core of interest to preserve highly sensitive ecosystems, the human wellbeing as an integral part of this ecosystem should be considered as the aim of sustainable development.

Balanced

Top-down and bottom-up: Consider both human and natural aspects of the ecosystem, the local communities should be an effective partner in the planning and development process, especially to overcome most of downsides and problems of the current development.

Supported by effective ES assessment tool

An effective ecosystem services/values assessment tool is required to support the planning and management process and enable well informed communication between science and decision making.

Flexible

To respond to complexity and uncertainty associated with this very dynamic ecosystem, the proposed framework should be flexible enough. Also, flexibility is required for wider applicability of this framework within different local contexts of coastal wetlands including; social, cultural, institutional, and data characteristics.

Moreover applying the 12 principles of the Ecosystem Approach, The EA-PMCW's objectives were formed to precisely reflect the

Main features of the EA-PMCW framework

- **Holistic:** To understand the multidimensional interactions, external, and global factors affecting coastal wetlands as open ecosystems, the proposed framework should take a holistic perspective.
- **Dynamic:** The proposed framework should be dynamic/ adaptive to reflect the continuous changes of the ecosystem.
- **Respecting the natural boundaries of the ecosystem:** The unity of the ecosystem and flow of its natural functions should be maintained to achieve sustainability.
- **Multi-temporal** (Long-term and Short-term): To achieve suitability in the long term and to respond to short term challenges and needs, the framework should include different time scales of plans and ensure that the short-term managements plans are well fitting within the long-term plans.
- **Multi-spatial scales** (local, regional, and global): Applying the MEA's conceptual framework especially coastal wetlands play vital roles on different scales; the proposed framework should consider the development objectives and effects on these scales.
- **Multi-dimensional** (Ecological, Social, Economic..): however the environmental dimension should be at the core of interest to preserve highly sensitive ecosystems, the human wellbeing as an integral part of this ecosystem should be considered as the aim of sustainable development.
- **Balanced** (top-down & bottom-up): Consider both human and natural aspects of the ecosystem, the local communities should be an effective partner in the planning and development process, especially to overcome most of downsides and problems of the current development.
- **Supported by effective ES assessment tool**: an effective ecosystem services/value assessment tool is required to support the planning and management process and enable well informed communication between science and decision making.

Flexible: To respond to complexity and uncertainty associated with this very dynamic ecosystem, the proposed framework should be flexible enough. Also, flexibility is required for wider applicability of this framework within different local contexts of coastal wetlands including; social, cultural, institutional, and data characteristics.

Figure 1: Main Feature of the EA-PMCW framework.

especial features of coastal wetlands, the challenges they are facing, and the main requirements of land use planning process, as follows [1]:

- Maintain/sustain biodiversity and ecosystem services for the long term.
- Protect the rare, unique, threatened species and their habitats.
- Protect the human cultural features as an integrated part of the ecosystem.
- Meet the local communities' objectives within the global national, and regional frameworks.
- Minimize impacts of human development.
- Minimize effects of global warming, sea level rise and other natural coastal hazards or expected extreme events.
- Respect the natural units.
- Ensure that environmental, social, and economic objectives are achieved at an acceptable cost to society.
- Enhance implementation through participatory planning, knowledge management and capacity building.
- Maintain continuity of effective planning–management process.

Stages and Steps of the EA-PMCW

This section briefly explains the main stages of the EA-PMCW framework [1]. The EA-PMCW can be divided into seven stages, which reflect a contiSSnuous nonlinear process, as it contains a few feedback loops for checking, updating, learning and adapting different proposals and actions of planning and management. Figure 2 illustrates these stages and their suggested steps, as follows [1]:

The pre-planning stage

This stage aims to determine the scale of the planning and management regime, and its essential requirements. It starts with looking at the coastal wetland area from global and national perspectives to obtain strategic reflections of these levels on the planning and management of this area. This stage includes:

- Identifying the class, level and scale of the coastal wetland.
- Identifying the external objectives and role within international, national, and regional strategies.
- Identifying financial resources.
- Identifying and preparing the institutional requirements.

The scoping stage

This stage aims to draw the direct and indirect boundaries of the planning and management regime. It is a multilevel stage which attempts to create a balance between the natural and the human aspects not only in the definition of spatial boundaries but also in the identification of the main issues of concern, goals, and the primary objectives.

The Analysis stage

This can be considered the most complex, multidisciplinary and critical stage of the process. It aims to combine the different current and future characteristics of the coastal wetland, and to draw integrated pictures, which will form the base of decision making in the planning and management stage.

The planning stage

The aim of this stage is to develop effective and efficient land-use plans, management plans, and emergency plans, in order to achieve the agreed objectives, using the results of all previous steps.

The implementation stage

To ensure efficiency of implementation, a time frame and measurements of progress should be set within each of the plans. Then the institutional arrangements, and other required resources should be prepared because implementation phases and programmes are very case specific, so they are not discussed in details here.

The monitoring (not limited to this stage)

Details of monitoring programmes and activities are also case specific, but generally regular monitoring programmes should be established and the required human and technical resources should be ensured from the beginning. Situation analysis, taking suitable actions, and resolving conflicts is a continuous set of steps which should run throughout the whole framework and not just be limited to the end of the plan making cycle to meet the dynamic and sensitive nature of coastal wetlands. Responding to this dynamic nature, the database should be regularly updated with a simple kind of trend analysis, and a conclusion of the situation should be shared with partners and stakeholders. It should be mentioned here that a suitable action at this stage could be applying one of the emergency plans, whenever its related scenario happens.

The evaluation and adaptation stage

Evaluation and adaptation again is a multi-temporal process, which differs in details, techniques used and outputs, from one time frame to another.

The Case Study

Egypt enjoys a vital strategic location between three continents. This gives it a special significance from the point of view of biodiversity. The Egyptian coastline extends 3,500 kilometres along the Mediterranean Sea and Red Sea in addition to the Suez and Aqaba gulfs [8]. The coastlines of Egypt are rich with wetland ecosystems, such as: coral reefs, mangroves, sand dunes, sea grass beds, estuaries, and coastal forests. At the same time, the Egyptian coasts are amongst the most densely populated areas in the MENA region [13]. Although the Egyptian coastal wetlands have attracted wide variety of the economic activities, most of them require especial protection due to their internationally recognized biodiversity. Indeed, coastal wetlands are facing main environmental challenges, including increased impacts of climate change all-over the world, they are under intensive pressure in Egypt [14]. This has caused water pollution, huge deterioration of their natural resources, and increased shoreline erosion. The socio-economic context of Egypt as a developing country has put extra pressures of development and impacts of unplanned activities on its coastal wetland [15]. Moreover some negative changes naturally driven by their location, including land subsidence along the northern coast due to the tectonic movement, which increases the effects of sea level rise. From here, the Egyptian case was selected as a rich case study for testing applicability of the EA-PMCW.

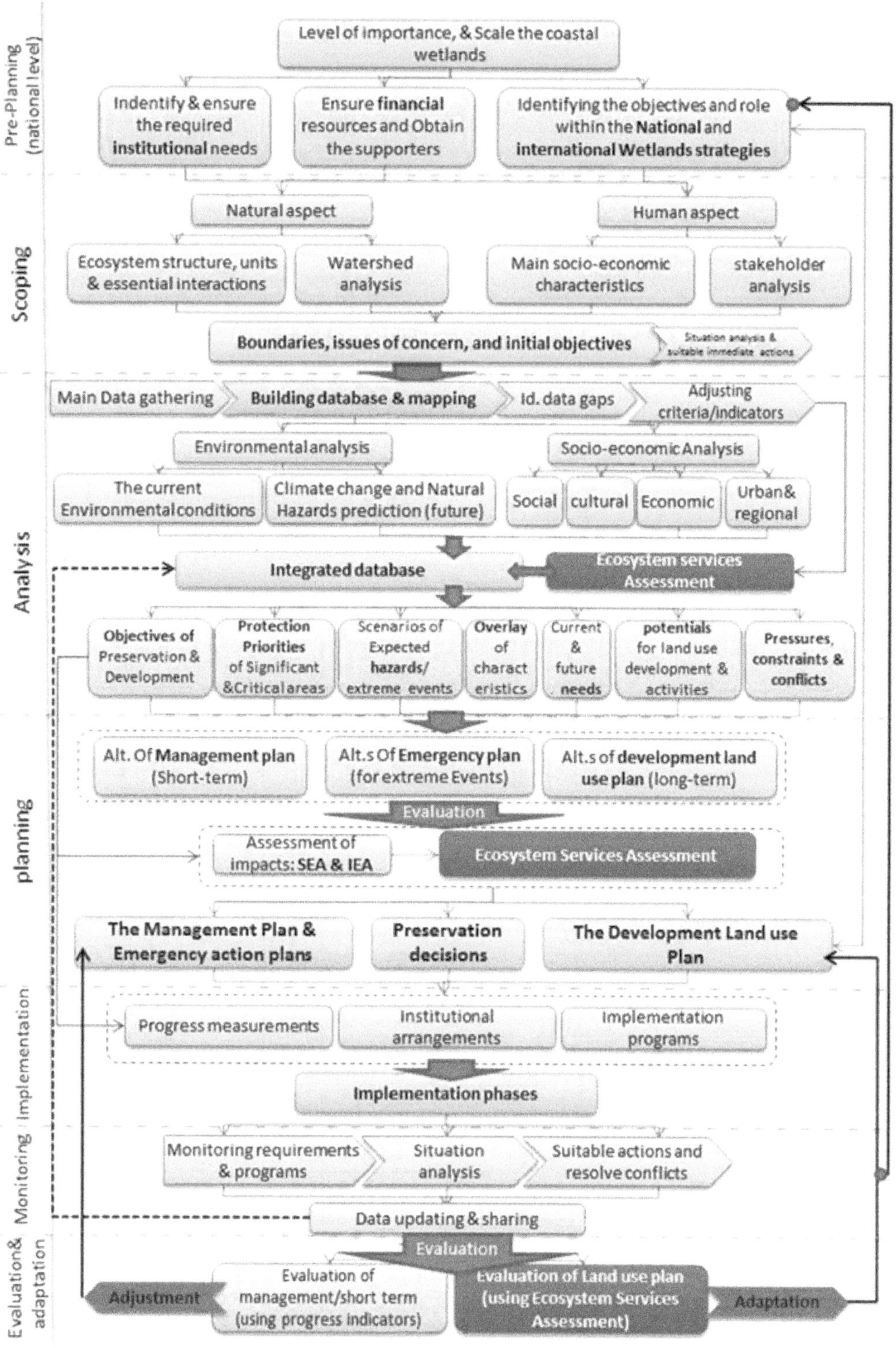

Figure 2: The EA framework for planning and management of coastal wetlands Source [1].

Overview the methods and techniques of data-collection

Several techniques of primary and secondary data collection were used here in order to increase accuracy and reliability of results. The Expert-Interviews and the Focus-Groups techniques were chosen as suitable empirical data-collection techniques for this research task [16]. The documentary data was used to build a suitable understanding of the local context of the case study (i.e. natural, social, economic, and institutional conditions). However the semi-structured interviews with the local experts mainly helped to test the applicability of the EA-PMCW (i.e. the expected potentials and challenges), the focus-groups technique with stakeholders and environmental managers was helped to form solid recommendations for more successful applications of the EA-PMCW within the Egyptian case [17-22].

Focus on the results

The results show a great acceptance of the EA-PMCW to be applied within the Egyptian context, as 100% of the interviewees evaluated it as a comprehensive methodology which meets the local and international objectives in coastal wetlands. However, they almost agreed that a few adaptation guidelines should be considered in order to overcome the especial challenges, which may face its application in Egypt and other developing countries. In more detailed vision, the focus of the interviews turned to explore potentials and challenges of applying the eight stages of the EA-PMCW framework. Then, recommendations and guidelines for successful applications were formed. Figures 3 and 4 draw an overview of the potentials and challenges of applying the framework, as seen by the interviewees.

Main potentials

Several significant potentials for applying the suggested framework in the Egyptian context were mentioned by the interviewees, as illustrated in Figure 3. Most of them are related to human, academic, and technical resources. For example, Egypt is the pioneer of several disciplines in the region such as through unique Faculty of Urban and Regional Planning, in Cairo, the Authority of Remote Sensing and Space Science, the development of Arabic GIS software. Also it has the longest history and leadership of most of the needed scientific disciplines in the region (e.g. The Institute of Oceanic Science, Water Research Institute). However concern about environmental issues comes in the second place of potentials in both levels of the government and in nongovernmental actions and they represent initial steps which need to be integrated and maximized. According to the stages of the suggested framework, the distribution of potentials tends to support most the planning stages, especially the scientific and technical potentials.

Main challenges

Figure 3 shows an overview of the challenges which may face application of the suggested EA-PMCW framework within the current local context, as mentioned by the Egyptian expert interviewees. Several weaknesses are related to integration and coordination of planning and management process, which are likely to significantly affect the second half of the process (e.g. planning, implementation, monitoring). Although technical challenges (e.g. data availability and quality, boundary determination) can affect the whole process, weakness of social participation and public awareness are also likely to be very significant throughout.

Extracted Recommendations

The previous section has explored the potentials and challenges of applying the EA-PMCW framework in Egypt as a case study and validated the findings of the interviews with the other available sources of evidence. This section presents some recommendations towards successful application based on tackling the challenges and maximizes the potentials within the local context. These recommendations have

Potentials	Pre-planning	Scoping	Analysis	Integrated Analysis	Planning	Implementation	Monitoring	Evaluation Adaptation
Active involvement with the international environmental movement	●	◑	◑	◔	◔	○	○	◔
Having a powerful central national environmental agency 'EEAA'	●	●	○	○	◑	◔	◑	◑
Specialized academic institutions supporting the required qualifications	●	●	●	●	●	◔	●	◔
The existing national movement towards applying principles of EA in sensitive areas	◑	◑	○	○	◑	○	◔	◔
Having a supportive National Strategy for wetlands	●	●	◑	◑	●	◕	◑	◑
Wide experience with the international environmental sponsors	◑	◑	○	○	◑	◑	◔	◔
The increased role of GIS in different institutions and levels of projects	○	◔	●	●	◑	◔	◕	●
Leading capabilities of remote sensing in the region	○	◑	◑	◑	◑	◔	◕	◔
Leading acadimic experience in environmental planning and physical planning in the region	◑	◑	●	◕	●	◑	◕	◕
The increased number and activities of social groups and NGOs interested in environment and sustainability	◔	◔	○	○	◑	◑	◔	◔
The increased social power and desire of participation in decision making	◔	●	○	○	◕	◑	◔	◑

Source: Author upon interviews with Egyptian experts

Frequency of the mentioned potentials by the interviewees

● the vast majority (>80%)
◑ around half of them (40-60%)
◕ the most of them (60-80%)
◔ few mentions (<40)

Figure 3: The EA framework for planning and management of coastal wetlands. Source [1].

Challenges	Pre-planning	Scoping	Analysis	Integrated Analysis	Planning	Implementation	Monitoring	Evaluation Adaptation
Insufficient integration between national institutions and local governance	○	◕	○	◔	◑	◕	●	●
Weak cross-sectoral links & lack of integration/communication between parties	◔	◑	◑	●	●	◑	◔	◕
Lack of Integration between disciplines/the boundaries of too narrowly focused disciplines	○	◑	◑	◕	◑	○	○	◑
Insufficient of monitoring and inforcement effort	○	○	○	○	○	●	●	○
Limited financial resources and insufficient equipment	●	○	◔	◔	◑	●	●	◑
Difficulty of identifying boundaries and units of management	○	●	◕	◕	◕	●	●	●
Weak participation of the local communities, stakeholders, and interest groups	○	◑	○	○	◕	◕	○	◕
Weak public awareness of environmental and sustainable issues	◔	○	○	○	○	◕	◕	◑

Source: upon interviews with Egyptian experts

Frequency of the mentioned challenges by the interviewees

● the vast majority (>80%) ◑ around half of them (40-60%)

◕ the most of them (60-80%) ◔ few mentions (<40)

Figure 4: Challenges which may affect successful applications of the suggested EAPM-CW framework in Egypt.

been gained through focus groups sessions with the local experts and decision makers.

Coordination and leadership

Establishing leadership or identifying a coordinating agency was strongly recommended by the focus group, as a first step towards solving problems of lack of integration. There was a primary agreement that the main responsibility of this lead agency should be to facilitate and control the quality of the planning and management process vertically through the different levels and horizontally among different parties. Although a few other agencies were mentioned as alternatives to take this responsibility, the discussions concluded that the Egyptian Environmental Affairs Agency (EEAA) is the most recommended one because it has several advantages over the other bodies in this regards, including:

- It is already responsible for drawing the National Wetland Strategy.
- It has already played a significant role of integration on the regional and global levels (e.g. UNFCCC, RAMSAR).
- It is responsible for the main environmental data drivers in Egypt (e.g. biodiversity annual reports).
- It is the central national body responsible for evaluating EIA studies.
- It is responsible for Egypt's sea and coastal zone policies, although different authorities also have different responsibilities in the coastal zones.
- It has a successful experience of managing protected areas, and it already has a network of local management units.
- The environmental law has given it higher authority in environmental related issues than any other governmental body.

Integration

weak integrations was identified as a main challenge facing the planning and management process in Egypt, so establishing a clear vision for integration was a key task and product of the focus group discussions. The suggested vision for effective integration can be illustrated as follows:

Vertical integration: A balanced relationship between top-down and bottom-up Integration between different levels of strategies and plans (i.e. local, regional, national, and global) was called for. The Egyptian Environmental Affairs Agency 'EEAA' already has valuable institutional units which can be linked and supported with the required capacity to play an effective role in integration between different levels of plans and strategies [15]. The EEAA is already coordinating the Wetlands Committee which draws the National Strategy of Wetlands, and at the same time it coordinates two ICZM committees for the Red Sea and the Mediterranean [8]. Moreover, the natural preservation sector in EEAA is managing all the preservation areas in Egypt and it has a local management and monitoring station in each one, including those in wetlands. Although currently all these committees and sections are located within and coordinated by the EEAA, an integration mechanism needs to be established and coordinated by EEAA. The existing units of EEAA can be used, after supporting them with the needed capacity, to manage the flow of information, guidelines, plans, and strategies vertically between different levels.

Horizontal integration on the national level: The research concluded that the EEAA has the capability to lead the integration between different development sectors and governmental agencies, and produce the final plan. On the other hand it was felt that the academic and scientific institutions should play very important role to support the EEAA and its committees and local units with the needed capacity of staff and physical capabilities as applicable. The suggested view of integration on the national level should be based on the role of EEAA as the central coordinator between national institutions and the local level. However as data sharing and management is a fundamental part of the integration and coordination tasks of EEAA, strong connection with the national data drivers through the IDSC 'Information and Decision Support Centre' is required.

Horizontal integration on the local level 'the wetland': The research emphasized the importance of establishing a powerful local management unit/committee for each coastal wetland. This unit would be the coordinator/leader of the process on the local level and it should be well linked with its related main committee in the EEAA. However the flow of information, guidance, technical support and feedback should be active and dynamic between them and in two directions. Institutionally other local governmental bodies representing the development sector would integrate information and facilities with the wetland unit through the governorate. Two strong wings would support effectiveness of the process: a) The local community (e.g. social groups, stakeholders, NGOs); and b) scientific and technical support (e.g. academic and research centres.

Current management boundaries and natural units of ecosystems

Current boundaries of governments, under any classification are not suitable for planning and management of the coastal wetlands. As discussed before, no clear boundaries have been drawn for development or management except limited boundaries of preservation areas. A positive step in this regard, has been taken by the NARSS using remote sensing capabilities to help determine the boundaries of main features in the northern coastal wetlands including; water basins, fish farms, and salt farms. A common suggestion by the focus groups was to declare the final produced boundaries together with the suggested local management unit of each coastal wetland as a new legal and institutional framework of these areas.

Local communities and public participation

Public awareness of environmental issues and sustainable development is growing but efforts are needed to prepare the local communities and social groups for positive participation in development plans and activities. Workshops and training programs are recommended, as persuading them to be a responsible partner in the planning and management process to achieve balanced the sustainable development is essential. Several mechanisms can be used such as: giving the local communities advantages and facilities to own or benefit from the proposed projects, creating new job opportunities for them, encouraging them to participate in data collection and other planning and management by employing them as assistants in the task.

Conclusion

The EA-PMCW - developed by Ghoneim [1] - transferred the wide principles of the ecosystem approach into a clear, step by step methodological framework to support land use and development decision making. However its steps was designed to fit the requirements and main stages of the planning process, it carefully consider the especial features of coastal wetlands and the global challenges facing them. The primary test of applicability shows promising results and gave solid evidence of acceptability of the EA-PMCW; however several points of consideration clearly formed to tackle the expected challenges, maximize the potentials, and achieve successful application within the local context of the Egyptian context.

References

1. Ghoneim SA (2012) An Ecosystem approach and an ecosystem service assessment technique for planning and management of coastal wetlands, PhD, University of Liverpool.
2. Ramsar (2016) Wetlands for our future: The 4th Strategic Plan 2016 - 2024, Adopted by the 12th Meeting of the Conference of the Parties at Punta del Este, Uruguay, 1-9 June 2015, through Resolution XII.2. Ramsar (Ramsar Convention Secretariat).
3. Ramsar (2010) Coastal management - Wetland issues in Integrated Coastal Zone Management, Gland, Switzerland, Ramsar Convention Secretariat.
4. Ramsar (2010) Managing wetlands, Gland, Switzerland, Ramsar Convention Secretariat.
5. Ramsar (2010) Wise use of wetlands: Concepts and approaches for the wise use of wetlands, Gland, Switzerland, Ramsar Convention Secretariat.
6. MEA (2005) Ecosystems and Human Well-Being: Wetlands and Water. In: Assessment Me (ed.) Washington. DC: World Resources Institute.
7. Ibrahim HS, Hegazy I (2014) Towards Applying Strategic Environmental Assessment (SEA) in the Egyptian Context. Asian Acd Research J of Social Sci and Humanities 21: 325-341.
8. Ibrahim HS, Ghoneim SA (2013) Integration as a significance factor in effective coastal management: Egypt as a case study. International Journal of Marine Science 46: 389-401.
9. Ehler C, Douvere F (2009) In: Dahl R (ed.) Marine Spatial Planning: a step-by-step approach toward ecosystem-based management, Paris UNESCO.
10. Randolph J (2004) Environmental land use planning and management, London, Washington, Island Press.
11. Shepherd G (2008) The Ecosystem Approach: Learning from Experience. Gland, Switzerland: IUCN p: 190.
12. Kidd S, Maltby E, Robinson L, Barker A, Lumb C (2011) The Ecosystem Approach to marine planning and management, London: Earthscan.
13. EEAA- Ministry of Environment (2016) Egyptian Biodiversity Strategy And Action Plan (2015 - 2030), UNDP, GEF, PIMS no: 4864.
14. El-raey M (2010) Egypt: Coastal Zone Development and Climate Change.
15. Ghoneim SA (2014) An Ecosystem approach for Planning and management of coastal wetlands, ECSEE 'The 2nd European Confference On Sustainability, Energy, And The Environment, Brighton, proceeding.
16. Bogner A, Littig B, Menz W (2009) Interviewing Experts, London: ECPR.
17. CBD (2016) About the Convention. Norway: Convention on Biological Diversity (CBD).
18. CBD-COP (2010) Strategic Plan For Biodiversity 2011-2020 and The Aichi Biodiversity Targets: Living in harmony with nature. In: DIVERSITY, COB (ed.) The Conference of the Parties Japan.
19. Gillespie AR, Miller BK, Johnson AKD (1995) Effects of Ground Cover on Tree Survival and Growth in Filter Strips of the Cornbelt Region of the Midwestern US, Texas A&M Blackland Research Center.
20. Ibrahim H (2013) Towards an effective framework for coastal zone management: the Egyptian experience. Journal of Coastal Conservation: Planning and Management 3: 601-613.
21. Kelly M, Tuxen K, Stralberg AD (2011) Mapping changes to vegetation pattern in a restoring wetland: Finding pattern metrics that are consistent across spatial scale and time. Ecological Indicators p: 11.
22. Maltby E (2009) The Changing Wetland Paradigm. In: Maltby E, Barker T (ed.) Wetlands Handbook. Blackwell Publishing Ltd.

11

A Re-evaluation of the Depositional Environments and Sedimentary Facies of Middle Miocene Paralic Deposits (Agbada Formation), Ewan and Oloye Fields, Northwestern Niger Delta

Durogbitan A Abimbola*

Petrfac IES, Research and Development, Manchester, UK

Abstract

This study analyses depositional environments and facies of the middle Miocene (Agbada Formation), northwestern Niger delta, based on cores and well log data, and incorporating ichonological data, that has led to a revision and re-evaluation of the facies within the study area. Log motif analysis, calibrated by lithology descriptions based on core photos, was used to define log facies associations and to identify stacking patterns and key surfaces. Six main wireline log facies associations have been recognized using gamma ray log motifs calibrated with core.

Within the cored interval two lithofacies associations have also been identified, based on composition, colour, sedimentary structures, and ichnofabric. They are: slope channel sand and shoreface sequences. The interpreted depositional facies model suggests an extensive development of marine dominated slope channel fill and shoreface deposits in the northwestern part of the delta. This suggests that in the studied area, the coastal barrier complex contains a large amount of sand that was originally deposited in river mouth bars, but was laterally removed by longshore current and re-deposited on the shoreline margin and basin ward through developed of canyons on the slope which serves as feeders and conducts for slope and basin fans.

The descriptive framework for the ichnofabrics analysis is based on bioturbation index (degree of bioturbation), burrow size, burrow frequency and ichonological diversity. Six ichnofabrics are recognized within the core namely: Planolites-Teichnichnus, Planolites-Phoebichnus, Planolites, Planolites-Thalassinoides, Teichichnus-Phoebichnus-Planolites and Palaeophycus-Planolites ichnofabrics. Observed tiering patterns are both simple and complex, which may reflect gradual aggradation or degradation of the substrate leading to overprinting as a stable ichnofauna structure develops. This type of tiering arrangement indicates a stable environment. The ichnofabrics are named after the dominant trace observed. Sedimentological analyses indicated that the depositional environments range from inner shelf to offshore.

Keywords: Niger delta; Miocene; Depositional environments; Lithofacies

Introduction

The Niger delta is an example of a mixed wave, tide and fluvial deltaic system [1]. It has long been recognised that deposits within the Miocene Agbada Formation of the Niger delta represent a prograding fluvial-deltaic depositional system intercalated by regionally extensive marine mudstone [1-5]. Sandstone units within the formation consist of fluvial (channel) or fluvio-marine (barrier bar), tidal channel and shoreface facies. Mineralogically, they consist of quartz with traces of feldspar, glauconite and zircon. Mudstones are generally fluvio-marine or lagoonal facies. The clay minerals are mainly mixtures of kaolinite, smectite and illites with some traces of pyrite, feldspar and carbonate [6] (Figure 1).

The reservoir intervals within the Agbada Formation have also been interpreted and documented to be deposited as highstand system tracts and transgressive system tract in a shelf delta setting [4]. The thickness of the reservoir intervals ranges from 45 feet to >150 feet. Most of the reservoirs observed within the shallow marine paralic section (Agbada Formation) have been described as point bars of a distributary channels, tidal channel fills, coastal barrier bars and shoreface deposits [2,7-11].

Aims and Objectives

The aims and objectives of the work were to re-evaluate the depositional environments and facies scheme which had been originally outlined by the previous authors by incorporating ichonological descriptions. This in turn will help to identifying the depositional conditions under which the sediment were deposited and preserved.

Data and Methods

In this study, well log analysis (gamma, density, neutron log etc.) and detailed core sedimentological descriptions, based on examination of core photographs have been undertaken, incorporating an ichonological description, constrained by seismic section. Log motif analysis, calibrated by core photographs and constrained with seismic were used to defining log facies associations which were used to identifying stacking patterns and key stratal surfaces.

Core sedimentology

The sedimentological descriptions from two wells, Ewan 1 and Oloye 3 (Figure 2) focused on the lithology, colour, grain size,

***Corresponding author:** Durogbitan A Abimbola, Principal Geoscientist, Petrfac IES, Research and Development, Manchester, UK
E-mail: adewoledurogbitan@yahoo.co.uk

Figure 1: Landsat Thematic Images of Niger Delta illustrating the mixed wave, tide and fluvial influence on the delta [1].

sedimentary structures ichnofauna, ichnofabric analysis and diagenetic overprint (ferruginisation and carbonate nodules). This allowed definition of lithofacies within the studied area, and the interpretation of the depositional environmental, identification of some key surfaces. Together this aided an interpretation of the physical, chemical and biological conditions within the delta at the time.

Lithofacies/Ichnofabric analysis

Lithofacies: Three main facies have been identified from the cored interval of the Ewan 1 and Oloye 3 wells, from the onshore area of the northwestern Niger delta. They are: mudstone, heterolithic (mixture of sandstone and mudstone) and sandstone. These were further subdivided into eleven lithofacies units based on sedimentary structures, indurations, composition, ichnofabric, colour and ichnofaunal diversity and degree of bioturbation (Table 1). Recognising the above features played a major role in the facies interpretation. The eleven facies are described below.

Mudstone facies: Mudstones are a detrital sedimentary rock or suspension fall-out deposit composed of clay-sized particles, often associated with low energy depositional environments. Three basic lithofacies were recognized. The level of bioturbation is very hard to assess using core photographs in mudstone, because of the lack of visual contrast between the host sediment and burrows.

- **Mudstone (MS1)-laminated mudstone (friable):** This friable lithofacies is dark greyish, parallel laminated and contains carbonaceous material. Parallel laminations are defined by the carbonaceous material (darker in colour) and streaks of sand (brownish in colour) (Figure 3). Within this facies, a Planolites ichnofabric has been identified, with small burrows.

The small burrow size could indicate environmental stress. The streak of sand indicates episodic high energy and could be some evidence of a neritic/freshwater influence. This influence may also be reflected in smaller sizes observed in Planolites and low diversity of the ichnofauna within the facies. The presence of carbonaceous materials suggests a terrestrial input. This facies is only observed in the Oloye 3 core in the upper part of the section. Their depositional environment is interpreted to be a low energy. This lithofacies is thought to be a deltaic lower shoreface sequences.

- **Mudstone (MS2) - laminated mudstone (indurated):** This lithofacies is characterized as being an indurated, light greyish mudstone with rare streaks of sand or silt bands (Figure 4). Laminations in the mudstone are not distinct as in case of MS1, this may be due to their cohesiveness caused by the increase in the amount of clay minerals or different clay in mineral type or increases in water depth. This facies is observed in both Ewan 1 and Oloye 3 with thickness of about 0.3 m. The streaks of sand have indistinct small scales cross lamination, possibly rippled, suggest episodic energy. The carbonaceous materials also suggest terrestrial input. In this lithofacies, a Planolites ichnofabric is recognised, with small burrows. The dominance of mudstones suggests low energy, with periodic higher energy condition and input of this, sometime rippled sandstone lenses. The ichnofabric suggests environmental stress possibly caused by above terrestrial and fresh water incursion. This lithofacies is thought to have been deposited in deltaic lower shoreface sequences.

- **Mudstone (MS3) - with lenticular bedded sandstones:** This facies is composed of bioturbated, lenticular bedded sandstones, carbonaceous rich mudstones. Synaeresis? desiccation cracks are observed (shrinkage crack). A Planolites ichnofabric is identified, and the burrows are elongate and aligned parallel with bedding. The lenticular bedding is defined by thin bands of sandstones. The thin bedded sandstones exhibit mm-scale ripples lamination (Figure 5).

This type of facies is only observed in Oloye 3 directly on top of an interpreted flooding surfaces (FS2); (Figure 2). The presence of synaeresis?/desiccation cracks may result from salinity fluctuation, suggesting tidal influences [12]. Also, the presence of lenticular bedding support tidal influence the carbonaceous material suggests terrestrial input. The Planolites burrows are also small in sizes, thus suggesting environmental stress. This lithofacies is thought to have been deposited in deltaic lower shoreface sequences.

Heterolithic facies: The heterolithic lithofacies were classified on the percentage of sandstone and mudstone and identified sedimentary structures. The alternation of sandier and muddy layer reflects deposition during fluctuating energy conditions (episodic). The sand layers were deposited during medium energy regime of current and storms while the muddy layers settled out from suspension. This

Figure 2: Core Sedimentological Descriptions of Ewan 1 and Oloye 3 Wells [30].

Facies	Lithology Descriptions	Thickness (m)	Trace fossils	Dominant Ichnofabric	BI	Sedimentary structures	Processes	Environmental interpretation
Laminated friable mudstone (MS1)	Bioturbated, friable, dark grey laminated carbonaceous mudstone with rare siderite.	0.3-0.6	*Planolites*	*Planolites*	4-6	Parallel lamination	Low sedimentation rate, suspension fall out deposit	Low energy, below fair-weather wave base: Lower shoreface-offshore
Laminated indurated mudstone (MS2)	Bioturbated, indurated, light grey mudstone with rare sand streak	0.3	*Planolites*	*Planolites*	4-6	Indistinct parallel lamination.	Low- medium sedimentation rate, suspension fall out deposit	Low energy, increase in water depth compare to MS1. Lower shoreface-offshore.
Lenticular Laminated mudstone (MS3)	Bioturbated, lenticular bedded (0.03 cm), carbonaceous mudstone with horizontal burrows and *synaeresis* crack.	0.6	*Planolites*	*Planolites*-defined the beddings	4-6	Lenticular beddings, desiccation crack? and parallel lamination	Low-medium sedimentation rate, suspension fall out deposit	Lower shoreface under fair-weather conditions. Presence of lenticular beddings reflects tidal indicator.
Thinly bedded disrupted heterolithic sandy mudstone (HE1)	Highly bioturbated, structureless to thinly bedded, highly disrupted	0.6-1	*Palaeophycus, Planolites*	*Palaeophycus-Planolites*	6	Poorly defined	Low-medium sedimentation rate and low energy regime, sediment reworking.	Lower shoreface associated with fair-weather deposition.
Muddy sandstone with *Planolites* ichnofabric (HE2)	Intensely bioturbated, disrupted, carbonaceous muddy sandstone with diverse suite of horizontal burrows.	2.2	*Planolites, Teichichnus, Palaeophycus, Ophiomorpha, Skolithos, Phoebichnus*	*Planolites-Teichichnus*	6	Lack of any discernable sedimentary structures due to complete bioturbation.	Low-medium sedimentation rate and low energy regime. Diversity of trace fossils reflect conducive environmental conditions	Marine dominated slope channel/fan deposits.
Mudstone dominated heterolithic (HE3)	Interbedded mudstone and sandstone. Interbedding is variable, ranging from 0.1 cm -0.4 cm thick. Mean mud bed thickness is 0. 7 cm and means sand bed thickness is 0.25 cm.	0.3	*Planolites, Palaeophycus, Phoebichnus.*	*Planolites-Palaeophycus*	5-6	Parallel lamination	Deposition during fluctuating energy conditions. Low-medium sedimentation rate	Lower shoreface below fair-weather wave base
Sandstone dominated heterolithic (HE4)	Intercalation of mudstone and sandstone. Interbedding is variable, ranging from 0.2 cm- 0. 6 cm thick. Mean mudstone bed thicknesses is 0.2 cm and mean sand bed thickness is 0.6 cm	0.6	*Planolites, Phoebichnus, Palaeophycus, Ancorichnus*	*Planolites-Phoebichnus*	4-5	Wave ripples and parallel lamination	Deposition during fluctuating energy condition. Low-medium sedimentation rate	Lower shoreface below fair-weather wave base
Highly bioturbated poorly bedded sandstone (SA1)	Fine grained brownish sandstone, with remnant beddings and ichnofabric texture.	0.3-2.2	*Teichichnus, Phoebichnus, Planolites, Skolithos, Ophiomorpha, Ancorichnus, Palaeophycus*	*Teichichnus-Phoebichnus - Planolites*	5-6	Remnant bedding structures of parallel and hummocky cross stratification	High diversity of ichnofauna indicates conducive environmental conditions. Sediment reworking	Shallow marine lower shoreface environment Presences of HCS indicate evidence of storm processes.
Completely bioturbated , sandstone (SA2)	Highly bioturbated, mottled, sandstone with rare iron stain and lag deposit	1-3.5	*Planolites, Thalassinoides Skolithos, Ophiomorpha, Phoebichnus, Palaeophycus,*	*Planolites -Thalassinoides*	6	Absent due to complete bioturbation.	High diversity and presence of *Thalassinoides* indicate break in sedimentation-episodic regime (firm ground).	Marine dominated conditions., marine dominated slope channel/ fan environment
Slightly bioturbated thinly bedded cross laminated sandstone (SA3)	Slightly bioturbated parallel ripples and cross lamination (0.6 cm), carbonaceous sandstone.	2	*Planolites, Phoebichnus, Skolithos, Ophiomorpha, Arenicolites, Ancorichnus*	*Planolites-Phoebichnus*	2-3	Cross and parallel lamination, presence of trough stratifications	High sedimentation rate and medium energy conditions, under current/storm conditions.	Low to medium energy settings: Marine dominated slope channel sand.
Carbonate nodules SA4	Carbonate nodules within bioturbated sandstone	0.3	*Planolites, Skolithos*	*Planolites*	2-4	Poorly defined, due to diagenesis	Nodules suggest a period of reduced clastic input.	Slope channel/fan environment. (Reduced salinity, stressful environment).

Table 1: Summary of the facies association from core studies.

a b c

Figure 3: Laminated friable carbonaceous mudstone, highly bioturbated and streak/laminae of sandstone (a, b, c). Occurred at depth 1394.2 m, 1393.9 m and 1392.7 m respectively within Oloye well.

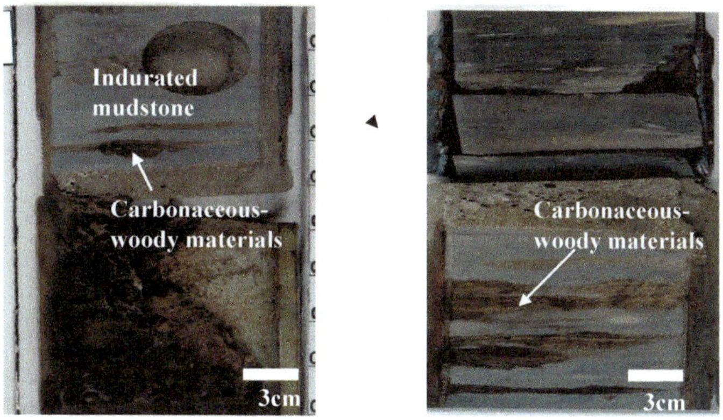

a b

Figure 4: Laminated indurated greyish mudstone, with streak of carbonaceous materials and rare streak of sand band at depth 1508.6 m and 1508.3 m respectively within Ewan well.

Figure 5: Lenticular mudstone, highly bioturbated. The bioturbation consist of *Planolites (Pl)* and ichnofabric, indicating that the burrows event occurred within a reducing environment. Presence of *lenticular bedding* reflects tidal evidence and salinity fluctuation. Depth 1394.5 m within Oloye 3 well (Above uncored section).

lithofacies association is subdivided into four lithofacies on the basis of the mean sand to mud content, bedding structures and bioturbation.

- **Heterolithic (HE1) - highly bioturbated, thinly bedded and disrupted:** This lithofacies is characterised as a light brownish in colour structureless to thinly bedded, highly disrupted sandy mudstone with carbonaceous rich material. This facies is highly bioturbated. This lithofacies contain a pervasive ichnofabric that consists of small (about 0.05 mm-0.1 mm) ovate and elongate Palaeophycus, Phoebichnus and Planolites sand filled-burrows with carbonaceous rich materials (Figure 6). The lack of distinct bedding which may be due to organism activity or the original homogeneous nature of the substrate. This facies is only observed in the Oloye 3 well. The presence of carbonaceous material suggests some terrestrial input. This deposit is thought to have been deposited in deltaic lower shoreface sequences.

- **Heterolithic (HE2) – highly bioturbated muddy sandstone with ichnofabric:** This facies is characterised by being intensively bioturbated, containing a composite ichnofauna of Planolites, Palaeophycus, Ancorichnus, Ophiomorpha, Teichichnus and Phoebichnus, dominated by Planolites, and Palaeophycus ichnofabrics. As well as interbedded sandstone and mudstone, it contains some carbonaceous material (Figure 7). Due to the high intensity and diversity of the bioturbation none of the original sedimentary structures are visible. The burrows are elongated and are aligned parallel; this parallel alignment may reflect the original bedding structures.

The facies is about 2 m thick (Figure 2) based on fining upward log motif observed from the core study integrated with wireline log. The top of this facies is interpreted to mark the initiation of deepening within the Oloye 3 well. Based on the core description and ichnofabric, constrained with shoreline position from the seismic survey (Appendix), the depositional environment is interpreted to be marine dominated slope channel sand. The presence of carbonaceous material indicates terrestrial input. It is observed to be more marine than MS1-3 due to high diversity of ichnofauna.

- **Heterolithic (HE3) - mudstone/sandstone:** This lithofacies is mudstone dominated (>60:40 mudstone/sandstone) and characterised by intercalation of indurated, laminated mudstone and sand, with some internal lamination. The scale of the lamination is variable, ranging 0.03 m -0.1 m thick. The mean mud bed thicknesses is 0.07 m and mean sand bed thickness is 0.03 m. The mudstone is characterised internally by parallel lamination while the sandstone is characterised by wave ripples (Figure 8). This lithofacies contains Palaeophycus, Ophiomorpha, Planolites and Phoebichnus.

The presence of parallel lamination, wave ripple and the intercalation of mudstone and sandstones is interpreted to indicate

Figure 6: Structureless, soft and highly distorted muddy sandstone with *Palaeophycus* (*Pal*), *Phoebichnus* (Phoe) and *Planolites (Pl)* ichnofabric occurred at depth 1393.6 m and 1393.3 m respectively within Oloye well.

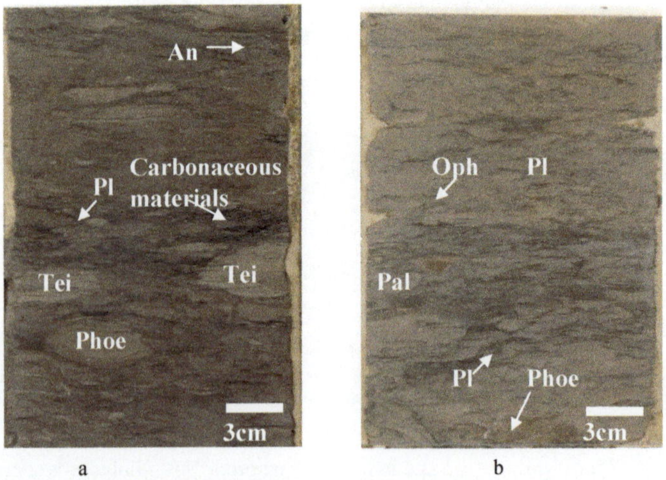

Figure 7: Highly bioturbated, carbonaceous rich muddy sandstone with elongated burrows of *Planolites, Teichichnus, Palaeophycus, Ophiomorpha, Ancorichnus* and *Phoebichnus* ichnofabrics aligned horizontal. A is at depth 1718.2 m while B is depth 1719.1 m.

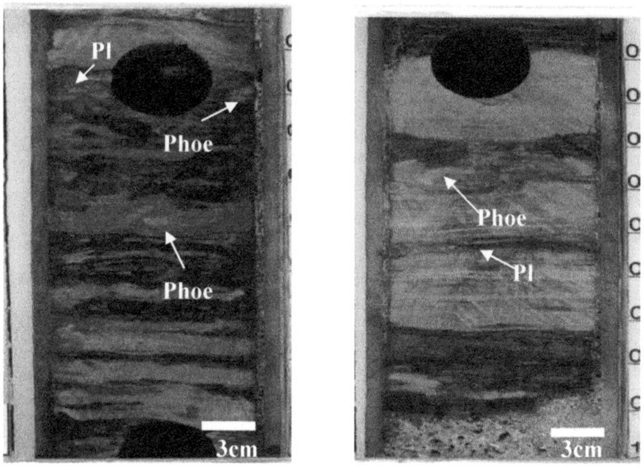

a b

Figure 8: Laminated sandy mudstone showing intercalation of sandstone and indurated mudstone layers 1498.7 m (a) and 1498.4 m (b) within Ewan well. *Planolites, Palaeophycus, Ophiomorpha* and *Phoebichnus* ichnofabric were also observed.

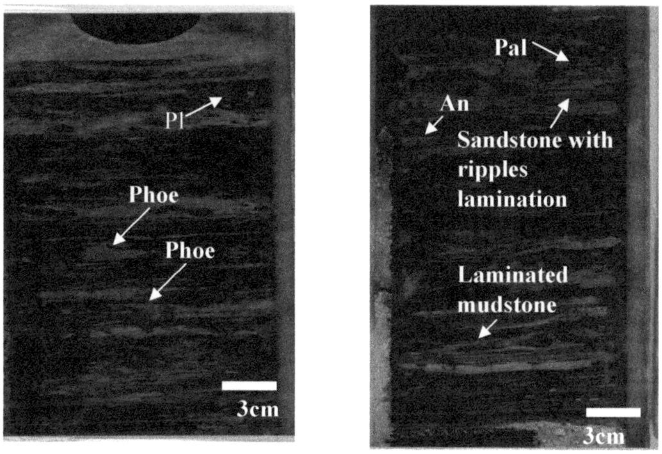

a b

Figure 9: Bioturbated laminated muddy sandstone showing intercalation of sand layer and indurated muddy layers. a and b occurred at depth 1502.2 m and 1502.4 m respectively within Ewan 1 well. *Planolites, Palaeophycus, Ancorichnus* and *Phoebichnus* ichnofabric were also observed.

deposition during fluctuating energy conditions. The environment of deposition is interpreted to be lower shoreface below fair-weather wave base.

- **Heterolithic (HE4) - mudstone/sandstone:** This lithofacies is sandstone dominated (>40:60 mudstone/sandstone), and characterised by the intercalation of indurated, laminated mudstone and sandstone (Figure 9). In this lithofacies, the scale of lamination is also variable, ranging from 0.1 cm- 3 cm thick. The mean mud bed thicknesses is 3 cm and mean sand bed thickness is 0.25 m. The mudstone is characterised internally by parallel laminations while the sands are characterised by horizontal, current or wave ripple lamination and lenticular bedding. The mudstone bands are generally cm scale. A Planolites, Palaeophycus, Ancorichnus and Phoebichnus ichnofabric were observed within the facies. The environment of deposition is interpreted to be lower shoreface below fair-weather wave base deposited during fluctuating energy condition.

Sandstone facies: The sandstones were classified into four lithofacies based on the degree of bioturbation, sedimentary structures and diagenetic imprint (ferruginisation and nodules) observed within the core. Bioturbation, primary structures and diagenetic imprint preserved within these cores have provided some useful information on the hydrodynamic conditions during deposition.

- **Sandstone (SA1) - intensively bioturbated sandstone with remnant bedding and ichnofabric textures:** This lithofacies occurred in both the Ewan 1 and Oloye 3 wells and is characterised as brownish sandstone, with remnant bedding structures (cross stratification- possibly hummocky cross stratification (HCS?) and planar lamination) (Figure 5), and a composite ichnofabric. The ichno-diversity is high with a Teichichnus, Ancorichnus, Palaeophycus, Planolites, Skolithos, Ophiomorpha and Phoebichnus ichnofauna with Planolites, Phoebichnus and Teichichnus as dominant ichnofabrics (Figure 10). The result is that the sediment is completely bioturbated and most of the original sedimentary structures are not visible. The style of burrows varies from horizontal, vertical and inclined to homogenised structures. The presence of cross lamination may indicate storm deposition [13,14]. The intensity of fair-weather burrowing is variable, depending upon the degree of storm dominance. High intensity and diversity of

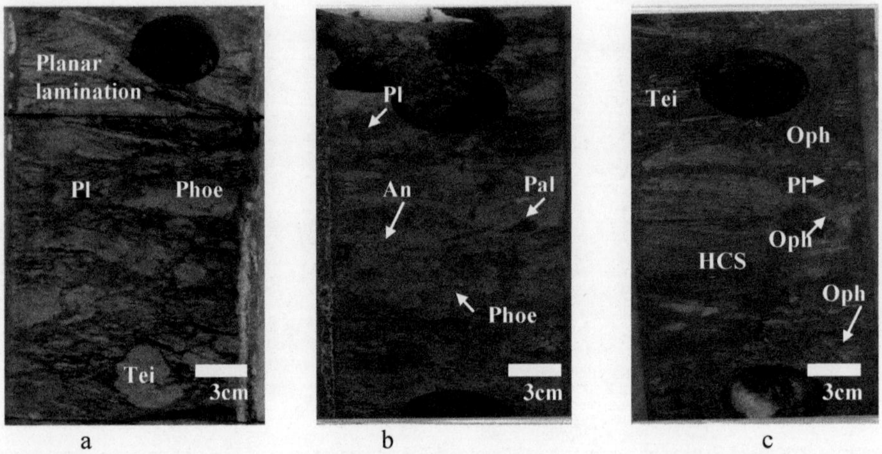

Figure 10: (a). Completely bioturbated carbonaceous rich sandstone, dominated by *Planolites* and *Teichichnus* ichnofabric. Oloye well, depth 1722.1 m.
(b)Highly bioturbated carbonaceous fined grained sandstone with remnant bedding and ichnofrabics of *Planolites* (dominant) and *Ancorichnus* (An) ichnofabric texture (Ewan 1 well, depth 1501.1 m).
(c) Highly bioturbated, carbonaceous fine grained sandstone with remnant of parallel, wavy bedding and hummocky cross stratification, *Ophiomorpha* (Oph) trace and *Planolites* (Pl) ichnofabric, (Ewan 1 well, depth 1497.8 m).

burrows in the facies favour a low intensity and low frequency storm event. Also, high diversity of ichnofauna indicated episodic flow of energy regime (low energy regime and low sedimentation rate). The environment of deposition is interpreted to be shallow marine (lower shoreface).

• **Sandstone (SA2) - intensively bioturbated with no bedding structures (mottled sandstone):** This lithofacies is characterised as a highly bioturbated, mottled sandstone. Primary sedimentary structures are not visible; this may be due to the original homogenous nature of the sediment or the intensity of the organism activity (Figure 11). Some of this lithofacies is brecciated (Figure 2) and lithified which may reflect high energy regime. High diversity of ichnofauna Skolithos, Ophiomorpha, Planolites, Phoebichnus and Palaeophycus, and the presence of Thalassinoides (firmground) indicate a break in sedimentation-episodic regime. The depositional environment is interpreted to be marine dominated condition, slope channel sand. Constrained with shoreline position and gross depositional environment (GDE) map from the seismic survey (Appendix)

• **Sandstone (SA3) – moderately-slightly bioturbated, thinly bedded sandstone with ripples and cross lamination:** This lithofacies is characterised a sandstone containing parallel lamination, moderate-heavy bioturbation, ripples and tangential cross lamination-TCS (Figure 12). Thin laminations of carbonaceous material are also recorded. Thin clay laminae pick out simple and wavy bedding. The occurrence of wavy bedding and thin clay laminae indicate a period of quiescence within a higher energy wave/current regime [15]. This facies was only observed in Oloye well. The diversity in ichnofabric is moderate with Phoebichnus, Skolithos, Ophiomorpha, Ancorichnus and Arenicolites. The Planolites are smaller in size, indicating environmental stress caused by the fresh water influence and terrestrial input.

The environment of deposition is interpreted to be low-medium energy setting, marine dominated slope channel sand (lower regime). Constrained with shoreline position and gross depositional environment (GDE) map from the seismic survey (Appendix).

• **Sandstone (SA4) - with carbonate nodules:** This lithofacies is characterised by sandstone containing carbonate nodules (1-3 cm diameter), Skolithos and clasts of remnant Planolites ichnofabric with no preserved bedding structures (Figure 13). The lack of bedding may be caused by reworking. The nodules are somewhat elongated in a near horizontal direction.

The presence of the nodules is indicative of diagenesis after deposition of the sediment i . a period of reduced clastic input. This suggests a period of lithification and diagenesis, which may be an indication of break in sedimentation caused by relative sea level fall, and possible proximity to the sediment-water interface. The presence of Planolites within the lithofacies and its smaller sizes indicate reduced salinity and stressful environment. The depositional environment is interpreted to be slope channel sand. Constrained with shoreline position and gross depositional environment (GDE) map from the seismic survey (Appendix).

Ichnofabric/Trace Fossil Analysis

Trace fossils and ichnofabric analysis can be employed effectively to aid in the recognition of various discontinuity types and assist in genetic interpretation [16]. By definition, trace fossils or ichnofossils are structures preserved in sedimentary rocks that record biological activity in the form of tracks, trails, burrows and borings. They are invaluable in sedimentological analysis and thus have great importance to the petroleum geologists, sedimentologists and palaeontologists as it helps to determine depositional environments. They are also proving to be important in delineating stratigraphically important boundaries related to sequence stratigraphy, allostratigraphy [17] and event (beds) stratigraphy [18-23]. Event layers are generally regarded as strata that are attributed to rapid, episodic deposition and include such diverse entities as: volcanic ash beds, or tephra deposits [24] beds resulting from seismic shocks, or seismites [25] and episodic sedimentation events such as turbidites [26]; tempestites, or storm deposits [27] phytodetritus pulses [28]; and inundites, or flood deposits [29]. Trace fossils are known to be significant features of most of these deposits.

Commonly, trace fossils are the only evidence of in-situ macrofaunal life in clastic sediments. When combined with the ichnofabric analysis, they can provide palaeonvironmental parameters, which are difficult to obtain by other means. To-date little attention has been paid to the ichnology of the deltaic section in the Tertiary Niger delta basin.

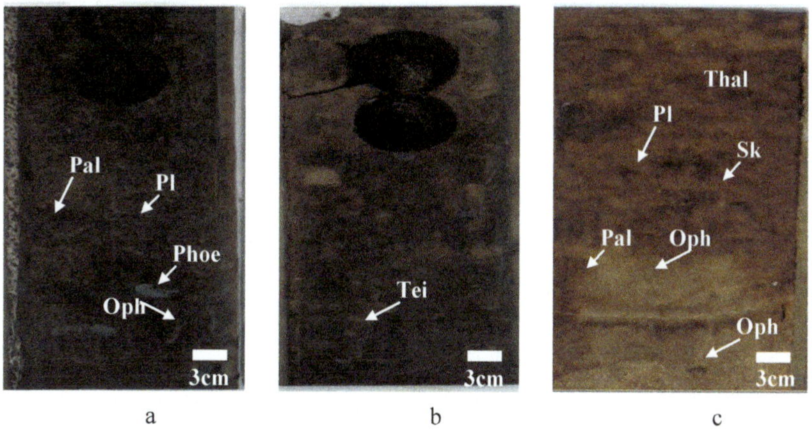

Figure 11: Fine grained sandstone consists of *Phoebichnus* ichnofabric *(Phoe)* and *Teichichnus* trace fossil (*Tei*) with no bedding structures a and b, occurred at depth 1506 m and 1506.4 m respectively within Ewan well.
C) Highly bioturbated, mottled ferruginised, homogeneous fine grained sandstone with *Thalassinoides (Thal) and Planolites (Pl)* ichnofabric, no visible bedding structures, occurred at depth 1723.6 m within Oloye well.

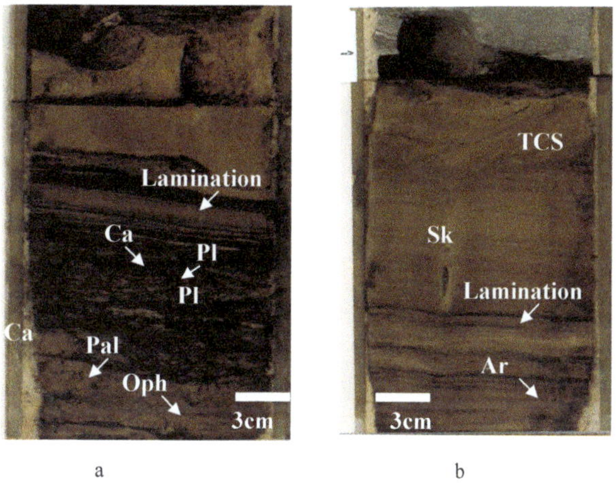

Figure 12: (a) Parallel laminated fine grained sandstone consists of carbonaceous materials (ca) dominated with *Planolites* ichnofabric (*Pl*) occurred at depth 1721.5 m within Oloye well.
(b) Laminated fine grained sandstone consist of carbonaceous materials (ca) occurred at depth 1719.7 m within Oloye well. Tangential cross stratification (TCS), *Skolithos (Sk)* (escaped burrows) and *Arenicolites (Ar)* trace fossils were observed.

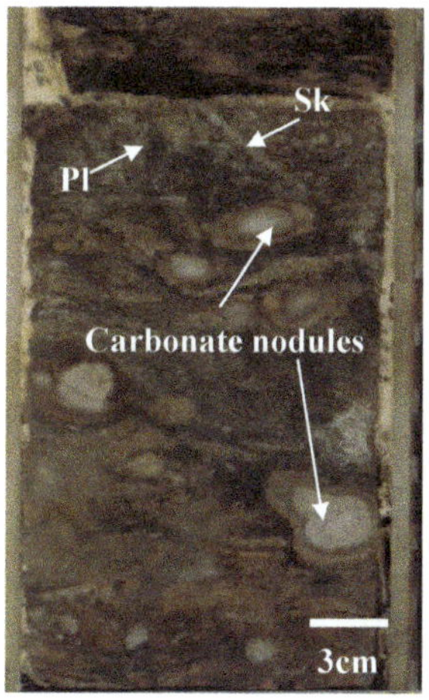

Figure 13: Fine grained sandstone consists of carbonate noodles (concretions) (1-3 cm) diameters and *Planolites* ichnofabric with no bedding structures occurred at 1721.8 m within Oloye well.

A re-evaluation of depositional environments of the Agbada Formation in the northwestern Niger delta (Ewan and Oloye fields) [30] was carried out, incorporating an ichnological description based on analysis of core photos. This has resulted in a revision of the facies scheme originally outlined by [1-5,11]. A summary of the facies scheme within the study area is given in Table 1.

The cored intervals studied within the Ewan 1 and Oloye 3 wells contain diverse assemblages of trace fossils that show a palaeoenvironmental control on their distribution. This has allowed the facies analysis to be refined, when combined with wireline logs and sedimentological interpretation. The diverse assemblages of trace fossils documented from the core studies are listed below. They are: - Planolites, Palaeophycus, Phoebichnus, Teichichnus, Thalassinoides, Skolithos, Arenicolites and Ophiomorpha.

The ichnofabric was assessed using several attributes, such as bioturbation index (Table 2), burrow size, burrow frequency and ichonological diversity. This has led to a greater understanding of the depositional conditions before, during and after deposition of the beds, and is useful in the recognition of surfaces of stratigraphic importance (sequence boundaries and flooding surfaces) e . [31-34]. It is a powerful tool for interpreting biogenic and depositional processes operating during basin fill, most importantly where there is complete bioturbation i . when bioturbation has destroyed all primary sedimentary structures. It is a descriptive technique for information relating to faunal colonization, palaeoenvironmental reconstruction, nature of stratal surfaces and other aspect of sequence stratigraphy [33,34]. By integrating ichnological and sedimentological data, one is able to produce a refined facies model with the aim to assess the ichnology of the study area provide diagnostic criteria for recognizing each ichnofabric through their morphology form core photos and palaeoenvironmental reconstruction.

The description of the ichnofabrics is based on the type examples comparison with corroborative literature using the core photos. Six (6) ichnofabrics are recognized and each of the ichnofabrics has been diagnosed by defining their diagnostic features, the lithofacies in which they occurred, the degree of bioturbation based on [32] (Table 2) and their environmental significance and cross cutting relationships. The degree of bioturbation can be related to the rate of sedimentation [33] and the relative rate of bioturbation have been assessed in comparison to slow, fast, discontinuous and continuous rates of sedimentation. Low or no bioturbation may be due to non preservation of the ichnofauna which may have been caused by rapid sedimentation rates or erosion which removes the colonized area. Moderate to high bioturbation (incomplete bioturbation) may reflect changes in sedimentation rate and fluctuation in the chemical (salinity, oxygen) and physical factors (depth and temperature). Complete bioturbation reflects the condition where bioturbation exceeds sedimentation rates under a conducive and favourable colonized area. Trace fossils are good indicators of oxygenation in sediment. Their diversity, composition and size change with oxygenation changes [35]. The degree of diversity have also been inferred with environmental stress and variability [36] and the trends in ichnodiversity with respect to salinity, oxygen and substrate consistency have been highlighted by Ekdale [35] and their controls on ichnodiversity (Figure 14).

Tiering analysis within the studied area was also assessed, i . the distribution of organisms (and their traces) below the sediment surface based on the tier arrangement proposed by Taylor and Goldring [33,37]. The preservation of a tiering profile relies upon rapid burial of the community. This is because, with continuous deposition, deeper burrows tend to overprint the shallower ones. Recognition of tiering patterns within the studied interval was based on the classification of tiering patterns by Taylor and Goldring [33,37]. There are six possible tier arrangements proposed by Taylor and Goldring [33,37]. They are: - primary colonization, successive colonization, simple tiering, complex tiering, gradational overlap and omission overlap (Figure 15). However, based on the tier arrangement by these authors, simple and complex tiering are the observed tiering patterns from the studied area. This reflects gradual aggradation or degradation of the substrate leading to overprinting as a stable ichnofauna structure develops. These types of tiering have been discussed by Wetzel [38] and Bromley [39] in deep-sea sediment. These types of tiering arrangement indicate a stable environment [33,37]. In this study the choice of name for the ichnofabrics reflects the dominant traces (Table 1). The identified fabrics have been interpreted to indicate a shallow marine setting during the basin fill (deposition).

In summary, six ichnofabric have been recognised from the core

Grade	% Bioturbation	Classification
0	0	No bioturbation
1	1-5	Sparse bioturbation: few discrete traces and/or escape structures
2	6-30	Low bioturbation: bedding distinct, low trace density, escape structures often common
3	31-60	Moderate bioturbation: bedding boundaries sharp, trace discrete
4	61-90	High bioturbation: bedding boundaries indistinct, high trace density with overlap common
5	91-99	Intense bioturbation: bedding completely disturbed limited reworking due to repeated overprinting
6	100	Complete bioturbation: sediment reworking due to repeated overprinting

Table 2: Bioturbation index (BI) based on the amount of reworking with respect to the original sedimentary fabric [32].

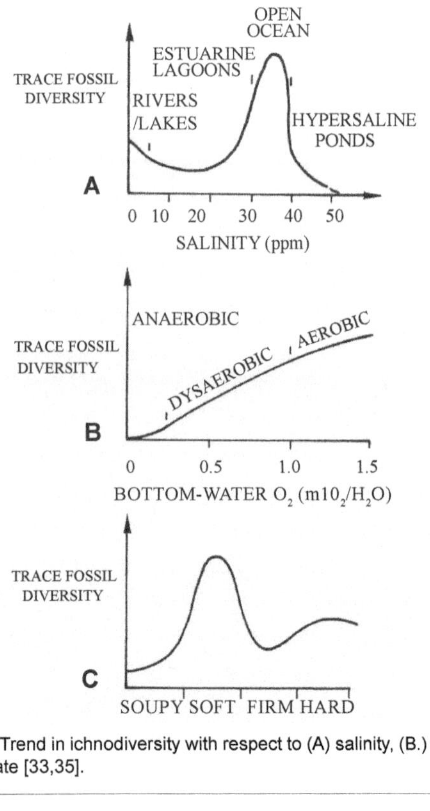

Figure 14: Trend in ichnodiversity with respect to (A) salinity, (B.) oxygen and (C.) substrate [33,35].

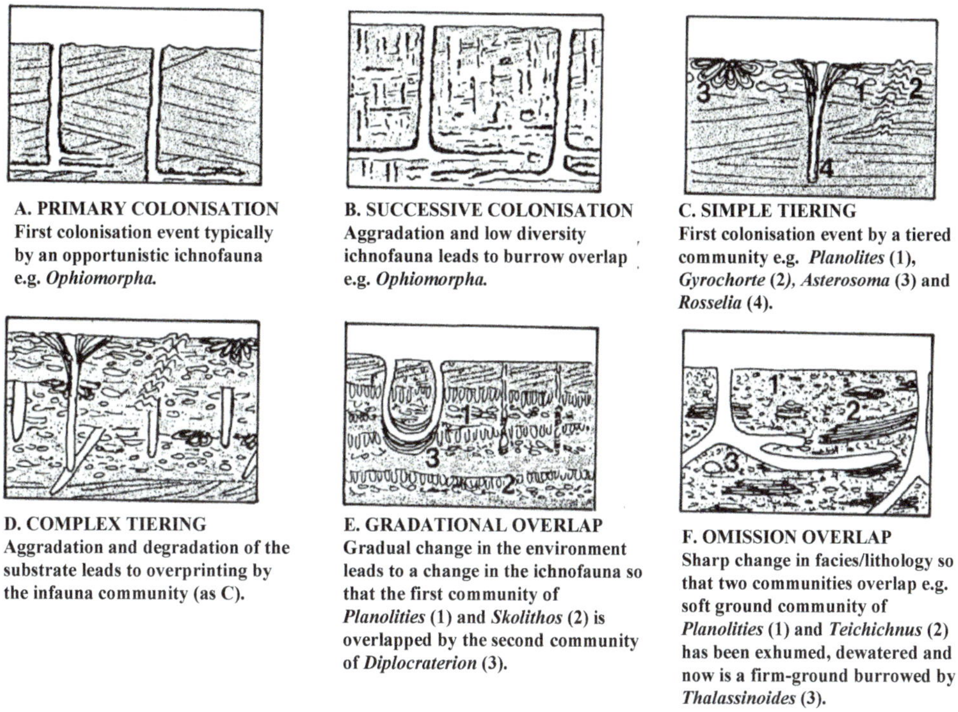

Figure 15: Diagrammatic illustration of the six possible styles of tiering (primary colonization, successive colonization, simple tiering, complex tiering, gradational overlap and omission overlap) as proposed by Goldring [37].

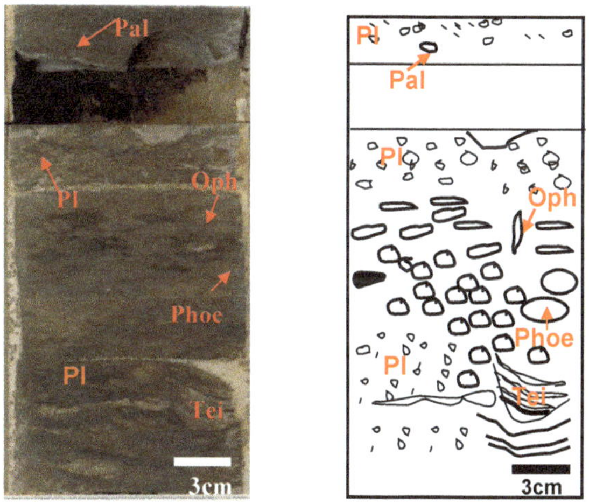

Figure 16: Completely bioturbated, carbonaceous muddy sandstone. This has been burrowed by 3cm irregular, retrusive *Teichichnus (Tei)*, and 0.05 - 0.1cm elongated burrows of *Planolites (Pl).*

intervals and are described below.

Planolites/Teichichnus ichnofabrics

Description: This composite ichnofabric is found in association with heterolithic facies (HE2) within slope channel sand. The beds are carbonaceous rich muddy sandstone with diverse assemblages of horizontal burrows showing disruption due to faunal activities, No distinct sedimentary structures are observed due to the intense bioturbation of trace fossils. The associated ichnofauna comprises an opportunistic colonization of Palaeophycus (Pal), Ophiomorpha (Oph), Skolithos (Sk) and Phoebichnus (Phoe) (Figure 16).

Ichnology: Bioturbation is complete (BI=6) with burrows commonly overlapping. The density of burrowing is high with five types of trace fossils recorded. The Planolites burrows show elongate forms (0.3-0.5 mm in diameter) and an irregular, retrusive Teichichnus (Tei) ichnofabric (1-3 cm). The associated trace fossils include Ophiomorpha which show thick pellet walled horizontal shaft (0.05-0.1 cm), Palaeophycus (Pal) (lined burrows) which are horizontal-oblique, elongated-ovate small burrows and Phoebichnus (Phoe) which is an elongated burrow. This ichnofabric is characterized by complex tiering.

Environmental interpretation: The high diversity suggests that this ichnofacies is more marine than facies MS1-3, which reflects the conducive environmental conditions, with the traces demonstrating a deposit-feeding mode of life. The presence of Skolithos (Sk) may indicate water movement/agitation and high sediment supply. The depositional environment is interpreted to be marine dominated slope channel fan based on the observed ichnofauna, constrained with shoreline position from the seismic survey (Appendix).

Planolites/Phoebichnus ichnofabric

Description: This ichnofabric contains Planolites (Pl)-Phoebichnus (Phoe) occurring in a heterolithic (HE4) and thinly bedded cross laminated sandstone facies (SA3) (Figure 17). The sediments are typically slightly–moderately bioturbated with a diverse ichnofauna consisting of fully a marine ichnotaxa of Skolithos (Sk), Ophiomorpha (Oph), Arenicolites (Ar) and Palaeophycus (Pal). Phoebichnus consist of elongated burrows (0.5-1.5 mm) while Planolites is represented by small unlined contrasting fill burrows (0.3-0.5 mm in diameter) (Figure 6).

Ichnology: This represents incomplete bioturbation (BI=3-5). The style of burrows is horizontal to elongate showing a high diversity, with four trace fossils recorded. The burrows of Planolites tend to produce horizontal to elongated burrows (0.05-0.1 cm in diameter) and elongated to cylindrical Phoebichnus ichnofabric (1-2 cm).

Environmental interpretation: Based on the above description, the depositional environment in which the ichnofabric is developed is inferred to be a marine setting with moderate to low rate of sedimentation in a lower shoreface below fair-weather wave base depositional environment.

Planolites ichnofabric

Description: The Planolites ichnofabric is composed of exclusively millimetres scale sand/mud filled Planolites, and is common in the sandstone facies of Ewan and Oloye wells. Bioturbation is characterised by incomplete bioturbation, burrowed by Planolites characterised as a small burrow with contrasting fill. Some Skolithos burrows are also observed. The level of bioturbation varies from low to moderate (BI=2-4) with few burrows overlapping. The burrow density can be interpreted as moderate consisting mostly of a mono-specific fauna, Planolites. The Planolites burrows are small size with a convoluted form. They are horizontal to oblique, simple flattened cylinders, which are 0.03-0.1 cm diameter. Carbonaceous materials are found as a thin bed within the lamination (Figure 18).

Ichnology: Diagnostically, Planolites here differs from Ancorichnus (An) in terms of the fill in the core of the burrows and from Palaeophycus in terms of morphology and wall structure. Palaeophycus tends to be infilled with sediment that is the same as the host matrix, whereas the sediment filling unlined Planolites tends to differ in composition [12]. The Planolites burrow fill differs from the host material indicating an active back filling of sediment by expulsion by a mobile deposit feeder that occurs in many environments and is probably produced by a number of different organisms [12,40].

Environmental interpretation: Based on the low diversity of trace fossils, the presences of carbonaceous rich material with cross and parallel lamination, the environment is interpreted to be medium-high energy settings, lower shoreface and slope channel fan complexes. It is noted that this ichnofabric is different from Planolites-Teichichnus ichnofabric in term of size. Planolites observed in this lithofacies are smaller and the small size of Planolites could be due to environmental stress. Low diversity of the ichnofauna may be due to influx of fresh water into the environment. The presence of Planolites within the diagenetic sandstone nodules suggests reduced salinity, and stressful depositional environment.

Planolites/Thalassinoides ichnofabric

Description: This composite ichnofabric is found within the intensive bioturbated sandstone facies with no remnant bedding structures observed (SA2). The sediment was colonized by Planolites and overprinted by Thalassinoides (Thal). They are characterized as completely bioturbated, reddish brown, mottled, fine grained sandstone with patches of carbonaceous material. It is difficult to discern sedimentary structures as the sandstone is homogeneous as a result of the bioturbation and ferruginisation. The trace fossil fabric is composed mainly of Planolites, comprising oval-elongated sand filled burrows with a mottled firm ground Thalassinoides ichnofabirc. In this core, the fabric of Planolites and Thalassinoides consists of elongated burrows ranging 0.05-1 cm diameter (Figure 19).

Ichnology: The bioturbation is complete (BI=6). The burrow

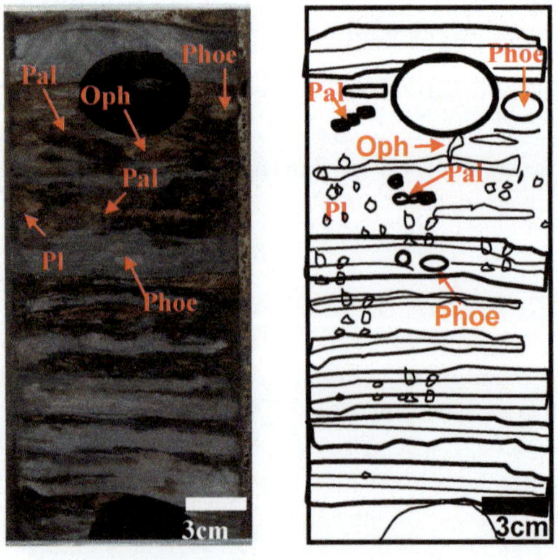

Figure 17: Incomplete bioturbated, carbonaceous mudstone dominated heterolithic. Burrowed by 1-2 cm diameter cylindrical burrows of *Phoebichnus* (*Phoe*) and 0.05-0.1 cm sand filled burrow of *Planolites* (*Pl*).

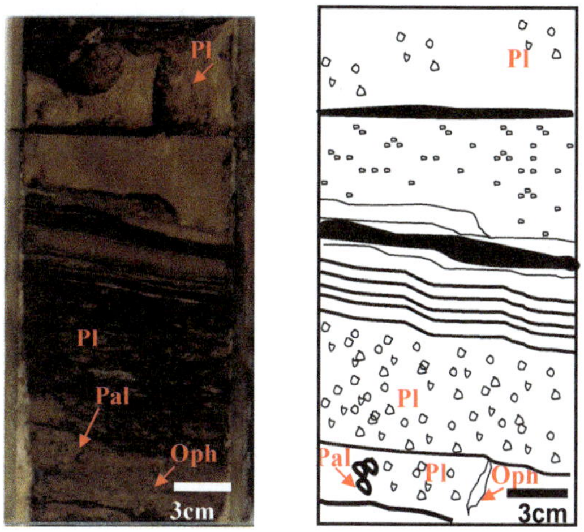

Figure 18: Incompletely bioturbated, carbonaceous-rich, fine grained sandstone which exhibits mm- scale planar laminations. This has been burrowed by 0.03 - 0.1 cm diameter, sand filled burrows of *Planolites* (Pl). This is an example of primary colonisation.

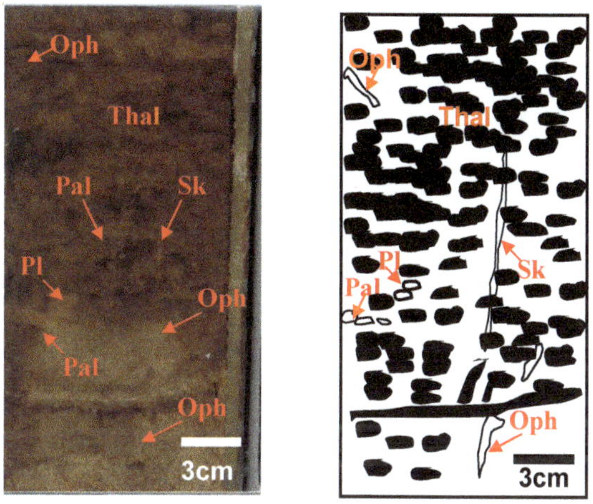

Figure 19: Completely bioturbated, mottled, ferruginised sandstones. Burrowed by 0.05 – 0.1cm sand filled *Planolites*, 0.1-0.2cm *Thalassinoides* and vertical burrow *Skolithos*. The presence of *Thalassinoides* indicate break in sedimentation and early cementation (firm ground). This suggests that the energy regime is episodic.

density and diversity is high, and consists of Planolites, Skolithos, Ophiomorpha, Phoebichnus, Palaeophycus and Thalassinoides. The diagnostic feature of Thalassinoides is the burrows have no fill structure which allows possible distinction between large Planolites that are an unlined actively filled burrow system [12].

Environmental interpretation: The presence of the Thalassinoides fabric results from processes operating in a marine environment and it occurs in a great variety of marine environments. The Thalassinoides ichnofabric has been recorded from fresh water fluvial conditions in fully marine shorefaces [41] in the upper Jurassic Fulmar Formation continental shelf [33], in the offshore transition zone below storm wave base level of the Nukul Formation, Suez Rift, Egypt [42] and is considered to be environmentally non diagnostic [43].

Based on this core, the associated lithofacies and context, the Planolites/Thalassinoides ichnofabric environment is interpreted to be marine dominated slope channel sand, constrained with shoreline position from the seismic survey (Appendix). This ichnofabric indicates increasing substrate cohesion from softground condition into firmground condition. High diversity of trace fossils within the facies and presence of Thalassinoides indicates a break in sedimentation that has led to early compaction (firmground), developed from initial softground. This suggests that the energy regime is episodic.

Teichichnus/Phoebichnus/Planolites ichnofabric

Description: This ichnofabric is found in association with highly bioturbated sandstones containing remnant bedding structures (SA1). The level of bioturbation is high and the sediments are characterised as being incomplete to complete bioturbated, greyish, fine-grained, carbonaceous muddy sandstone. It is dominated by an irregular, retrusive Teichichnus ichnofabric, comprising cylindrical burrows of Phoebichnus and elongated burrows of Planolites. They are associated with variety of other trace fossils which include, Palaeophycus, Ancorichnus, Ophiomorpha, Skolithos and Planolites. It is difficult to recognize any original sedimentary structures either due to the high level of bioturbation or the original homogeneous nature of the sediment (Figure 20).

Ichnology: The level of the bioturbation can be classified as complete bioturbation with burrow overlap (BI=6). The burrow density

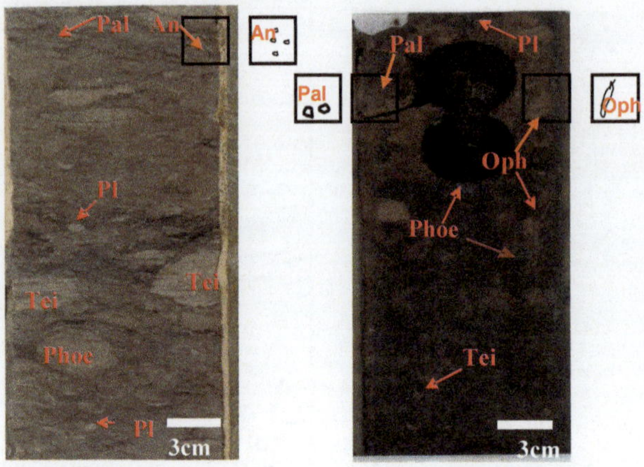

a b

Figure 20: (a) Completely bioturbated, carbonaceous muddy sandstone, (b) incompletely bioturbated fine grained sandstone. This has been burrowed by 0.5-2 cm irregular, retrusive *Teichichnus*, 0.5-2 cm cylindrical burrows of *Phoebichnus* and 0.05-0.1 cm elongated burrows of *Planolites*.

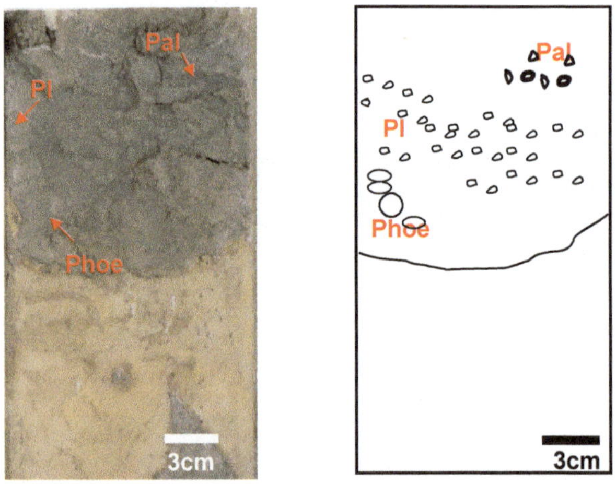

Figure 21: Completely bioturbated, structureless and highly disrupted sandy mudstones. This has been burrowed by small ovate–elongated 0.3-0.5 mm diameter sand filled burrows of *Planolites* and *Palaeophycus*.

is high with a high level of diversity of the ichnofauna. The style of burrows varies from horizontal, vertical and inclined to homogenised structures.

Environmental interpretation: This type of ichnofabric has been recorded within the sand siltstone event beds from the middle Jurassic Not Formation, Norwegian continental shelf [33]. But in this work, these ichnofabrics are recorded in the lower shoreface. The intensity of bioturbation and the diverse nature of the assemblages suggests episodic flow of energy regime and sediment reworking.

Palaeophycus-planolites ichnofabric

Description: Completely bioturbated, disrupted, greyish sandy mudstone with high density of Palaeophycus, Phoebichnus and Planolites showing low ichnofauna diversity. The sandy mudstone lacks any discernable sedimentary structures due to the intensity of bioturbation. Morphologically, Palaeophycus has been differentiated from Planolites by the burrow infill because of their similarity in burrows style. The Palaeophycus infill tends to be the same material as the host matrix, and is interpreted as an open tube produced by carnivorous of omnivorous invertebrates, mostly polychaetes, whereas that of sediments filling unlined Planolites tends to differ [12]. Within these beds, the Palaeophycus burrows are oriented horizontal-oblique, elongated-ovate small size burrows with fill similar to the host (Figure 21).

Ichnology: This facies is classified as complete bioturbated (BI=6). Burrows intensity is high but diversity is low, being restricted to Palaeophycus, Phoebichnus and Planolites.

Environmental interpretation: Palaeophycus ichnofabric has been recorded in the middle Jurassic Not Formation, Norwegian continental shelf, Palaeophycus trace fossil has also been recorded in the offshore transition zone below storm wave base level of Nukul Formation, Suez Rift, Egypt [42]. In this study area, Palaeophycus and Planolites have been encountered in the lower shoreface environment.

Summary Interpretation of Ewan 1 Well Cores

The thickness of the cored interval is 15 m. The sediments are characterized as moderately to intensively bioturbated, fine grained muddy sandstones, sandy mudstones and mudstone that contain a pervasive and irregular ichnofabric composed of burrows of Skolithos, Ophiomorpha, Arenicolites, Planolites, Palaeophycus, Ancorichnus,

Thalassinoides and Teichichnus. Based on the bioturbation index (Table 2), the core interval can be classified as incompletely-completely bioturbated (BI=3-6) [31,32].

This cored interval consists of dominantly sandstone facies (heavily bioturbated sandstone with remnant bedding structures (hummocky cross stratification, wavy bedding and plannar lamination) and ichnofabric texture (SA1), heavily bioturbated homogeneous/mottled sandstone with no remnant bedding (SA2), laminated mudstone dominated heterolithic facies (HE3), laminated sandstone dominated heterolithic (HE4) and laminated indurated mudstone (MS2). The lithofacies within this core are moderately to completely bioturbated. Where they are highly bioturbated the intensity suggest that the sand bodies could be shoreface attached rather than sand shoals i . high energy regime where currents energy are prevailing. This would allow enough time for sediment to be reworked by the ichnofauna as a result of the reduced energy regime.

In this cored interval, the bioturbated sandstones facies (SA1) are interbedded with bioturbated mudstones (MS2) and heterolithic facies (HE4) (Figure 2). The sandstone beds are commonly 1-2.5 m thick with remnant bedding structure overprinted by ichnofabric textures. The cross stratifications, wavy bedding and planar cross beddings observed are of small scale (about 0.6 m bed set thickness). No medium or large scale cross bedding was observed this may be due to the dimension of the core. This may also be a function of grain size control, with cross bedding forming mainly in medium–and coarse grained sand [44].

The cored interval, when located on the wireline log is characterised by being within a relatively continuous coarsening upward facies succession (average 3 m thick), dominated shoreface deposits. There is some degree of facies variability, both sedimentologically and ichnologically. Variability of facies and trace fossils may reflect a combination of water turbidity, water salinity, fluctuating/episodic depositional rate and storm energy condition [45].

Intercalation of sandstone facies with the heterolithic facies reflects deposition during fluctuating energy conditions and their repetitive pattern likely reflects relative changes of sea level controlling the sedimentation. The occurrence of parallel and cross bedding populated by shallow marine ichnofauna also suggests shallow marine sediment deposited within lower shoreface zone under wave and storm influence. The presence of carbonaceous material indicates a terrestrial input. A flooding surface was recognised within this interval on top of facies SA1 at a depth of 1508.4 m (Figure 2), overlain by facies MS2. This surface suggests an increase in water depth (deepening) as a result of relative sea level rise.

Summary Interpretation of Oloye 3 Well Cores

The overall thickness of this cored interval is 11.4 m. The lithofacies are characterized as slightly to completely bioturbated, fine to muddy sandstone, sandy mudstone and mudstone, with medium to high bioturbation (BI= 2-6). The identified trace fossils and ichnofabrics texture are Planolites, Thalassinoides, Palaeophycus, Phoebichnus, Teichichnus Skolithos, Arenicolites and Ophiomorpha. Inability to decipher the ichnofossils in some of the core interval within this well is due largely to the sediment distortion caused by intensive bioturbation and the homogeneous nature of the substrates. These cores show a general upward-fining motif.

The basal part of the core log (Figure 2), shows a brecciated, intensively bioturbation sandstone with Planolites, Palaeophycus, Phoebichnus, Teichichnus, Thalassinoides and Skolithos trace fossils dominated by Planolites and Thalassinoides ichnofabrics. High diversity of ichnofauna and the presence of Thalassinoides indicate a break in sedimentation-episodic regime (firmground). There is some iron staining but no distinct bedding structures (SA2). The infill within the Planolites burrows differs from the host materials, which may indicate an active backfilling of the sediment by injection and expulsion by the mobile deposit feeders [12]. The burrows are about 0.2 mm-0.75 mm in diameter. The brecciation indicates early lithification followed by erosion in a high-energy deposition environment. Some of the observed burrows are elongate and aligned parallel to bedding; this may suggest compaction after deposition. The thickness of this unit (SA2) is about 3.3 m.

On top of this is a 0.9 m thick sandstone (SA1) characterised by a composite ichnofabric that formed as a result of the activities of many organisms. The dominant trace fossils observed are Planolites, Teichichnus and Phoebichnus ichnofabric and they range between 0.25 mm-2 cm in diameter with Planolites the smallest diameter. The interval is completely bioturbated (BI=6) and none of the original sedimentary structures are visible. This is overlain by unit (SA4) which is about 0.3 m thick and characterised by containing carbonate nodules. The occurrence of these nodules (0.5 mm-1.5 cm in diameter) indicates a diagenetic overprint which may suggest a period of reduced clastic input and accommodation initiated by sea level fall. The large-scale stratigraphic setting of this unit suggests that nodule formation is linked to clastic sediment bypass. The lack of accommodation caused by the relative fall in sea level in this setting is supported by this presence of carbonate nodules. Based on these observations the top of this facies is interpreted as a candidate sequence boundary (~ depth 1721.5 m) (Figure 2).

Above the sequence boundary about 2 m of sandstone facies (SA1) is characterized by thin beds (0.3 m) of highly bioturbated sandstone with remnant cross-bedding structures and Teichichnus-Phoebichnus-Planolites ichnofabric (0.1 mm-2 cm in diameter). This is followed by about 1.7 m of slightly bioturbated carbonaceous, ferruginised, homogeneous, ripples and cross laminated sandstone facies with Planolites, Palaeophycus-Phoebichnus, ichnofabric (SA3). At the top of this facies, gently dipping trough cross stratifications occurs.

Above SA3 is about 2 m of highly bioturbated muddy sandstone with trace fossils (HE2) of Teichichnus, Palaeophycus, Planolites, Ophiomorpha and Phoebichnus, dominated by Planolites-Teichichnus ichnofabric. The top of the 2 m thick HE2 facies is marked by a flooding surface (transgression) denoted by a bioturbated, laminated indurated mudstone with rare sand streaks (MS2).

From the base of the sequence boundary, the stacking pattern shows a fining upward trend from sandstone to muddy sandstone.

A 330 m section with no core intervals was recorded at the top of this laminated indurated mudstone (MS2).

The top of the uncored interval is marked as 1395.4 m. It is overlain by highly bioturbated, thinly bedded, carbonaceous, disrupted sandy mudstone of about 1 m (HE1). The top of this facies marks another flooding surface (deepening) overlain by about 1.2 m of mudstone facies MS3 (lenticular bedding and laminated mudstones with desiccation cracks?). The presence of lenticular bedding and desiccation cracks within this facies is interpreted to provide evidence of tidal conditions and salinity fluctuation [12].

Overlying this mudstone facies is a repetition of (HE1) follow by bioturbated, laminated, friable, carbonaceous mudstone facies with rare siderite (MS1). Intercalation of sandy mudstone and mudstone facies that occurred at the upper part of the core log reflects deposition during fluctuating energy conditions. The presence of the following ichnofauna (Skolithos, Planolites, Thalassinoides, Palaeophycus Ophiomorpha, Teichichnus and Phoebichnus), sedimentary structures (parallel and cross lamination), suggest that the sediment is deposited in a marine dominated environment. Abundant carbonaceous materials indicate terrestrial input.

In the upper part of the core i . the interval overlying the uncored section, highly bioturbated lenticular mudstone followed by thinly bedded, disrupted sandy mudstones facies displays ferruginisation and iron staining, which suggests that burrowing occurred within a reducing environment. The level of ferruginisation is higher than that observed in the Ewan 1 well. Intensive bioturbation within this core interval indicates medium-high sedimentation rate where the organisms are preserved from re-working. The deposition environments of the core interval below the uncored interval are interpreted as marine dominated slope channel sand. This is constrained with shoreline position and gross depositional environment (GDE) map from the seismic survey (Appendix). The core interval above the uncored section is interpreted as upper shoreface environment with some tidal/fluvial incursion (Figures 22-24).

Wireline Log Facies and Depositional Environment

Six main gamma ray facies association have been recognised (Figure 25) using the wireline gamma ray log motif constrained with the help of core photographs within the interval 1496-1512 m in well Ewan 01 and 1392-1395 m, 1717-1726 m in well Oloye 3 (Figure 2). Each log facies is linked to its depositional environment and thus provides the basis for understanding its three-dimensional geometry and connectivity. Detailed core sedimentological descriptions carried out on core photos within the intervals from the two wells have helped to constrain the paleoenvironments. These investigations have also provided useful information on the structures, diagenetic imprint, trace fossils and ichnofabric textures within the sediments.

Six major wireline log facies association were recognised. They are:-

- Blocky and upward fining
- Blocky and upward cleaning
- High gamma serrated/erratic
- Low gamma serrated log motif
- Blocky sharp base
- Gradual cleaning upward

Wireline log facies descriptions (Gamma)

Several publications have shown the log shapes expected in various facies [46-50]. They all depend on the relationship between log shape and grain size trends in sandstone bodies. Most interpretation of log shapes are based on the gamma logs. The gamma ray measures radioactivity radioactive isotopes within clays, organic matters or

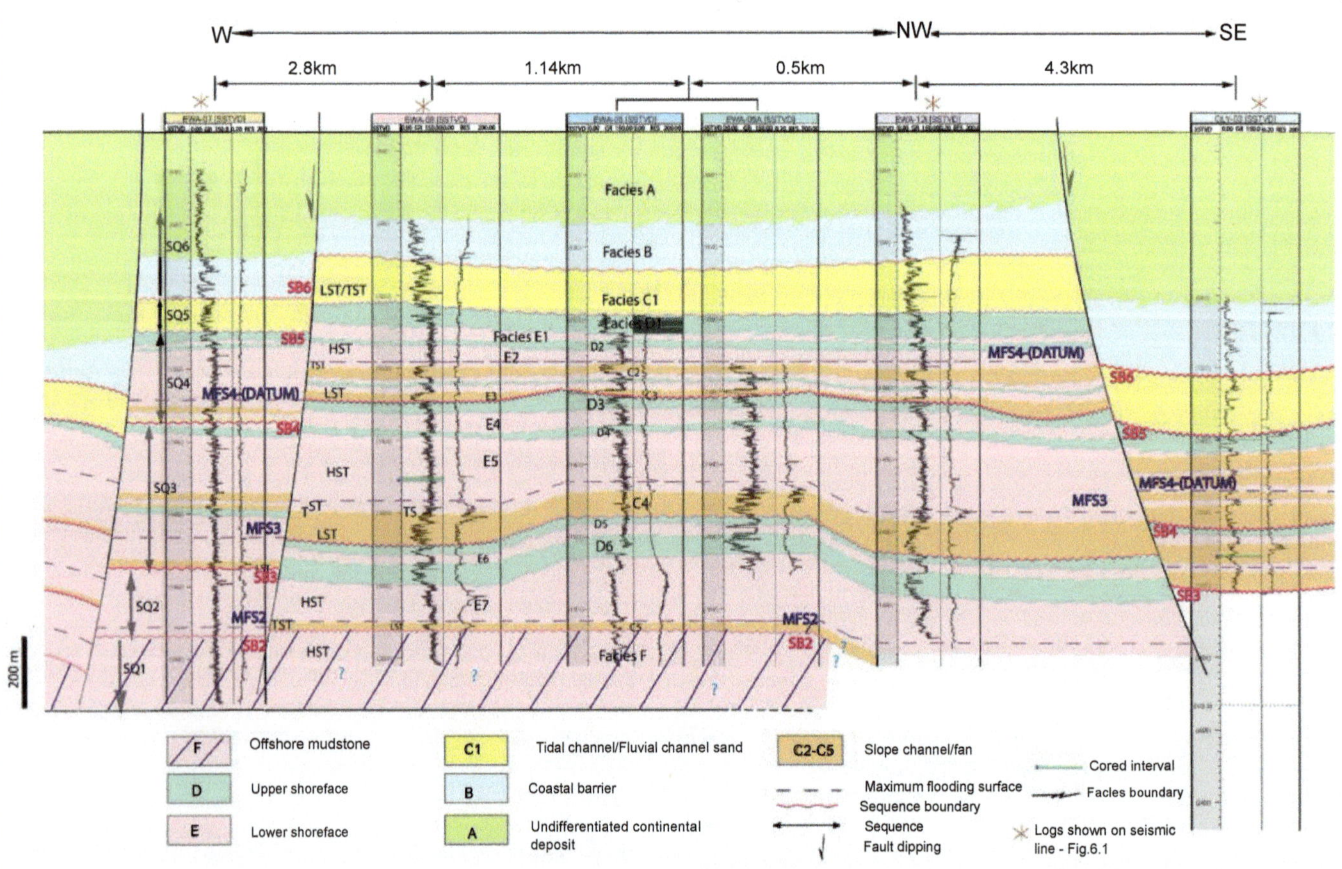

Figure 22: Correlation panel of Ewan and Oloye logs illustrating sequences component, parasequence sets and possible depositional environments from wireline log signatures within the middle Miocene Agbada Formation. Wells location is shown in Figure 23 [30].

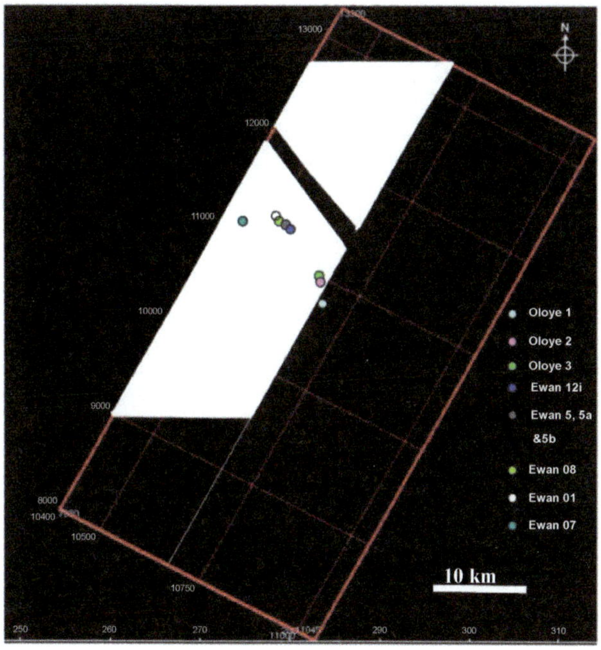

Figure 23: Seismic survey showing area where seismic section covers and position of wells.

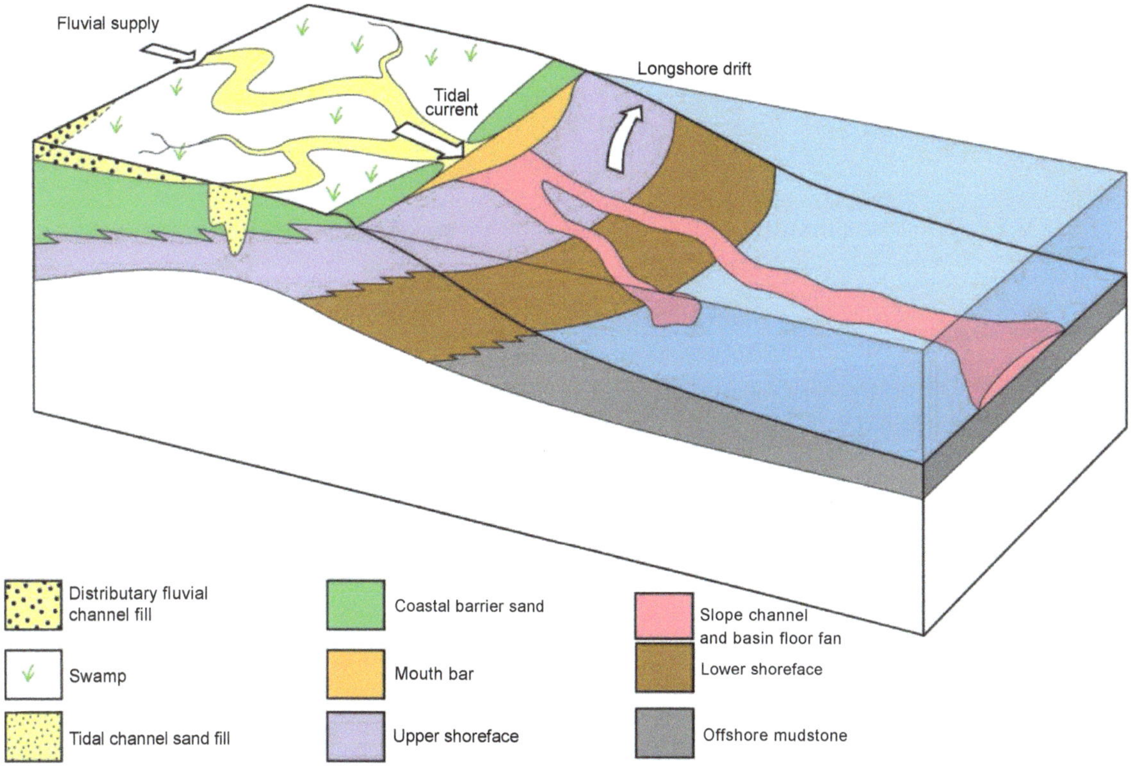

Figure 24: Conceptual Lithofacies Relationship in the Study Area. Based on [2,7].

heavy minerals. From this we may interpret the grain size, and thus deduce log stacking i . the fining/coarsening upward trend etc. A bell shape indicates a fining up sequence which may be an alluvial/ fluvial channel. A funnel shape is a coarsening up sequence which may record deltaic progradation or a shallow marine progradation. The six log facies are described below.

Log facies A: Log facies A is characterised by vertically-stacked, low gamma, serrated wireline log motifs (Figure 22). On a small scale the patterns have a funnel-shaped geometry. The shale baseline is given

	Well log signatures	Log patterns/ motifs 0 API 150	Lithofacies Descriptions	Lithofacies Interpretations
A	Low serrated gamma	I 20m	Stacked up of bell shape (coastal barrier bar) and funnel shape (fluvial/tidally influence channel sand	Undifferentiated Continental deposits (Benin Fomation)
B	Blocky and cleaning upwards	I 20m	Funnel shape morphology with average thickness of about 50m bounded at the basal part by candidate sequence Boundary or transgressive surface	Coastal barrier bar
C1	Blocky and dirtying upwards	I 20m	Bow/ bell shape morphology, shows some high gamma values anormalies, the top units marked the transgressive surfaces while the basal marked the sequence boundaries.	Tidal/Fluvial channel fills
C2-5	Blocky sharp base	I 20m	Blocky and sharp base shape morphology, shows some high gamma anomalies within the clean sand, The top and the basal contacts correspond to transgressive surfaces and sequence boundary respectively	Slope channel/Channel sand
D	Gradual cleaning upwards	I 20m	Funnel shape morphology and less blocky than coastal barrier bar. The upper contact is defined by the transgressive/ flooding surfaces while the basal part is defined by shallow marine deepening.	Upper shoreface sand
E & F	High serrated gamma	I 20m	High serrated gamma ray values with aggrading stacking patterns this can differentiated form each other by their neutron and density log characters	Offshore mudstone/ Lower shoreface deposit

Figure 25: Gamma ray facies association from well log pattern used in defining depositional environments within the study area [30].

at a value of 95 API. The gamma ray values for this log facies range from 20 to 35 API. This facies is sandy as revealed from the relatively low gamma ray responses (25-30 API), and commonly contain a series of relatively stacked coarsening-upward facies succession. These stacked coarsening-up units are usually less than twenty five metres thick. In this log facies, no abrupt changes were observed. The spikey gamma ray value within this interval may be caused by the presence of mica (heavy minerals) within the sandstone bodies.

This is facies was not observed within the cored interval that would allow correlation with core photos. The environment of deposition is interpreted to be a continental fluvially/tidally influenced deposits (denoted as undifferentiated continental sand in this study), the overall geometry exhibiting a coarsening-up, progradational stacking pattern.

Log facies B: Log facies B is characterised by a blocky and cleaning-upward log motif with some high serrated gamma values within the cleaning-upward sand (Figure 22). The combined thickness of this lithofacies is about 110 m. vertically stacked. The morphology of the log pattern is characterised by a funnel shape. The gamma ray values range from 20 to 80 API with the shale baseline marked at 95 API. There is abrupt change of gamma ray values within the cleaning-upward log motif, this changes may be related to lithology e . changes in mica content within the sandstone bodies. Upwardly, this lithofacies transits gradually into stacked up fluvial/tidally influence channels sands (undifferentiated continental deposit). This is facies was not observed within the cored interval that would allow correlation with core photos Based on the log motif and stacking patterns, this lithofacies has been interpreted to be coastal barrier sand. The overall geometry is pradational.

Log facies C: This log facies is characterised by a dirtying upward log signature with a bow/bell shape and a blocky gamma ray log signature (Figure 22). This bow shape may give a clue on the position of maximum deposition which may be at the centre of the bow shape [51]. Five units of this log facies were characterized and interpreted from given gamma ray logs within the study area. They are denoted by facies C1, C2, C3, C4 and C5. The thickness of each unit is 100 m, 25 m, 20 m, 60 m and 20 m respectively. Gamma ray values range from 20-80 API. The log shape of log facies C1 and upper part of unit C4 are characterised by bow/bell shape while C2, C3 and C5 are characterised by a blocky sharp base. Gamma ray log signature anomalies, i . high gamma within this clean sand may be due high mica content within the sandstone facies.

The geometry and relationships of the basal part log facies C shows a sharp contact followed by concave-up pattern, which is interpreted to define an erosion surface. The base of units C1, C2, C3, C4, C5 are all interpreted to be a candidate sequence boundary, and apart from unit C1, all have a transgressive surface at the top. Unit C1 is cut by a major candidate sequence boundary, and probably the transgressive surface has been eroded at this location.

Correlation of this log facies with the Oloye 3 core photos at interval 1717.3-1725.5 m confirms the lithofacies as a micaceous fine-grained sandstone. Based on core sedimentology and the log stacking pattern, constrained with shoreline position from the seismic survey (Appendix), log facies C2-C5 is interpreted as slope channel sand with a possible erosive base. While log facies C1 is interpreted as tidal/fluvial channel sand with possible erosion base.

Log facies D: Log facies D1, D2, D3, D4, D5 and D6 are characterised by a cleaning-upward log motif. The average thickness of this log facies is about 40 m. The log pattern is characterised by a funnel shape. The shale baseline is given at a value of 95 API with gamma ray values ranging from 20 to 45 API, so this facies is interpreted to be sandy. There are some small gamma ray log signature anomalies i . high serrated gamma within the log facies may indicate the presence of mica or feldspars within the sandstone.

Based on correlation with core sedimentology from interval 1392.7-1395 m in well Oloye 3, which is the lateral equivalent of unit D1 in the Ewan wells, (Figure 5), this log facies has been interpreted to be a prograding-aggrading shoreface (upper-lower shoreface).

Log facies E: The log facies E1, E2, E3, E4, E5, E6 and E7 are characterised by having a sharp base and a highly serrated gamma log signature, with gamma ray values of about 90 API. The shale baseline is about 95 API (Figure 22); therefore this is interpreted as a heterolithic (muddy sandstone). The average thickness varies from well log to the other. Gamma ray log signature anomalies i . excessive high serrated gamma values within the log facies E may also indicate the mica and feldspars content within the facies.

As observed from the well logs, Maximum Flooding Surfaces are located within this log facies (Figure 22). This log facies can be correlated with core photos from Ewan 1 well, which is a lateral equivalent of Ewan 08 well (interval 1496.6-1511.5 m) (Figures 22 and 23). Based on the wireline log signatures and core descriptions, the lithofacies have been interpreted to be a muddy sandstone.

This is interpreted to be prograding shoreface (upper shoreface) and aggrading shoreface (lower shoreface) deposits that are characteristic of highstand system tracts [44]. It is observed from the well data that the sub-lithofacies Es thins seaward as observed from Oloye 3 which may imply a coastal environment/flooding event (retrogradation).

Log facies F: This log facies is characterised by a high serrated gamma ray log motif. This facies is mudstone as revealed from the relative high gamma ray responses (> or = 95 API). This log signature can be distinguished from Log Facies E as it has a higher gamma ray reading, higher that or equal to 95 API that defines the shale base line for this interpretation. The neutron and density log signature values are also higher (between 95-105 API) compared to the ones observed from Log facies E.

Where core photos are available for this lithofacies, it should be possible to differentiate from the lower shoreface by palaeoentological and petrological data. Only wells Ewan, 07, 08 and 05 penetrated this facies. The spiky gamma signature within this interval may indicate the presence of streaks and bands of siderite within the lithofacies. The sub-environment of deposition is interpreted as offshore mudstone.

Log facies distribution

In summary, from the wireline log analysis, six log facies association were identified partly constrained by core sedimentology. They are:- undifferentiated continental sand which is defined as Benin Formation, coastal barrier sand, tidal/fluvial channel- fill sand, upper shoreface, lower shoreface, slope channel sand and prodelta/offshore mudstone.

Synthesis of Lithofacies Distribution, Sedimentary Processes and Environments from Core

Description of core photographs from selected cored intervals within the studied area allows interpretation of depositional environments from specific intervals from wells Ewan 1 and Oloye 3. A detailed lithofacies classification has been established and two lithofacies associations identified. They are: slope channel sand sequences and lower-shoreface sequences (Table 1).

Slope channel sandstone

This lithofacies association in the Oloye 3 well (1718-1726 m) Figure 2, is characteristically thickly bedded (bedding about 8 m thick), fine grained and moderately to highly bioturbated with trace fossils of Skolithos, Ophiomorpha, and Planolites. It comprises poorly sorted, fine grained sandstone lithofacies SA2, SA3, SA4 and heterolithic lithofacies HE2. This unit shows a fining upward to blocky shape on gamma ray log profiles (Figure 22). Stacking pattern based on log character cross bedding structures, the presence of carbonate nodules and degree of bioturbation suggest that they were deposited by low to high energy traction currents in a shallow marine slope channel setting with some fluvial. The fluvial influence is supported by carbonaceous fragments, carbonaceous laminae recognised within the lithofacies. The overlying highly bioturbated muddy sandstone has a composite ichnofabrics of Teichichnus, Palaeophycus, Planolites (dominant), Ophiomorpha and Phoebichnus (about 2 m thick) i . at the top of sandstone faces, below the uncored intervals (Figure 2) indicate upward decrease in flow velocity as a result of channel migration, channel abandonment and/or decrease in discharge.

The large amount of carbonaceous (woody) materials within this facies may suggest a terrestrial input. However, the high preserved ichnofauna within this lithofacies suggest little or no reworking within a marine dominated environment, where there was deoxygenation caused by biological action.

The carbonate nodules and ferruginisation sometimes observed at the top of the facies (in Oloye 3) probably formed shortly after deposition (diagenetic) and indicates reducing (anoxic) conditions resulting from consumption of available oxygen by aerobic bacteria.

Environmental interpretation: The above characteristics are interpreted to record sub aqueous confined flows in a meandering slope channel setting with fluvial influence. This unit is characterised as fining upward sequence of cross bedded and fine laminated sands, with an erosional base underlain by carbonate nodules. Petrophysically, it is characterised by a blocky to fining upward profile.

Lower shoreface

The thick, highly bioturbated sandstone unit composed of fine-grained SA1, MS1, MS2, MS3, HE1, HE3 and HE4. The unit is characterised by a coarsening-upward log motif and contains ripples, wavy, planar lamination and cross stratification. The presences of wave ripples and cross lamination suggests moderate to high energy. There is also an abundance of carbonaceous materials in the sandstone facies, which can be related to terrestrial input. These sandstones are generally more compacted and cemented than above described slope channel sandstones. This may reflects the finer-grained nature of the sands which allows more compaction and the availability of mud and calcium carbonate within the pore water. The alternation of parallel bedded sandstone in which burrows are rare to absent with densely burrowed zones suggests considerable variation in sedimentation rates, probably associated with storm/fair-weather deposition. Bedding is defined by thin laminae of carbonaceous material and reddish brown streaks of sandstone. The streak of sandstone reflects episodic high energy condition. Bioturbation has largely destroyed the original sedimentary structures.

Graded mudstone facies characterised generally with thin laminated basal layer are a common feature of lower shoreface-transitional zone facies [49]. Alternating thinly bedded, disrupted sandy mudstone and mudstone lithofacies within the Oloye 3 well in the upper part of the core (section above the uncored interval) in which original sedimentary structures were largely destroyed by the intensive bioturbation are interpreted to be upper shoreface environment.

Environmental interpretation: Based on the trace fossils, ichnofabric textures and the observed sedimentary structures, the interpretation of this unit reflects a lower shoreface depositional environment under storm/wave dominated conditions. Proximal mouth bars on top of delta slope deposit may be the most likely sub environment for this above described unit. They are formed in a wave dominated lower shoreface environment.

Core and Wireline Log Facies Model

Core sedimentology and wireline log analysis, constrained with shoreline position from the seismic survey have been integrated to produce a facies scheme and interpret the depositional environment. The results are compared with that suggested by previous workers [7-11]. A summary of the new facies scheme is given in Table 1, and the depositional trend of the facies in the studied area is illustrated in Figures 2 and 24. Previous works have shown that Niger delta is an excellent example of a mixed wave, tide and fluvial deltaic system [1,52] (Figure 1), in which the prograding delta has been modified by wave action creating an arcuate coast with beach deposits, back barrier lagoons and chenier ridges. The coastline of the lower delta plain of the Niger is bordered by thick mangroves, incised into this coastline with numerous tide dominated coastal estuaries that are gradually infilling with sediment during the present-day transgressive to highstand system tract (Figure 1). Therefore, understanding the configuration of the different and complex depositional systems in modern deltaic analogues is the key for interpreting and understanding the geometries, scales and processes that formed the ancient system. The complex interplay of physical processes along the present-day Niger delta is apparent in the preserved sediments of the Ewan and Oloye fields (northwestern delta). On the field reservoir scale it is difficult to decipher the large-scale deltaic features of the Niger Delta and what is apparent are sedimentary bodies representing smaller scale distributary fluvial channel fill, tidal channel fill, coastal barrier bar, fan deposit (mouth bar sand), shoreface, slope channel fan and offshore mudstones.

Based on the core sedimentological descriptions and the wireline log interpretations, it is also proposed that the modern delta represents an excellent analogue based on its scale, depositional systems and stratigraphic setting (Figures 22 and 24). Generally, the middle Miocene succession within the studied area (northwestern Niger delta) is a complex mosaic formed by a series of discrete deltaic deposition events at different times. The succession comprises fine grained sandstone continental deposits of delta plain facies, tidal/fluvial channel sandstones and coastal barrier bar deposits; mouth bar and shoreface sandstones of delta front facies, slope channel sand and offshore mudstone (Figure 22).

Sharp-based sandstones are interpreted as slope channel and tidal/fluvial channel sand; and gradational-based sandstones as lower shoreface deposit respectively. The slope channel and tidal channel sands are fine grained sandstone, fine-upward, and are poorly sorted, often carbonaceous, micaceous, ferruginised. The lower shoreface units are typically fine grained poorly sorted and muddy sandstone, carbonaceous, stratified ferrugineous. Interpretation from these observation shows that sands were originally deposited in river- and mouth bars, with tidal currents producing inlet channels that maintained openings along coastline. Subsequent erosion and coast parallel transport by wave and tide induced currents (longshore drift currents) lead to deposition of considerable amounts of sand along an elongated shoreface and basinal environments.

The lithofacies associations interpreted from wireline log motifs and core photographs show that middle Miocene paralic sub environments range from inner shelf to offshore (Figure 22). This agrees with previous works which showed that the Agbada Formation consists of stacked fluvial/tidal influenced channels whose main lithofacies include distributaries channel fills, levees, crevasse splays, coastal barrier bars, shoreface, slope channel/fan and offshore mudstones sediments [7-11].

Discussion

Lithofacies interpretation forms the primary tool for identifying the depositional conditions under which the sediment were deposited and preserved. Based on wireline interpretation and core sedimentological description within the studied area, it can be observed that the Niger delta prograded and regressed several times in response to fluctuation in relative sea level; and the deposits have been modified by wave and tidal processes producing a complex suite of continental deposits, beaches, and estuaries along the delta margin (Figure 1). These processes have implications on the interpretation of depositional sequences at the field scale, but on the seismic scale, the depositional model reflects the delta progradation and recession. However, on the parasequence to parasequence set scale (well log), the depositional environments comprises, continental sand, tidal/fluvial channels/ mouth bar sand, coastal barrier bars, shoreface, slope channel/fan and offshore mudstone systems.

Ichnofabric analysis has proved very useful in this study to constrain the depositional processes, environments and help refine the facies model. The study of ichnofabric and the diagenetic imprint has also helped to delineate some key surfaces. For example the presence of Thalassinoides ichnofabric and diagenetic nodules within the core of Oloye 3 well. Thalassinoides is usually considered a firmground indicator in clastic substrates [53-56]. The occurrence of Thalassinoides reflects cessation of deposition, with minimal erosion (subaerial exposure) or burial and exhumation manifested by firmground formation [57]. Within this unit, firmground development seems to have occurred episodically and gradationally. It is commonly believed that sharp breaks in the vertical stratigraphic record may signify changes in depositional environment and initiation of new sedimentation cycles. However, gradational facies contacts are generally observed within the core interval. This suggests or implies a more gradual shift in depositional condition. It may be speculated that intense bioturbation could be responsible for destruction of the erosion surfaces through biogenic homogenization, making the contact between the superposed facies appear gradational. The presence of carbonate nodules is interpreted to be evidence of such surfaces (sequence boundary) within the study core interval (Figure 2). The recognition of Planolites-Phoebichnus, Planolites-Thalassinoides and Planolites-Teichichnus-Phoebichnus ichnofabric within the shallow marine slope depositional environment, coupled with the differences in the three dimensional geometry morphology, wall structure (burrow infill) and sizes of Planolites and Palaeophycus-Planolites ichnofabric has provided more details about the environment of deposition within the stratigraphic interval i how stressful the environment of deposition may have been.

Marine flooding surfaces develop as a result of increasing accommodation and/or accompanied reduced sediment input into the system during a sedimentary hiatus. This produces geochemical conditions which are conducive for preservation of trace fossils a high diversity of ichnofossil and precipitation of early bacterially mediated carbonate cement for their preservation. The presence of firmground and hardgrounds should also be one of the criteria in demonstrating these surfaces, but identification of firm/hard grounds was not possible because the analysis is based on core photos.

The main controls on the development of ichnofabrics include, water turbidity, water oxygenation, nature of the relict sediment, water salinity, episodic depositional rate and storm/wave energy condition. Individual marine flooding surfaces and sequence boundary exhibit different ichnofossils colonising different environments. For example, the interpreted FS 1 (Figures 2 and 22) in upper shoreface settings is dominated by Palaeophycus and Planolites (Table 1 and Figure 2). In FS 2 (Figure 2), it is marked by Planolites and Teichichnus ichnofabric with high ichno-diversities of Planolites, Palaeophycus, Ophiomorpha, Teichichnus, Skolithos and Phoebichnus. This environment of deposition is interpreted as marine dominated slope channel/fan.

The lower shoreface complex reflect some degree of facies variability, both sedimentologically and ichnologically. This facies ranges from vertically unburrowed HCS, wavy beddings and planar lamination sandstone to thoroughly bioturbated muddy sandstone. Variability in the preserved trace fossil record may locally reflect combination of water turbidity, water salinity, fluctuating/episodic depositional rate and storm/wave energy condition. Reduced oxygenation also has a profound effect upon the trace assemblages [50,57,58] and do appear to exert a control on the facies themselves within the study area. From this observation storms and to a lesser degree, fair-weather waves are the dominant physical processes operating on the lower shoreface [59].

Wireline log stacking patterns has been used to delineate log facies associations in this study. The use of only gamma ray log stacking patterns in predicting deposition environment has posed some difficulties, but through integration of identified key surfaces within the core; constrained with shoreline position from the seismic survey, a robust constrained interpretation is obtained. The scale of facies succession is a very important criterion constraining the interpretation of the log pattern.

Difficulty in interpreting of log pattern within the study area occurs where units amalgamate e . the stacked clastic progradation continental sandstones (e . Facies A). In this case, the succession generates a continuously upward cleaning gamma ray pattern. Within facies A is not possible to easily distinguish and sub units.

From the wireline log, correlation, the trend and nature of facies variation can be observed to be lateral extensive (~10 km wide) within a stratigraphic horizon (middle Miocene). Their facies relationship can be used to understand strata stacking patterns and to estimate lateral extent of reservoir facies. Vertically, the facies trend shows both small scale transgression and regression cycles. From observation in the study area, the geometry of major sand bodies produced in deltas is strongly related to the time of sea level fall.

The model developed within this study is similar to the existing Quaternary delta system model [7,58-60] and the coastal plain/shallow marine deposition system model of Oboh [2]. This suggests that in the studied area, the coastal barrier complex contains a large amount of sand that was originally deposited in river mouth bars, but was laterally removed by longshore current and re-deposited on the shoreline margin and basinward through developed of canyons on the basin slope which serves as a conduits for transporting sand basinward. Variability in sedimentological depositional processes within the studied area may be due to numerous variables that influence sedimentation rate (river, tide and wave), mode of deltaic sedimentation and salinity fluctuation.

Conclusion

The lithofacies, ichnofacies and depositional environments of the middle Miocene; northwestern Niger delta has been evaluated using core photographs and well logs. Log motif analysis, calibrated by core photographs and constrained with seismic was used to define log facies associations which were used to identify stacking patterns and key stratal surfaces. Identified lithofacies associations are slope channel sand and upper-lower shoreface sequences.

The ichnofabric succession allows recognition of gradatioal transitions between ichnofabrics, the identification of stratal units, composed of genetically related ichnofabrics, and key stratal surfaces e . sequence boundary and marine flooding surfaces. Integration of core and wireline sedimentology; constrained with shoreline position from the seismic survey, have provided a depositional model for lower delta plain to offshore (prodelta) environments within a tide/wave/storm influenced delta. Within each depositional complex, some sub-environments have been differentiated through a combination of physical, ichnofabric and biogenic features. The lower shoreface complex is strongly affected by both storm and fair-weather conditions, and hence, the preserved record shows a great degree of variability. This degree of variability is attributed to variations in overall storm intensity, storm frequency and relative water depth. The marine dominated slope channel sand complex is also quite variable in terms of ichnology.

Well logs and core data indicate that the depositional environments range from inner shelf to offshore (prodelta). Constrained with shoreline position and gross depositional environment (GDE) map from the seismic survey (Appendix).

References

1. Doust H, Omatsola E (1990) In: Edwards and Santogrossi PA (Ed) Niger Delta in JD Divergent/Passive Margin Basin AAPG 48: 201-238.
2. Allen JRL (1965) Late Quaternary Niger delta, and adjacent areas- sedimentary environments and lithofacies: AAPG Bull 49: 547-600.
3. Weber KJ, Daukoru EM (1975) Petroleum Geology of the Niger Delta Proc Ninth World Petrol. Congress 2: 209-221.
4. Evamy BD, Haremboure J, Kimberling P, Knapp WA, Molloy FA, et al. (1978) Hydrocarbon habitat of Tertiary Niger delta. AAPG Bull 62: 1-39.
5. Reijers TJA (1995) Reservoir geological core description using standardised lithofacies and tier associations in the Tertiary Niger Delta. Nigerian Asso of Petroleum Explorationists Bull 10: 27- 39.
6. Idowu AO (1993) Reservoir Fluid Differentiation Case Study from an Onshore Niger Delta Fields. Nigerian Ass of Petrl Geologists 10: 157-168.

7. Oomkens E (1974) Lithofacies relations in the Late Quaternary Niger Delta complex. Sedimentology 21: 195-222.

8. Adedokun OA (1981) Petrology, Provenance and Depositional environment of the Reservoir Sandstone of OSSU-1ZOMBE Oilfield, Imo State, Nigeria. J of Petroleum Geology 4: 25-56.

9. Petters SW (1984) An ancient submarine canyon in the Oligocene-Miocene of the western Niger Delta. Sedimentology 31: 805-810.

10. Amajor LC, Agbaire DW (1989) Depositional history of the Reservoir Sandstones, Akpor and Apara Oilfields Eastern Niger Delta. J of Petroleum Geology 4: 453-464.

11. Oboh EA (1992) Middle Miocene palaeonvironments of the Niger Delta. Palaeogeography, Palaeoclimatology, Palaeoecology. Elsevier Science Publishers BV, Amsterdam 92: 55-84.

12. Pemberton SG, Frey RW (1982) Trace fossil nomenclature and the *Planolites-Palaeophycus* dilemma. J Paleont 56: 843-881.

13. Dott RH Jr, Bourgeois J (1982) Hummocky stratification: Significance of its variable bedding sequences: Geological Society of America Bulletin 93: 663-680.

14. Duke WL (1985) Hummocky cross-stratification, tropical hurricanes, and intense winter storms: Sedimentology 32: 167-194.

15. Reineck HE, Singh IB (1980) Depositional Sedimentary Environment. Springer, Berlin p: 551.

16. Pemberton SG, MacEachern JA (1995) The sequence stratigraphic significance of trace fossils: examples from the Cretaceous foreland basin of Alberta, Canada: *in* JC Van Wagoner and G Bertram eds. Sequence stratigraphy of foreland basin deposits- outcrop and subsurface examples from the Cretaceous of North America: AAPG Memoir 64: 429-475.

17. Pemberton SG, MacEachern JA, Frey RW (1992a) Trace fossil facies models: environmental and allostratigraphic significance: *in* RG Walker and NP James eds. Facies Models: Response to Sea Level Change. Geological Association of Canada, Geotext 1: 47-72.

18. Seilacher A (1981) Towards an evolutionary stratigraphy: Acta Geologica Hispanica 16: 39-44.

19. Seilacher A (1982a) General remarks about event beds: *in* G Einsele and A Seilacher eds., Cyclic and Event Stratification. Springer-Verlag, Berlin p: 161-174.

20. Seilacher A (1982b) Distinctive features of sandy tempestites: *in* G. Einsele and A. Seilacher eds., Cyclic and Event Stratification. Springer-Verlag, Berlin p: 333-349.

21. Wheatcroft RA (1990) Preservation potential of sedimentary event layers: Geology 18: 843-845.

22. Pemberton SG, MacEachern JA, Ranger MJ (1992b) Ichnology and event stratigraphy: The use of trace fossils in recognizing tempestites: *in* SG Pemberton ed. Application of Ichnology to Petroleum Exploration, A Core Workshop. Society of Economic Paleontologists and Mineralogists, Core Workshop Notes 17: 85-117.

23. Frey RW, Goldring R (1992) Marine event beds and recolonization surfaces as revealed by trace fossil analysis: Geological Magazine 129: 325-335.

24. Pedersen GK, Surlyk F (1983) The Fur Formation, a late Paleocene ash-bearing diatomite from northern Denmark: Bulletin of the Geological Society of Denmark 32: 43-65.

25. Seilacher A (1969) Fault-graded beds interpreted as seismites: Sedimentology 13: 155-159.

26. Seilacher A (1962) Paleontological studies in turbidite sedimentation and erosion. J of Geology 70: 227-234.

27. Aigner T (1985) Storm Depositional Systems: Berlin; Springer-Verlag p: 174.

28. Rice AL, Billett DSM, Fry J, John AWG, Lampitt RS, et al. (1986) Seasonal deposition of phytodetritus to the deep-sea floor: Royal Society of Edinburgh Proceedings 88B: 205-279.

29. Leithold EL (1989) Depositional processes on an ancient and modern muddy shelf, northern California: Sedimentology 36: 179-202.

30. Durogbitan, AA (2010) Seismic, Sequence Stratigraphic and Structural Analysis of Ewan and Oloye Fields (Middle Miocene), Northwestern Niger Delta. PhD Thesis unpubl. Earth Science Department, The University of Manchester p: 337.

31. Taylor A, Gawthorpe RL (1993) Application of sequence stratigraphy and trace fossil analysis to reservoir description: examples from the Jurassic of the North Sea. In: Parker JR (Ed), Petroleum of Northwest Europe: Proceedings of the 4th Conference. Geological Society, London p: 317-335.

32. Taylor A, Goldring R (1993) Description and analysis of bioturbation and ichnofabric. J of the Geological Society, London 150: 141-148.

33. Taylor A, Goldring R, Gowland S (2003) Analysis and application of ichnofabrics. Earth-Science Reviews 60: 227-259.

34. Taylor A (1991) Trace Fossil Fabric Analysis in the Sub-surface Exploration of Jurassic Sequences from the North Sea Basin. University of Manchester, PhD thesis unpublished p: 489.

35. Ekdale AA (1988) Pitfalls and palaeobathymentric interpretation based on trace fossil assemblages. Palaios 3: 464-472.

36. Staff GM, Powell EN (1988) The paleoecological significance of diversity: the effect of time averaging and differential preservation on macro invertebrate species richness in death assemblages. Palaeogeogr, Palaeoclimatol., Palaeoecol 63: 73-89.

37. Goldring R (1995) In Bosense DWJ and Allinson PA (Ed) Organism and the subtrate: response and effect. Marine Palaeoenvironmental Analysis from Fossils. Geological Society, London 83: 151-180.

38. Wetzel A (1983) Biogenic structures in modern slope to deep-sea sediments in the Sulu Sea Basin (Philippines). Palaeogeogr Palaeoclimatol, Palaeoecol 42: 285-304.

39. Bromley GG, Asgaard U (1979) Triassic freshwater ichnocoenoses from Carlsberg Fjord, East Greenland. Palaeogeogr Palaeoclimatol Palaeoecol 28: 39-80.

40. Keighley DG, Pickerill, RK (1995) The ichnotaxa *Palaeophycus* and *Planolites*; historical perspectives and recommendations. Ichnos 3: 301-309.

41. Ager DV, Wallace P (1970) The distribution and significance of trace fossil in the uppermost Jurassic rock of the Boulonnais, northern France Geol J Spec Issues 3: 1-18.

42. Malpas JA, Gawthorpe RL, Pollard JE, Sharp IR (2005) Ichnofabric analysis of the shallow marine Nukhul Formation (Miocene), Suez Rift, Egypt: implication for depositional process and sequence stratigraphic evolution. Palaeogeogr Palaeoclimatol, Palaeoecol 215: 239-264.

43. Ekdale AA, Bromley RG (1984) Comparative ichnology of shelf-sea and deep-sea chalks. *J Paleont* 58: 322-332.

44. Walker RG, Plint AG (1992) Wave and Storm dominated Shallow Marine Systems. In Walker RG and James NP (eds). Facies Models: Response to seal level changes. Geological Assoc of Canada p: 219-238.

45. Pemberton SG (1992) Applications of Ichnology to Petroleum Exploration: A Core Workshop. SEPM (Society of Economic Paleontologists and Mineralogists), Core Workshop No. 17, Calgary p: 429.

46. Galloway WE (1968) Depositional Systems in the lower Wilcox group, North-Central Gulf Coast basin. Gulf Coast Soc Trans 18: 275-289.

47. Fishers WL (1969) Facies characterisation of Gulf Coast basin delta systems with some Holocene analogues. Gulf Coast Soc Trans 21: 239-261.

48. Coleman JM, Prior DB (1982) Deltaic environment. In: Scholle PA, Spearing D (Ed.) Sandstone Depositional Environment. American Assc of Petr Geol. Memoir 31: 139-178.

49. Selley RC (1978) Concepts and methods of subsurface facies analysis: American Association Petroleum Geologists, Continuing Education Course Notes Series 9 p: 82.

50. Escalona A, Mann P (2006) Sequence-stratigraphic analysis of Eocene clastic foreland basin deposits in central Lake Maracaibo using high resolution well correlation and 3-D seismic data. AAPG Bull 90: 581-623.

51. Emery D, Myers KJ (1996) Sequence stratigraphy: Oxford, Blackwell Science p: 297.

52. Allen JRL (1970) In: J.P Morgan (Ed.) Sediments of the modern Niger delta: a summary and review. Deltaic Sedimentation. Modern and Ancient. SEPM Spec Publ 15: 138-151.

53. Seilacher A (1964) Sedimentological classification and nomenclature of trace fossils. Sedimentology 3: 253-256.

54. Pemberton SG, Frey RW (1985) The *Glossifungites* ichnofacies: modern examples from the Georgia coast, USA. In: Curran HA (Ed), Biogenic structures: their use in interpreting depositional environments, Spec Publ Soc Eon Paleontol Mineral 35: 237-259.

55. Mac Eachern JA, Raychaudhuri I, Pemberton SG (1992) In. Pemberton, SG. (Ed.) Stratigraphic applications of the *Glossifungites* ichnofacies: Delineating discontinuities in the rock record. Application of Ichnology to Petroleum Exploration, a core workshop, Soc Econ Paleontol Mineral Core Workshop 17: 169-198.

56. Gingras MK, Pemberton SG, Saundners T (2001) Bathymetry, sediment texture and substrate cohesiveness; their impact on modern Glossifungites trace assemblages at Willapa Bay, Washington. Palaeogeogr Palaeoclimatol Palaeoecol 169: 1-21.

57. Bromley RG, Ekdale AA (1984) Chondrites: a trace fossil indicator of anoxia in sediments. Science 224: 872-874.

58. Ekdale AA, Bromley RG (1984) Sedimentology and ichnology of the Cretaceous-Tertiary boundary in Denmark: implications for the causes of the terminal Cretaceous extinction. J of Sedimentary Petrology 54: 681-703.

59. Greenwood B, Mittler PR (1985) Vertical sequence and lateral transitions in the facies of a barred nearshore environment: Journal of Sedimentary Petrology 55: 366-375.

60. Weber KJ (1971) Sedimentological aspect of oil fields in the Niger Delta. Geol Mijnbouw 50: 559-576.

12

Physico-Chemical Analysis of Marine Soils From Different Ecosystem Palk Strait Coastal Regions of Tamilnadu, India

Manikandan R* and Vijayakumar R

Research Department of Microbiology, Bharathidasan University Constituent College, Kurumbalur, Perambalur District, Tamil Nadu, India

Abstract

Many researchers focus on the marine coastal zones. Present study was physical and chemical characteristics of soil in three different marine ecosystem of six different places sea shore (Point Calimere [Kodiyakarai], Adirampattinam, Mallipattinam and Manora), saltpan (Vedaranyam), mangroves Forest (Muthupet) Palk Strait coastal regions of Tamilnadu, India, was performed. The marine soil were selected for the following boundaries like Soil texture, Calcium Carbonate, Electrical conductivity, Power of hydrogen, Macronutrients like (Organic carbon, Nitrogen, Phosphorus, Potassium), Micronutrients like (Iron, Manganese, Zinc, Copper) and others Caution exchange capacity, Magnesium, Sodium were studied. At the end of the soil collected from various stations showed differences in all analyzed features.

Keywords: Mangrove; Sediment soils; Physico-chemical analysis; Palk Strait regions

Introduction

Soil is one of the major and costly properties of the nature. Every living is straight and lastly dependent on soil for the human food is derivative from the earth. Creation plan for have strong and soil is essential to human being survival. Soil has composite role which is useful to human and other living organism. Marine environmental conditions such as temperature, pH, salinity, nutrients, sediments and soil texture find out composing the biota [1]. Increased temperature [2], low salinity [3], sediment nutrients and availability of organic matter Nair et al. [4] have been found to favor the growth and distribution of microorganisms. Microbes like fungus, bacteria and other atomic life form available surrounded by the soil are also important and therefore soil is an active medium made up of mineral deposits, macrobiotic matter, aquatic, atmosphere and microorganisms. The natural history of soil first and advanced depend on its nonstop change under the cause of physical reasons like the close relative substance, instance, the weather, the organic in it etc. [5].

The physical-chemical features like power of hydrogen, electrical conductivity, moisture content, organic matter, nitrogen, and potassium were studied, tabulated with in brief discussed Das and Bindi [6]. The physical-chemical study of soil samples under revision show erratic concentration of different features. Potent distributions of micronutrients were recorded in the present investigation which may be recognized to the added fertilizers during the crop formation [7-8].

Human being behavior with metal mining, smelting and dying, coal fire, refuse burning, fossil fuel burning and agricultural practices have at large numerous to excess of trace elements into the environment [9]. With them Think, during many research for the study idea and the main goal for many years, in the waters naturally large high salinity, climate change and the sea caused a volcanic eruption, earthquake, natural disasters die in the soil decayed the marine life, plants, animals and micro-organisms. Therefore, the present study aims to record the Physical-chemical features marine soil from different marine ecosystem like seashore, hypersaline salt pan and mangrove environments of Palk Strait region of Tamil Nadu, India.

Materials and Method

Collection and processing of marine soil Sample

Soil sample were collected from six different locations of three different ecosystems like sea shore [Point Calimere (Kodiyakarai)], Adirampattinam, Mallipattinam and Manora, saltpan (Vedaranyam), mangrove (Muthupet) marine environmental soils of Palk Strait coastal regions of Tamil Nadu, India (Figure 1). Surface contaminated soils were removed using a spade to a depth of 2 cm [10]. Using V-shaped holes in the soil to a depth of 20 cm thick piece of sample collection is taken in a plastic bucket digging collected. The soil was collected after the registered name of the place. Merrily soil collected in the dryer and the wind blew them into cubes and store in clean plastic bags secured was studied in laboratory conditions (Figure 1) [11].

Inspection of marine soil texture, $CaCO_3$, EC, CEC, pH, macro and micro nutrients

Comparative amount of different range soil is an important physical limit to find out soil texture. The percentage of the soil particles were determined by 10 g air dried soil sample was mixed with 20 mL distilled water. The mix was heated for 3 min to break the large sized soil. 5 mL hydrogen peroxide was added to the mix. The solution was filtered through common filter paper. The soil balance of filter paper was then treating with 125 mL of 3 Normality HCl and left it for immediately. The solution was filtered again and soil filtrate dried in air. The dried soil was passed through 0.2 mm pore sized technical filter. The filter soil was transferred into a flask and sufficient amount of 1 Normality

***Corresponding author:** Manikandan R, Research Scholar, Research Department of Microbiology, Bharathidasan University Constituent College, Kurumbalur, Perambalur District, Tamil Nadu, India E-mail: rmani612@gmail.com

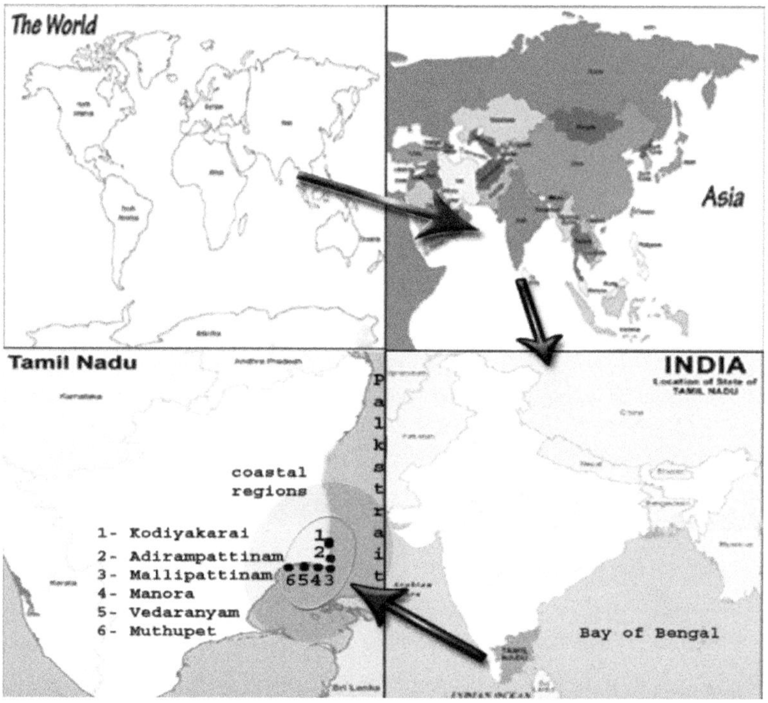

Figure 1: Sea coastal area of soil collected by the Palk Strait in southern India's Tamil Nadu Map.

sodium hydroxide was added to make the inside alkaline. After shaking the container for 6 h on the mechanical shaker, the content was move to 500 mL measuring jar and the volume was made up to 250 mL by distilled water. It was stopper and was allowed to settle after shaking. The supernatant was free into a separate beaker and soil was washed again and again till the delay was clear. This was transferred to another tarred dish, heated and weighted, which was the reading of fine sand [12-15].

Calcium carbonate was extracted with neutral 1 N ammonium acetate and the available calcium carbonate in the extract was determined by Smith and Doran [15]. Electrical conductivity (EC) of the soluble salts in the soil contents of the ion current fixing reveals. If any one of the electrical conductivity of the soil samples are included in 20 g soil suspension 40 mL distilled water and stirred for 30 min after the soluble salts had been fully used by an Equiptronics digital electrical conductivity was detected and recorded. Cation exchange capacity (CEC) of the soil was determined by using 1 Normality ammonium acetate solution as give details by Jackson [13].

Power of hydrogen values were determined by pH meter. 10 g soil samples were mixed with 20 mL distilled water in. The suspension was stirred continuously with glass rod for 15 min and left for 1 h. The electrode was added into supernatant and pH was recorded. Power of hydrogen value as calculates of the hydrogen ion action of the soil water and state the acidity and alkalinity of the soil. It is an important property of soil as it fixes the accessibility of nutrients and physical condition of soil [14-15].

The Macronutrients like Organic carbon, Nitrogen, Phosphorus, Potassium was determined by Jackson [16-17]. The Micronutrients like Iron, Manganese, Zinc, Copper and others Magnesium, Sodium were studied [18-20].

Result and Discussion

The marine soil is measured like the mainly significant issue in the marine atmosphere that influences the growth, replica and metabolic actions of biotic components including microbes. In the current study, the physico-chemical factor including Soil texture, Calcium Carbonate, Electrical conductivity, Power of hydrogen, Macronutrients like (Organic carbon, Nitrogen, Phosphorus, Potassium), Micronutrients like (Iron, Manganese, Zinc, Copper) and others Caution exchange capacity, Magnesium and Sodium of the Palk Strait coastal soil samples were study (Table 1).

The soil textures are shown in Table 1. The results showed that various marine environmental soils change from Sands, Sands and Clay, Sandy Loams, and Clay Loams in soil value were reported (Figures 2a-2f).

The physicochemical characterization of seven marine sediment samples was investigated and collected from different zones with Thondi coastal sediments along the southeast coast of India. The study found that variation in physical parameters and mud contents were due to differences in sediment source. Results of Station 3 revealed the calculated average values as follows: Bulk density (BD) (1.3 g/cm^3), particle size distribution of-sand (50.5%), silt (31.5%), and clay (18%). Further, the water holding capacity (WHC) was observed to be 65.3% [21]. The soil parameters like texture, calcium carbonate content, electrical conductivity (EC), pH, organic carbon (OC), available nitrogen, available phosphorus, available potassium iron, manganese, zinc, copper and cation exchange capacity were studied by the standard methods. Further, the correlation coefficient analysis between the physical and chemical boundaries of the marine soil samples and total actinobacterial isolates were performed [22].

The minimum Calcium Carbonate (5.56 mg/kg), Magnesium (3.15 mg/kg) and Sodium (2.21 mg/kg) was present in Adirampattinam, Mallipattinam soil samples were recorded, the maximum Calcium Carbonate (8.25 mg/kg), Magnesium (9.10 mg/kg) and Sodium (3.17 mg/kg) was present in Vedaranyam, Muthupet soil respectively (Figure 3a).

Name of the Parameters	Soil samples from different ecosystems (Session: July 2015)					
	sea shore				*saltpan*	*mangrove*
	KS	**AS**	**MPS**	**MS**	**VS**	**MUS**
Soil texture (%)						
Sands	64.2	28.2	61.0	57.8	45.7	12.7
Sands & Clay	05.1	53.9	25.4	24.9	36.2	27.9
Sandy Loams	20.7	12.0	27.9	35.1	35.4	38.2
Clay Loams	02.3	28.7	12.4	32.0	12.9	50.6
$CaCO_3$ (mg/kg)	7.44	5.56	7.73	7.33	8.25	7.94
EC (dsm-1)	2.10	11.0	2.57	6.12	12.3	16.2
pH	5.25	8.23	6.62	5.35	7.12	6.13
Organic carbon (%)	0.22	1.25	1.01	0.57	0.20	2.30
Nitrogen (kg/ha)	25.1	59.2	32.2	57.3	33.2	98.4
Phosphorus (kg/ha)	7.12	3.45	5.81	5.01	3.23	7.51
Potassium (kg/ha)	150	392	52	277	387	887
Iron (ppm)	4.3	11.0	7.01	11.2	7.47	7.64
Manganese(ppm)	4.71	15.2	4.31	1.57	9.52	11.3
Zinc (ppm)	1.30	0.21	0.14	0.17	0.12	1.21
Copper (ppm)	2.23	1.21	1.05	1.24	2.52	1.42
CEC (C.mole proton/kg)	24.4	19.2	17.3	13.2	53.4	14.8
Magnesium (mg/kg)	6.71	5.12	3.15	4.22	5.86	9.10
Sodium (mg/kg)	2.36	2.42	2.21	2.52	3.17	2.70

KS: Kodiyakarai Soil; AS: Adirampattinam Soil; MPS: Mallipattinam Soil; MS: Manora Soil; VS: Vedaranyam Soil; MUS: Muthupet Soil; $CaCO_3$: Calcium Carbonate; EC: Electrical Conductivity; pH: Power of Hydrogen; CEC: Caution Exchange Capacity

Table 1: Detect of various physico-chemical parameters in marine soil samples from Palk Strait.

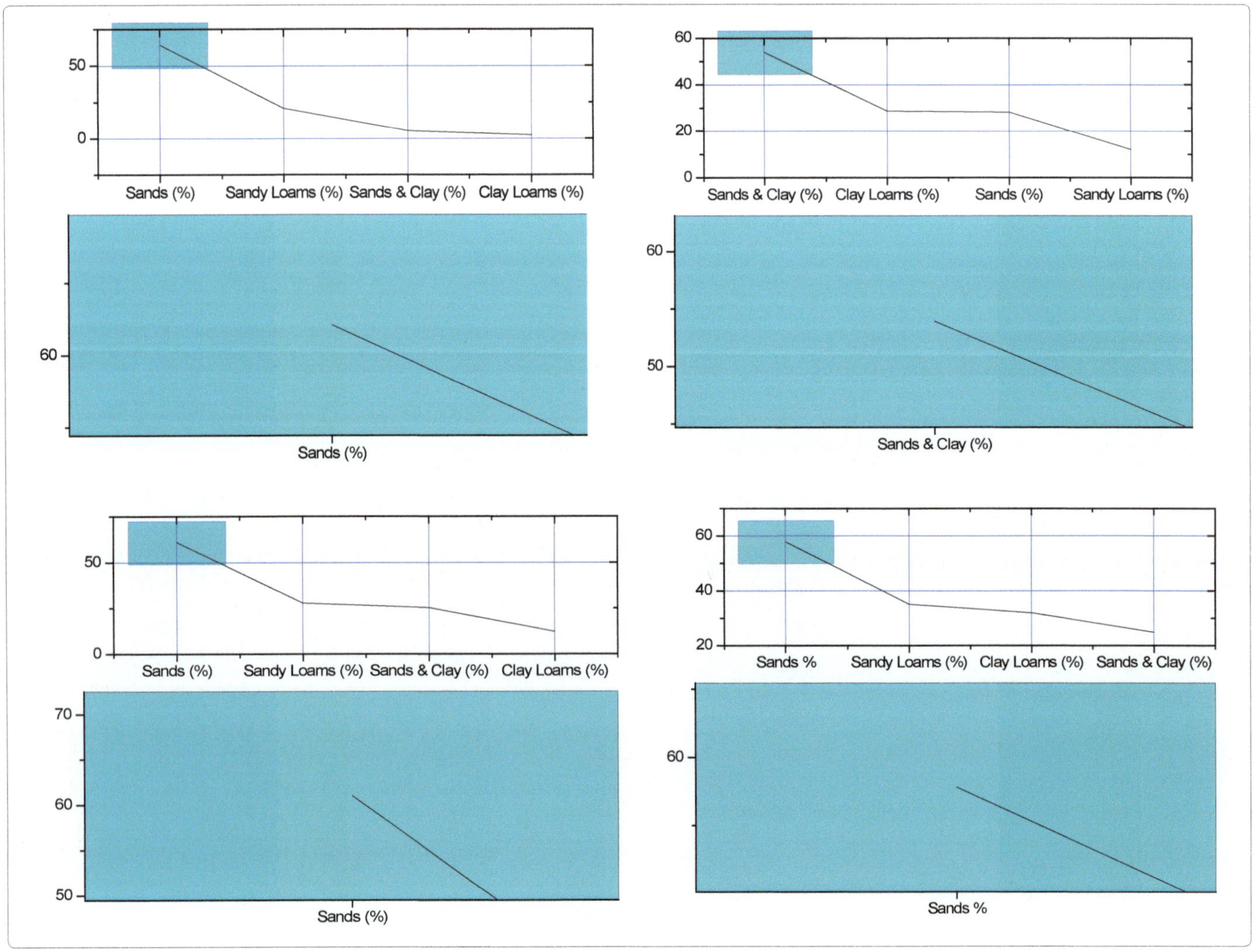

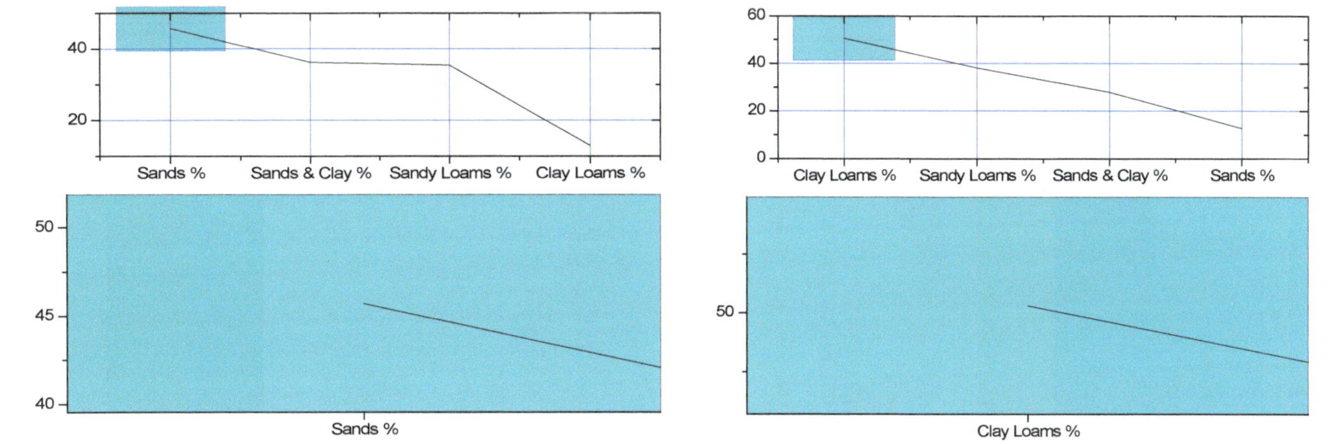

Figure 2: Marine Soil Texture group in Palk Strait (a) KS-marine soil texture percentage, (b) AS-marine soil texture percentage, (c) MPS-marine soil texture percentage, (d) MS-marine soil texture percentage, and (e) VS-marine soil texture percentage, MUS-marine soil texture percentage.

The minimum Caution exchange capacity (13.2 C. mole proton/kg), Electrical conductivity (2.10 dsm-1) and pH (5.25) Manora, Kodiyakarai soil samples were study, the soil maximum Caution exchange capacity (53.4 C. mole proton/kg), electrical conductivity (16.2 dsm-1) and pH (8.23), Vedaranyam, Muthupet mangrove and Adirampattinam soils at the produce, respectively (Figure 3b).

The macronutrients like Organic carbon (41%), Nitrogen (98.4 kg/ha), Phosphorus (7.51 kg/ha) and Potassium (887 kg/ha), the present study were recorded maximum at Muthupet mangrove soil. The minimum contents of the macronutrients organic carbon (3.6%), Nitrogen (25.1 kg/ha), Phosphorus (3.23 kg/ha) and Potassium (52 kg/ha) recorded at soil samples collected from Vedaranyam, Point Calimere, Mallipattinam respectively (Figures 3c-3d).

The micronutrients like Iron (11.2 ppm), Manganese (15.2 ppm), Zinc (1.30 ppm) and Copper (2.52 ppm) the present investigated were recorded maximum at Manora, Adirampattinam, Kodiyakarai, Vedaranyam soil. The minimum contents of Iron (4.3 ppm), Manganese (15.2 ppm), Zinc (1.30 ppm) and Copper (2.52 ppm) were documented at Kodiyakarai, Manora, Mallipattinam soil (Figure 3e). Sea, soil, characteristics, properties, pH-acidic natures have been reported previously [23].

Soil characteristics such as pH 7.22-7.69, electrical conductivity 0.41 dSm-1 to 0.54 dSm-1, cation exchange capacity 16.3-23.8 C. mole proton+/kg, organic carbon 0.36-0.60%, nitrogen 2.74-3.29 mg/kg, phosphorus 1.1-1.28 mg/kg, potassium 4.13-5.34 mg/kg, zinc 0.53-0.89 ppm, copper 0.53-1.97 ppm, iron 2.46-8.54 ppm, manganese 3.10-3.69 ppm, calcium 10.5-12.8 C. mole Proton+/kg, magnesium 6.7-9.2 C. mole Proton+/kg, sodium 1.33-2.98 C. mole Proton+/kg and potassium 0.14-0.28 C. mole Proton+/kg, showed variation during different seasons [18].

It is clear that the threat of ocean acidification on marine ecosystems and species represents a priority for future investigations and large-scale investments in green-energy sources. The most important challenge of research is to identify the vulnerability of some physiological processes of key marine species but also rate of tolerance and adaptation capability to the global climate change [23-26].

Conclusion

The physico-chemical analysis of soil samples under evaluation showed variable concentrations of various boundaries. Irregular

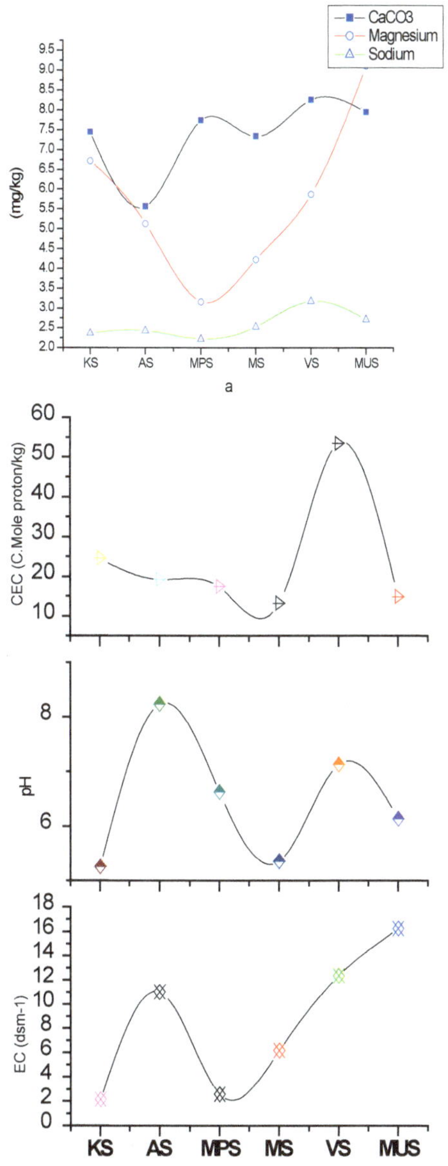

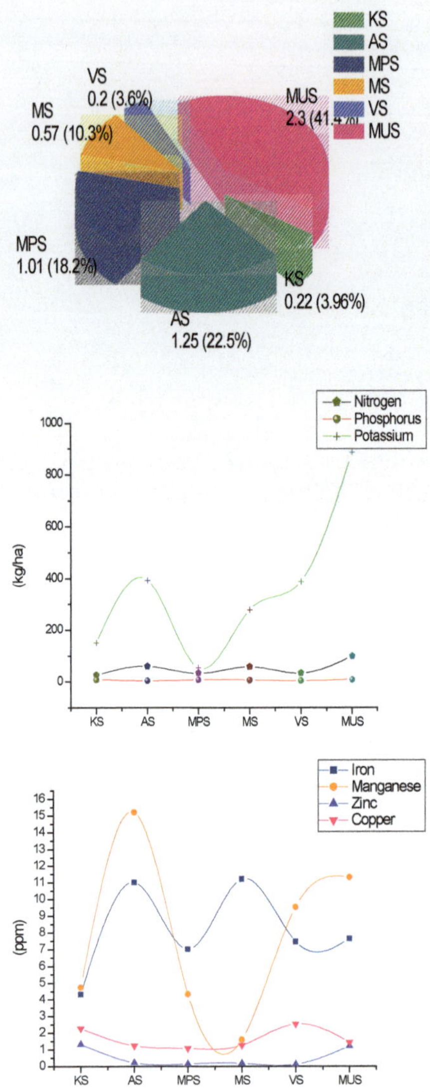

Figure 3: Graphical structure of marine soil parameters from Palk Strait (a) $CaCO_3$, Mg, Sodium, (b) CEC, pH, EC, (c) Organic carbon, (d) N, P, K, and (e) Fe, Mn, Zn, Cu.

distributions of macronutrients and micronutrients were recorded during the present investigation which may be recognized to the further fertilizers for the period of the produce development. The survey of Marine soil ecosystems (sea shore, saltpan, and mangrove) displays of micronutrients in the soils occasionally an able to assess the qualitative and quantitative of the metal concentrations. From here the marine soil strength is recommended as a way of data quality. It is data to take up technique, sample testing, just before gain peak in series appearance from a coast-effective resulting outside program.

Acknowledgment

This study was supported in part by grants from Tamil Nadu State Council for Science and Technology, Chennai-600025, Tamil Nadu (Project ref. No. TNSCST/S and T Projects/VR/ES/2013-2014/468 dated 16.04.2014).

References

1. Galal M, Zaiad (2010) Physico-Chemical Analysis of Soils in Al-Khums city, Libya. J Applied Sci Res 6: 1040-1044.
2. Anzecc, Armcanz (2000) Australian and New Zealand Guidelines for Fresh and Marine Water Quality. National Water Quality Management Strategy 1: 1-314.
3. Lucena F, Bostch A, Jofre J, Schwarlzbrod L (1985) Identification of Viruses isolated from sewage, riverwater and coastal seawater in Barcelona. Water Res 19: 1237-1239.
4. Nair NB, Azis PKA, Dharmaraj K, Arunachalam M, Krishnakumar K, et al. (1983) Ecology of Indian Estuaries: Part I- Physico-chemical features of waters and sediments-nutrients of Ashtamudi Estuary. Indian J of Marine Sci 12: 143-150.
5. Wagh GS, Chavhan DM, Sayyed MRG (2013) Physicochemical Analysis of Soils from Eastern Part of Pune City. Universal J Env Res Tech 3: 93-99.
6. Das B, Bindi (2014) Physical and Chemical Analysis of Soil Collected from Jaisamand. Universal J Env Res Tech 4: 260-264.
7. Bimashankar R, Pirgonde (2014) Analysis and Physical-Chemical Study of Heavy Metal Concentration in Soil. Indian Streams Res J 4: 1-6.
8. Velmurugan S, Govindaraj R, Gokulakumar B, Ravi S (2012) Physico-chemical parameters and elemental analysis of the soils of sunflower (*Helianthus annus* L) Growing field with different manure treatment. Asian J Plant Sci Res 2: 473-477.
9. Nriagu JO, Pacyna JM (1988) Quantitative assessment of worldwide contamination of air, water and soils by trace metals. Nature 333: 134-139.
10. Gupta PK (2007) Soil, plant, water and fertilizer analysis (2nd edn.) Dr. Updesh Purohit for Agrobios, Jodhpur, India.
11. Nalebuff EA (1975) The recognition and treatment of tendon ruptures in the rheumatoid hand. In: AAOS Symposium on Tendon Surgery in the Hand. Mosby, St Louis, MO pp: 255-269.
12. Piper CJ (1964) Soil and plant analysis Adelaide. The University Adelaide pp: 1-368.
13. Jackson ML (1973) Soil chemical analysis. Prentice Hall of India Pvt. Ltd., New Delhi.
14. Jackson ML (1967) Soil Chemical analysis. Prentice Hall of India, Pvt. Ltd., New Delhi p: 498.
15. Smith JL, Doran JW (1996) Measurement and use of pH and electrical conductivity for soil quality analysis. In Methods for assessing soil quality. Soil Science Society of America Special Publication 49: 169-182.
16. Jackson MC (1962) Soil chemical analysis. Prentice Hall inc. Engle Wood Cliffs, New Jersey, USA.
17. Bray RH, Kurtz LT (1945) Determination of total, organic, and available forms of phosphorus in soils. Soil Sci 59: 39-45.
18. Karthikeyan P, Senthilkumar G, Panneerselvam A (2013) Isolation, characterization and identification of actinobacteria of Mangrove ecosystem Ennoor, east coast of Tamil Nadu, India. Adv Appl Sci Res 4: 296-301.
19. Lindsay WC and Norvell A (1978) Development of DTPA soil test for Zn, iron, manganese and cu. Proc Sci Soc IS 42: 421-428.
20. Barnes H (1959) Apparatus and methods of Oceanography, Part I Chemical, George allen and Unwin, London.
21. Sankaran G, Jothivel S, Govindasamy V, Nateesan M (2013) Geochemical and textural characterization of minerals in Thondi coastal sediments along the southeast coast of India. J Oceanograph Marine Sci 5: 37-44.
22. Manikandan R, Vijayakumar R (2015) Marine actinobacterial diversity from Palk Strait coastal region of Bay of Bengal. Wide Spectrum 3: 25-34.
23. Vaijayanthi G, Vijayakumar R (2014) Physico-chemical analysis of marine sediment soils from Muthupet mangrove environments in Tamilnadu, India. Int J Adv Life Sci 7: 417-423.
24. Guinotte JM, Fabry VJ (2008) Ocean acidification and its potential effects on marine ecosystems. Ann NY Acad Sci 1134: 320-342.
25. Byrne M (2012) Global change ecotoxicology: Identification of early life history bottlenecks in marine invertebrates, variable species responses and variable experimental approaches. Mar Environ Res 76: 3-15.
26. Gallo A, Tosti E (2016) Adverse Effect of Ocean Acidification on Marine Organisms. J Marine Sci Res Devop 6: 139.

13

Effects of Harvest and Post-Harvest Processing Methods on Quality of Beche-de-mer in Fiji Islands

Ravinesh Ram[1*], Roveena Vandana Chand[2] and Paul C. Southgate[1]

[1]*Centre for Sustainable Tropical Fisheries and Aquaculture, College of Marine & Environmental Sciences, James Cook University, Townsville, Queensland 4811, Australia*
[2]*School of Biological and Chemical Sciences, Faculty of Science Technology and Environment, University of the South Pacific, Suva, Fiji Islands*

Abstract

The sea cucumber fishery and processing of sea cucumbers into bêche-de-mer (BDM) supports livelihoods and generates significant export income in Fiji. Harvesting and processing of sea cucumbers has been done in Fiji for over two centuries and the processing technique has changed little over this time. Processing of sea cucumbers involves a number of steps: first boiling, slitting and gutting, smoke drying and sun drying. This study investigated the effects of harvesting and post-harvest handling techniques on the quality of BDM produced in Fiji and documented difficulties faced by the BDM processors in producing a good quality product. Our results show that to generate income as soon as possible, sea cucumber fishers in Fiji hastened BDM processing by eradicating or shortening necessary processing steps. This resulted in production of lower quality product with lower value and reduced acceptance. The first boiling and smoking steps of processing were particularly problematic for fishers. A well processed BDM product has good odour, appearance, shape, color and size and has moisture content appropriate for storage without spoilage. Processed BDM sold by fishers to the main exporters in Fiji were generally of low quality because of a number of consistent characteristics that greatly impacted on quality and product value. Common problems with product quality included improper cleaning, contamination with sand, peeling skin resulting from over-boiling and burning resulting from an inappropriate smoking technique. The results of our study have prompted immediate awareness and education activities for sea cucumber fishers and processors in Fiji supporting improvements in product quality and improved income from this important coastal livelihood.

Keywords: Bêche-de-mer; Processing; Quality; Harvest; Post-harvest processing

Introduction

Sea cucumbers have been harvested and utilized commercially for at least 1000 years in the Indo – Pacific region [1]. It is a product that is consumed in many tropical and subtropical countries [2,3] and in Fiji was first exploited for commercial purposes by Chinese exporters in the early 1820's [4]. Sea cucumbers are processed into a dried and non-perishable product known as 'bêche-de-mer' (BDM), and processing of BDM offered a source of income for Fijian communities subsequent to the collapse of the sandalwood trade in the early 1800's [4]. Still today, harvesting and producing appropriately processed and presented BDM offers good income opportunities for fishers in Fiji. Consumption of BDM is considered to be beneficial for human health as it is a good source of animal protein, vitamins, minerals, micronutrients and essential fatty acids [5-8].

Literature on the processing methods used for BDM and their effects on resulting quality is very limited. Processing BDM involves a number of steps including boiling, gutting, smoking and drying [9,10]. Each contributes to the resulting quality of the final product which determines the suitability of the processed product for Asian markets [2,9-11] and its value. Small developing nations such as Fiji are reliant on marine resources as their basis for income generation and livelihoods. Nevertheless, due to a lack of equipment required to optimize the quality of BDM, such as kerosene burners, smoking sheds and drying amenities, BDM is produced in Fiji using simple uncomplicated customary technologies principally through fire smoking and air or sun drying [9,12].

Bêche-de-mer processing entails an uncomplicated sequence of actions to accomplish a product that is non – perishable if stored in dry, dark conditions. Post-harvest steps include first boiling, slitting and gutting, second boiling, smoking and finally sun-drying [2,9,13,14]. Although these steps are uncomplicated, processing requires continuous attention to obtain a standard dry product. Failure to do so can result in reduced quality and value of the final product [9,10].

The main objectives of this study were to investigate the effects of harvesting and post-harvest handling techniques of Fijian communities and the main BDM processing agents on the quality of BDM produced in Fiji, and to identify difficulties faced by the BDM processors in producing a good quality product. The economic impacts of inadequate processing of BDM are detailed in a related study (Ram et al., in press) [15].

Materials and Methods

Study site

This study was conducted in the Republic of the Fiji Islands. The sites of study were eastern and western Viti Levu, northern and southern Vanua Levu and two islands from outer Lau group (Vanua Balavu and Lakeba Island). The latter two islands have two main BDM buyers who also buy product from other islands in the Lau group (Figure 1). On this basis the data from Vanua Balavu and Lakeba Island represents the Lau Group as a whole. Catch analyses and interviews with sea cucumber fishers were done when the fishers visited the main

***Corresponding author:** Ravinesh Ram, Centre for Sustainable Tropical Fisheries and Aquaculture, College of Marine & Environmental Sciences, James Cook University, Townsville, Queensland 4811, Australia
E-mail: ravinesh.ram@my.jcu.edu.au, ravineshram@gmail.com

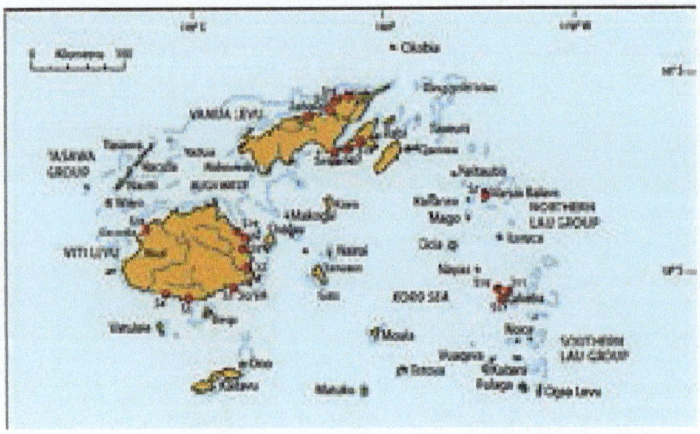

Figure 1: Sites visited in Fiji for collection of data (source: Ram, 2008)

buyers on Vanua Balavu and Lakeba Island to sell their products in either fresh or dry form. However, fishers based on the main islands of Fiji (Viti Levu and Vanua Levu) were visited directly in their villages to collect data on harvesting and processing.

Data collection

Data on processing sea cucumbers was acquired through formal and informal interviews using a structured questionnaire given to fishers at the various study sites around Fiji (n=21; Figure 1). The key people targeted for interviews and collection of catch analysis data were artisanal fishers (n=86), middlemen (n=8) who purchase sea cucumbers and process them for a higher return from the exporters, the main agents for BDM exporters (n=5) located in Suva, Fiji and relevant fisheries officers at the Ministry of Fisheries and Forests, Fiji.

The questionnaire was designed to derive information about the species of sea cucumbers that were commonly harvested, total catch of each species harvested, methods used to store sea cucumbers in the sea after harvest, methods used for processing, difficult steps in the processing sequence, and sea cucumber species that were difficult to process or handle. A sea cucumber identification card and pictures of various sea cucumber species were shown to the fishers and buyers to assist with accurate sea cucumber identification [9].

Harvesting and processing of sea cucumber

The daily harvest of the fishers was recorded by species and size. Sea cucumbers were left out of the water for approximately 5 min before length and weight were determined. The length of harvested sea cucumbers was measured using a standard ruler to the nearest 0.1 cm and their weight was measured using an analytical balance to the nearest 0.1 g. Details of the processing techniques (harvesting, boiling, gutting, smoke drying and sun drying) used by the fishers were noted through critically evaluating all steps involved. Measurements (length and weight) of individuals were taken between steps in the processing sequence to determine water loss during processing and to assess whether the processed BDM satisfied Fiji's export size limit of 7.62 cm or 3 inches [16] before shipment. Data on processing methods were analyzed using Logistic regression analysis and a Hosmer and Lemeshow Goodness-of-Fit Test was done to determine the relationships between processing variables (harvesting, smoking, first cook, cutting and gutting) at $P<0.05$. The Hosmer and Lemeshow Goodness-of-Fit Test was performed on the different steps of processing methods such as harvesting, boiling, cutting and gutting, smoking and drying so that the main step of the processing step could be determined leading to major quality losses. All data were expressed as percentages plus or minus standard error and were analyzed using SPSS software.

Results and Discussion

Problems with harvesting

Collection of all accessible sea cucumbers was done during low tide by communities involved in the BDM trade. A total of 19 species were captured and recorded during this study (Table 1). Daily harvests were stored in 20 litre drums that were cut in half. Our results revealed that fishers collected a large number of juvenile sea cucumbers that were below the regulated minimum size limit of 7.62 cm (Table 1) [16]. The harvested species with the highest proportions of undersized individuals were Curryfish, Greenfish, Lollyfish, Sandfish (currently banned as a target species in Fiji), Curryfish 'D' grade (local name 'Geci': *Stichopus horrens*) (no size limit due to its small size), Stonefish, Surf Redfish, Tigerfish and sea cucumber (local name 'Mundra': *Bohadschia marmorata*) (Table 1). Harvesting of undersized individuals indicates over-exploitation of these sea cucumber species in Fiji. However, the Fisheries Division in Fiji lacks monitoring and a proper management plan for the sea cucumber fishery and this poses a significant threat to the future of Fiji's BDM fishery.

During the harvesting of sea cucumbers, villagers also targeted other species for sale or for local consumption including crabs, lobster, clams, fish and other shellfish. With the exception of Sandfish, sea cucumbers usually have a fragile and soft skin that is easily damaged as a result of improper handling [9,10]. Villagers in Fiji tend to mix sea cucumbers with other seafoods after collection and this causes deep scratches in the sea cucumber skin [10] (Figure 2). These scratches are sometimes severe and may affect the quality of the product after final drying [9,10,12].

High value species collected by fishers in Fiji are also affected by poor storage in the sea. Instead of being kept separately, White Teatfish is mixed together with other sea cucumber species and usually also with fish and crabs. Proper storage techniques for sea cucumbers have been described previously [9,10], yet our results show that 45% of harvested White Teatfish are damaged due to poor storage (Table 1 and Figure 2). Curryfish (*Stichopus hermanni*) and Greenfish (*S. chloronotos*) also showed a high proportion of damage during harvesting (Table 1) resulting in skin deterioration (Figure 3).

Sea cucumber species with handling issues

Two species of sea cucumber, Curryfish (*Stichopus hermanni*) and Greenfish (*Stichopus cloronotos*) were determined to be the most difficult species for handling and storage in the sea by the fishers interviewed. This is because of skin and flesh deterioration which occurs as soon as they are removed from seawater and mixed with other sea cucumber species (Figure 3). The highly fragile skin and flesh of Curryfish and Greenfish discourages fishers from harvesting these two species for processing. This is despite them being ranked as medium to high value species in Fiji and a good revenue earner. Further research on skin and flesh disintegration needs to be done to determine the causes of this problem and to provide a basis for developing more appropriate harvesting and handling methods for these species.

Post harvest processing

Sea cucumber processing is accomplished using two different techniques; the salting and unsalting methods (Figure 4). The salting

Common name	Fijian name	Value	% undersized species harvested	% species damaged during harvest	% over cooked	% poorly cutting and gutting	% smoke burnt	% semi dried product	% spoilt
Amberfish	*Dri volavola*	Low	11.2 ± 2.3	14.4 ± 2.6	10.6 ± 2.2	4.3 ± 1.5	8.0 ± 2.0	6.9 ± 1.9	0.0 ± 0.0
Blackfish	*Dri loa*	Medium	12.5 ± 1.4	26.6 ± 1.9	23.0 ± 1.8	50.9 ± 2.1	15.0 ± 1.5	7.9 ± 1.1	17.0 ± 1.6
Black teatfish	*Loaloa*	High	17.4 ± 2.0	14.6 ± 1.9	5.7 ± 1.2	1.7 ± 0.7	7.1 ± 1.4	0.0 ± 0.0	2.3 ± 0.8
Brown sandfish	*Vula*	Low	1.3 ± 0.3	23.1 ± 1.2	35.5 ± 1.4	72.8 ± 1.3	18.9 ± 1.1	31.6 ± 1.3	59.2 ± 1.4
Curryfish	*Lokoloko; Laulevu*	Low – Medium	36.3 ± 1.2	36.0 ± 1.2	30.6 ± 1.1	42.6 ± 1.2	32.4 ± 1.2	26.7 ± 1.1	44.0 ± 1.2
Deepwater redfish	*Dritabua*	Low – Medium	0.0 ± 0.0	13.4 ± 1.9	4.5 ± 1.2	10.5 ± 1.7	10.8 ± 1.8	0.0 ± 0.0	0.0 ± 0.0
Elephants trunkfish	*Dairo ni cakau*	Low	0.0 ± 0.0	8.8 ± 2.2	9.4 ± 2.3	13.1 ± 2.7	8.8 ± 2.2	7.5 ± 2.1	0.0 ± 0.0
Greenfish	*Dri votovoto; Tarase*	Medium - High	46.1 ± 1.1	32.4 ± 1.0	14.5 ± 0.7	11.4 ± 0.7	14.7 ± 0.8	2.1 ± 0.3	18.3 ± 0.8
Lollyfish	*Loliloli*	Low	46.5 ± 1.1	4.7 ± 0.5	14.2 ± 0.8	39.3 ± 1.1	27.6 ± 1.0	29.4 ± 1.0	12.5 ± 0.7
Pinkfish	*Dri damu*	Low	9.0 ± 1.7	0.0 ± 0.0	12.0 ± 1.9	0.0 ± 0.0	19.3 ± 2.3	14.7 ± 2.0	0.0 ± 0.0
Prickly redfish	*Sucudrau*	Medium-High	8.1 ± 1.6	16.3 ± 2.1	0.7 ± 0.5	4.4 ± 1.2	0.0 ± 0.0	6.8 ± 1.5	0.0 ± 0.0
Sandfish **[banned]**	*Dairo*	High	41.2 ± 2.7	0.0 ± 0.0	32.5 ± 2.6	0.0 ± 0.0	45.5 ± 2.8	0.0 ± 0.0	0.0 ± 0.0
Curryfish 'D' grade*	*Laulevu; Geci*	low	100.0 ± 0.0	0.0 ± 0.0	10.7 ± 0.8	0.0 ± 0.0	0.0 ± 0.0	0.0 ± 0.0	0.0 ± 0.0
Stonefish	*Dri vatu*	Medium	34.2 ± 2.3	12.5 ± 1.6	0.0 ± 0.0	2.4 ± 0.7	8.3 ± 1.3	0.5 ± 0.3	0.0 ± 0.0
Surf redfish	*Tarasea*	Medium	21.9 ± 1.9	4.1 ± 0.9	9.0 ± 1.3	16.2 ± 1.7	38.4 ± 2.2	9.4 ± 1.3	3.3 ± 0.8
Tigerfish	*Vula ika*	Low – Medium	32.9 ± 1.5	25.9 ± 1.4	11.8 ± 1.0	38.7 ± 1.6	22.3 ± 1.3	23.9 ± 1.4	19.7 ± 1.3
White teatfish	*sucuwalu*	High	12.6 ± 1.9	45.0 ± 2.8	3.2 ± 1.0	12.9 ± 1.9	4.2 ± 1.1	0.0 ± 0.0	3.9 ± 1.1
Sea cucumber	*Mundra*	Low	31.8 ± 1.2	14.8 ± 0.9	20.1 ± 1.0	16.4 ± 0.9	21.7 ± 1.0	10.3 ± 0.8	12.3 ± 0.8
Snake fish	*Loliloli*	Low	0.0 ± 0.0	0.0 ± 0.0	21.2 ± 2.0	0.0 ± 0.0	6.3 ± 1.2	0.0 ± 0.0	0.0 ± 0.0

*Species with no export size limit of 7.62 cm.

Table 1: The main species of sea cucumbers processed in Fiji and the percentages of each influenced by factors affecting product quality and value.

Figure 2: Sea cucumbers damaged during harvesting as a result of mixing with other seafood.

Figure 3: Flesh deterioration observed in Curry fish (*Stichopus hermanni*) after harvesting

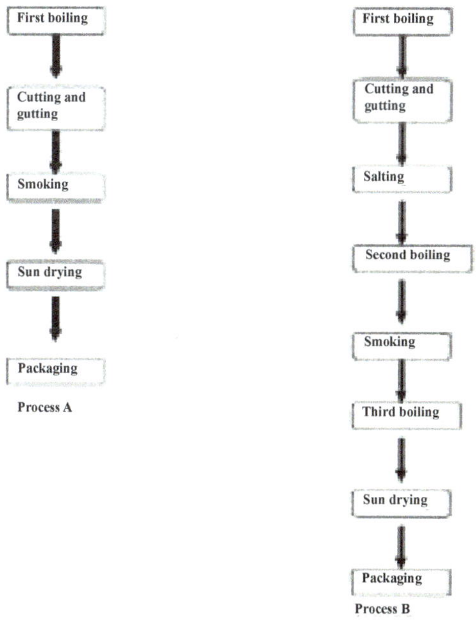

Figure 4: Sea cucumber processing by fishers and the main agents in Fiji Islands. Process A is the 'unsalting' method and process B is the 'salting' method. (Source: Ram, 2008)

method is generally used by the main agents of bêche-de-mer exporters in Fiji and the middlemen engaged in this trade. The simpler unsalting method is generally used by fishers in Fiji because of their lack of required processing resources. Some major product quality issues relating to production of a 'grade A' BDM are linked to the unsalting method of processing.

Bêche-de-mer processing and problems (post-harvest processing)

Analysis of data using logistic regression analysis and a Hosmer and Lemeshow Goodness-of-Fit Test showed that in addition to first cook ($P<0.05$), cutting and gutting ($P<0.06$) and (smoking) ($P<0.04$) were the major issue leading to loss of BDM quality as a result of poor processing at the fisher's level (Table 2). Improperly smoked products have a tendency to deteriorate rapidly due to case hardening. When sold to the marine product agents these products are generally rejected or receive reduced grading with lower value.

The major issue affecting BDM product quality and value in Fiji is that fishers do not follow the standard procedure for processing bêche-de-mer [9]. In general they process sea cucumbers with little regard for quality to obtain a quicker source of income [17]. Similar research in Madagascar revealed that, to generate quick income, sea cucumber fishers missed fundamental steps in the processing procedure which badly affected the quality of the final product [18]. If the standard processing steps for BDM are followed, good quality product will result that earns greater income for fishers and communities than poorly processed product [12].

First boiling

Our interviews revealed that 100% of fishers boil all the sea cucumber species of all sizes together without any sorting or grading before processing. Larger sizes of sea cucumber require ample time for cooking during first boil [9,10]. However, attention to this component of processing by Fijian fishers is poor and a high proportion of sea cucumbers are overcooked (Table 1). Overcooked sea cucumbers often burst affecting product quality [10].

Inappropriate cooking also results in skin peeling of high value sea cucumber species such as White Teatfish and Black Teatfish, when cooked in boiling water instead of warm water (50–65°C). In addition, Blackfish should be cooked at a high temperature (90–95°C) but not boiled. Introducing commercial sea cucumbers into boiling water results in skin peeling that affects their appearance after drying and reduces value [9,10].

Cutting and gutting

After first boiling sea cucumbers are cut and gutted to remove the viscera. The cutting methods used by fishers in Fiji are generally not of standard quality [9]. Instead of making small cuts in the anal region for species that do not require cuts in the body wall (such as Tigerfish, Stonefish, Deepwater Redfish and Brownspotted Sandfish) [9], very long cuts are made. These cuts are also generally not straight in the case of high value species. Chinese markets prefer BDM products with neat cuts which enhance their appearance; this is usually not considered during local processing in Fiji. Products may also be incompletely gutted resulting in a product containing gut contents that are contaminated with sand, ash and dust (Table 1). The data for cutting and gutting shown for Sandfish in Table 1 (0%) reflects the fact that this species is generally harvested and sold frozen to the main agents. Processing of Sandfish is currently banned in Fiji and our data are incomplete because the main agents are very secretive regarding Sandfish export data. There could still be significant harvesting of Sandfish in Fiji and, without monitoring and harvesting records, this could have serious consequences for the recovery of Sandfish stocks in Fiji.

Smoking

Smoking is performed after cutting and gutting (Figure 4). In the salting procedure (Fiji's current standard procedure), after cutting and gutting, the products are usually salted in grade 11 coarse solar salt for 2-3 days before they are boiled again and smoked (Figure 4). Instead of producing a gentle smoke for the smoking step, fishers in Fiji generate fire that either roasts the products (Figure 5) or generates a hard skin on the product known as "case – hardening" [19]. "Case – hardening" is described as burnt skin resulting from high temperature that prevents moisture loss from the product resulting in spoilage. The proportion of burnt product resulting from smoke drying during sea cucumber processing by fishers in Fiji is shown for all harvested species in Table 1.

Usually the smoking process is done for 24 – 48 hours and the resulting product is sun dried for four to five days to accomplish a 'grade A' product [9,10]. However, to generate rapid income, village processors in Fiji generally perform the smoking and sun drying process together for a day and sell the product to middlemen or the main agents. The products contain 25-35% moisture and due to poor storage, a high proportion of the product may spoil as a result of insufficient drying (Table 1). Insufficient drying is a major issue in local processing of BDM in Fiji. This problem results because sea cucumbers are generally piled on top of each other during processing

		B	S.E.	Wald	df	Sig.	Exp(B)
Step 1(a)	Smoking	2.132	1.038	4.218	1	.040	8.431
	Harvesting	.690	.426	2.627	1	.105	1.993
	First cook	1.214	.625	3.775	1	.052	3.367
	Cutting and gutting	1.296	.689	3.542	1	.060	3.655
	Constant	-1.652	1.345	1.510	1	.219	.192

Table 2: Logistic Regression analysis and Hosmer and Lemeshow test for beche-de-mer processing techniques in Fiji.

Figure 5: Various examples of smoke burnt sea cucumber species.

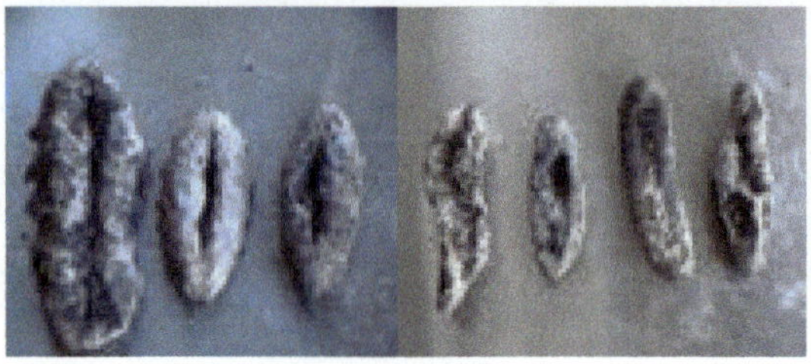

Figure 6: Poorly processed sea cucumbers for export from Fiji

Common name	Grades of product	Percentage undersized species exported	Percentage damaged species exported	Percentage over dried species exported	Percentage discarded as waste
White teatfish	A+	0	4 ± 7.4	0	0
	B	0	17 ± 5.3	0	0
	C	0	26 ± 6.2	0	0
	D	14 ± 3.5	43 ± 5.0	0	0
Black teatfish	A	0	17 ± 5.3	0	0
	B	0	16 ± 5.2	0	0
	C	10 ± 3.0	42 ± 4.9	0	0
Sandfish	Banned	Banned	Banned	Banned	Banned
Prickly redfish	1 grade	0	12 ± 4.6	0	0
Stonefish	1 grade	82 ± 3.1	0	9.0 ± 2.9	1.3 ± 1.5
Surf Redfish	A	4.0 ± 1.9	12 ± 3.2	0	0
	B	40 ± 5.0	36 ± 4.8	0	1.0 ± 1.3
Blackfish	A+	0	12 ± 4.6	0	0
	B	44 ± 4.9	13 ± 3.4	0	0
	C	86 ± 3.5	29 ± 4.5	7.0 ± 2.6	2.2 ± 1.9
Elephant Trunkfish	1 grade	0	0	0	0
Greenfish	1 grade	80 ± 4.0	5 ± 2.2	11 ± 3.1	1.7 ± 1.7
Curryfish	A	0	5 ± 3.1	0	0
	B	12 ± 3.2	10 ± 3.0	3.0 ± 1.7	0
	C	48 ± 5.0	30 ± 4.6	12 ± 3.2	4.0 ± 2.5
	D (*Geci*)	100	0	0	0
Amberfish	1 grade	0	15 ± 5.0	0	1.0 ± 1.3
Brown sandfish	1 grade	8.0 ± 2.7	8 ± 2.7	0	0.7 ± 1.1
Deep water redfish	1 grade	0	10 ± 4.2	0	0
Lollyfish	A	36 ± 4.8	0	1.0 ± 1.0	0
	B	80 ± 4.0	0	3.0 ± 1.7	0.5 ± 0.9
Tigerfish	A	0	23 ± 6.0	0	0
	B	16 ± 3.7	44 ± 5.0	2.0 ± 1.4	1.7 ± 1.7
Snake fish	1 grade	17 ± 3.7	0	0	2.5 ± 2.0
Pinkfish	1 grade	26 ± 4.4	4 ± 2.0	0	0
Sea cucumber (*Mundra*)	1 grade	98 ± 1.4	20 ± 4.0	26 ± 4.4	0.8 ± 1.2

Table 3: Percentages of poor quality bêche-de-mer products exported by the main buyers in Fiji.

which hinders moisture loss. The products are kept like this for days and this result in spoilage (Table 1).

Bêche-de-mer quality standard at the main buyer's warehouse before export

The main BDM exporters in Fiji either process sea cucumbers themselves to a marketable standard before exporting to Asian markets, or purchase product processed by local fishers and reprocess prior to export. Reprocessed product is given a low grade and has reduced value. Data generated from the main agent's ware house revealed that nearly all the bêche-de-mer (all species) exported from Fiji to Asian markets were of poor quality (Table 3). Factors that influenced this grading included:

- Undersized (less than 7.62 cm) (Table 1)
- Damaged (skin damage, flesh ripped, skin peeled and holes in the tissue)
- Product had poor appearance (shape distorted, gut cavity contaminated with sand, dust and remaining gut contents, poorly cut for gutting) (Figure 6)
- Products were over-dried (Table 3) and burnt as a result of smoking at higher temperatures.

When BDM products are traded by village processors to the main agents or exporters, they are sometimes rejected due to extremely bad odour resulting from spoilage, insufficient drying and burning during the smoking process. Buyers however, do purchase the product but when graded by size and quality, extremely poor quality products that cannot be reprocessed for the market are rejected or discarded as waste (Table 3). High value species such as White Teatfish, Black Teatfish are not rejected if they are processed by the fishers but instead purchased and reprocessed as a low grade product (e.g. 'Grade D') and exported to Asian markets for low value. Because populations of the high value species of sea cucumbers are declining rapidly in Fiji [20], buyers tend to purchase the product in perishable state from the fishers for little value and process themselves to obtain a higher return from Asian markets (Table 3). The results in Table 3 show that 43% of White Teatfish (a high value species) is exported as damaged product that was processed by fishers. As a result, processors now prevent fishers from processing the higher value species.

The percentages of damaged individuals are high for higher value species (Table 3) due to flesh and skin damage caused during harvesting and processing. In addition, high value species are generally purchased raw from the villages and extended exposure of the sea cucumbers to the air or to sun may result death or very poor condition. The skin may also peel as a result of the drying effect from the air or sun [9]. Data for the percentage of damaged individuals shown in Table 3 also includes smoke burnt products that were processed by village processors.

The proportion of over-dried product for the major exported species is also shown in Table 3. Over-dried BDM has reduced nutritive value [21,22] and, especially after over boiling, may have skin and flesh that is fragile and may easily break leading to serious product quality issues [9,10]. Greenfish, Amberfish, Surf Redfish and Curryfish, in particular, show a high proportion of damaged product (Table 3) because of the fragility of their skin and flesh and their susceptibility to damage during harvesting.

Conclusion

Bêche-de-mer processing in Fiji is generally done rapidly to obtain income as soon as possible. As a result essential processing steps are shortened or missed completely and this has serious consequences for the quality of the final product. Processing steps in Fiji involved first boil after harvest, slitting and gutting, followed by salting (main agents only), smoke drying and finally sun-drying. Fishers found smoke drying and first boiling to be difficult steps during processing and this result in poor quality products being produced in the Fiji. The final dried product generally has poor exterior appearance, distorted or twisted shape, and particulate matter present in the gut cavities and, perhaps of most importance, a high rate of spoilage due to improper storage. These factors reduce the grade of the final product from 'A' to 'B-D' grade. Further research is required into the impacts of processing methods on the quality of BDM produced in the Fiji and to develop ways to add value to BDM exported from Fiji. The results of our study have prompted immediate awareness and education activities for sea cucumber fishers and processors in Fiji supporting improvements in product quality and improved income from this important coastal livelihood.

Acknowledgement

The authors wish to thank the main buyers and village people in Fiji for their assistance with data gathering and provision of information. Involvement of officials from the Ministry of Fisheries and Forest of Fiji is also thankfully acknowledged.

References

1. Bumrasarinpai R (2006) Overview of issues on concerned aquatic species under the convention on international trade in endangered species of wild fauna and flora (CITES), in Regional technical consultation on international fisheries related issues: WP03.
2. Conand C (1990) The fishing resources of the Pacific Island Countries. Part 2: Holothuria, in Overview of Bêche-de-mer industry in Milne Bay Province, Papua New Guinea, J. Kinch, Editor. SPC Bêche-de-mer information bulletin. p. 2-15.
3. Anon (1994) Sea cucumbers and Bêche-de-mer of the tropical Pacific: a handbook for fishers, no 18, in Bêche-de-mer production in Tuvalu, K. Belhadjali, Editor. 1994, SPC Bêche-de-mer information bulletin. p. 2-3.
4. Ward P (1972) The Pacific Bêche-de-mer with special reference to Fiji. In Pp 2 – 15. , in Bêche-de-mer and sandalwood trade in Fiji. Master's Thesis, M. Dough, Editor. 1972: Macquarie University. Department of History. School of general studies: 2-15.
5. James D (1988) Production, Consumption, and Demand in Fish Drying and Smoking; production and quality, P.E. Doe, Editor Technomic Publishing Co Inc.1-4.
6. Sikorski Z, Haard N, Motohiro T, Pan BS (1998) Quality, in Fish Drying and Smoking; production and quality, P.E. Doe, Editor., Technomic Publishing Co Inc:88-116
7. Ferdouse F (1999) Bêche-de-mer markets and utilisation. SPC Bêche-de-mer information bulletin,11: 3-9.
8. Bordbar S, Anwar F, Saari N (2011) High-value components and bioactives from sea cucumbers for functional foods--a review. Mar Drugs 9: 1761-1805.
9. SPC (1994) Sea cucumbers and beche-de-mer of the tropical pacific. Handbook No. 18 Noumea: SPC. 51.
10. Sachithananthan K, Osman SS, Mlay M, Schoemaker R n (1985) Report on the national workshop on fish handling in Zanzibar, in United Republic of Tanzania and report on the Tanzania/ SWIOP national workshop on Bêche-de-mer processing. 1985, National workshop on fish handling. 45-83.
11. Purcell SW (2014) Value, market preferences and trade of Beche-de-mer from Pacific Island sea cucumbers. PLoS One 9: e95075.
12. Seeto J (1999) Bêche-de-mer processing – a little more effort to get much more money while saving precious resources. SPC Bêche-de-mer information bulletin, 11: 2-3.
13. Schoppe S (2000) Sea cucumber fishery in Philippines. SPC Bêche-de-mer information bulletin, 13:10-12.
14. Jun A (2002) Trepang exploitation in the Philippines: updated information. SPC Bêche-de-mer information bulletin, 17:17-21.
15. Ram R, Chand R.V, Southgate PC (2014) Effects of Processing Methods on the value of bêche-de-mer from the Fiji Islands. Journal of Marine Science: Research and Development ,in press.
16. Bruckner AW, Johnson KA, Field JD (2003) Conservation strategies for sea cucumbers: Can a CITES appendix II listing promote sustainable international trade. SPC Bêche-de-mer information bulletin, 18: 24-33.
17. Kinch J (2002) Overview of Bêche-de-mer industry in Milne Bay Province, Papua New Guinea. SPC Bêche-de-mer information bulletin, 2002. 17:2-15.
18. Irwing MA (1994) Fishing Bêche-de-mer in Madagascar. SPC Bêche-de-mer information bulletin, 6:10-11.
19. Venugopal V (2006) Seafood processing: adding value through quick freezing, retortable packaging, and cook – chilling. United States of America, CRC press, Taylor and Francis group.
20. Pakoa K, Saladrau W, Lalavanua W, Valotu D, Tuinasavusavu I et al, (2013) The status of sea cucumber resources and fisheries management in Fiji. SPC :51.
21. Chen J (2001) Present status and prospects of sea cucumber industry in China, in Advances in sea cucumber aquaculture and management .A. Lovatelli et al, FAO Fisheries Technical Paper. No. 463: Rome; FAO:25-38.
22. Ram R (2008) Impacts of harvest and post-harvest processing methods on quality and value of bêche-de-mer in Fiji Islands. Master of Science Thesis, University of the South Pacific. 191.

14

Design and Prototype Development of a 1-DOF Flapping-foil & Gliding UUV

Ding Hao*, Song Bao-wei and Cao Yong-hui

College of Marine Engineering, Northwestern Polytechnical University, Xi'an 710072, China

Abstract

The main objective of this paper is to investigate an underwater unmanned vehicle (UUV) which has both flapping-foil and gliding function. Firstly, the paper introduced the advantages of the flapping-foil & gliding UUV, and described the flapping-foil motion model of bionic object turtle in detail. Then, by using the dynamic meshing technology in the code Fluent, the numerical computations of unsteady hydrodynamic characteristic for single degree of freedom (1-DOF) flapping-foil movement was carried out, which demonstrated the mechanism of flapping-foil. During the computation, turbulence effects were taken into account, and the turbulent stressed was evaluated by means of the Realizable k-epsilon model. Finally, the kinematics simulation of the UUV was performed. The design on virtual prototype of the UUV was completed in UG software, and the prototype was processed in the digital factory. Moreover, we presented the corresponding solutions aiming at the problems of the assembly prototype.

Keywords: Underwater unmanned vehicle (UUV); Bionic; Flapping-foil; Gliding; 1-DOF

Introduction

The traditional propeller used by most of the UUVs has a broad field of application of the ship. However, the maneuverability and speed performance of the propeller still need to be improved. Considering the complicated environment of the seabed in marine engineering, such as undercurrent, wave, bay area, etc., the traditional propeller is not able to meet the requirements of the modern marine operations [1].

In recent years, the flapping-foil technology and underwater gliding technology have been developed independently, and some achievements are also obtained, which can partially overcome the subsistent problems of the traditional propeller well.

The physics of beat and flying focus on the considerable theoretical, numerical, and experimental work. Experimental work with live fish [2-4] and biorabotic devices modeled on the blue-fin tuna [5], and small-mouth bass [6] have been studied extensively. One of the motivations has been to improve the design and performance of underwater vehicles, recognizing that in many ways there is an extraordinary gap between the abilities of man-made machines and other marine animals. Quantitative comparisons between the hydrodynamics of fish and small underwater vehicles have been formulated [7]. Extensive testing of oscillating foils has been performed in the Massachusetts Institute of Technology (MIT) Ship Model Testing Tank, resulting in a solid understanding of the fundamental parameters of thrust production in foils [8-10], and design and projected performance of the flapping-foil AUV [11,12] has also been studied.

The concept of underwater glider was envisioned in 1989 [13], then three operational gliders have been developed, including the Slocum Battery/Thermal glider [14], the Spray slider [15] and the Sea glider [16], which change their volume and buoyancy to cycle vertically in the ocean and use lift on wings to convert the vertical velocity into forward motions [17].

However, the research on combining technology of these two propulsion technologies is few in the literatures. It will become an important direction that how to compromises the merits of this two technologies. In this paper, we present a 1-DOF flapping-foil & gliding UUV, which imitates the swimming mechanism of some marine animals such as turtles, penguins and seals. It is a new concept UUV combines flapping-foil technology with underwater gliding technology, which has many advantages including high efficiency, high stability, high maneuverability, long range and so on.

As part of the development of the UUV, this paper carried out numerical simulations to investigate the mechanism of flapping foil propulsion. Besides, virtual design and the prototype manufacture were introduced. Finally, motion simulations were proposed to evaluate the motion performance of the vehicle.

Motion Mode of Bionic Object

In this paper, the UUV was designed with green turtles as bionic object. Swimming in the water, the turtle generates thrust mainly relying on the forelimb stroking, while after-limb only plays the role of the rudder. There are three basic swimming forms of the turtle, such as horizontal straightline movement, turning and heaving. The most typical movement pattern is the horizontal straightline, and the other two movements are just auxiliary motions basing on the motion of horizontal straightline. That is, the forelimb has different flapping frequency and amplitude to cooperate with the motion of the after-limb in the turning and heaving motions.

In order to give an intuitive description of the forelimb motion mode when the turtle moving in a horizontal straightline, we assign a Cartesian coordinate system O-XYZ fixed to the turtle body with its origin at the root point of the forelimb, as shown in Figure 1. Let axis Y parallel the long axis of the turtle (positive in the direction of the turtle's nose), and axis Z, which points to the top, lie in the longitudinal plane

***Corresponding author:** Ding Hao, College of Marine Engineering, Northwestern Polytechnical University, Xi'an 710072, China, E-mail: hao_shreck@163.com

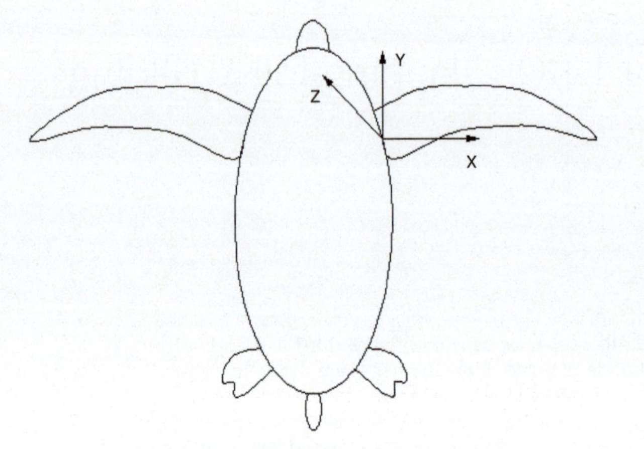

Figure 1: The coordinate frame of turtle forelimb motion.

and perpendicular to axis Y. Axis X points in the direction orthogonal to Y-Z plane [18].

In generally, swimming straight and level, the forelimb motion mode contains three kinds of rotation around X-axis, Y-axis and Z-axis, respectively. The rotation around Y-axis makes turtle forelimb beat water up and down; the rotation around X-axis makes the forelimb pitch and change angle of attack of the forelimb; the rotation around Z-axis makes the forelimb stroke forward and backward, while the range is relatively narrow. All these three cooperative rotations make the turtle get a continuous forward thrust, and the motion mode has three degrees of freedom (3-DOF).

Numerical Simulation of 1-DOF Flapping-foil Motion

Mathematical model of 1-DOF flapping-foil motion

This paper simplified the complex 3-DOF flapping-foil motion of turtle to 1-DOF flapping mode, and considered that the foil was rigid in free boundary flow. Defining the foil flapping according to the sine law, the expression of the flapping angle displacement [19] is:

$$\theta(t)=\theta_0 sin(2\pi ft+\psi)+\theta_{bias} \quad (1)$$

Where θ_0 is the foil flapping amplitude, f is the frequency, ψ is the initial phase angle. θ_{bias} is used to change the equilibrium position of the foil flapping, which leads to a change of the direction of the thrust produced by the flapping-foil motion.

The expression of the angle velocity of flapping-foil can be derived by differentiating with respect to $\theta(t)$:

$$\omega(t)=2\pi f\theta_0 \cos(2\pi ft+\psi) \quad (2)$$

Computational domain and grid system

Size of the computational domain:

$1500mm\times1500mm\times1500mm$

Inlet boundary conditions: velocity inlet;

Outlet boundary conditions: outflow;

Other boundary conditions: wall.

Using the dynamic meshing technology in the code Fluent, the numerical simulation of unsteady hydrodynamic characteristic for 1-DOF flapping-foil motion can be carried out by solving the incompressible Navier-Stokes equations numerically in a non-inertial reference frame. During the computation, turbulence effects were taken into account, and the turbulent stressed was evaluated by means of the Realizable k-epsilon model.

Since the dynamic meshing technique should base on the unstructured grid system, we discretized the computational domain into the unstructured grid using the ANSYS ICEM software. In the process of computation, the unstructured grids were updating with every time step by spring-based smoothing method and local remeshing method. To improve the quality of the mesh updating, a spherical interface was set to divide the computational domain into two parts. Moreover, to increase the calculation accuracy, the meshes near the surface of foil wall were refined. The grid system is shown as Figure 2.

Calculation results and analysis

For the motion object, the foil is three-dimensional and rigid, and the airfoil profile of section is NACA0010. Its root chord is $c_{max}=70mm$, and tail chord is $c_{min}=1/3c_{max}$, i.e., the chord shrinks in a proportion 3:1 in the direction of span. To explore the change principle of the thrust produced by the flapping-foil motion with time increasing, numerical simulation was carried out in the conditions that the coming flow velocity is $U=0.01m/s$, the amplitude of beat is $\theta_0=30^o$, and the

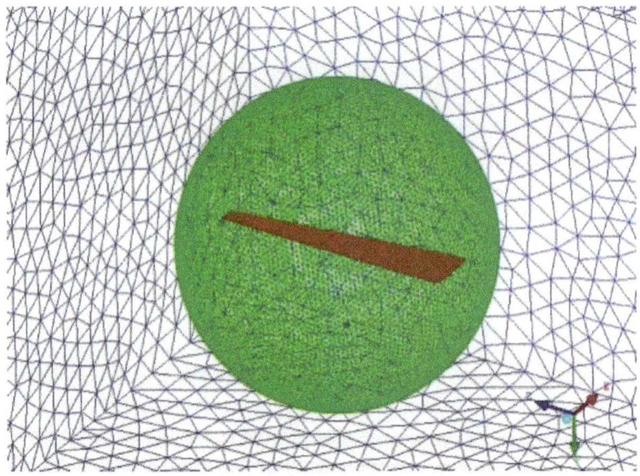

Figure 2: Unstructured grid system around the foil.

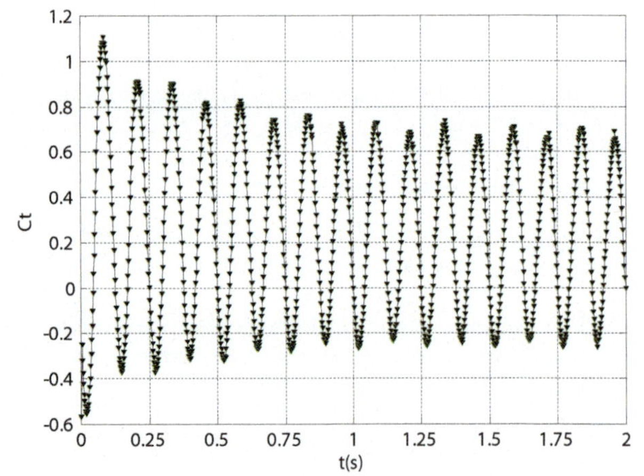

Figure 3: Curves of thrust coefficient changing with time.

frequency is f=*4Hz*. Figure 3 shows the curve of the thrust coefficient *Ct* changing with time.

The time step of this numerical example was set as 0.002s with total 1000 steps. The flapping-foil motion lasted 2s from start to end and had eight full periods. After the third period (0.75s), the flow field was tending to steady. It can be seen from Figure 3:

(1) At the beginning of the flapping-foil motion, thrust coefficient is negative, which shows the drag is produced at this moment;

(2) In one period of the motion, two peak values of the thrust and two peak values of the drag ($Ct<0$) are observed;

(3) In one period, the average thrust coefficient is positive, which means the continuous flapping-foil motion can produce a continuous thrust for the UUV.

(4) Periodicity has been found from the curve of the thrust coefficient changing with time, and its frequency is twice as much as that of the flapping-foil motion.

Influence of motion parameters on thrust

Relationship between frequency and thrust: When coming flow velocity $U=0.2m/s$, and the amplitude $\theta_0=30°$, thrusts are calculated in the cases when the frequency f are *1Hz, 1.5Hz, 2Hz, 2.5Hz, 3Hz, 4Hz, 5Hz* and *6Hz*, respectively.

Figure 4 shows that in the case when other parameters are identical, the average thrust is approximately a square function of flapping frequency, $T\alpha f^2$. That is, since the frequency increases, the work done on the fluid is also increases, which leads to a raise of thrust.

Relationship between amplitude and thrust: When coming flow velocity $U=0.2m/s$, and flapping frequency, $f=2Hz$ thrusts are calculated in the cases when the amplitude θ_0 are 10°, 20°, 30°, 40°, 50°, 60°, 70°, 80°, 85° and 90°, respectively.

As can be seen in Figure 5, thrust gets larger as the amplitude increases, but the changing rate slows down when θ_0 >50°, and the thrust approximately stayed the same when θ_0 >85°. This can be explained as when the amplitude increases, the resistance on the foils evidently increases and the thrust generated mainly contributes to counteract the resistance.

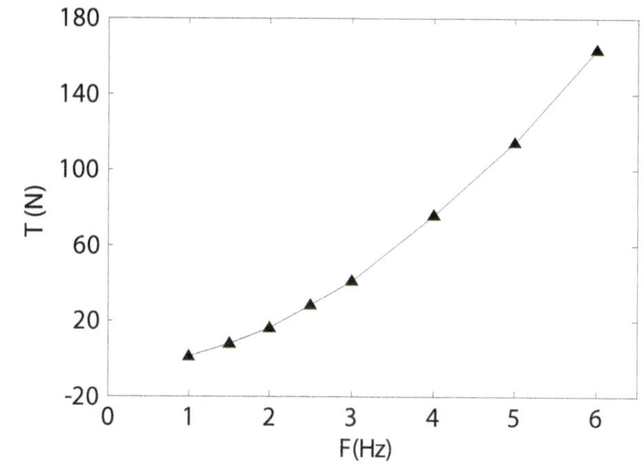

Figure 4: Curve of thrust changing with frequency.

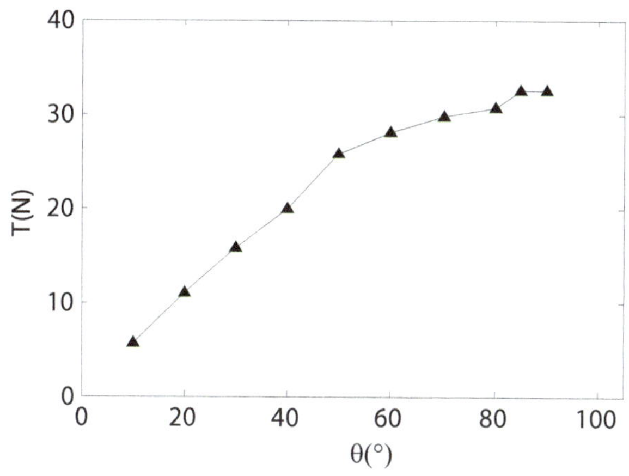

Figure 5: Curve of thrust changing with amplitude.

Virtual Design of 1-DOF Flapping-foil & Gliding UUV

The overall structure of the 1-DOF flapping-foil & gliding UUV designed in this paper mainly included five parts: (1) main body and accessories; (2) 1-DOF flapping-foil thruster; (3) glide control system; (4) steering gear structure; (5) control and communication system. The idea of modular design was adopted for each part, which has many advantages such as high reliability, easy for assembly and disassembly, excellent expansibility. Figure 6 shows the integrated layout of the UUV [20].

Main body

The design of the main body can be divided into the following parts: configuration design of pressure shell, material selection of the shell, design of seal, and optimization of interior space, etc.

From a bionic standpoint, we selected ellipsoid as the configuration of the pressure shell in consideration of the maneuverability, running resistance, internal installation space of the UUV. Except the foil, the designed UUV had a total length of 600 mm, width of 420 mm, height

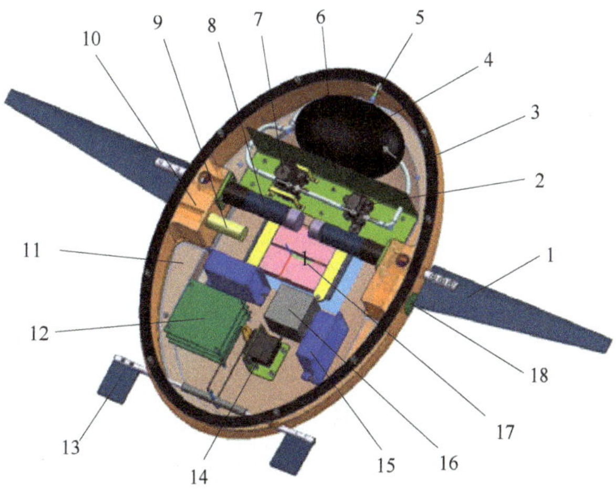

1-foil; 2-solenoid valve; 3- sealing gasket; 4-water bag; 5-cover for draining; 6-clapboard; 7-pump; 8-motor component; 9-depth sensor; 10-nether shell; 11-mounting plate; 12-PC104; 13-rudder; 14-steering gear; 15-motor actuator; 16-attitude sensor; 17-battery component; 18-cover for checking

Figure 6: Schematic drawing of the integrated layout of the UUV.

of 220 mm, and minimum wall thickness of 10 mm, hence the external offsets of configuration is:

$$\frac{x^2}{300^2}+\frac{y^2}{210^2}+\frac{z^2}{110^2}=1 \quad (3)$$

The shell consisted of two parts, where the joint of the upper and nether was sealed with rubber gasket, and the shafts with foil and rudder were sealed with O-shape seal ring. The good sealing properties ensured that the security of the internal space to install various devices and apparatuses. In order to save space and make the center-of-gravity position of the UUV easier to control, a mounting plate fixed on the nether shell was designed for laying up most of the devices. Considering the fabrication, procedure and work environment of the UUV, we selected cemented aluminum alloy as the material of pressure shell.

Swimming in the water, the turtle promotes thrust mainly relying on the forelimb stroking, and the after-limb only plays the role of rudder. According to this point and the requirement of gliding movement, the 1-DOF flapping-foil thruster was laid at both sides of the UUV away from the centerline 50 mm forward, and the rudder 250 mm backward, as shown in Figure 7.

1-DOF flapping-foil thruster

When the flapping-foil & gliding UUV was sailing underwater, the thruster can be used as a machine to generate power by flapping-foil motion to make the UUV in high mobility, or a glider to provide enough lift for the UUV gliding. Thus, the design of flapping-foil thruster is primary.

The 1-DOF flapping-foil thruster consisted of multiple components such as foil, shaft, coupling, speed reducer, motor etc. Relying on lithium batteries to provide power source, the motor drove the speed reducer to transfer the rotation to shaft through the coupling, and the foil was flapping with the shaft. The schematic diagram of thruster structure is shown in Figure 8. Then, the UUV could realize the function of gliding, when the foil was fixed in a prescribed station by motor self-lock and the centre-of-gravity adjustment device was working. This design reduced the driving medium from motor to the foil, therefore the transmission efficiency and accuracy were improved dramatically. It made the thruster in a very compact form and service life longer, however, this form required the high coaxiality of the motor shaft, foil shaft and mounting hole in the shell, which increased the difficulties in processing and assembly. In this paper, we ensured the high coaxiality requirement by increasing a shaft shoulder in the mounting motor flange.

To select the suitable motor for driving the foil, nominal power and maximum torque of the motor should be estimated. The foil was reduced to a trapeziform plate, and we selected a rectangular differential element with an area of dA on the plate in the direction of span. When the foil flapping, the drag on the element can be written as:

$$dF_D=\frac{1}{2}C_D\rho V_\infty{}^2 dA \quad (4)$$

Where, $C_D=1.2$ is the coefficient of viscosity of rectangular plate, $\rho=1000kg/m^3$ is the density of water, and l is the span of foil. Since $dA=cdl$, $V_\infty=\omega(t)l$, we have

$$dF_D=\frac{1}{2}C_D\rho(\omega(t)l)^2 cdl \quad (5)$$

The torque that the element on shaft is:

$$dM=\frac{1}{2}C_D\rho\omega^2(t)l^3 cdl \quad (6)$$

The total torque of the motor overcoming the drag can be obtained by integrating dM in the direction of span, and the power of motor is:

$$P=M\omega \quad (7)$$

Taking gravity of the foil into consideration, we substitute the maximum frequency $f_{max}=5$ and the maximum amplitude $\phi_0=\pi/2$ into the equation, then the nominal power and maximum torque of the motor were calculated as 42.6w and 2.3Nm, respectively. Finally, we selected the DC servo motor RE30 manufactured by Maxon Corporation in Switzerland.

This 1-DOF flapping-foil thruster makes the UUV designed have many advantages including high efficiency, high maneuverability and long range. However, the thruster has the following limitations: (1) it applies only to small UUVs and provides a relatively lower velocity, (2) the structure is complicate and (3) the control system is complex.

Glide control system and steering gear structure

Glide control system depended on the center-of-gravity adjustment device to control the angle of attack of the UUV, and it consisted of bump, solenoid valve, controller, switch, water bag and so on. Glide is an important function of the flapping-foil & gliding UUV. Cooperating with the flapping-foil motion, it can both improve the high mobility and save the energy source to ensure a long range. Generally, sliding straight at low speed, the value and the center of buoyancy of the UUV are constant. Thus, in the situation without power, using the device to change the gravity and the relative position with the center of buoyancy, the lift for gliding is produced when the angle of attack of the UUV changed [21]. The steering gear structure laid on the rear end of the UUV can be used to give a fine-adjustment to attitude of the UUV, which makes it slide according to a predetermined trajectory.

Control and communication system

Control and communication system consisted of hardware

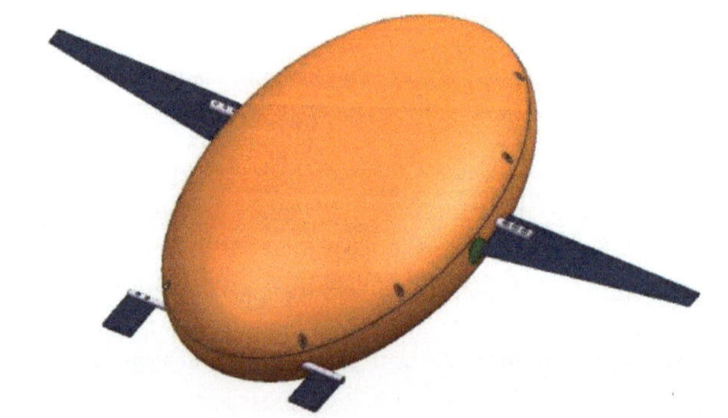

Figure 7: The 3D view of the UUV configuration and layout of foil and rudder.

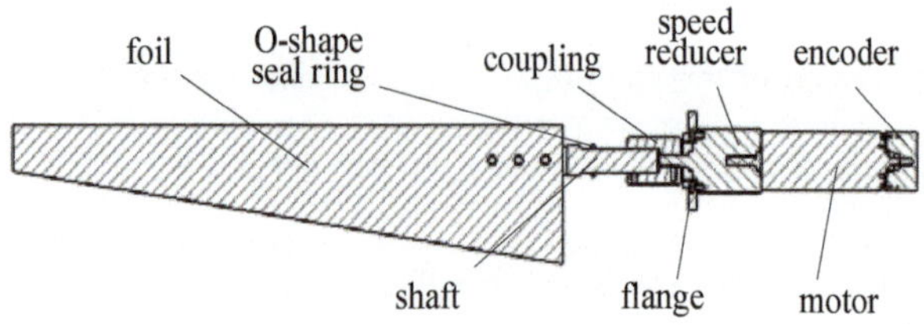

Figure 8: Schematic diagram of thruster structure.

circuitry and matching software. For processing various kinds of information effectively, a high performance central processing unit PC104 and buffer circuit were used, including various I/O ports and internal Bus. All sensors and devices, including depth sensor, motor actuator, and attitude sensor and so on, were connected to the Bus through the I/O ports. With the help of the software to conduct the real time information interchange, each module of the flapping-foil & sliding UUV can work independently or in parallel (Figure 9).

Kinematics Simulation of the 1-DOF Flapping-foil & Gliding UUV

In this section, by constructing a kinematics model for the designed 1-DOF flapping-foil & gliding UUV, the analysis of kinematical performance was studied when the UUV was cruising underwater in flapping-foil propulsion way and gliding way.

Kinematics simulation in flapping-foil working state

The kinematics simulation of the UUV diving with a negative angle of attack of 30° had performed in the conditions that the two thrusters are working together with the same frequency $f=5Hz$ and amplitude $\phi_0=\pi/4$.

From Figure 10, we can see that the UUV has dived for 60m when going forward for 100m. The UUV has a bumpy movement at the initial stage, but after a short time, it dives in a straightline with zero angle between the equilibrium position of the foil flapping and the horizontal axis plane of the UUV.

To make the UUV turn left for 60°, we set left foil flapping in frequency 3Hz and right foil in a changing frequency 4Hz~0Hz. As shown in Figure 11, the UUV has turned 60° to the left within less than 10s, and then it is sailing with a yaw angle for 60°. This simulation result illustrates that the UUV designed in this paper has a high motility when it works in flapping-foil state.

Kinematics simulation in sliding working state

The gliding kinematics simulation of the UUV had performed in the initial conditions that delivery speed for 1m/s and the mass of water bag for 0.2kg.

It can be seen from Figure 12 (a) that the UUV has submerged for 180m when going forward for 150m in no power state. Since the mass changing of the water bag has a relative long time process, in the initial stage, the UUV has a changing angle of attack and a bumpy movement as shown in Figure 12 (a)~(b). After about 120s, the angle of attack is fixed at about 37°, and at this time the UUV begins to slide down in a approximate straight line.

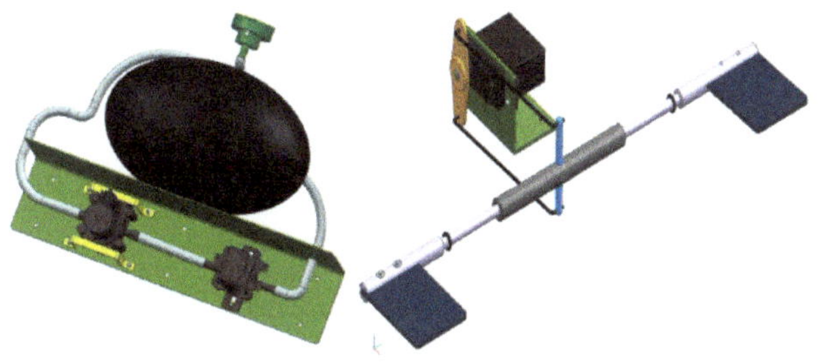

Figure 9: Center-of-gravity adjustment device and steering gear structure.

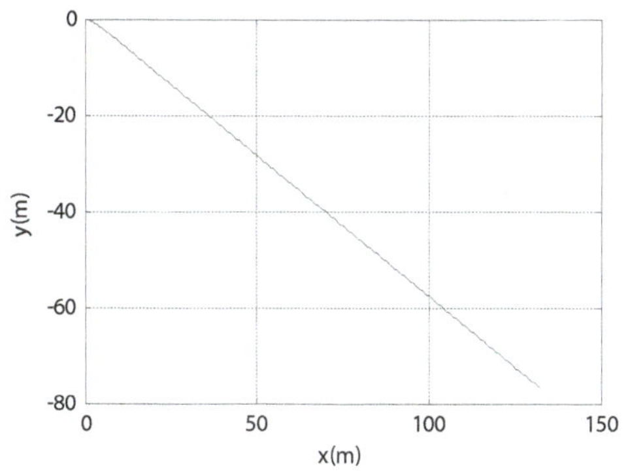

Figure 10: Diving depth changing with forward range.

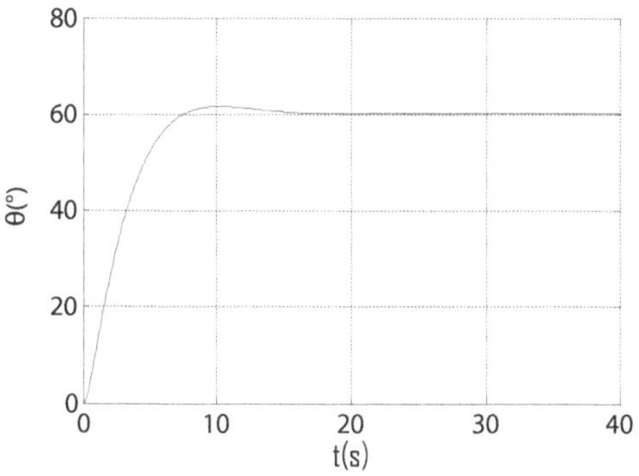

Figure 11: Yaw angle changing with time.

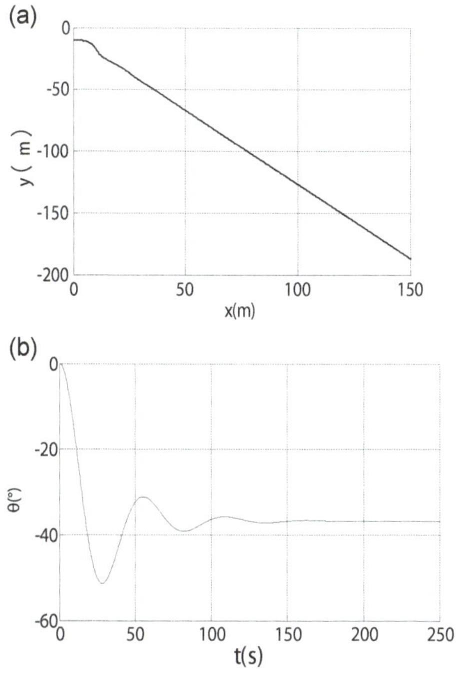

Figure 12: Sliding simulation of the UUV. (a) Diving depth changing with forward range. (b) Angle of attack of the UUV changing with time.

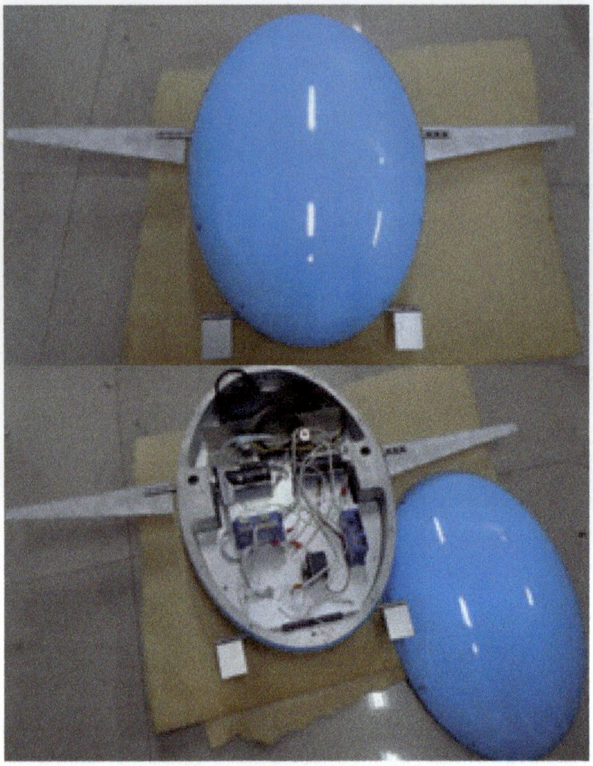

Figure 13: Appearance of the vehicle.

Prototype Design and Manufacture

Figure 13 shows the appearance and inner structure of the vehicle. All the parts of the vehicle were processed in a NC machine. Schedule drawings of the parts were drawn by means of a commercial 3-D modeling code NX v6.0. Several problems arose during the test, and were finally solved. Among all the problems, the following ones seemed the most significant and typical:

(1) Because of the accuracy error in processing and set up, the motor shaft, foil shaft and body installing hole were not concentric, giving rise to an increase of friction force on the flapping foil. This problem was successfully solved by adding sealed shims between the flanges.

(2) Seal between the upper and nether shell. Since the preload force on the fastening screw was not balance, seal problems appeared. By adding silica gel between the rubber blanket and the shells, we achieved a good seal performance.

(3) The vehicle had been designed to be zero buoyancy, however, due to the large size, the prototype was positive buoyancy after assembling. We succeeded balanced buoyancy and gravity of the vehicle by mounting lead block inside the vehicle.

Conclusion

A novel UUV, the 1-DOF Flapping-foil & Gliding UUV was developed in this paper. Numerical simulation, virtual design with the aid of CAD software, kinematics simulation, high-accuracy process on NC machine and vehicle assembling were presented. During the test, the vehicle showed good performance. Since the control algorithm is still under developing, underwater trials will be the next step of work. Moreover, validation of the vehicle characteristics also depends on the future experiments.

Acknowledgement

The authors gratefully thank the Basic Research Foundation of Northwestern Polytechnical University and the Graduate Starting Seed Fund of Northwestern Polytechnical University.

References

1. Chi DG, Yan GZ (2001) Biommetic robot research and its perspective. Robot 23: 476-480.
2. Fish F (1993) Power output and propulsive efficiency of swimming bottlenose dolphins. J Exp Biol 185: 179-193.
3. Lauder B (1995) Speed effects on midline kinematics during steady undulatory swimming of largemouth bass, Micropterus salmoides. J Exp Biol 198: 585-602.
4. Videler J (1992) Fish Swimming. Chapman and Hall London UK.
5. Barrett DS, Triantafyllou MS, Yue DKP, Grosenbaugh MA, Wolfgang MJ (1999) Drag reduction in fish-like locomotion. J Fluid Mech 392: 183-212.
6. Kato N (2000) Control performance in horizontal plan of fish robot with mechanical pectoral fins. IEEE J Oceanic Eng 25: 121-129.
7. Bandyopadhyay P, Castano J, Rice J, Phillips J, Nedderman R, et al. (1997) Low speed maneuvering hydrodynamics of fish and small underwater vehicles. J Fluids Eng 119: 136-144.
8. Anderson J (1996) Vorticity control for efficient propulsion. Ph D dissertation Dept Oceanographic Eng Massachusetts Inst Technol Cambridge.
9. Read D (1999) Oscillating foils for propulsion and maneuvering of ships and underwater vehicles. M S thesis, Dept Naval Architect, Marine Eng Massachusetts Inst Technol Cambridge.
10. Haugsdal O (2000) Motion control of oscillating foils for steady propulsion and starting maneuvers. M S thesis, Massachusetts Inst Technol Cambridge.
11. Licht S, Polidoro V, Flores M, Hover FS, Triantafyllou MS (2004) Design and projected performance of a flapping foil AUV. IEEE J Oceanic Eng 29: 786-794.
12. Schouviler L, Hover FS, Triantafyllou MS (2005) Performance of flapping foil propulsion. J Fluids Struct 20: 949-959.
13. Stommel H (1989) The Slocum Mission. Oceanography 2: 22-25.
14. Webb DC, Simonetti PJ, Jones CP (2001) Slocum: an underwater glider propelled by environmental energy. IEEE J Oceanic Eng 26: 447-452.
15. Sherman J, Davis RE, Owens WB, Valdes J (2001) The autonomous underwater glider 'Spray'. IEEE J Oceanic Eng 26: 437-446.
16. Eriksen CC, Osse TJ, Light RD, Wen T, Lehman TW, et al. (2001) Seaglider: a long range autonomous underwater vehicle for oceanographic research. IEEE J Oceanic Eng 26: 424-436.
17. Davis RE, Eriksen CC, Jones CP (2003) Autonomous buoyancy-driven underwater gliders. The Technology and Applications of Autonomous Underwater Vehicles, 37-58.
18. Liu XB, Zhang MJ (2007) Research on bionic turtle hydrofoil propulsion technology of autonomous underwater vehicle. Harbin: Harbin Engineering University.
19. Schouveiler L, Hover FS, Triantafyllou MS (2005) Performance of flapping foil propulsion. J Fluids Str 20: 949-959.
20. Singh SN, Simha A, Mittal R (2004) Biorobotic AUV maneuvering by pectoral fins: Inverse control design based on CFD parameterization. IEEE J Oceanic Eng 29: 777-785.
21. Liu K (2007) Conceptual design and study of hybrid autonomous underwater vehicle. Tianjin, Tianjin university.

15

Isolation and Identification of A Novel Aporphine Alkaloid SSV, An Antitumor Antibiotic from Fermented Broth of Marine Associated *Streptomyces* sp. KS1908

Sunanda Kadiri[1]*, Nagendra Sastry Yarla[2] and Siddaiah Vidavalur[1]

[1]*Department of Organic Chemistry and FDW, Andhra University, Visakhapatnam, 530 003, India*
[2]*Department of Biochemistry and Bioinformatics, School of Life Sciences, Institute of Science, GITAM University, Visakhapatnam, 530045, India*

Abstract

A marine actinomycete, *Streptomyces* sp. KS1908 was isolated from a marine sediment sample collected in the Bay of Bengal, India and identified based on morphological, cultural, physiological, biochemical characteristics, along with the cell wall analysis and 16S rRNA gene sequence analysis. Strain KS1908 has more similarities with *Streptomyces albus* but some variations were observed. A novel aporphine alkaloid SSV was isolated according to bioactivity guided fractionation of fermented broth through solvent extraction and chromatography. Chemical structure of the aporphine alkaloid SSV was elucidated on the basis of spectroscopic analysis, including two-dimensional (2D) NMR and HR-ESI-MS data. Aporphine alkaloid SSV showed significant antibacterial, antifungal activity and also possesses considerable anticancer activity against human larynx (HEp-2), cervical (HeLa), leukemia HL-60 and MCF-7 breast cancer cell lines.

Keywords: Antitumor antibiotic; *Streptomyces* sp. KS1908; Aporphine alkaloid SSV; Fermented broth

Introduction

Streptomycetes have been shown to possess the ability to synthesize antibacterial, antifungal, insecticidal, antitumor [1-5], anti-inflammatory, anti-parasitic, antiviral, anti-infective, antioxidant and herbicidal compounds [1,2,6-9]. Hence, these are widely recognized as industrially important microorganisms [10]. Moreover, approximately 60% of the antibiotics discovered in the year 1990 and most of the antibiotics are from the genus *Streptomyces* [11]. These characteristics make this genus an important research area. Earlier literature suggests that many antimicrobial molecules have been isolated from *Streptomyces albus*. Salinomycin, a new polyether antibiotic was produced by strain of *Streptomyces albus* ATCC 21838 [12]. A new macromolecular peptide antibiotic, named AN-1 was isolated from the culture broth of *Streptomyces albus* AJ 9003 [13]. An antibiotic complex identical to Paulomycins A and B active against multiple resistant strains of staphylococci and other gram-positive bacteria was isolated from cultures of *Streptomyces albus* G [14]. In present study, bioactive actinomycete was collected from marine sediment and identified as *Streptomyces* sp. KS 1908 further responsible antibiotic was isolated and spectroscopic assignment of structure was demonstrated here.

Isolation and Taxonomy

Marine sediment samples were collected at Bay of Bengal near Gangavaram Coast, Visakhapatnam, India. Sample serially diluted to isolate as pure culture on a starch casein agar (SCA-soluble starch, 10.0 g; vitamin free casein, 0.3 g; KNO_3, 2.0 g; NaCl, 2.0 g; K_2HPO_4, 2.0 g; $MgSO_4{\bullet}7H_2O$, 0.05 g; $CaCO_3$, 0.02 g; $FeSO_4{\bullet}7H_2O$, 0.01 g; agar, 20.0 g; sterilized natural aged sea water, 1.0 L; pH, 7.2; supplemented with rifampicin 25 µg/ml and cycloheximide 75 µg/ml to inhibit bacterial and fungal contamination, respectively) plate, which had been seeded with a sediment sample suspension and incubated at 28°C for 14 days [10]. The isolated actinomycete colonies being filamentous, compact, often leathery giving a conical appearance, and maintained on YEME (Yeast Extract Malt Extract) slants at 4°C and as a glycerol suspension (20%, v/v) at −20°C [15]. This pure culture was later used for taxonomic and bioactivity studies.

Taxonomic identification was done by physiological conditions, biochemical tests, chemo taxonomic investigations and molecular characterization. Morphological observation through macroscopic based on growth pattern on different media like yeast malt extract agar (ISP-2), Oatmeal agar (ISP-3), Inorganic salts starch agar (ISP-4), Glycerol asparagines agar (ISP-5), tryptone yeast glucose agar, peptone agar, and nutrient agar. The color of aerial mycelium, substrate mycelium and soluble pigment were observed by naked eye. Optical and Scanning electron microscopies ((JSM-6610LV, JEOL Ltd.) were used for microscopic observations. Organism growth conditions were studied on SCA at various pH, Temperature and NaCl levels for physiological characterization [16]. Organism was biochemically characterized using the tests viz., Enzymes, H_2S production tests, Carbon and nitrogen utilization tests using different substrates12. Cell wall chemical composition was demonstrated according to the procedures of Lechevalier [17] for chemotaxonomic investigation17. Molecular characterization was done by 16s rRNA gene sequencing [16,18,19].

Fermentation

Sporulated isolate prepared with sterile water according to 0.5Mc Farland standard. The resulted spore suspension at 10% level was transferred aseptically into a 250 ml Erlenmeyer flask containing 45 ml of the inoculation medium [8]. The flasks were inoculated and incubated at 28°C for 48 hrs at 120 rpm. Thoroughly washed pellet containing cell

***Corresponding author:** Sunanda Kadiri, Department of Organic Chemistry and FDW, Andhra University, Visakhapatnam, 530 003, India
E-mail: sunandakadiri65@gmail.com

mass was suspended in sterile 0.9% sodium chloride solution and used as inoculum was then transferred to the modified production medium [8] and incubated at 30°C for 96 hrs at 160 rpm on a rotary shaker. The fermented broth was used for extraction of the active principle.

Bioactivity guided fractionation and purification

The fermented broth was aseptically collected in a sterile centrifuge tube and centrifuged at 4000 rpm for 15 min at 4°C. The culture filtrate (supernatant) and mycelial pellet obtained were extracted separately for identification of the active principle source. Antibiotics from the cell mass were isolated usually by extraction with polar and non-polar solvents while that from the fermented medium were extracted by solvent extraction only when the antibiotic has a reasonably high degree of solubility in non-polar organic solvents. Bioautography was performed to identify bioactive fraction. The other alternative technique for the separation of bioactive principle from the culture filtrate is the adsorption of the compound on some inert material like silica [20].

Compound identification and Structure elucidation

Thin Layer Chromatography (TLC) was analyzed on the glass percolated silica gel plates GF254, and spots were checked by UV light, Iodine and spraying with 10% sulfuric acid in methanol followed with heating. The melting point was determined on Fisher-Johns melting point apparatus. FT-IR spectra were recorded on a Perkin-Elmer spectrophotometer with KBr pellet. The sample was scanned between 400 and 4000 cm^{-1} wave number. High Resolution Mass Spectrum (HRMS) was recorded on QSTAR XL, HYBRID MS System and EI MS was recorded on VG 7070H (70 eV). The NMR data of purified compound was acquired using an AMX-400 spectrometer (Bruker, Rheinstetten, Germany). ^{1}H NMR spectra were obtained at 400.13 MHz and ^{13}C NMR spectra were obtained at 100.6 MHz. All NMR spectra were recorded in DMSO $-d_6$. The chemical shifts were expressed in δ (ppm) using DMSO-d_6 as solvent and TMS as internal reference. Dragendorff's reagent test was used for the identification of alkaloid [21].

Antimicrobial assay

Anti-microbial studies were carried out on clinical isolates of human pathogenic bacteria and dermatophytic fungi, *Salmonella typhi, Vibrio cholerae, Shigella dysenteriae, Enterococcus faecalis* are gastrointestinal pathogens, which were collected at King Gerorge Hospital, Visakhapatnam, India. Proteus vulgaris NCIM 2813, *Pseudomonas aeruginosa* NCIM 5031 cultures were collected from NCL, Pune, India. *Staphylococcus aureus* MTCC 7443, *Bacillus subtilis* MTCC 8141, *Aspergillus niger* MTCC 6484, *Aspergillus awamori* MTCC 7711, *Candida albicans* MTCC 1346, *Trichophyton rubrum.* MTCC 3272 cultures were collected from Indian microbial technology, India. Clinical isolate of *Candida albicans* was collected from skin lesions. Zone of inhibitions were determined using agar well diffusion method ad Minimum Inhibitory Concentration (MIC) was done by broth dilution assay. Microbial broth cultures (Mueller Hinton broth for bacteria, Sabouraud Dextrose broth for fungi) were adjusted the absorbance to 0.6 (Optical Density at 620 nm) in Spectrophotometer according to CLSI guidelines. These cultures were used as Inoculums. The agar plates were prepared by pour plate method using 20 ml of sterilized agar medium (MH agar for bacteria, SD agar for fungi). The sterile agar medium was cooled to 45°C and mixed thoroughly with 1ml of growth culture of concerned test organism (inoculum) and then poured into the sterile petri dishes and allowed to solidify. Wells of 6 mm size were made with sterile cork borer and test compounds were added. The agar plates were incubated at for 4 days at 28°C for fungi wile 24 hours at 37°C for bacteria. Zone of inhibitions were measured by Himedia milli meter zone reader. MIC was performed on broth media (10 ml) containing 1000-1 μg/ml of test compound prepared by 10 fold dilution. 0.1 ml of culture inoculums was added. The MIC was determined at which concentration of compound causes nil absorbance (no growth) in the spectrophotometer at 620 nm. All the experiments were conducted according to Clinical Laboratory Standard Institute. Ciprofloxacin (for bacteria) and fluconazole (for fungi) were antibiotics used as positive control [22-26].

MTT assay for anticancer activity

The cytotoxic activities of the compound (SSV) isolated from *Streptomyces* sp. KS1908, were tested against human larynx carcinoma cells *HEp*-2, cervical cancer (HeLa), human leukemia HL-60 and MCF-7 breast carcinoma cell lines. This cell line was obtained from TRIMS and National Centre for Cell Sciences (NCCS), India further cultured at 37°C with 5% CO_2, using MEM(minimum essential medium) medium. MTT assay [8] was used to study the cytotoxic properties of the sample. 200 μl of cell culture (2×10^4 cells/ml) was added in each well containing 100 μl MEM medium in a 96 well plate. After 24 hrs of incubation 20 μl of different test concentrations like 1000 μg /ml, 500 μg /ml, 250 μg /ml, 125 μg /ml, 62.5 μg /ml were added to the respective wells. After incubation of 4 days at 37°C temperature and 5% CO_2 in a CO_2 incubator. 20 μl of MTT (3-(4, 5- dimethyl-2-thiozolyl)-2,5-diphenyl-2H-tetrazolium bromide) reagent 5mg/ml concentration, was added to each well and incubated for 4hrs. Then the medium was carefully discarded and the Formazan complexes formed in the cells were dissolved in 200 μl of DMSO (Dimethyl sulphoxide). The rate of color appearance was measured at 570 nm in a spectrophotometer. The percent of cell inhibitions were calculated based on control and test absorbance values. Results were expressed as IC_{50} means concentration of compound where 50% of cancer cell growth inhibition occurred. The experimental measurements were made in five replicates. Camptothecin was used as the standard [27,28].

Molecular docking

Topoisomerase I (PDB ID 1T8I) and IIA (PDB ID 2XCT) crystal structures of proteins were obtained from Protein Data Bank. Co-crystallized ligands and water molecules are removed from target protein using Argus lab. Ligands are prepared using Chemoffice (Cambridge). Energy minimization was done using molecular mechanics. The minimized was executed until root mean square value reached smaller than 0.001 Kcal/mol. Such energy minimized ligands and receptor used for docking studies using Molegro Virtual Docker [29].

Results

Isolation and Taxonomy of isolated marine actinomycete

The isolated bioactive actinomycete colonies being filamentous, compact, often leathery giving a conical appearance, dry surface on SCA, which can easily be distinguished from fungi and non filamentous bacteria. Morphological and cultural observations of the isolate grown on different ISP media given in Table 1, revealed that vegetative mycelium showed yellow-brown color, aerial hyphae were abundant, well-developed with white color on different test media and substrate mycelium with pale yellow color. It didn't produce any pigments but faint yellow color pigmentation on Yeast-malt extract agar (ISP-2). The scanning electron micrograph of the strain KS1908 revealed that aerial mycelia were monopodially branched with compact spirals of sporophore terminating in long open coils. Each spore chain consisted of 8-20 white, oblong to cylindrical shaped spores, 0.6 ~ 0.7 x 0.8 ~ 0.9 µm in size, having smooth surface (Figure 1). The chemotaxonomic investigations revealed that the cell wall peptidoglycan of isolate contained L-diaminopimelic acid and glycine. This indicates that isolate belongs to cell wall type I which is characteristic of the genus *Streptomyces*.

16S rRNA gene sequence analysis (Genbank accession no. KC556777) and other cultural, biochemical physiological, chemotaxonomic characteristics revealed that Strain KS1908 has close similarities with *Streptomyces albus* [12] (Figure 2 and Tables 1 and 2) but some variations were observed so named as *Streptomyces* SP. KS 1908.

Bioactivity guided fractionation and purification

Fermented broth of sterptomyces sp. KS1908 was extracted using various solvents but only ethyl acetate extract showed bioactivity. Then ethyl acetate extract was run by chromatography using silica gel to obtain bioactive fraction II (255 mg), which was further fractionated with Sephadex LH-20 column and separated into five major fractions that included with fraction IIc. Preparative reverse phase HPLC was used to get light brown colored pure bioactive compound SSV in FIIc2 fraction. Schematic representation of detailed fractionation and purification was given in Figure 3.

Structural elucidation of aporphine alkaloid SSV

Compound SSV was obtained as light brown needles having melting point 68-72°C; [α] [25 $_{]D}$ -28.0 (c = 1.5, MeOH). Dragendorff's reagent test showed positive [16]. The molecular formula of SSV was assigned as $C_{19}H_{19}NO_4$ from its elemental and mass spectral analyses (HRMS: *m/z* 326.1386 $[M+H]^+$, 348.1206 $[M+Na]^+$). This was corroborated by the decoupled [13] C-NMR spectrum which showed signals for all the nineteen carbons of the molecule (Table 3). The mass fragmentation pattern of compound SSV showed typical of aprophine alkaloid [17]. The IR absorption bands at 3430 (OH), 945 (-O-CH2-O-) cm^{-1} indicated the presence of hydroxyl and methylenedioxy groups.

The proton NMR spectrum of compound SSV showed three, one proton singlet peaks at δ 7.55, 6.75 and 6.57 corresponding to H-11, H-8 and H-3 respectively of an aporphine alkaloid (Table 3). The two, one proton singlet peaks at δ 6.11 and 5.97 indicated the presence of one methylenedioxy group on C1-C2 and two, three proton singlet peaks at δ 3.77 and 2.42 attributed to one methoxyl group and one N-methyl group, respectively. The proton NMR spectrum also showed a one proton singlet peak at δ 9.19 attributed to hydroxyl group. Further the correlations observed in HMQC, HMBC and HSQC confirmed that methylenedioxy group was present at C-1 and C-2 carbons, methoxyl group was present on C-9 and free hydroxyl group was present on C-10 (Table 3). Thus from the foregoing spectral studies, the structure of compound SSV was established as 10-hydroxy-9-methoxy-1,2-methylenedioxy-6-methyl-4,5,6,6a-tetrahydro-*7H*,6-azabenzanthrene (Figure 4).

Medium	Characteristics				
	Growth	Vegetative mycelia	Aerial mycelia	Spore mass	Soluble pigment
Nutrient agar	Good	Moderate, pale yellow	Moderate, white	Poor, white	Nil
Yeast – malt extract agar (ISP-2)	Abundant	Moderate, yellow	Abundant, white	Moderate, white	Faint yellow
Oatmeal agar (ISP-3)	Abundant	Good, pale-yellow	Abundant, white	Moderate, white	Nil
norganic salts starch agar (ISP-4)	Good	Moderate, yellow	Good, white	Poor, white	Nil
Glycerol asparagines agar (ISP-5)	Good	Moderate, yellow	Good, white	Moderate, white	Nil
Tyrosine agar (ISP-7)	Moderate	Moderate, pale-yellow	Good, white	Poor, white	Nil
Yeast – starch aga	Moderate to good	Good, pale yellow	Good, white	Moderate, white	Nil
Sucrose – nitrate agar	Moderate	Moderate, pale yellow	Moderate, white	Poor, white	Nil

Table 1: Cultural characteristics of the isolated *Streptomyces* sp. KS1908.

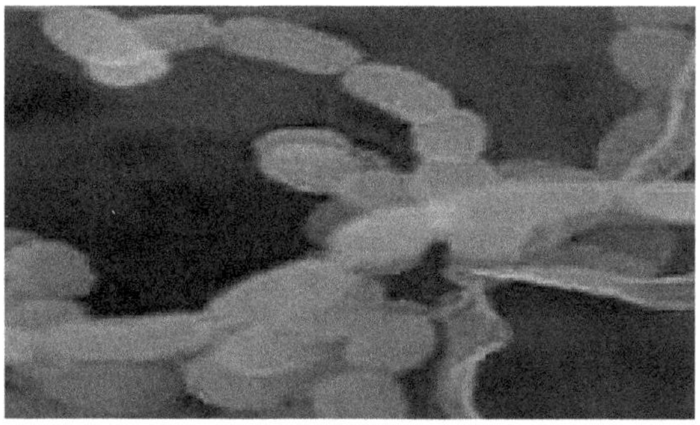

Figure 1: Scanning electron micrograph of *Streptomeces* sp. KS1908.

Antimicrobial and anticancer activities of aporphine alkaloid SSV

Aporphine alkaloid SSV showed good antimicrobial activity especially on multi drug resistant clinical isolates including bacteria and fungi. Aporphine alkaloid SSV was more effective against bacteria than fungi.

As shown in Table 4, zone of inhibition found to be between 9-14 mm at 30 µg of compound. MIC range found to be between 1-100 µg/ml. 14 mm was the inhibitory zone showed by SSV on *S. typhi* and with lowest MIC of 1 µg/ml. Aporphine alkaloid SSV showed potent antibacterial activity against both gram positive and gram negative bacteria. Compound SSV showed very effective activity against gastrointestinal pathogenic bacteria (*S. typhi, V. cholerae, E. faecalis* and *E. coli*). Dermatophytic fungi (*T. rubrum* and *C. albicans*) showed slight resistance. Aporphine alkaloid SSV showed comparable antimicrobial potency with ciprofloxacin and flucanozole (antibiotics).

Anticancer activity of aporphine alkaloid SSV

Anticancer activity of compound SSV performed by MTT assay and found significant cytotoxic activity on human larynx (HEp-2), cervical (HeLa), human leukemia HL-60 and MCF-7 breast cancer cell lines with the IC_{50} values of 1.10, 1.13, 2.85 and 4.44 µg/ml, respectively. Aporphine alkaloid SSV showed comparable potency with camptothecin (Table 5).

Docking of aporphine alkaloid SSV

Molecular docking studies of aporphine alkaloid SSV were on topoisomerases using Molegro Virtual Docker. Docked energy or binding energy was inversely proportional to affinity of compounds towards enzyme. Lower binding energy indicated higher binding affinity. -98.4 kcal was the binding energy of aporphine alkaloid SSV on topo II, which was less than docked score of ciprofloxacin (-79.7 Kcal/mol). Aporphine alkaloid SSV showed comparable binding energy (-78.6 Kcal/mol) with camptothecin (-77.1 Kcal/mol) on topoisomerase I. Figure 4, demonstrated that Compound SSV binding interactions

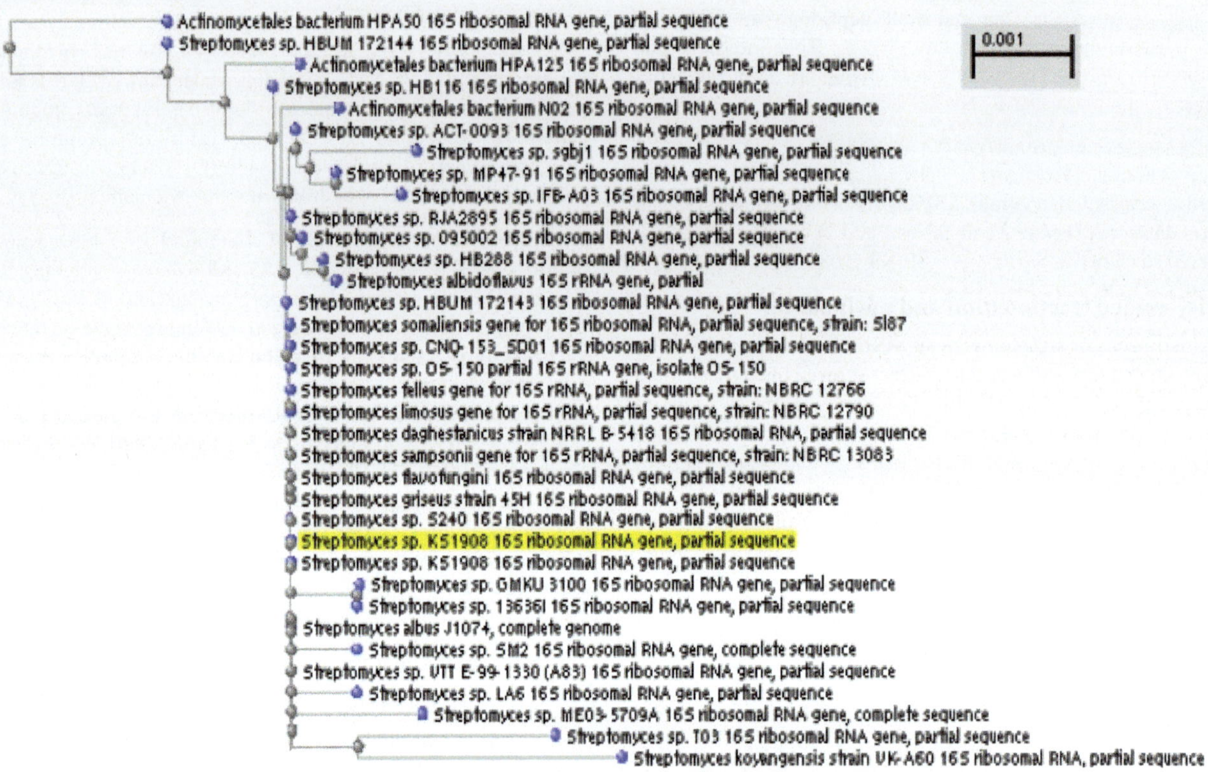

Figure 2: Phylogenetic tree of obtained by distance matrix analysis of 16s rRNA gene sequence of closely related *streptomyces* sps. obtained from BLAST results and constructed using neighbor-joining method, showing phylogenetic position of *Streptomyces* sp. KS1908 with yellow marking.

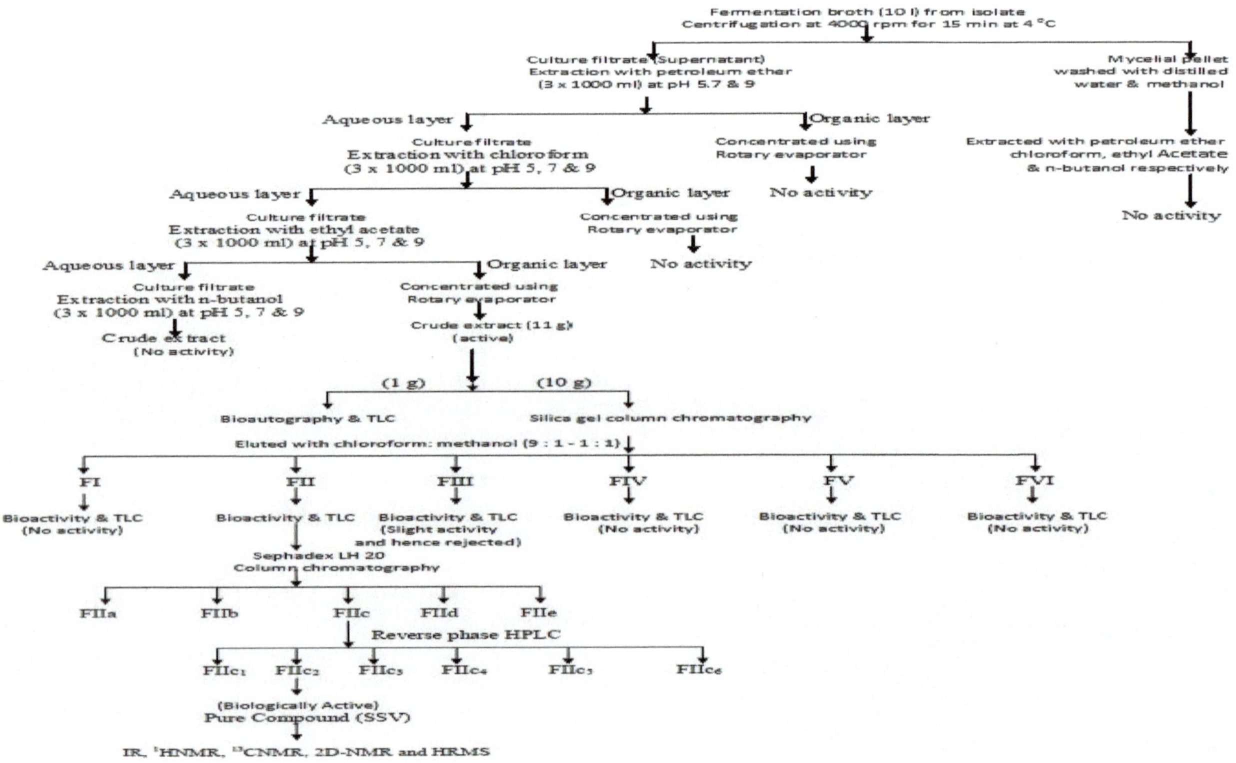

Figure 3: Isolation and purification procedure of compound SSV.

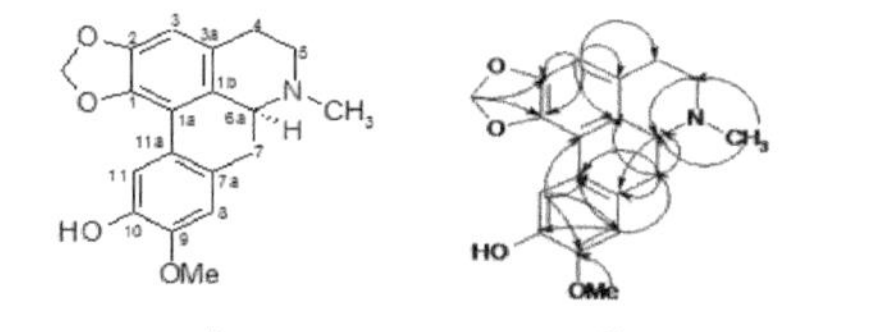

A B

(A) Structure of aporphine alkaloid SSV and (B) HMBC correlations.

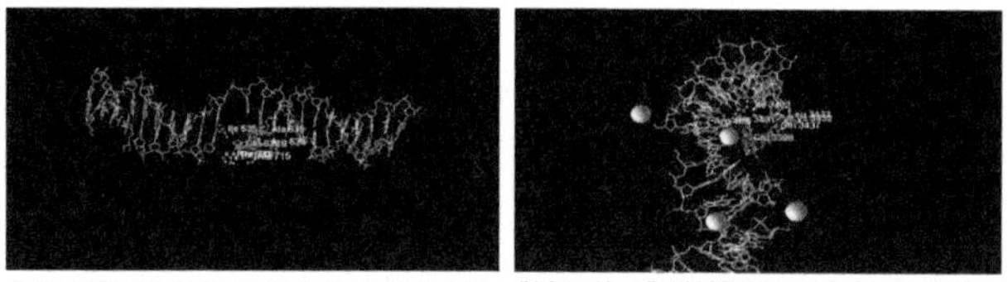

(a) Aporphine alkaloid SSV binding interactions with DNA catalytic amino acids of Topo I (PDB ID 1T8I)

(b) Aporphine alkaloid SSV binding interactions with DNA catalytic amino acids of Topo IIA (PDB ID 2XCT).

Figure 4: Molecular docking studies of aporphine alkaloid SSV on Topoisomerases.

with DNA catalytic residues of topoisomerase I and II enzymes.

Discussion

Multidrug resistance is the one of major problem in treatment of various microbial infectious diseases and cancer. There is a necessity to develop novel drugs with high potency and less toxic. Most of the antitumor antibiotics have been isolated from marine microbes especially *Streptomyces* sp. These antitumor antibiotics are commonly interacting with DNA to cause cell death [5]. Anthracyclines are a special class of antitumor antibiotics which act through topoisomerase II inhibition [30]. Pimprinine, an extracellular alkaloid has been isolated from the culture filtrate of *Streptomyces* CDRIL-312 [31]. Alkaloid group of aporphine antibiotics are topoisomerase I inhibitors from *Streptomyces* sp [32].

In conclusion, bioactive *streptomyces* sp. KS1908 was isolated and characterized from marine associated actinomycetes further novel aporphine alkaloid SSV is an antitumor antibiotic which was isolated from bioactive fraction of fermented broth and chemically characterized through advanced spectroscopic data. Therefore, the isolated aporphine alkaloid SSV can be promising agent for treatment of cancer and microbial infections.

Acknowledgement

Sunanda Kadiri thankful to Department of Science and Technology (DST) for the financial assistance by sanctioning a project under the women scientist scheme with Grant No. WOS-A/LS-74/2009. We acknowledged Bioserve, India for help in gene sequencing and Phylogenetic tree construction.

References

1. Berdy J (2005) Bioactive microbial metabolites. J Antibiot 58:1-26.
2. Sacramento DR, Coelho RRR, Wigg MD, Linhares LFTL, Santos MGM, et al. (2004) Antimicrobial and antiviral activities of an actinomycete (*Streptomyces sp.*) isolated from a Brazilian tropical forest soil. World J Microbiol Biotechnol 20: 225-229.
3. Prabavathy VR, Mathivanan N, Murugesan K (2006) Control of blast and sheath blight diseases of rice using antifungal metabolites produced by *Streptomyces sp.* PM5. Biol Control 39: 313-319.
4. Ebrahimi ZM, Shahidi BGH, Padasht DF, Ayatollahi MSA, Rashid FP, et al. (2009) Biological control of rice blast (*Magnaporthe oryzae*) by use of *Streptomyces sindeneus* is isolate 263 in greenhouse. Am J Appl Sci 6: 194-199.
5. Galm U, Martin HH, Steven GVL, Jianhua J, Jon ST, et al. (2005) Antitumor Antibiotics: Bleomycin, Enediynes, and Mitomycin. Chem Rev 105: 739-758.
6. Pimentel-Elardo SM, Kozytska S, Bugni TS, Ireland CM, Moll H, et al. (2010) Antiparasitic compounds from *Streptomyces* sp. strains isolated from Mediterranean sponges. Mar Drugs 8: 373-380.
7. Rahman H, Austin B, Mitchell WJ, Morris PC, Jamieson DJ, et al. (2010) Novel Anti-Infective Compounds from Marine Bacteria. Mar Drugs 8: 498-518.
8. Teppei K, Miho I, Misa O, Hideki Y, Masayuki H, et al. (2012) JBIR-94 and JBIR-125, Antioxidative Phenolic Compounds from *Streptomyces* sp. R56-07. J Nat Prod 75: 107-110.
9. Sousa CS, Soares ACF, Garrido MS (2008) Characterization of *Streptomycetes* with potential to promote plant growth and biocontrol. Sci Agric (Piracicaba, Braz) 65: 50-55.
10. Ramesh S, Mathivanan N (2009) Screening of marine actinomycetes isolated from the Bay of Bengal, India for antimicrobial activity and industrial enzymes. World J Microbiol Biotechnol 25: 2103-2111.
11. Ramesh S, William A (2012) Marine actinomycetes: An ongoing source of novel bioactive metabolites Microbiol Res 167: 571-580.
12. Yohko K, Shojiroh M, Yukiko O, Akio S (1980) Antibiotic no. 6016, a polyether antibiotic. J Antibiot 33: 1437-1442.
13. Shigeyoshi M, Takao K, Hiroshiro S, Tsuyoshi S (1983) The fermentation, Isolation and Characterization of a Macromolecular peptide antibiotic: AN-1. J Antibiot 37: 27-32.
14. Majer J, Chater KF (1987) *Streptomyces albus* G produces an antibiotic complex identical to paulomycins A & B. J Gen Microbiol 133: 2503-2507.
15. Williams ST, Cross T (1971) Methods in Microbiology. New York: Academic press.
16. Lu J, Ma Y, Liang J, Xing Y, Xi T, et al. (2012) Aureolic acids from a marine-derived *Streptomyces sp.* WBF16. Microbiol Res 67: 590-595.
17. Lechevalier HA, Lechevalier MP (1970) A critical evaluation of the genera of Aerobic actinomycetes. 393- 405.
18. Saitou N, Nei M (1987) The neighbor-joining method: a new method for reconstructing phylogenetic trees. Mol Biol Evol 4: 406-425.
19. Sharma AD, Singh J (2005) A non enzymztic method to isolate genomic DNA from bacteria and actinomycetes. Anal Biochem 337: 354-356.
20. Venkata RRKD, Murali KKM, Murali YN, Sri RRD (2012) Novel Pyridinium compound from marine actinomycete, Amycolatopsis alba var. nov. DVR D4 showing antimicrobial and cytotoxic activities in vitro. Microbiol Res 67: 346- 351.
21. Coe FG, Parikh DM, Johnson CA (2010) Alkaloid presence and brine shrimp (Artemia salina) bioassay of medicinal species of eastern Nicaragua. Pharm Biol 48: 439-445.
22. Mackie, Mac cartney (1969) Practical Medical Microbiology, Churchill Livingstone Publisher (Elsevier); 14: 141.
23. Taplin D, Zaias N, Rebell G, Blank H (1969) Isolation and recognition of dermatophytes on a new medium (DTM). Arch Dermatol 99: 203-209.
24. Perez C, Pauli M, Bazevque P (1990) An antibiotic assay by the agar well diffusion method. Acta Biologiae et Medicine Experimentalis 15: 113-115.
25. Clinical Laboratory Standards Institute Methods for Dilution Antimicrobial Susceptibility Test for Bacteria That Grow Aerobically; Approved Standard-Ninth Edition. M07-A9, 32: 12-20.
26. Pfaller MA, Castanheira M, Diekema DJ, Messer SA, Moet GJ, et al. (2010) Comparison of European Committee on Antimicrobial Susceptibility Testing (EUCAST) and E test Methods with the CLSI Broth Microdilution Method for Echinocandin Susceptibility Testing of *Candida Species.* J Clin Microbiol 48: 1592-1599.
27. Mosmann T (1983) Rapid colorimetric assay for cellular growth and survival: application to proliferation and cytotoxicity assays. J Immunol Methods 65: 55-63.
28. Govinda Rao Duddukuri, Nagendra Sastry Y, Kaladhar DSVGK, Ajay Babu P, Kamalakara Rao K, et al. (2011) preliminary studies on *in vitro* anti-tumor activity of tender seed coat extract of *borassus flabellifer* l. On HeLa cell line. International Journal of Current Research 3: 075-077.
29. Rene T, Mikael HC (2006) MolDock: A New Technique for High-Accuracy Molecular Docking. J Med Chem 49: 3315 -3321.
30. Fujiwara A, Hoshino T, Westley JW (1985) Anthracycline Antibiotics. Critical Reviews in Biotechnology 3: 133-157.
31. Naik SR, Harindran J, Varde AB (2001) Pimprinine, an extracellular alkaloid produced by *Streptomyces* CDRIL-312: fermentation, isolation and pharmacological activity. J Biotechnol 88: 1-10.
32. García MT, Blazquez MA (2011) New alkaloid antibiotics that target the DNA topoisomerase I of *Streptococcus pneumoniae.* J Biol Chem 286: 6402-6413.

16

Distribution and Seasonal Variation of Heavy Metal in Surface Sediments from Arvand River, Persian Gulf

Abdolah Raeisi Sarasiab[1], Zohreh Mirsalari[2] and Mehdi Hosseini[3*]

[1]*Department of Marine Ecology, Farhangian Science University of Iran, Iran*
[2]*Department of Biodiversity, Science and Research Branch, Islamic Azad University, Ahvaz, Iran*
[3]*Department of Marine Biology, Faculty of Biological Science, Shahid Beheshti University, Tehran, Iran*

Abstract

Distribution and seasonal variation of heavy metals copper (Cu), nickel (Ni), cadmium (Cd), cobalt (Co), lead (Pb) and mercury (Hg) were measured in the surface sediments samples taken from Arvand river in the northeast Persian Gulf. Concentration of heavy metals varied depending on different season and sampling sites. The concentration order of heavy metals in sediment were Ni > Cu > Co > Pb > Cd > Hg. There was significant difference ($p<0.05$) in metals levels between different seasons and stations. The highest concentration of Ni, Cu, Cd, Pb and Hg was detected in S3 in July and the highest concentration of Co was detected in S3 in April. The percentage of mercury associated with different fractions in the sediment from was in the order of residual > organic matter > easily and exchangeable > acid reduction. I_{geo} values calculated for Cu (4.5) and Ni (3.6) showed higher values in the July month in the upstream region of the river than in the other seasons. BAFs of Ni were extremely higher than other metals followed by Cu, Co, Pb, Cd and Hg, respectively. These results indicate that Ni and Cu are more easily available than other elements. The analysed heavy metals were found in sediment samples at mean concentration in the sediment quality guideline proposed by NOAA and ROPME, except for Ni concentration in some cases.

Keywords: Heavy metal; Surface sediment; Seasonal variation; Arvand River

Introduction

Heavy metals can be classified as potentially toxic elements such as cadmium, lead and mercury [1] and essential elements such as copper, zinc and iron [2]. Toxic metals naturally occur in aquatic environments in very low concentrations, but their concentration levels have increased due to anthropogenic pollutants [3] over time. Industrial activities as well as agriculture and mining create a potential source of metals pollution in aquatic environment [4]. As a result, living things inhabiting contaminated waters show rather high metal concentrations. It is well known that metals accumulate in sediments of aquatic environments and hence metals measured in the sediments can reflect past exposures [1,5,6].

Sediments are important sinks for various pollutants such as heavy metals and also play a useful role in the assessment of heavy metal contamination [1,5,6]. Sediments, particularly surficial sediments, may serve as a metal pool that can release metals to the overlaying water via natural anthropogenic processes, causing potential adverse health effects to the ecosystems because of their serious toxicity and persistence [7,8]. Bottom sediments are known to act as a sink for heavy metals introduced to the marine environment [9]. Therefore, benthic organisms, which occupy the bottom layer of the water column, tend to accumulate the highest levels of heavy metals in comparison to pelagic organisms [9-11].

A study of the distribution of trace metals in the sediments is important to the assessment of the probable influence of the mining of iron ore in this region on the estuarine environment. The measurement of trace element concentrations and distribution in the marine environment leads to better understanding of their behaviour in the aquatic environment and is important for detecting sources of pollution [6]. Studies of various physicochemical, biological and mineralogical aspects of the Persian Gulf have been carried out, but limited data are available on the metal concentrations and seasonal variation in the sediments of the Persian Gulf. The trace metal data provide new information on the surface sediment geochemistry in the region whilst the study of element interrelationships, including organic carbon, gives information on their possible origin [12]. The distribution of metals in surficial sediments in the Arvand river, northwest of Persian Gulf has been examined, therefore, in order to enhance the data inventory for the region, characterize the geochemistry of Cd, Co, Cu, Ni, Pb and Hg in surficial sediments and to help understand the influence of anthropogenic activities in this tropical estuarine system.

Materials and Methods

Four sampling sites were selected along the northwest coasts of the Persian Gulf (Figure 1). The first one (S1) was located near the Khoramshahr city, the second one (S2) was located near Minoo island, the third one (S3) was located near the Abadan city, and the last one (S4) was located near Arvand river's mouth. The Arvand river, the border between Iraq and Iran, is the biggest river in the Persian Gulf. It passes three main cities including Al-Basre in Iraq, Abadan and Khoramshahr in Iran. For people of these cities, the Arvand river is considered as a main resource of seafood and drinking water. For this reason, striking quantities of marine organisms in the markets of these cities are caught from Arvand river.

Surface sediments were collected in Spring (April), Summer (July) and Autumn (October) of 2010 and Winter (January) of 2011, using a Van Veen Grab from 4 stations in the Persian Gulf (Figure 1) across

***Corresponding author:** Mehdi Hosseini, Department of Marine Biology, Faculty of Biological Science, Shahid Beheshti University, Tehran, Iran
E-mail: smhbio@yahoo.com

the upper, mid, and lower regions, covering a distance of 50 km and a range of salinities from freshwater in the upstream area to seawater at the mouth. Subsamples were taken from the uppermost layer of the sediment taking care to minimize contamination. The samples were frozen after collection and later thawed, dried at 50-60°C in an oven and disaggregated in an agate mortar, before chemical treatment for total metal analysis. For each sample a known quantity (1 g) of sediment was digested with a solution of concentrated $HClO_4$ (2 ml) and HF (10 ml) to near dryness. Subsequently, a second addition of $HClO_4$ (1 ml) and HF (10 ml) was made and the mixture was evaporated to near dryness. Finally, $HClO_4$ (1 ml) alone was added and the sample was evaporated until white fumes appeared. The residue was dissolved in concentrated HCl and diluted to 25 ml [13].

Figure 1: Map of Persian Gulf showing study area, Arvand river.

Concentrations of Cd, Cu, Ni, Pb and Hg were measured with a GBC (Dandenong, Vic. 3175, and Australia) Flam Atomic Absorption Spectrometer (Savant AA sigma) with hyper pulse deuterium background corrector. Mercury concentration was determined by a cold vapor atomic absorption spectrometer (Unicam, model 919).

The accuracy of the analytical procedures was assessed using the certified reference material BCR-1 and yielded results within the reference value range [14]. The recovery means for Cd, Cu, Co, Ni, Pb and Hg were 91.4%, 90.5%, 93.5%, 94%, 92.1% and 108.4% respectively.

All data were tested for normal distribution with Shapiro-wilk normality test. Significant differences between heavy metals concentration in the samples of various stations and between seasons were determined using One-Way analysis of variance (ANOVA) fallowed by Duncan post hoc test.

Results and Discussion

The mean concentrations of studied metals in sediment are given in Table 1. The heavy metals concentration in sediments at all sampling sites occurs in descending order of Ni > Cu > Co > Pb > Cd > Hg, except for station 1 where Cu concentration was higher than Ni.

Ni

In the sediments at all sampling station, concentration of Ni ranged

Metal	Station	Season			
		April	July	October	January
Cu	S 1	20.3 ± 0.01[c]	22.5 ± 0.02[c]	18.1 ± 0.05[c]	14.4 ± 0.05[a]
	S 2	12.1 ± 0.02[c]	12.4 ± 0.01[b]	11.3 ± 0.01[c]	10.2 ± 0.06[a]
	S 3	15.5 ± 0.03[b]	16.2 ± 0.04[c]	14.2 ± 0.05[c]	13.4 ± 0.05[b]
	S 4	10.4 ± 0.04[b]	15.2 ± 0.40[c]	17.2 ± 0.02[c]	13.8 ± 0.02[b]
Ni	S 1	34.6 ± 0.01[c]	38.2 ± 0.01[c]	34.1 ± 0.01[b]	20.8 ± 0.01[b]
	S 2	40.4 ± 0.05[c]	51.4 ± 0.04[c]	42.2 ± 0.05[c]	35.2 ± 0.06[b]
	S 3	51.5 ± 0.01[c]	64.5 ± 0.06[c]	45.1 ± 0.04[c]	42.5 ± 0.05[c]
	S 4	23.1 ± 0.[b]	28.2 ± 0.04[c]	35.3 ± 0.06[b]	21.5 ± 0.01[b]
Co	S 1	17.2 ± 0.05[b]	15.1 ± 0.06[c]	16.3 ± 0.01[c]	17.5 ± 0.01[c]
	S 2	18.4 ± 0.03[b]	17.1 ± 0.05[c]	16.5 ± 0.01[c]	16.2 ± 0.01[b]
	S 3	25.1 ± 0.02[c]	23.5 ± 0.01[c]	22.2 ± 0.01[c]	21.2 ± 0.06[c]
	S 4	16.2 ± 0.01[b]	14.4 ± 0.04[c]	15.1 ± 0.01[b]	14.5 ± 0.04[b]
Pb	S 1	7.1 ± 0.02[b]	8.2 ± 0.04[c]	6.2 ± 0.04[c]	3.31± 0.06[c]
	S 2	3.4 ± 0.06[c]	5.1 ± 0.03[c]	2.4 ± 0.01[c]	1.22 ± 0.03[c]
	S 3	7.1 ± 0.03[b]	6.1 ± 0.01[c]	3.1 ± 0.05[c]	0.71 ± 0.04[a]
	S 4	2.2 ± 0.05[c]	4.2 ± 0.03[c]	1.1 ± 0.03[c]	0.52 ± 0.09[a]
Cd	S 1	0.36 ± 0.01[a]	0.28 ± 0.04[a]	0.19 ± 0.02[a]	0.21 ± 0.05[c]
	S 2	0.21 ± 0.06[c]	0.18 ± 0.03[a]	0.14 ± 0.01[b]	0.15 ± 0.04[b]
	S 3	0.30 ± 0.04[c]	0.19 ± 0.04[a]	0.16 ± 0.01[b]	0.15 ± 0.03[b]
	S 4	0.15 ± 0.03[b]	0.12 ± 0.01[b]	0.15 ± 0.06[a]	0.11 ± 0.04[c]
Hg	S 1	0.03 ± 0.03[c]	0.07 ± 0.05[a]	0.04 ± 0.01[a]	0.04 ± 0.03[a]
	S 2	0.20 ± 0.02[b]	0.30 ± 0.01[c]	0.10 ± 0.02[c]	0.08 ± 0.01[b]
	S 3	0.30 ± 0.06[b]	0.60 ± 0.03[c]	0.20 ± 0.01[c]	0.10 ± 0.06[b]
	S 4	0.01 ± 0.07[a]	0.05 ± 0.02[b]	0.02 ± 0.04[a]	0.01 ± 0.03[a]

S1 (Khoramshahr), S2 (Minoo), S3 (Abadan) and S4 (Arvand River mouth); a, b, c show significant differences among seasons

Table 1: Mean concentration of heavy metals (µg g-1) in sediments during different seasons from Arvand river, northwest of Persian Gulf.

from of 21 to 64.5 μg g^{-1}. The highest concentration of Ni (64.5 μg g^{-1}) was detected in station 3 in July. There was significant difference ($p<0.05$) in Ni levels between different seasons. It is likely that this high concentration of Ni was related to the oil tankers traffic in this site [15]. This area is surrounded by Abadan petrochemical complex and petroleum refinery.

Cu

In the sediments at all sampling station, concentration of Cu ranged from 10.3 to 22.5 μg g^{-1}. The highest concentration of Cu (22.5 μg g^{-1}) was detected in site 1 in July. The high concentration of lead at site 1 was situated at the pH of the sediment was the lowest. Samecka-Cymerman and Kempers [16] have showed that the highest concentrations of copper were found in lakes which had the lowest sediment pH. Tokalioglu et al. [17] have indicated that copper concentrations in the lake samples, decreased with increasing pH, because the metal binding abilities are decreased with decreasing pH due to proton binding. A decrease in pH will increase the competition between metals and hydrogen ions for binding sites and may dissolve metal complexes, releasing free metal ions into the sediment [18]. There was significant difference ($p<0.05$) in Cu levels between different seasons. It may be mentioned here that there is no significant relation between salinity and the copper concentration for sediment along the Arvand River.

Co

The concentration of Co ranged from 14.3 to 26.81 μg g^{-1}. The highest concentration of Co (26.81 μg g^{-1}) was detected in station 3 in April. There was no significant difference ($p<0.05$) in Co levels between different seasons. No strongly elevated concentrations of Co were observed, although Co is apparently enriched above natural values. It should be noted that the choice of reference value for normalization alters the degree of enrichment.

Pb

The concentration of Pb ranged from 1.03 to 8.11 μg g^{-1}. The highest concentration of Ni (8.11 μg g^{-1}) was detected in station 1 in July. There was significant difference ($p<0.05$) in lead levels between different seasons. Nolting et al. [19] observed that the low and constant Pb concentrations indicating the minor importance of anthropogenic input from the Laptev Sea in contrast to other areas of world as anthropogenic inputs are considered to be the major source of elevated Pb concentrations in marine sediments [19]. The high concentration of lead at site 1 was situated at the boat stations where high boating activities takes place and in located where the pH of the sediment was the lowest. Byrd and Perona [20] have showed that the highest concentration of lead was found in the Turlok Lake when there were high boating activities.

Cd

In the sediments at all sampling station, concentration of Cd ranged from 10.3 μg g-1 to 22.5 μg g-1 with an average of14.8 μg g-1. The highest concentration of Cd (22.5 μg g-1) was detected in station 3 in July. There was significant difference ($p<0.05$) in Cu levels between different seasons. The high concentrations of cadmium in sediment at sampling site 3, seems to corresponds to low pH sediment in the former site while the latter site showed agricultural runoff entering river.

Hg

In the sediments at all sampling station, concentration of Hg ranged from 0.01 to 0.60 μg g-1. The highest concentration of Hg (0.60 μg g^{-1}) was detected in station 3 in July. There was significant difference ($p<0.05$) in Hg levels between different seasons. As similar to Cd, there is a general trend observed that for mercury decreasing its concentration in July. Mercury concentrations in the Arvand river sediments showed generally lowest values than other metals. Mercury can enter the aquatic environment from a number of sources including hazardous substance spills from various refineries, petrochemical and oil industries [15].

Correlation analysis also showed relationship between individual elements including Cu and Pb ($r=0.65$), Pb and Cd ($r=0.71$) and between Cd and Cu ($r=0.76$). Pearson correlation showed that there was no significant correlation among these metals ($p>0.05$).

The comparison between the sampling stations showed that the amount of metals varied from site to site, and the variation could be related to variability in the sources of metals input. Maximum concentration of Hg, Ni, Co and Cd were observed in sit 3 and Maximum concentration of Cu and Pb were observed in sit 4. The site 3 is the nearest station to Abadan petrochemical complex and petroleum refinery [15]. Site 3 was relatively rich in Ni concentration. It is likely that this high concentration of Ni was related to the oil tankers traffic in this site. About 800 offshore platforms and 25 major oil terminals as well as oil supertankers traffic exert tremendous stress by oil pollution in the Persian Gulf [15,21]. Therefore, site 3 receives different types of pollution such as heavy metal from the surrounding areas. Maximum concentration of Cu was observed in site 1. The concentrations of Cu were higher than the other metals in site 1, suggesting that there is an influence of the industrial activities in surroundings of the sampled stations and which confirmed the increased anthropogenic activities with time [21].

Minimum concentration of metals was observed in site 4. This station was located near the mouth of Persian Gulf. In the area river water with water of Persian Gulf to be mixed and concentration of metals was low. As the interactions between salt water and fluvial water in the mouth of rivers could lead to sediment deposition, a considerable quantity of the sediment is deposited in the mouth of the Arvand river [21,22]. Also, there is no industrial activity near this station, which is a relatively remote area compared to other stations.

This study can provide information on possible chemical forms of heavy metals in sediments. The effects of heavy metals in the environment depend to a large extent on whether they occur in forms that can be taken up by plants or animals. According to Elith and Garwood [18], lead may be strongly adsorbed onto sediment particles and therefore largely unavailable, while cadmium ions can be directly absorbed to sediment and it is known to be most mobile among the other metals [23]. A wide range of values for heavy metal concentrations was observed for the sediments. On the average, the percentage of cadmium associated with different fractions in the sediment from all sites was in the order of residual (49.4%) > organic matter (31.6%) > easily and exchangeable (11.2%) > acid reduction (7.8%). The percentage of copper from all sites was in the order of residual (70.3%) > organic matter (27.4) > easily and exchangeable (1.7%) > acid reduction (0.6%). The percentage of nickel was in the order of residual (73.5%) > organic matter (22.5%) > easily and exchangeable (3.6%) > acid reduction (0.4%). The percentage of mercury was in the order of residual (80.6%) > organic matter (17.5%) > acid reduction (1.3%) > easily and exchangeable (0.6%).The percentage of lead was in the order of residual (63.5%) > organic matter (30.5%) > easily and exchangeable (4.4%) > acid reduction (16.5%). The percentage of cobalt was in the order of residual (60.5%) > organic matter (29.5%) > easily and exchangeable

(6.8%) > acid reduction (3.2%). Assuming that bioavailability is related to solubility, then metal bioavailability decreases in the order of exchangeable forms > acid reduction forms > organic forms > residual forms [6,24]. The residual forms are not expected to be released under normal conditions in nature [6] and could be considered as an inert phase [24].

Metal availability is also influenced by other characteristics of the sediment system such as pH and organic matter content, which control the solubility and therefore the availability of metals [25]. Among the non-residual fractions, the organic matter fraction was much higher than other fractions in all sites. The percentage of cadmium in the non-residual fractions was greater than the residual fraction. About 73% of the cadmium in sediment was associated with the exchangeable, acid reduction and oxidation fractions. Forstner [26] reported that cadmium was characteristically enriched in the more mobile fractions and more mobile than most of other heavy metals [27]. The present results indicate that cadmium, cobalt and lead have greater potential for mobilization from the sediments than copper, nickel and mercury because of their higher concentration at the acid reduction fraction.

Organic matter has a high specific storage capacity for heavy metals [18]. Acidic pH condition is known to influence the sorption of lead by organic matter fraction in sediments [6]. Analysis of pH and percentage of organic matter present in sediments concluded that there were strong correlations between each of these factors and the concentrations of copper and lead. Results showed that the concentration of heavy metals at oxidation fraction increased with increasing pH levels and percentage of organic matter present in the sediments.

A direct correlation was carried out between organic matter percentages, pH and concentration of heavy metals at the oxidation fraction. There was no significant correlation between concentration of cadmium with organic matter percentage and pH ($p>0.05$). The result of Pearson correlation (r) for organic matter for copper was 0.75 ($p<0.001$), which means that there was a strong linear correlation with organic matter. There was a linear correlation between lead and the percentage of organic matter in the sediment ($r=0.55$ and $p<0.01$), copper ($r=0.64$ and $p<0.01$). Thus, organic matter shows a significant to bind copper and lead the sediment in Arvand river. There was a linear correlation between copper and pH ($r=0.69$ and $p<0.01$) and lead and pH ($r=0.80$ and $p<0.01$).

There was significant difference ($P<0.05$) between the levels of heavy metals during different seasons. Relatively high concentrations were observed in sediments collected during July. It may be mentioned here that there is significant relation between salinity and temperature and the heavy metals concentration for sediments along the coast of Persian Gulf [15,28]. Temperature could increase metal concentration in summer than winter [12,19]. Also, this variation could result variation in bioavailability of metal in environment [29]. Such this condition was happened during present study and metals levels in different sites showed higher in summer season compared to other seasons. Similarly in the present study, more metals levels uptake was showed during summer season. According to different studies the heavy metal levels in sediments showed higher in summer season [15,29]. It was revealed that sediments all show similar seasonal patterns in metal concentrations it would seem likely that environmental factors (discharges to the estuary, pH, salinity, suspended matter, etc.) are having a greater overall influence on seasonality of metal concentrations.

Index of geoaccumulation

Possible sediment enrichment of metals was evaluated in terms of the I_{geo} of Muller [30]. The formula used for the calculation of I_{geo} is $\log_2 (C_n/1.5B_n)$, where C_n is the measured content of element "n", and B_n the element's content in "average shale" [31]. The geoaccumulation index (I_{geo}) was originally defined by Muller [30] for a quantitative measure of the metal pollution in aquatic sediments [32] and is shown for the Arvand river sediments in (Figure 2). Elevated I_{geo}

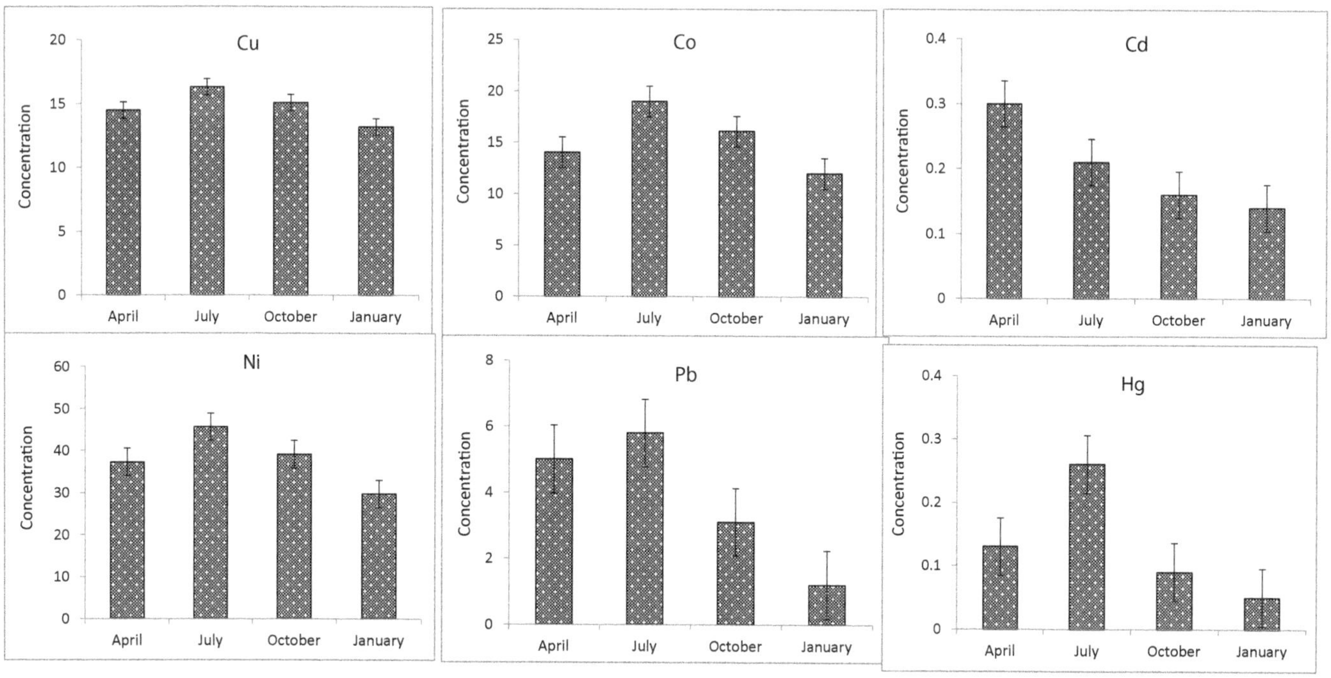

Figure 2: Concentration of heavy metals during different season.

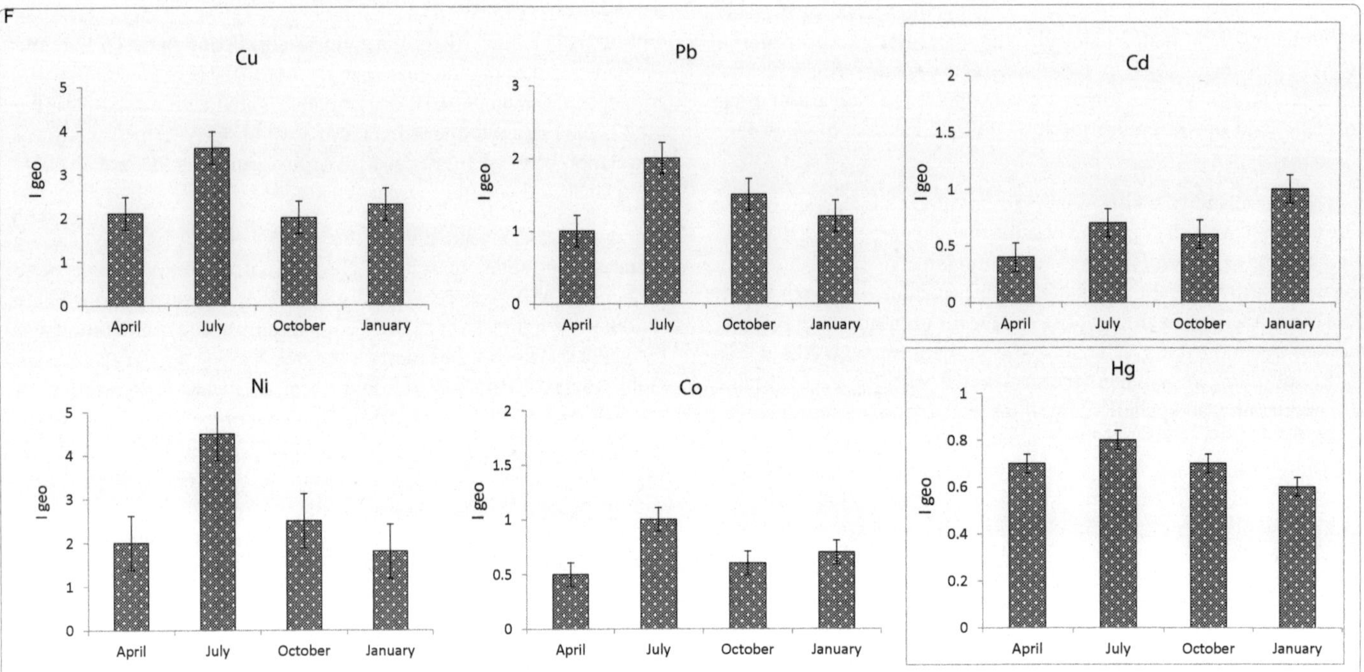

Figure 3: Plots showing I_{geo} seasonal values for each metal.

values identified for Cu and Ni in the same areas indicate that surface sediments are contaminated to some extent, most probably as a result of anthropogenic activities. I_{geo} values calculated for Cu (4.5) and Ni (3.6) showed higher values in the July month in the upstream region of the river than in the other seasons. This trace metal enrichment is attributed to iron ore processing in the higher reaches.

Bioaccumulation factors (BAFs)

BAFs significantly differed between heavy metals and sediments. BAFs of Ni were extremely higher than other metals followed by Cu, Co, Pb, Cd and Hg, respectively. These results indicate that Ni and Cu are more easily available than other elements. The bioavailability of metals in sediments could be controlled by species of metals and physicochemical factors such as pH, salinity, temperature, hardness, alkalinity, oxidation and reduction potential, particulate matter, organic carbon content, cation exchange capacity and differences in the grain size [12,33-36]. The present study also showed that the highest BAFs of essential and non-essential metals occurred in the station 3 followed by station 1, 2 and 4, respectively (Figure 3).

Conclusion

This study provides new information on the distribution and seasonal variation of metals in surface sediments along the Arvand river from northwest of Persian Gulf. The results showed that the concentration of metals varied among station and season sampling. The heavy metal concentration in the sediments are described in the descending order of Ni> Cu > Co > Pb > Cd > Hg at all sampling sites, except for station 1 where Cu concentration was higher than Ni. Results of this study also showed sediments from S3 showed greater concentration of the mercury than those from the other areas. The high concentration of metals in sediments at the S3 sampling could result from industrial effluents. BAFs showed that the risk of Ni is higher than the risk of other metals in this study. Therefore, the Arvand river as a major source of sediments and a source of metals can affect the concentration of metals in sediment of the area. Therefore, in order to enhance the data inventory for the region, characterize the geochemistry of meals in surficial sediments and to help understand the influence of anthropogenic activities in this tropical estuarine system.

Acknowledgement

The authors would like to thank all the field technicians helping in the field. This work was funded by Khoramshahr University of Marine Science and Technology and Environmental Protection of Tehran, Iran.

References

1. Clements W, Newman M (2002) Community ecotoxicology. New York, Wiley.
2. Tuzen M (2009) Toxic and essential trace elemental contents in fish species from the Black Sea, Turkey. Food Chem Toxicol 47: 1785-1790.
3. Kargın F, Donmez A, Cogun HY (2001) Distribution of heavy metals in different tissues of the shrimp Penaeus semiculatus and Metapenaeus monocerus from the Iskenderun Gulf, Turkey: seasonal variations. Bull Environ Contam Toxicol 66: 102-109.
4. Unlu E, Akba O, Sevim S, Gumgum B (1996) Heavy metal levels in mullet, Liza abu (Heckel, 1843) (Mugilidae) from the Tigris River, Turkey. Fresenius Environmental Bulletin 5: 107-112.
5. Chen MH (2002) Baseline metal concentrations in sediments and fishes, and the determination of bioindicators in the subtropical Chi-ku Lagoon, S. W. Taiwan. Mar Pollut Bull 44: 703-714.
6. Ikem A, Egiebor NO, Nyavor K (2003) Trace elements in water, fish and sediment from Tuskegee Lake, Southeastern USA. Water, Air, and Soil Pollution 149: 51-75.
7. Davies PJ (1974) Arsenic in sediments on the continental shelf of southeast Australia. Searches 5: 394-397.
8. McCready S, Birch GF, Long ER (2006) Metallic and organic contaminants in sediments of Sydney Harbour, Australia and vicinity-a chemical dataset for evaluating sediment quality guidelines. Environ Int 32: 455-465.
9. Yi Y, Wang Z, Zhang K, Yu G, Duan X (2008) Sediment pollution and its effect on fish through food chain in the Yangtze River. International Journal of Sediment Research 23: 338-347.
10. de Mora S, Fowler S, Wyse E, Azemard S (2004) Distribution of heavy metals in marine bivalves, fish and coastal sediments in Persian Gulf and Gulf of Oman. Mar Pollut Bull 49: 410-424.

11. Dalman O, Demirak A, Balci A (2006) Determination of heavy metals (Cd, Pb) and trace elements (Cu, Zn) in sediments and fish of the Southeastern Aegean Sea (Turkey) by atomic absorption spectrometry. Food Chemistry 95: 157-162.

12. Alagarsamy R (2006) Distribution and seasonal variation of trace metals in surface sediments of the Mandovi estuary, west coast of India. Estuarine, Coastal Shelf Science 67: 333-339.

13. Tessier A, Campbell PGC, Bisson M (1979) Sequential extraction procedure for the speciation of particulate trace metals. Analytical Chemistry 51: 844-851.

14. Flanagan FJ (1973) 1972 Values for international geochemical reference samples. Geochimica Cosmochimica Acta 37: 1189-1200.

15. Abdolahpur Monikh F, Safahieh A, Savari A, Doraghi A (2012) Heavy metal concentration in sediment, benthic, benthopelagic, and pelagic fish species from Musa Estuary (Persian Gulf). Environ Monit Assess 185: 215-222.

16. Samecka-Cymerman A, Kempers AJ (2001) Concentrations of heavy metals and plant nutrients in water, sediments and aquatic macrophytes of anthropogenic lakes former open cut brown coal/mines differing in stage of acidification. Sci Total Environ 281: 87-98.

17. Tokalioglu Ş, Karta Ş, Elçi L (2000) Speciation and determination of the heavy metals in lake waters by atomic absorption spectrometry after sorption on Amberlite XAD- 16 Resin. Analytical Sciences 16: 1169-1174.

18. Elith M, Garwood S (2001) Investigation into the levels of heavy metals within Manly Dam Catchment. In: Freshwater ecology report 2001. Sydney: Department of Environmental Sciences, University of Technology.

19. Nolting RF, van Dalen M, Helder W (1996) Distribution of trace and major elements in sediment and pore waters of the Lena Delta and Laptev Sea. Marine Chemistry 53: 285-299.

20. Byrd JE, Perona MJ (1980) The temporal variations of the lead concentration in a freshwater lake. Water, Air, and Soil Pollution 13: 207-220.

21. Sheppard C, Al-Husiani M, Al-Jamali F, Al-Yamani F, Baldwin R, et al. (2010) The Gulf: A young sea in decline. Mar Pollut Bull 60: 13-38.

22. ROPM (1999) Overview on Land-based Sources and Activities Affecting the Marine Environment in the ROPME Sea Area. UNEP/GPA Coordination Office & ROPME 127.

23. Kabata-Pendias A, Pendias H (2001) Trace elements in soils and plants (3rd ed.) Boca Raton, FL: CRC Press.

24. Xiangdong L, Zhenguo S, Onyx WHW, Yok-sheung L (2000) Chemical partitioning of heavy metal contaminants in sediments of the Pearl River Estuary. Chemical Speciation and Bioavailability 12: 17-25.

25. Vogel-Mikus K, Drobne D, Regvar M (2005) Zn, Cd and Pb accumulation and arbuscular mycorrhizal colonization of pennycress Thlaspi praecox Wulf. (Brassicaceae) from the vicinity of a lead mine and smelter in Slovenia. Environ Pollut 133: 233-242.

26. Forstner U (1985) Chemical forms and reactivities of metals in sediment. In R. Leschber, R. D. Davids, L'. Hermite (Eds), Chemical methods for assessing bio-available metal in sludges and soils (pp. 1–30). London: Elsevier.

27. Kong IH, Liu SH (1995) Determination of heavy metals distribution in the anoxic sediment slurries by chemical sequential fraction. Ecotoxicol Environ Saf 32: 34-38.

28. Karbassi AR, Amirnezhad R (2004) Geochemistry Ni, Cu, Zn, Pb, Cd, Co, Mn,V, Fe, Al and Ca in sediments of North Western part of the Persian Gulf. Intl J Env Studies 54: 205-212.

29. Bordin G, Mc-Court J, Rodriquez A (1992) Trace metal in the marine bivalve Macoma balthica in the westerschelde estuary the Nerthlands. Part 1:analysis of total copper, cadmium, zinc and iron concentrations; local and seasonal variations. Science of the Total Environ 127: 255-280.

30. Muller G (1979) Schwermetalle in den sedimenten des Rheins e Veranderungen seit 1971. Umschau 79: 778-783.

31. Turekian KK, Wedepohl KH (1961) Distribution of the elements in some major units of the earth's crust. Geological Society of America Bulletin 72: 175-192.

32. Ridgway J, Shimmield G (2002) Estuaries as repositories of historical contamination and their impact on shelf seas. Estuarine, Coastal and Shelf Science 55: 903-928.

33. Caussy D, Gochfeld M, Gurzau E, Neagu C, Ruedel H (2003) Lessons from case studies of metals: investigating exposure, bioavailability, and risk. Ecotoxicol Environ Saf 56: 45-51.

34. Sekhar KC, Chary NS, Kamala CT, Suman Raj DS, Sreenivasa Rao A (2003) Fractionation studies and bioaccumulation of sediment-bound heavy metals in Kolleru lake by edible fish. Environ Int 29: 1001-1008.

35. Zhang W, Yu L, Hutchinson SM, Xu S, Chen Z, et al. (2001) Chinas Yangtze Estuary: I. Geomorphic influence on heavy metal accumulation in intertidal sediments. Geomorphology 41: 195-205.

36. Bellucci IG, Moumni El, Collavini B, Frignani FM, Albertazzi S (2003) Heavy metals in Morocco Lagoon and river sediments. Journal de Phys 107: 139-142.

Polychlorinated Dibenzo-P-Dioxins (PCDD), Polychlorinated Dibenzofurans (PCDF) and Dioxin-Like Polychlorinated Biphenyls (Dl-PCB) in the Baltic and Arctic Fish and the Further Trophic Transfer of these Pollutants to Seabirds

Elżbieta Niemirycz[1], Joanna Szlinder-Richert[2], Ott Roots[3,4], Lucyna Falkowska[1], Ksenia Pazdro[5], Agata Zaborska[5], Matti Verta[6], Grażyna Sapota[7], Maria Witt[1], Andrzej R Reindl[1] and Marta E Kobusińska[1*]

[1]University of Gdansk, Institute of Oceanography, Department of Marine Chemistry and Environmental Protection, 46 Marszalka Pilsudskiego Str., Gdynia 81-378, Poland
[2]National Marine Fisheries Research Institute, Department of Food and Environmental Chemistry, Kollataja Str, Gdynia 81-332, Poland
[3]Estonian Environmental Research Centre, Marja 4D, 10 617 Tallinn, Estonia
[4]University of Tartu, Estonian Marine Institute, Mäealuse 10A, 12 618 Tallinn, Estonia
[5]Institute of Oceanology of the Polish Academy of Sciences, 55 Powstancow Warszawy Str, Sopot 81-712, Poland
[6]Finnish Environment Institute (SYKE), P.O. Box 140 Helsinki, Finland
[7]Maritime Institute in Gdansk, Department of Environmental Protection, 41/42 Dlugi Targ Str, Gdansk 80-830, Poland

Abstract

The study presents the analysis of polychlorinated dibenzo-p-dioxins, dibenzofurans (PCDD/F) and dioxin-like polychlorinated biphenyls (dl-PCB) in fish and seabirds from the Baltic Sea area and the Arctic region (the Svalbard Archipelago). Long-time studies revealed no significant temporal trends in PCDD/F and dl-PCB concentrations in fish from the Baltic Sea. Trace amounts of PCDD/F were detected in the Arctic fish as an evidence of their anthropogenic sources from the temperate zone, however congener profile in the cod was entirely different than in the Baltic fish. On the contrary, both concentrations and profiles of dl-PCB were comparable in cod from the Baltic Sea and from the polar region. PCDD/F`s congener profile in African penguin`s tissues breed in captivity and fed only by herring exhibited a clear resemblance to the profiles measured in herring. This indicates a relevant impact of diet (fish) on PCDD/F`s level in seabirds. Diet impact on the further transfer of these xenobiotics in the trophic chain, evidenced in the presented study, appears to be of the particular relevance due to the human health.

Keywords: Dioxins; Congener profiles; Marine organisms; Trophic chain; Diet impact; Climatic zones

Introduction

Properties of halogenated organic compounds, such as persistence, toxicity, bioaccumulation and potential for long range atmospheric transport on airborne particles pose a threat to marine organisms as well as human, so their emission must be reduced. The Stockholm Convention [1] and the Convention on Long-Range Transboundary Air Pollution (UNECE) [2] created a lists of hazardous substances and solutions aimed at mitigation of a global environmental risk Most of these compounds are produced commercially therefore, the control of their emission is controllable. However, polychlorinated dibenzo-p-dioxins and dibenzo-p-furans (PCDD/F) are a subject of particular concern since they are produced unintentionally in industrial and non-industrial combustion processes or chemical reactions occurring in the environment, as well as in natural processes like volcanic eruptions or forest fires. The list of unintentionally produced persistent organic pollutants (UPOPs) has been doubled during recent years [3]. Due to the typical properties of dioxins and other POPs as hydrophobicity and low vapor pressure, these contaminants are characterized by relatively limited mobility in environmental compartments. Nevertheless, a vertical migration into groundwater, as well as horizontal transport by air can be achieved easily [4-6]. Increased mobility can be caused by the presence of other contaminants, e.g. the group of dense non-aqueous phase liquids (DNAPLs), including pesticides and organic solvents, but also creosote and soot. These impurities may result in the formation of colloidal particles of mobile POPs or facilitate the sorption process in the solid phases of atmospheric aerosols. Sites contaminated with halogenated organic compounds revealed presence of 70% of DNAPLs in a groundwater (US EPA) [7]. Thus, the ability of POPs to migrate in environmental compartments, particularly via atmosphere has become an important criterion for the environmental risk assessment. In aquatic ecosystems, pollutants originating from industrial activities, transport, and other human activities enter the sea through river inflow, direct discharge from the land, and atmospheric inputs. In the marine environment, organic pollutants such PCDD/F/dl-PCB exists as dissolved in the water, and as adsorbed on the suspended particles in the water column. Due to the high hydrophobicity, concentrations of organic pollutants in waters are relatively low, while the sediments constitute the significant reservoir for organic contaminants, posing a threat for the benthic organisms and for the water column due to the resuspension or remobilization processes, dependent on the particular compound properties and environmental conditions. These compounds are known to induce a number of adverse changes in the ecosystem

***Corresponding author:** Marta E Kobusińska, University of Gdansk, Institute of Oceanography, Department of Marine Chemistry and Environmental Protection, 46 Marszalka Pilsudskiego Str., Gdynia 81-378, Poland
E-mail: marta.kobusinska@ug.edu.pl

while have been introduced to the trophic chain. Concentrations of pollutants in the sediments or water do not reflect their bioavailability. Therefore, biota is useful indicator for environmental risk assessment, representing a load of contaminants to the environment. Fish are a good indicator for assessment of the state of the environment, especially due to their trophic relationships to birds and other consumers as well as too human. Thus the risk assessment of aquatic ecosystems is very important.

Since the human remain exposed to PCDD/F/dl-PCB mainly via consumption, the EU has established their acceptable limits in several food groups, i.a. in fish. The latest European Commission Directive (EC) [8] establishes the maximum limit of PCDD/F in fish as 3.5 WHO-TEQ pg · g^{-1}w.w. and the maximum limit of the sum of PCDD/F and dl-PCB for 6.5 WHO-TEQ pg g^{-1}w.w.

The study aimed at comparison of levels and congener patterns of PCDD/F and dl-PCB in fish from the Baltic Sea- the basin particularly sensitive due to unique hydrological and geographical conditions, with the polar region, the Spitzbergen Island on the Arctic Ocean (the Svalbard Archipelago), distant from point sources of pollutants

The phenomenon of migration of pollutants is of the major relevance in the assessment of contamination of regions previously considered as unpolluted. Detection of hazardous substances in the Arctic region confirms a transboundary migration, suggesting a relevant threat to the organisms inhabiting there and also indicates insufficient decision-making in environmental protection of industrialized temperate zone. Furthermore, PCDD/F profiles in seabird (bred in captivity) and herring as its food was compared to assess the possible impact of diet.

Materials and Methods

Sampling

Fish: Concentrations of PCDD (7 congeners from tetra to octa), PCDF (10 congeners from tetra to octa) and 12 dl-PCB (non-ortho 77, 81, 126 and 169 and mono- ortho: 105, 114, 118, 123, 156, 157, 167 and 189) have been analysed in the muscle tissue of five fish species: herring (*Clupea harengus membras*), salmon (*Salmo salar*), sprat (*Sprattus sprattus balticus*), flounder (*Platichthys flesus*) cod (*Gadus morhua callarias*), collected in 2002-2014 from the Baltic Sea and a cod (*Gadus morhua morhua*) from the Kongsfiord on the Svalbard Archipelago (Figure 1), the Arctic.

Baltic fish were caught offshore by trawl and in the coastal area using gillnet tools. Arctic cod was caught using fishing rods from r/v "Oceania" board within "AREX 2013" campaign fishing area and the number of samples is given in Table 1. Some of the results presented have been yet published by Szlinder-Richert et al. [3,8-10].

Seabirds: PCDD/F were analyzed in muscle and liver tissues in three species of seabirds: wild living gulls- herring gull (*Larus argentatus*) and great great black-backed gull (*Larus marinus*), inhabiting the coastal zone of the Gulf of Gdansk (Poland) and the African penguin breed in captivity (*Spheniscus demersus*), from the culture the Zoological Garden in Gdansk. Three individuals of herring (*Clupea harengus membras*) from the Gulf of Gdansk were analyzed (whole fish as one sample) as a seabirds` diet component.

Sample Processing and Analysis: Each sample of herring, sprat and flounder consisted of a combined pooled muscle tissues from several individuals, classified by fish length. One individual of salmon, cod and bird, respectively, prepared as one sample Herring investigated as a diet component for African penguin was analyzed as a whole organisms. Homogenized, freeze-dried tissues were extracted by accelerated solvent extraction (ASE). The quantification of the studied compounds was based on the use of ^{13}C labeled internal standards that were spiked into the sample extracts before extraction. Lipids were removed by passing the extract through a multilayer silica gel column eluting with n-hexane.

Analyses of 17 congeners of PCDD/F and 12 congeners of dl-PCB have been performed according to norm: PN-EN 1984, US EPA 1613, and US EPA 1668 in accredited laboratories (National Institute of Nutrition and Seafood Research in Copenhagen, Denmark, National Reference Laboratory for Analysis of POPs in Ostrava, Czech Republic, Eurofins GfA Gmbh Laboratory in Hamburg, Germany, National Institute for Health and Welfare in Helsinki, Finland, National Public Health Institute in Kuopio, Finland, Laboratory for Trace Organic Analyses at Krakow University of Technology, Poland).

The limits of detection (LOD) for PCDD/F and dl-PCB congeners were congeners dependent and varied in different samples between 0.1 and 1 pg g^{-1}w.w. for PCDD/F and from 0.5 to 45 pg g^{-1}w.w. for dl-PCB. Concentrations below the limits of quantification (LOQ) were equated to half of the LOQ. The uncertainty of measurements varied from 30% to 35% for sum of PCDD/F and dl-PCB, depending on the laboratory. The recoveries of the internal standards ranged between 74% and 126% for PCDD/F and 56–54% for dl-PCB.

Expression of results: PCDD/F and dl-PCB determined in fish muscle tissue were expressed in pg g^{-1} w.w. (wet weight). Total toxicity of PCDD/F and dl-PCB (TEQ) was calculated by multiplying the individual congener levels by the respective Toxic Equivalency Factor (TEF), (WHO 2005) [11] and expressed as a WHO-TEQ pg g^{-1} w.w. PCDD/F and dl-PCB in birds` muscle and liver tissues were expressed in pg g^{-1}l.w. (lipid weight).

Quality Assurance/Quality Control: QA/QC was performed through the analysis of procedural blanks, duplicate samples and standard reference materials for each sample set with a relative standard deviation (RSD) below 15% for all detected compounds.

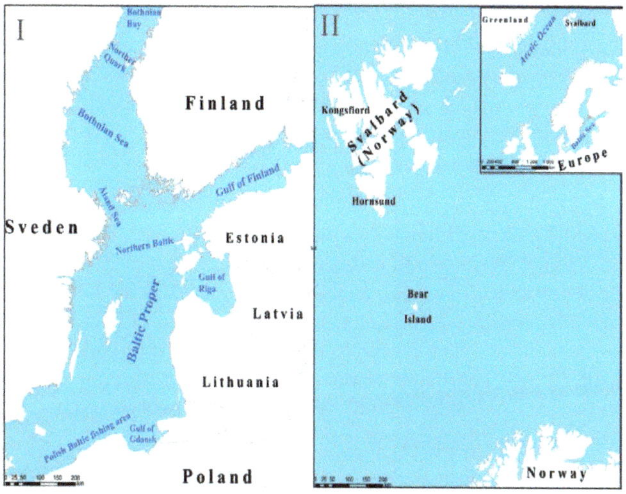

Figure 1: Study area: I) the Baltic Sea region (Table 1), 418 500 km^2, II) the Arctic (the Svalbard Archipelago).

Results and Discussion

Concentrations and congener profiles of PCDD/F and dl-PCB in Baltic and Arctic fish

The Baltic Sea, as a relatively shallow and small basin, characterized by limited water exchange through the Danish Straits, remains sensitive to the anthropogenic land pollution discharged via riverine outflow and via atmospheric deposition [12-17]. Currently, the concentration of the most hazardous halogenated organic substances such as PCDD/F, are higher in the Baltic Sea compared to other seas at the same latitude [18].

Anthropogenic pollution in the northern Europe increasingly affects the polar areas. Substances from the group of POPs unlikely undergo metabolic reactions and due to strong sorption capacity, are transported on aerosols over long distances. Due to their hydrophobicity they accumulate along trophic levels and induce multiple adverse effects in organisms, including developmental effects, immunotoxicity, and reproductive effects [19].

The level of POPs in marine organisms (zooplankton, fish) depends *i.a.* on seasonal changes in lipid content connected with physiological cycles (*e.g.* reproduction), age, feeding habits and the environmental conditions [18, 20-27] have been investigating the impact of seasonal changes in the concentrations of halogenated organic compounds in the Baltic salmon, due to the spawning run phenomenon, which remains insufficiently recognized.

The bioaccumulation and elimination of organochlorine contaminants in fish is complex and several routes for these processes have been proposed. The bioaccumulation defined as an increase in organisms/water chemical concentration ratio due to direct uptake from water (transport across respiratory surface and dermal absorption) and consumption. The elimination includes: respiration, egestion and metabolism (Mackay and Fraser, 2000). Development of analysis and synthesis of autocatalytic chemical systems, metabolomics and proteomics, facilitates identification of the possible migration routes of toxic substances in organisms [28,29]. Vuorinen et al. [30] proved that during vitellogenesis, estradiol-17β inhibits detoxification and influences accumulation of PCB in Baltic salmon roe to 68%.

Concentration of PCDD/F and dl-PCB in studied fish (331 samples) are presented in Table 1.

Any noticeable decreasing trend of PCDD/F/dl-PCB concentrations in Baltic fish during the decade (2002-2014) has been observed, despite the implementation of legal restrictions on the POPs emissions. This fact may suggest a relevance of uncontrolled diffuse sources and low emission (municipal combustion, traffic etc.) contribution, not to discard the physiological processes (e.g. spawning run process) as well as individual characteristics (age and sex).

PCDD/F and dl-PCB, as well as dl-PCB/PCDD/F ratio in the study revealed significant interspecies and spatial differences. Twelve of the non-ortho-PCB and mono-ortho-PCB (dl- PCB), have been calculated together with PCDD/F for establishing the TEQ (WHO 1998) and frequently considered as a one group. However, PCDD/F and PCB are characterized by different source of origin, thus PCDD/F concentration remains higher in a winter season, while dl-PCB predominates in the summer [31] and consequently, environmental fate of these chemicals proceeds differently, what was also observed in this study.

Minor interspecies differences in PCB concentration were observed in contrast to PCDD/F. Among the Baltic fish, the highest contribution of dl-PCB in PCDD/F/dl-PCB was noticed in cod, salmon and flounder. However the most evident difference was observed in the Baltic and the Arctic cod. Concentration of dl-PCB was about 5-fold higher than PCDD/F in the cod from the Baltic, whereas in the cod from the polar region concentrations of dl-PCB were 16-fold higher than PCDD/F. The highest concentrations of investigated compounds, particularly dl-PCB, were observed in salmon, which may be explained by the intensified bioaccumulation due to the high lipid content and biomagnification, considering the trophic chain level that is represents. The highest average concentration (12.6 WHO-TEQ pg g^{-1}w.w.) in salmon`s muscle tissue has been observed in the individuals from the southern part of the Baltic Sea (Table 1). Vuorinen et al. [18], observed halved level of contamination in the northern regions such as the Aland Sea and the Bothnian Sea about ten years later (6.7 WHO-TEQ pg g^{-1}w.w.and 6.6 WHO-TEQ pg g^{-1}w.w., respectively). These results exceed the acceptable level for both PCDD/F (3.5 WHO-TEQ pg · g^{-1}w.w.) and the sum PCDD/F/dl-PCB (6.5 WHO-TEQ pg g^{-1}w.w.) [8].

Concentrations determined in herrings and sprats were prominently lower. All averaged sums remain compliant with the acceptable limits for fish with one exception of elevated concentration in herring caught in 2008 and the maximum values have been exceeded occasionally (Table 1). Individuals of Baltic flounder and cod revealed the lowest PCDD/F contamination, lower than 1.05 WHO-TEQ pg g^{-1}w. w. of PCDD/F/dl-PCB sum.

Levels of the tested xenobiotics revealed correlation with fat content and the body weight of fish. Herring, belonging to the fatty fish that live only in marine waters is the most representative for the Baltic Sea, compared with fatty salmon which life cycle assumes the existence in both salt and fresh waters.

It was confirmed, that the biological factor as age, plays a relevant role for the contamination of the fish with toxic organohalogenated compounds such as PCDD/F. It was observed that fish older than 6 years and more than 17 cm in length had higher PCDD/F concentration than the internationally permitted threshold [4,32,33].

Uptake of contaminants by fish is strictly related to their diet, therefore, spatial variation of PCDD/F and dl-PCB level observed in organisms reflect the quality of their habitats. The best foraging conditions for fish prevail in the Baltic Proper and the Northern Quark, where their major diet components, species-dependent, are zooplankton and sprats. According to Peltonen et al. [18,27] the sum of PCDD/F and dl-PCB concentrations in the fish muscle did not exceed 4.5 WHO-TEQ pg g^{-1}w.w. The less preferred conditions are observed in the northern Baltic Sea (the Åland Sea, the Bothnian Sea, the Bothnian Bay) and in the southern Baltic Sea (the Bornholm Basin), which are dominated by fatty herring and sprats, major diet component for salmon and cod.

The theory of congener profiles of PCDD/F was created and introduced by Rappe thirty years ago as a method for presentation of PCDD/F congeners (isomers) associated with different sources of emissions, so the particular congener profile pattern may be applied for emission source identification (Figure 2) [34].

The congener profiles in fish from the different parts of the Baltic Sea area revealed resemblance in most studied cases [18,25]. Among congeners of PCDD/F, in Baltic fish, TCDF and PeCDF predominated, what was confirmed by the results of presented study (Figure 3a). The 1,2,3,7,8-PeCDD followed by 2,3,4,7,8-PeCDF are the main contributors to the sum of toxicity (WHO-TEQ) in Baltic fish and at the same air deposition indicators [12,25]. In Atlantic cod, only HpCDD and OCDD were identified but surprisingly, at comparable levels as the

Fish species	Sampling locations	Year	WHO-TEQ pg ·g^{-1}w.w.					
			ΣPCDD/F $\bar{X}$ ± SD	Σdl-PCB $\bar{X}$ ± SD	Total PCDD/F-dl-PCB	Ratio Σdl-PCB/PCDD/F	Min.–Max. ΣPCDD/F ± U	Min.-Max. Σdl-PCB ± U
Salmon n=59	Polish Baltic fishing area	2003	2.62 ± 0.75	5.12 ± 1.39	7.74*	1.95	0.75 ± 0.23 - 4.30* ± 1.29	1.39 ± 0.42 - 8.27 ± 2.48
		2004	3.43 ± 0.66	9.17 ± 2.84	12.60*	2.67	2.48 ± 0.74 - 4.32* ± 1.30	5.86 ± 1.76 - 13.06 ± 3.92
		2005	2.10 ± 0.62	3.96 ± 0.89	5.97	1.89	4.27* ± 1.28 - 7.07* ± 2.12	4.27 ± 1.28 - 7.07 ± 2.12
		2006	3.12 ± 0.7	5.49 ± 0.86	8.62*	1.76	1.20 ± 0.36 - 3.50* ± 1.05	2.60 ± 0.78 - 5.40 ± 1.62
		2010-2011	3.26 ± 2.53	7.05 ± 5.68	10.31*	2.16	0.86 ± 0.26 - 5.26* ± 1.58	2.67 ± 0.80 - 10.87 ± 3.26
		2010-2011	3.46 ± 1.27	7.43 ± 3.43	10.89*	2.15	1.35 ± 0.41 - 5.26* ± 1.58	3.85 ± 1.16 - 10.87 ± 3.26
	Gulf of Finland	2005	3.13 ± 0.41	3.94 ± 0.68	7.07*	1.26	2.67 ± 0.52 - 3.47 ± 0.69	3.50 ± 0.7 - 4.73 ± 0.95
Herring n= 134	Polish Baltic fishing area	2002	2.19 ± 0.91	1.77 ± 0.57	3.96	0.81	0.34 ± 0.10 - 3.77 ± 1.13	0.45 ± 0.14 - 2.69 ± 0.81
		2003	1.23 ± 0.37	1.51 ± 0.38	2.74	1.23	0.76 ± 0.23 - 2.05 ± 0.62	1.16 ± 0.35 - 2.05 ± 0.62
		2004	2.01 ± 0.55	2.11 ± 0.54	4.12	1.05	1.34 ± 0.40 - 3.28 ± 0.98	1.03 ± 0.31 - 3.02 ± 0.91
		2005	1.75 ± 0.57	1.83 ± 0.53	3.58	1.05	0.98 ± 0.29 - 2.60 ± 0.78	1.16 ± 0.35 - 2.61 ± 0.78
		2006	1.89 ± 0.47	2.12 ± 0.49	4.01	1.12	0.96 ± 0.29 - 2.60 ± 0.78	1.32 ± 0.40 - 3.01 ± 0.74
		2011	1.01 ± 0.38	1.51 ± 0.57	2.52	1.50	0.52 ± 0.16 - 2.01 ± 0.60	0.25 ± 0.08 - 2.47 ± 0.74
	Gulf of Riga	2003	3.23 ± 2.81	2.61 ± 1.00	5.84	0.81	1.19 ± 0.30 - 8.61 ± 0.34	1.34 ± 0.34 - 3.41 ± 0.85
		2004	1.63 ± 0.57	1.03 ± 0.38	2.66	0.63	1.23 ± 0.31 - 2.04 ± 0.55	0.75 ± 0.19 - 1.30 ± 0.33
		2007	3.42 ± 0.60	1.91 ± 0.31	5.33	0.56	2.83 ± 0.71 - 4.03 ± 0.98	1.63 ± 0.41 - 2.25 ± 0.56
		2008	1.94 ± 0.14	1.30 ± 0 0.04	3.24	0.67	1.83 ± 0.46 - 2.14 ± 0.53	1.27 ± 0.32 - 1.35 ± 0.34
		2009	1.83 ± 1.16	2.15 ± 1.97	3.98	1.17	1.11 ± 0.28 - 3.19 ± 1.23	0.97 ± 0.24 - 4.42 ± 1.11
	Gulf of Finland	2005	0.85 ± 0.24	0.72 ± 1.11	1.57	0.85	0.20 ± 0.05 - 1.16 ± 0.29	0.48 ± 0.12 - 0.80 ± 0.20
		2007	1.35 ± 0.08	0.94 ± 0.72	2.29	0.70	1.29 ± 0.32 - 1.44 ± 0.36	0.86 ± 0.22 - 1.01 ± 0.25
		2009	0.95 ± 0.36	0.99 ± 0.45	1.94	1.04	0.58 ± 0.15 - 1.37 ± 0.34	0.57 ± 0.14 - 1.54 ± 0.39
		2010	1.51 ± 0.33	1.01 ± 0.25	2.52	0.67	1.31 ± 0.33 - 2.00 ± 0.50	0.89 ± 0.22 - 1.27 ± 0.32
	Northern Baltic	2003	2.77 ± 1.00	2.86 ± 1.00	5.63	1.03	2.07 ± 0.52 - 3.46 ± 0.91	2.08 ± 0.51 - 3.63 ± 0.92
		2006	1.59 ± 0.24	1.41 ± 0.23	3.00	0.89	1.35 ± 0.30 - 1.79 ± 0.41	1.18 ± 0.28 - 1.63 ± 0.43
Sprat n= 104	Polish Baltic fishing area	2002	2.56 ± 0.47	3.28 ± 0.65	5.84	1.28	0.47 ± 0.14 - 3.19 ± 0.99	0.65 ± 0.20 - 4.23 ± 1.27
		2003	2.11 ± 0.89	2.94 ± 1.08	5.05	1.39	0.74 ± 0.22 - 2.80 ± 0.84	1.00 ± 0.30 - 3.98 ± 1.19
		2004	2.59 ± 0.65	3.52 ± 0.91	6.11	1.36	1.35 ± 0.41 - 3.14 ± 0.94	1.79 ± 0.54 - 4.69 ± 1.41
		2005	1.88 ± 0.56	2.73 ± 0.59	4.61	1.45	0.56 ± 0.17 - 2.04 ± 0.61	0.58 ± 0.17 - 3.80 ± 1.14
		2006	2.77 ± 0.35	3.60 ± 0.51	6.37	1.30	0.56 ± 0.17 - 3.01 ± 0.90	0.59 ± 0.18 - 4.30 ± 1.29
		2011	1.34 ± 0.51	2.28 ± 0.91	3.62	1.70	0.59 ± 0.18 - 2.18 ± 0.65	0.97 ± 0.29 - 0.51 ± 0.15
	Gulf of Finland	2003	2.89 ± 0.46	2.79 ± 0.22	5.68	0.97	2.49 ± 0.50 - 3.56* ± 0.89	2.48 ± 0.62 - 3.07 ± 0.77
		2004	1.15 ± 0.47	1.77 ± 0.72	2.92	1.54	0.76 ± 0.27 - 1.85 ± 0.46	0.84 ± 0.21 - 2.60 ± 0.65
		2007	1.60 ± 0.09	1.65 ± 0.26	3.25	1.03	1.45 ± 0.36 - 1.76 ± 0.44	1.36 ± 0.34 - 2.07 ± 0.52
	Northern Baltic	2004	1.80 ± 0.62	2.27 ± 1.23	4.07	1.26	1.18 ± 0.30 - 2.61 ± 0.65	0.96 ± 0.24 - 3.35 ± 0.84
		2006	1.51 ± 0.08	1.57 ± 0.09	3.08	1.04	1.42 ± 0.36 - 1.57 ± 0.39	1.47 ± 0.37 - 1.65 ± 0.41
		2010	1.86 ± 0.15	1.64 ± 0.17	3.50	0.88	1.73 ± 0.43 - 2.02 ± 0.51	1.47 ± 0.37 - 1.81 ± 0.45
Flounder n=28	Polish Baltic fishing area	2011	0.26 ± 0.15	0.63 ± 0.29	0.89	2.42	0.15 ± 0.05 - 0.77 ± 0.23	0.22 ± 0.07 - 1.55 ± 0.47
	Northern Baltic	2004	0.35 ± 0.14	0.70 ± 0.47	1.05	2.00	0.19 ± 0.04 - 0.45 ± 0.09	0.33 ± 0.07 - 1.22 ± 0.31
Cod n=6	Polish Baltic fishing area	2008-2009	0.18 ± 0.08	0.97 ± 0.45	1.05	5.39	0.03 ± 0.01 - 1.16 ± 0.35	0.65 ± 0.31 - 1.53 ± 0.51
	The Arctic Sea (the Svalbard Archipelago)	2014	0.03 ± 0.03	0.59 ± 0.15	0.62	16.83	0.002 ± 0.001 -0.063 ± .0126	0.30 ± 0.06 - 0.59 ± 0.118

Table 1: PCDD/F and dl-PCB (WHO-TEQ pg g^{-1}w.w.) in marine fish caught in the Baltic Sea and the Arctic region during 2002-2014. Total number of samples (n) equal of 311, average concentration ($\bar{X}$) and standard deviation (SD), range (Min.-Max.) with expanded uncertainty (U).

Baltic fish (Figure 2a). In terms of environmental risk, it is relevant, that PCDD/F congeners found in the Atlantic cod represents the lowest TEF, consequently- the lowest toxicity. This phenomenon could be explained by more efficient air transfer of highly chlorinated compounds, which are known to absorb on particulate matter to the higher extent than less chlorinated PCDD/F congeners [35-37]. On the contrary, the profile of dl-PCB was similar in the Baltic fish and the Atlantic cod from the Arctic region, what potentially may pose a threat to organisms from the higher trophic levels, e.g. polar bears. In both Baltic and Arctic fish, the PCB118 mono-ortho congener dominates among dl-PCB (Figure 2b). All non-ortho congeners (PCB77, PCB126, PCB169) have been detected in Baltic herring and Atlantic cod.

Fish as a source of pollutants to the higher trophic levels: Fish due to the high content of protein, minerals, fatty acids and vitamins are extremely valuable food for birds and other predators, but they are also a source of hazardous chemicals accumulating through trophic chains. In 2012, the University of Gdansk performed a study of the African penguin inhabiting the Zoological Garden in Gdansk, fed only by studied herring, as well as herring gull and great black-backed gull fed in the natural environment. The aim of the research was to determine

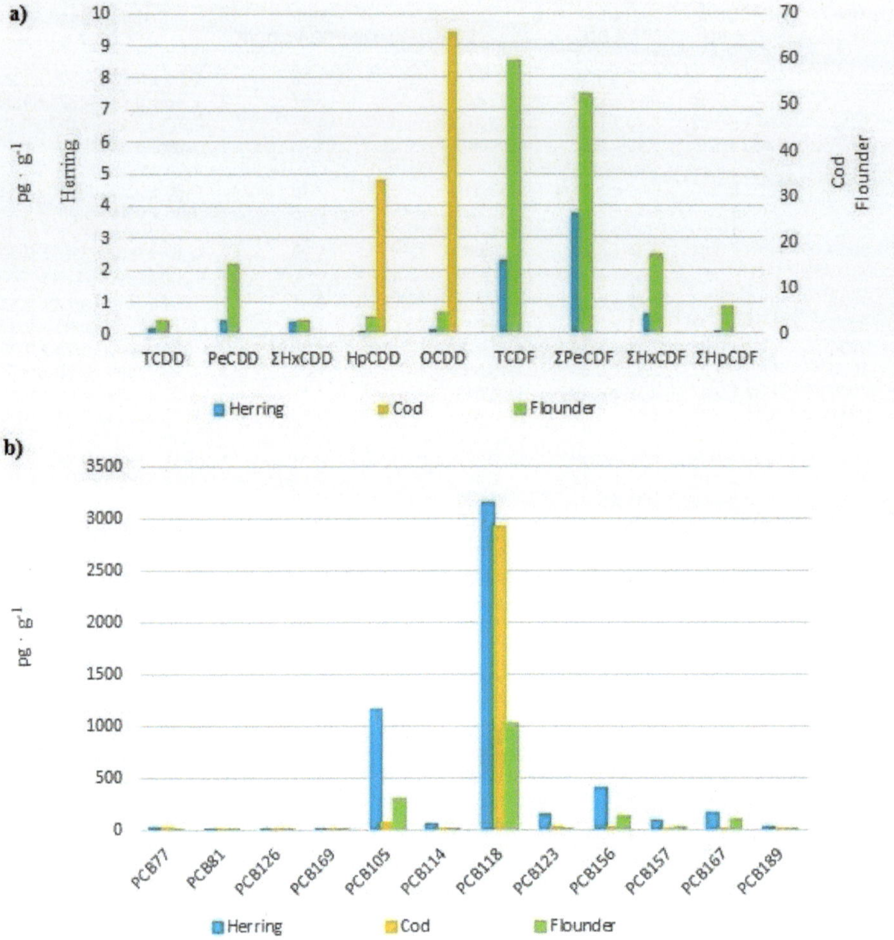

Figure 2: Congener profiles of a) PCDD/F and b) dl-PCB expressed in pg·g-1w.w. in fish collected from the southern Baltic Sea (flounder), the Gulf of Riga (herring) and the Svalbard Archipelago (cod).

the effects of the diet on the PCDD/F contamination of studied birds. PCDD/F congener profile in the whole herring was similar to those measured in herring muscles tissue. Profiles of PCDD/F in the whole herrings have a clear resemblance to the muscle tissue and the liver glandular of penguin (Figure 3a), which proves that diet determines the load of toxic substances to the consumer tissues. The highest concentration of congeners PeCDF and TCDF in herring dominated also in the liver and muscles of penguin.

Herring gulls are the most abundant bird (approx. 70% of individuals) in the southern area of the Baltic Sea (the Gulf of Gdansk) [37] while black-backed gulls represent the largest group of gulls occurring on islands and coastal areas of the northern Atlantic hemisphere (from the Great Lakes to the White Sea). The typical gulls` diet consists of fish, fish guts delivered from fishing boats, invertebrates, eggs and chicks of other birds, while the penguin`s diet includes only herring. The differences in the obtained congener profiles in the penguin`s tissues and the both species of seagulls indicate the influence of the specific diet`s composition on the birds` contamination (Figure 3).

The sum of PCDF congeners in liver is higher than the sum of these congeners in muscles, while the sum of PCDDs remains higher in muscles than in liver. Particularly high concentrations of TCDD- the most toxic congener (TEF=1), have been recorded in the black-backed gull, which may be related to the randomly acquired, significantly contaminated food.

Conclusion

Toxic substances from the POPs` group are spread over the long distances originating from the point sources located in the industrialized countries of the temperate latitudes, up to the Arctic region. The low mobility, nonetheless the strong chemical bonding with other pollutants from the DNAPL group, enhances their potential to be transported.

Presented paper includes an extensive and comprehensive data on PCDD/F and dl-PCB concentrations in various fish species from the Baltic Sea sub- basins and comparatively, the Arctic region. Concentrations of PCDD/F and dl-PCB in fish, in the majority, have met the stringent requirements established by European Commission in 2011 (3.5 WHO-TEQ pg g^{-1}w.w. for PCDD/F and the sum of PCDD/F and dl-PCB for 6.5 WHO-TEQ pg g^{-1} w.w.) and do not exhibit any noticeable fluctuations in the recent decade. The exceedances have been observed mostly in fatty individuals as salmon.

Dl-PCB/PCDD/F ratio in investigated Baltic fish remains comparable to the literature`s reports, however, the remarkably raised

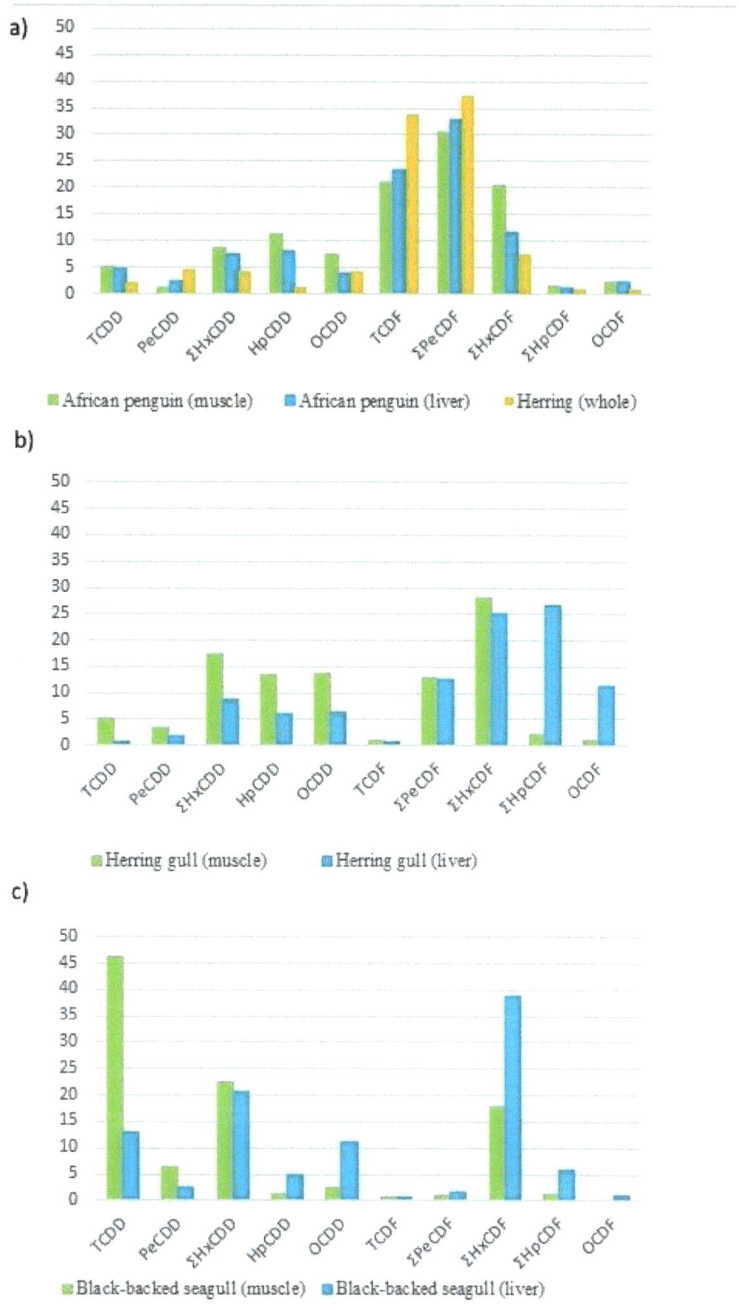

Figure 3: Congener profiles of PCDD/F in the whole Baltic herring, in the liver and muscles of African penguin (a), great black-backed gull wintering in the Gulf of Gdansk (b) and herring gull with breeding colony on the Gulf of Gdansk (c), expressed as a percentage contribution in the ΣPCDD/F (pg g^{-1} l.w).

ratio observed in the Arctic cod may indicate more efficient long range transport of dl-PCB. Nevertheless, the observed phenomenon requires a further examination.

Comparative analysis of PCDD/F congener profiles in the selected seabirds, the African penguin fed on the homogenous diet as herring and the two species of seagulls, fed on the randomly acquired food, proved the apparent trophic transfer of selected halogenated contaminants from fish to seabirds.

The diet impact on the further transfer of these xenobiotics in the trophic chain, evidenced in the presented study, appears to be of the particular relevance due to the human health.

References

1. The Stockholm Convention on Persistent Organic Pollutants (2001) United Nations Environment Programme (UNEP), Stockholm Sweden.

2. The Convention on Long-Range Transboundary Air Pollution: The Protocol on Persistent Organic Pollutants (1998) United Nations Economic Commission for Europe (UNECE), Aarhus, Denmark.

3. Weber R, Gaus C, Tysklind M, Johnston P, Forter M, et al. (2008) Dioxin- and

POP-contaminated sites-contemporary and future relevance and challenges. Environ Sci Pollut Res Int 15: 363-393.

4. Roots O, Lahne R, Otsa E, Simm M, Schramm KW (2003) Dioxins in the Baltic herring and sprat in Estonian coastal waters. Organohalogen Compounds 62: 201-203.
5. Persson Y, Shchukarev A, Oberg L, Tysklind M (2008) Dioxins chlorophenols and other chlorinated organic pollutants in colloidal and water fractions of groundwater from a contaminated sawmill site. Environ Sci and Pollut Res 15: 463-471.
6. Venier M, Ferrario J, Hites R (2009) Polychlorinated dibenzo-p-dioxins and dibenzofurans in the atmosphere around the Great lakes. Environ Sci and Tech 43: 1036-1041.
7. US EPA (United States Environmental Protection Agency) (2014) Estimating exposure to dioxin-like compounds, 1-3. Office of Research and Development, Washington.
1. EC (2011) European Commission Directive 2011/516/EU amending Regulation (EC) 1881/2006 as regards maximum levels for dioxins dioxin-like PCB and non-dioxin-like PCB in foodstuffs and Commission Recommendation of 23 August 2011 on the reduction of the presence of dioxins furans and PCB in feed and food. Official Journal of the European Union, I 218: 823-825.
8. Szlinder-Richert J, Barska I, Usydus Z, Ruczynska W, Grabic R (2009a) Investigation of PCDD/F and dl-PCB in fish from the southern Baltic Sea during the 2002–2006 period. Chemosphere 74: 1509-1515.
9. Szlinder-Richert J, Barska I, Mazerski J, Usydus Z (2009b) PCB in fish from the southern Baltic Sea: levels bioaccumulation features and temporal trends during the period from 1997 to 2006. Marine Pollution Bulletin 58: 85-92.
10. Van den Berg M, Birnbaum LS, Denison M, De Vito M, Farland W, et al. (2006) The 2005 World Health Organization Re-evaluation of Human and Mammalian Toxic Equivalency Factors for Dioxins and Dioxin-like Compounds. Toxicol Sci 93: 223-241.
11. Verta M, Salo S, Korhonen M, Assmuth T, Kiviranta H, et al. (2007) Dioxin concentrations in sediments of the Baltic Sea – a survey of existing data. Chemosphere 67: 1762-1775.
12. Roots O, Zitko V, Kiviranta H, Rantakokko P, Ruokojarvi P (2009) Polybrominated diphenyl ethers in Baltic herring from Estonian waters 2006-2008. Russian J of General Chem 80: 2721-2730.
13. Falkowska L, Bełdowska M (2011) Deposition of chemical substances from the atmosphere. In Sz Uścinowicz (Ed), Geochemistry in the Surface Baltic Sediments, Warsaw, Ministry of Environment.
14. Geeraerts C, Belpaire C (2010) The effects of contaminants in European eel: a review. Ecotoxicology, 19: 239-266.
15. Niemirycz E (2011) River discharges of chemical substances. In Sz. Uścinowicz (Ed). Geochemistry in the Surface Baltic Sediments, Warsaw, Ministry of Environment.
16. Niemirycz E, Jankowska D (2011) Concentration and profiles of PCDD/F in sediments of major polish rivers and the Gdansk Basin – Baltic Sea. Chemosphere 85: 525-532.
17. Vuorinen PJ, Kiviranta H, Koistinen J, Pöyhönen O, Ikonen E, et al. (2014) Organohalogen concentrations and feeding status in Atlantic salmon (Salmo salar L) of the Baltic Sea during the spawning run. Sci of the Total Environ 468-469: 449-456.
18. Geeraerts C, Focant JF, Eppe G, De Pauw E, Belpaire C (2010) Reproduction of European eel jeopardized by high levels of dioxins and dioxin-like PCB. Science of the Total Environment 209: 4039-4047.
19. Hites RA (2006) Persistent Organic Pollutants in the Great Lakes Part N Handbook. In Environ Chem, DC: Springer-Verlag Berlin and Heidelberg.
20. Szlinder-Richert J, Barska I, Usydus Z, Grabic R (2010) Polybrominated diphenyl ethers (PBDEs) in selected fish species from the southern Baltic Sea. Chemosphere 78: 695-700.
21. Szlinder-Richert J, Usydus Z, Malesa-Ciećwierz M, Polak-Juszczak L, Ruczynska W (2011) Marine and farmed fish on the Polish market: Comparison of the nutritive value and human exposure to PCDD/F and other contaminants. Chemosphere 85: 1725-1733.
22. Szlinder-Richert J, Usydus Z, Drgas A (2012) Persistent organic pollutants in sediment from the southern Baltic: risk assessment. J of EnvironMonitoring 14: 2100-2107.
23. Piskorska-Pliszczynska J, Maszewski S, Warenik-Bany M, Mikolajczyk S, Goraj L (2012) Survey of persistent organochlorine contaminants (PCDD, PCDF and PCB) in fish collected from the Polish Baltic fishing areas. The Scientific World Journal 973292: 1-7.
24. Vuorinen PJ, Keinanen M, Kiviranta H, Koistinen J, Kiljunen M, et al. (2012) Biomagnification of organic halogen in Atlantic salmon (Salmo salar) from its main prey species in three areas of the Baltic Sea. Sci of the Total Environ 421-422: 129-143.
25. Strucinski P, Piskorska-Pliszczynska J, Maszewski S, Górlaczyk K, Warenik-Bany M, et al. (2013) PCDD/F and dl-PCB intake from fish caught in Polish fishing ground in the Baltic Sea- characterizing the risk for consumers. Environ Sci and Technol 30: 2294-2304.
26. Peltonen H, Ruokojarvi P, Korhonen M, Kiviranta H, Flinkman J, et al. (2014) PCDD/F PCB and PBDEs in zooplankton in the Baltic Sea- spatial and temporal shifts in the congener specific concentrations. Chemosphere 114: 172-180.
27. Mackay D, Fraser A (2000) Bioaccumulation of persistent organic chemicals: mechanisms and models. Environ Poll 110: 375-391.
28. Jakóbkiewicz-Banecka J, Gabig-Ciminska M, Banecka-Majkutewicz Z, Banecki B, Węgrzyn A, (2014) Factors and processes modulating phenotypes in neuronopathic lysosomal storage diseases. Metabolic Brain Disease 29: 1-8.
29. Vuorinen PJ, Vuorinen M (1985) Effect of bleached kraft mill on effluent on reproduction of brown trout (Salmo trutta L) on a restricted diet. Finnish Fisheries Research 6: 92-105.
30. Niemirycz E, Witt M, Kobusińska M (2016) Polychlorinated Dibenzo-P-Dioxins and Dibenzofurans (PCDD/F) Measurements in Ambient Air over the Northern Poland. Intern J of Chemical Eng and App 7: 32-35.
31. Roots O, Schramm K-W, Simm M, Henkelmann B, Lankov A (2006) Polychlorinated dibenzo-p-dioxins and dibenzofurans in Baltic herring and sprat in the north-eastern part of the Baltic Sea, Proc. Proceedings of the Estonian Academy of Sciences, Biol and Ecol 55: 51-60.
32. Pandelova M, Henkelmann B, Roots O, Simm M, Jarv L, et al. (2008) Levels of PCDD/F and dioxin-like PCB in Baltic fish of different age and gender. Chemosphere 71: 369-378.
33. Rappe C, Bergqvist PA, Kjeller LO, Swanson S, Belton T, et al. (1991) Levels and patterns of PCDD and PCDF contamination in fish brabs and lobsters from Newark Bay and New York Bright. Chemosphere 22: 239-266.
34. Falandysz J (1999) Polichlorowane bifenyle (PCB) w środowisku: chemia analiza toksyczność stężenia i ocena ryzyka. Gdańsk, DC: Fundacja Rozwoju Uniwersytetu Gdanskiego (in polish) 54: 657-668.
35. Lohmann R, Belkin I (2014) Organic pollutants and ocean fronts across the Atlantic Ocean: A review. Progress in Oceanography 128: 172-184.
36. Meissner W, Betleja J (2007) Skład gatunkowy liczebność i struktura wiekowa mew Laridae zimujących na składowiskach odpadów komunalnych w Polsce. Notatki Ornitologiczne 48: 11-27.

18

Statistical Analysis of Sea Surface Temperature and Chlorophyll-*a* Concentration Patterns in the Gulf of Tadjourah (Djibouti)

Youssouf MO[1,2]*, Laurent M[2] and Xavier C[3]

[1]*Centre for Study and Research in Djibouti Airport Road, PO Box 486, Djibouti*
[2]*UMR6539 UBO/CNRS/IRD/Ifremer European University Institute of the Sea (IUEM) Technopole Brest Iroise, 29280 Plouzane, France*
[3]*University of Western Brittany (UBO) UMR 6523 CNRS/IFREMER/UBO Physics Laboratory of the Ocean, IUEM, Plouzané, France*

Abstract

The sea surface temperature (SST) and chlorophyll-*a* concentration (CHL-*a*) were analysed in the Gulf of Tadjourah from two set of 8-day composite satellite data, respectively from 2008 to 2012 and from 2005 to 2011. A singular spectrum analysis (SSA) shows that the annual cycle of SST is strong (74.3% of variance) and consists of warming (April-October) and cooling (November-March) of about 2.5C than the long-term average. The semi-annual cycle captures only 14.6% of temperature variance and emphasises the drop of SST during July-August. Similarly, the annual cycle of CHL-*a* (29.7% of variance) depicts high CHL-*a* from June to October and low concentration from November to May. In addition, the first spatial empirical orthogonal function (EOF) of SST (93% of variance) shows that the seasonal warming/cooling is in phase across the whole study area but the southeastern part always remaining warmer or cooler. In contrast to the SST, the first EOF of CHL-*a* (54.1% of variance) indicates the continental shelf in phase opposition with the offshore area in winter during which the CHL-*a* remains sequestrated in the coastal area particularly in the south-east and in the Ghoubet Al-Kharab Bay. Inversely during summer, higher CHL-*a* quantities appear in the offshore waters. In order to investigate processes generating these patterns, a multichannel spectrum analysis was applied to a set of oceanic (SST, CHL-*a*) and atmospheric parameters (wind speed, air temperature and air specific humidity). This analysis shows that the SST is well correlated to the atmospheric parameters at an annual scale. The windowed cross correlation indicates that this correlation is significant only from October to May. During this period, the warming was related to the solar heating of the surface water when the wind is low (April-May and October) while the cooling (November-March) was linked to the strong and cold North-East winds and to convective mixing. The summer drop in SST followed by a peak of CHL-*a*, seems strongly correlated to the upwelling. The second EOF modes of SST and CHL-*a* explain respectively 1.3% and 5% of the variance and show an east-west gradient during winter that is reversed during summer. This work showed that the seasonal signals have a wide spatial influence and dominate the variability of the SST and CHL-*a* while the east-west gradient are specific for the Gulf of Tadjourah and seem induced by the local wind modulated by the topography.

Keywords: Sea surface temperature; Chlorophyll *a* concentration; Singular spectrum analysis; Empirical orthogonal function; Multichannel singular spectrum

Introduction

The Gulf of Tadjourah (GOT) extends from the Gulf of Aden and penetrates deep into the Horn of Africa (Figure 1). It is almost completely surrounded by the Djibouti land masses, except for the south eastern portion which prolongs into Somalia. This semi-enclosed sea has specific geomorphological features that distinguish it from the Gulf of Aden. In the south-east, the continental shelf is wider and includes two islands (Moucha and Maskali) and many sandbanks are scattered throughout the entrance of the GOT. In the North, the GOT is crossed by a large rift (Tadjourah rift) extending westward toward the Ghoubet El Kharab Bay and the Assal Lake (150 m below sea level), one the hottest and most volcanically active areas in the world [1,2].

The extreme climate conditions of the Afar depression and the arid lands surrounding the GOT seem to have an important impact on the water properties, in particular on surface temperature and salinity. The volcanic formations in the west may also enhance marine primary production by providing minerals such as iron. On the oceanic side, GOT is located on the confluence of the Red Sea and the Indian Ocean. The climate is mainly governed by the Indian Ocean monsoon regime; alternation of North-Eastern and South-Western monsoon winds regulates surface circulation. Given these aspects, the GOT constitutes a unique marine ecosystem with many ill-understood but remarkable phenomena.

From an economic perspective, the GOT includes most of the exclusive economic zone of Djibouti (EEZD) and provides almost all local fish resources. The coastal area also provides seafood and services to the inhabitants of the surrounding arid lands. These resources could be affected by changes in physical and biogeochemical properties due to the global warming [3,4]. It is therefore important to establish a fishing policy that is based upon scientific knowledge of the ecosystem and its exploitation potential. Surprisingly, very little information is available on oceanographic conditions and bio-productivity of the GOT. The area has been very little studied.

The few data available come from research ships passing through Djibouti. In the pioneering study and using merchant ships, Donguy, 1970, 1972 and 1974 [5-7] collected data on sea surface temperature, salinity and wind speed between Djibouti port and Cap Guardefui.

***Corresponding author:** Youssouf MO, Centre for Study and Research in Djibouti Airport Road, PO Box 486, Djibouti
E-mail: assoum_omar42@yahoo.fr

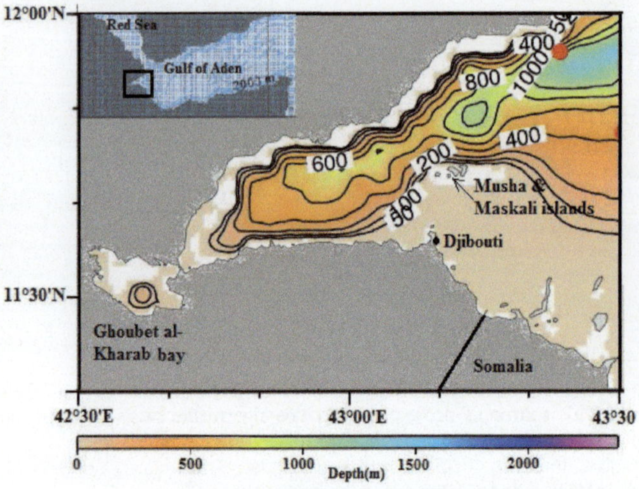

Figure 1: Map of the Gulf of Tadjoura showing geographic location and bathymetric features. The small rectangle (in the top left insert) shows the localization of the Gulf of Tadjoura within the Gulf of Aden and southern Red Sea region (source: Ifremer/sismer).

Limited to the Eastern part of the GOT, these data indicate the existence of two marine seasons. The highest temperatures were observed between March and May in the offshore waters of Djibouti and the lowest ones from November to February. Subsequently, some descriptive ecological studies provide limited data on climatic and oceanographic conditions of the GOT. In 1974-1976, Perez and Braud [8] collected data on temperature, turbidity and wind speed around the Moucha and Maskali Islands. These data showed that from January to April, surface water temperatures remain stable and spatially homogenous (24-26C). However, from May to November, temperatures show high frequency oscillations and are spatially heterogeneous. Robineau and Rose [9] reported that temperatures fluctuate between 25C during winter and 29°C during summer.

However, there is very limited information on primary production of the GOT. The descriptive works [10,11] on fish assessment relate the summer abundance of pelagic fish to the phytoplankton bloom. More recently, as part of the monitoring of the quality of the coastal water, Awaleh et al. [12] collected data on 26 stations located close to Djibouti city. During this survey, very low chlorophyll *a* concentrations (0.006 to 0.06 µg/l) were observed at sampling sites. On a regional scale, the Gulf of Aden is considered as an oligotrophic area during summer, with a surface chlorophyll bloom which only occurs during winter due to deep convective mixing [13,14].

Understanding surface patterns of temperature and primary production is of great importance, particularly when looking at their implications for the ecosystem and the influences on the local fisheries. Unfortunately, the above referenced large-scale studies in the Gulf of Aden are spatially limited. The descriptive studies at a local small scale are restricted to certain areas or are temporally sporadic. They do not provide a synoptic and clear picture of the variability of the two parameters as they evolve throughout the year.

The present work aims to improve the understanding of the changes in thermal structure and primary production and their patterns of temporal and spatial variability in the GOT. As this study is essentially based on a statistical approach, the robust statistical analysis methods, as well the techniques to separate signal and noise through decomposition are briefly described. Results focus on the most important modes of variability of two studied parameters (SST and CHL-*a*). The relationship between these two parameters and also the link with atmospheric parameters is studied through covariance analysis.

Data and Methodology

Data

The challenge was to find satellite data with the same temporal and spatial resolutions for SST and CHL-*a*, in our study area. Subsequently, the two parameters were first studied separately, using data at different spatial resolutions and over different periods. The purpose here is to decompose data and identify the mean patterns of variability for each parameter.

Data on sea surface temperature

SST AVHRR_MetopA: For temporal and spatial variability analysis, we used the daily SST derived from the METOP/Advanced Very High Resolution Radiometer (AVHRR) for the period January 2008 to December 2012. These data were provided by the CERSAT (Centre ERS d'Archivage et de Traitement - *French ERS Processing and Archiving Facility)* at IFREMER (Institut français de Recherche et d'exploitation de la mer). The data have spatial resolution of 0.05°× 0.05° which corresponds to approximately to 16 grid points in latitude direction and 20 grids points in longitude direction for the GOT, so we have 16×20 grid points and 225 observations (2008-2012), at each grid point to describe the state of GOT SST. Details on this product can be found in [15,16]. The validation of data for July 2007 to December 2013 at a global scale was made using the Matchup Data Base and also by comparison with data from the other sensors [17]. This indicates a negligible error of - 0.05 K and a standard deviation of 0.48 K for night time data against 0.9 K and 0.58 K for daytime data [18]. We used in this study only the night time data. To test the validity of these data in our area of study, we used *in situ* daily collected data using a temperature-conductivity sensor (Hobo U24) at near-surface (2 m depth) in the Gulf of Tadjourah from April to July 2013. The comparison of the SST AVHRR_MetopA daily-averaged data in the vicinity of the temperature-conductivity sensor showed that the two data evolve linearly during this short period. The mean difference was estimated at 0.6°C.

Ocean colour data: The CHL-*a* distribution in the GOT was analysed using the merged ocean colour product (Chlorophyll-*a*, case I water: CHL_1) from the Moderate Resolution Imaging Spectroradiometer (MODIS) and Medium Resolution Imaging Spectrometer (MERIS) with a 1 km by 1 km resolution. These data are 8-day composites and were provided by GlobColour/ST-ACRI (see: http://hermes.acri.fr/) from 2005 to 2011. Thus, in our area of study, the data consist of 81×61 grid points and 303 observations at each grid point.

Atmospheric parameter data: Atmospheric parameters data were reprocessed from "Daily Turbulent Flux" product provided by CERSAT/Ifremer. This product contains daily data on turbulent flux air temperature (Tair), specific air humidity (Qa10), and wind speed (W10) and also the sea surface temperature (SST). The data are gridded over the global ocean at a spatial resolution of 0.25° by 0.25° [19]. They cannot be used alone to analyse spatial variability in the GOT because of their low resolution. Therefore, the limited grid points were averaged to perform time series from 2005 to 2009 in order to analyse the relationship between SST, CHL-*a* and the atmospheric parameters.

Methodology

Pre-processing: As shown in Figure 1, the study area of study is limited between 42°30E - 43°30E and 11°20N - 12°00N. We have extracted a set of satellite images for this area for the above described products of SST and CHL-*a*. The daily data for SST were first transformed to 8-day composites by simple averaging over the entire area of study. This transformation has allowed the reduction of missing data and to provide with have the same temporal resolution as the CHL-data. For the latter, as the chlorophyll-*a* concentration is approximately log normally distributed in nature [20] a logarithmic scale (base 10) was used.

Temporal variability analysis: For each field, from each image, "sea pixels" were retrieved using land masking. Then time series was performed through a simple averaging of the "sea pixels" of each image. These time series can be presented as shown in (1): X is a time dependent physical variable, t is the equally spaced record time and *N*: the length of the sample. The objective here is to separate signals from noise as showed in (2). This was done using singular spectrum analysis decomposition.

$$X_m(t) = x_t, x_{t+1}, x_{t+2}, \ldots\ldots x_n \quad (1)$$

$$X(t) = \text{Signal}(t) + \varepsilon(t) \quad (2)$$

Singular spectrum analysis (SSA): SSA is designed to extract useful information from short and noisy time series [21]. In simple terms, the SSA method can be visualised as a sliding window of length M, moving down a time series X(t) of length *N*. SSA decomposes time series into eigenvectors or temporal EOFs. The projection of these EOFs on the original time series yield temporal principal components or PCs Each PCs captures a type of variation representing trend, periodic oscillations, and structure less variations or noises [22-26].

Separability, identification and grouping of the temporal principal components: Following decomposition, the resulting additive components (PCs) must be separated and identified. This is firstly based on eigenvalues spectrum inspection. Thus, the oscillatory component is generally characterised by a pair of nearby equal eigenvalues and the associated PCs are in phase quadrature. After identification, the entire time series, or parts of it that correspond to trends, oscillatory modes or noise, can be reconstructed by linear combinations of the PCs and EOFs, providing the reconstructed components (RCs). The dependence between RCs is measured using the weighted correlation (See [21] for the calculation of w-correlation matrix). The main steps of the SSA method are described in detail in [27,28]. Here, the Caterpillar SSA 3.4 software, (available at http//www.gistagroup.com, see also [29]) was used to implement all steps of computation. The window length (M) determination is a critical step of the SSA method. The choice of this parameter impacts SSA results directly, and can be a source of numerical noise [30]. To determine this parameter, we first applied a simple spectral analysis to the time series in order to find the main period (T). Then, as recommended by [27], M was set as a multiple of this period, but less than a third of the length of the time series (M = *k*T; M< *N*/3). Furthermore, we have tested different values of M, and the root mean square error (RMSE) - between the original time series and the reconstructed ones was computed for each case. Thus, the value of M that corresponds to the lowest RMSE is retained.

Monte Carlo SSA test: As mentioned above, selection of the RCs are empirically based on the SSA reparability theory [27]. Unfortunately, this approach is limited in the presence of the autocorrelation. As pointed out in reference [31], the interdependence between observations introduces red noise that could lead to RCs with high ranks in the eigenspectrum. In order to identify either the oscillatory components associated with red noise or to objectively separate signal from noise, the Monte Carlo SSA (MC-SSA) was applied against the null hypothesis H∅: data are derived from red noise, in particular autocorrelation at lag 1 or AR(1). This was done using simulated realisations (surrogate data) and the comparison of their eigenspectrum with the eigenvalues of the original data. The test is performed in three main steps:

1- The surrogate data were generated from original data using linear autoregressive model (AR1).

$$Xt = aX_{t-1} + \varepsilon_t \quad (3)$$

2- Then, the covariance matrix (*C*) is computed from all surrogate data. This matrix is projected onto the eigenvalue basis (**EX**) of the original data (4).

$$\Lambda_R = E_x^T C E_x \quad (4)$$

3- The comparison between the given surrogate data and the original data is quantified by computing the values of the discriminating statistics (confidence interval) of the diagonal elements of Λ_R via Monte Carlo simulation. The null hypothesis is rejected where the eigenvalue falls outside the red noise confidence interval [31].

Harmonic analysis: Harmonic analysis aims to decompose periodic time dependent data into a sum of sinusoidal functions. Here, this method was applied to supplement SSA analysis in order to isolate and clearly define periodic components in SST and CHL-a time series. From a discrete data set (1), a periodogram was obtained using Fast Fourier Transform (Matlab fft). From the periodogram, the fundamental frequency (ω_0) is determined graphically. Then, the Fourier coefficients A_n and B_n were estimated by integrating equations (6) and (7) between zero and the fundamental period T. The amplitude and phase of each harmonic were calculated using equations 8 and 9.

$$F(t) = A_n\cos(n\omega_0 t) + B_n\sin(n\omega_0 t) \quad (5)$$

$$A_n = \frac{2}{T}\int_0^T X(t)\cos(n\omega_0 t)\,dt \quad (6)$$

$$B_n = \frac{2}{T}\int_0^T X(t)\sin(n\omega_0 t)\,dt \quad (7)$$

$$C_n = \sqrt{A^2_n + B^2_n} \quad (8)$$

$$(\phi_n) = \arctan\left(\frac{B_n}{A_n}\right) \quad (9)$$

Empirical orthogonal analysis: While the SSA technique is suitable for one-dimensional time series analysis, empirical orthogonal function (EOF) is helpful in identifying spatial components (spatially uncorrelated) varying in time [32,33]. The EOF analysis was applied to a set of 8-day composites images of SST and CHL-*a*, after the long-term average was removed from each grid point (10).

$$Z'(t,x) = Z(t,x) - \frac{1}{N}\sum_{t=1}^{N} Z(t,x) \quad (10)$$

$$CE = E_n(x)\lambda i \quad (11)$$

$$a_n(t) = \sum_{x=1}^{M} Z'(x,t)\times E_n(x) \quad (12)$$

The resulting anomalies Z'(t,x) are then decomposed through eigenanalysis (or diagonalisation). This produced the eigenvectors E_n and eigenvalues (λi) as shown in (11). The temporal coefficients $a_n(t)$ are obtained by projecting data on the associated eigenvectors. The a_n

(t) represent amplitudes of the eigenvector at each sampling time t. To facilitate physical interpretation, $a_n(x)$ were normalised to unit variance and regressed against the original data set. This makes it possible to obtain regression maps (EOF_n) in the same units as the initial data. In addition, individual EOF must be evaluated for uniqueness. Following [34], it is possible to assign a "standard error" δλ, to each eigenvalue. This error depends on the size of sample *N* or, more precisely, on the number of effective independent samples *N'* (13): the total number of observations reduced by the autocorrelation time scale (Tscale) as in (14). The latter is estimated from ACF (autocorrelation functions) using the formula proposed by Thomson et al. [35].

$$\delta\lambda_i \approx \lambda_i \times \sqrt{\frac{2}{N'}} \Rightarrow \frac{\delta\lambda_i}{\lambda_i} = \sqrt{\frac{2}{N'}} \quad (13)$$

$$N' = N/\text{Tscale} \quad (14)$$

$$C(\tau) = \frac{1}{c(0)} \sum_{k=0}^{m-1} \frac{\Delta\tau}{2} [c(\tau_k + \Delta\tau) + c(\tau_k)] \quad (15)$$

One mode (*i*) is considered significant if the error $\delta\lambda i$ of its associated eigenvalue λi is smaller than the distance $\Delta\lambda i$ between λi and its nearest eigenvalue.

Multichannel singular spectrum analysis: Multichannel spectrum analysis (M-SSA) is an extension of the SSA method (described in section 2.2.3) adapted to more than one time series or channel. The aim of the M-SSA is to isolate and distinguish patterns of similar temporal behaviour [36]. Here, in order to extract oscillatory and persistent common modes, the M-SSA is first applied to a trajectory matrix of normalised 8-day time series of all the oceanic and atmospheric parameters.

Results

Temporal variability

SST: Figures 2a and 2b shows the spatially 8-day averaged SST from January 2008 to December 2012. From this time series, the annual cycle is discernible. The average SST varies from a minimum of approximately 25-26°C between December and February to a maximum of 31-32C achieved in May and September. Between these two maxima, a sharp drop during July-August was discerned.

Choosing M (window length = 90) and performing SSA decomposition of the trajectory matrix, provided us with 90 eigenvalues, of which the first 50 are shown in Figure 3a. The first two pairs of eigenvalue (ranks 1 to 4) are well separated from the remainder. The plotting of the related eigenvectors (not shown here) indicates periodic components from rank 1 to 6. From rank 7, eigenvalues are very close and asymptomatically linear. The highly correlated components (rank 1-2, rank 3-4 and rank 5-6) were grouped to perform the three first RCs (Figure 3b).

The first RCs (Figure 4a, thick blue line) accounts for 74.3% of the total variance and show an increase in temperature from March to May-June, followed by a decrease until December. The second RCs (red line) account for 14.6% of the variance and emphasises the drop in temperature in July. The third component (green line) accounts only for 3% and shows three maxima and minima of low amplitudes (Figure 4b). The reconstructed time series from the three RCs contribute at about 91.9% of temperature variance. The RMSE between the original and the reconstructed time series remains low (RMSE ≈ 0.79). The residuals (Figure 4c) have no apparent signal but indicate the extrema, particularly in 2008 and 2011.

CHL-*a*: The spatially 8-day averaged CHL-*a* is illustrated in Figure 2b. This time series is characterised by a pronounced and amplitude variable peak occurring every year in July-August. During the rest of the year, CHL-*a* remains low. The SSA decomposition (window length M=86) of logarithm (base 10) of the CHL-*a* results in 86 eigenvalues whose first 50 are shown in Figure 3a). The first pair and the three following eigenvalues are well separated. The first RCs corresponding to the first pair of eigenvalues (blue thick line) cover 29.7% of variance and present a maximum in July-August (Figure 4a). The second RCs associated with the three eigenvalues (rank 3, 4 and 5) explain 19% of the variance and indicate a first peak in February-March and a second

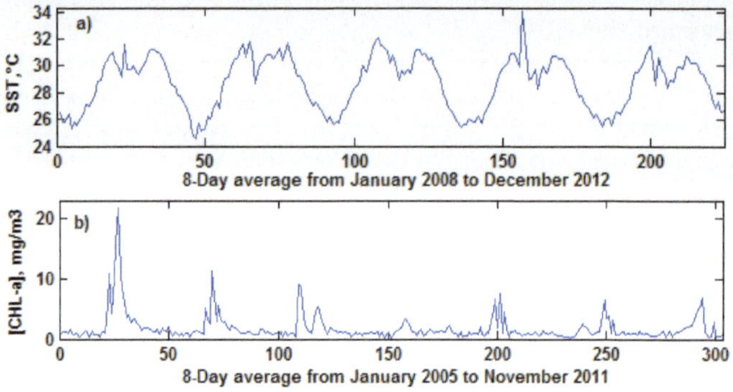

Figure 2: Time series of spatially 8-day averaged SST and CHL-a from 2008 to 2012 and from 2005 to 2011 respectively. Spatial averaging is performed over the entire study area (see Figure 1).

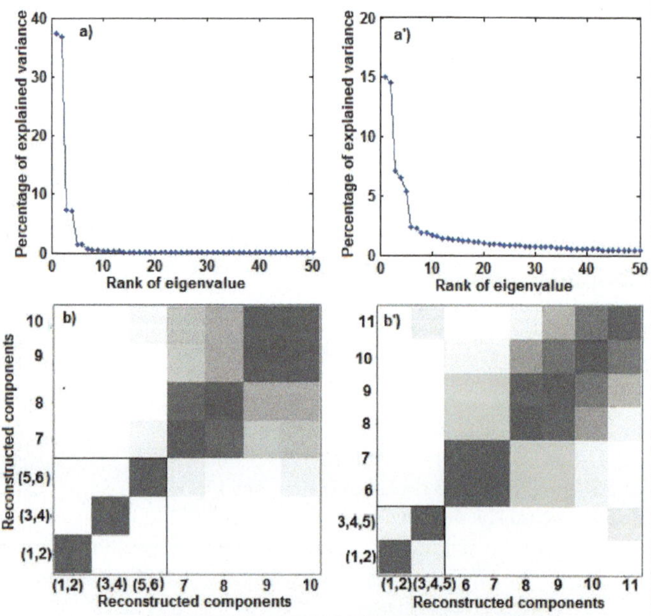

Figure 3: Top panel: Percentage of explained variance of the first 50 eigenvalues of SST and CHL-*a* (a and a'). Low panel: Matrices of weighted correlation for the 10-11 reconstructed components of SST and CHL-*a* (b and b').

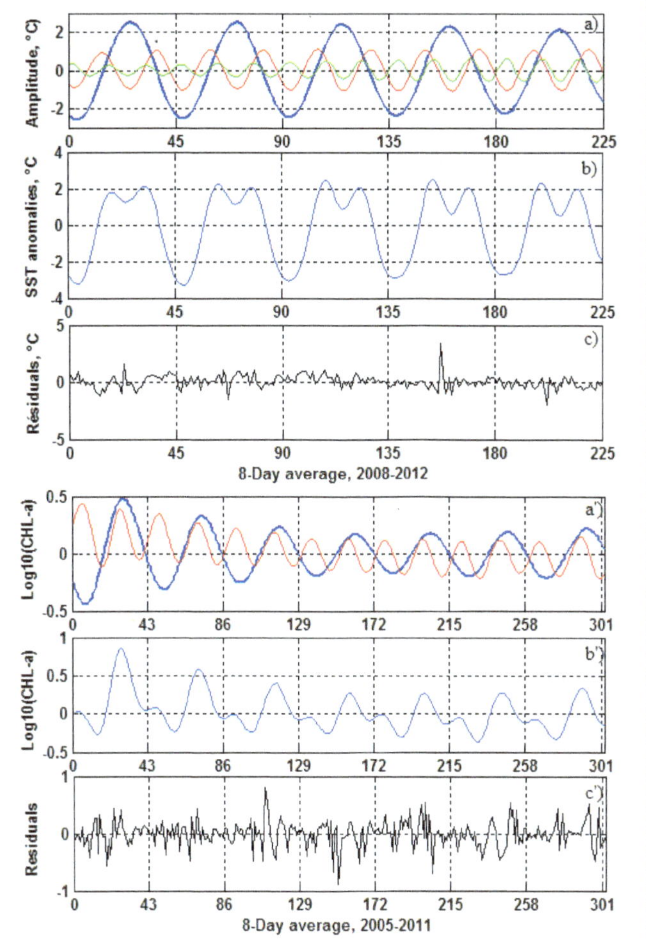

Figure 4: Top: SSA reconstructed components, reconstructed time series and residuals for SST (a,b,c) and the same for CHL-a (low panel).

in July-August (Figure 4a). The reconstructed time series from these two RCs (Figure 4b) fit well with the original time series (RMSE= 0.22).

To characterise the cyclic components described above, spectral analysis based on Fast Fourier Transform (FFT) was applied to the original time series of SST and CHL-*a*. This resulted in three peaks of 45, 22.5 and 15, 8-day cycle (360, 180 and 120 days) for SST (Figure 5a) and two peaks of 43 and 22, 8-day cycle (344 and 176 days) for CHL-*a* (Figure 5b). The same analysis applied to each SSA isolated RCs of SST and CHL-*a* shows identical periods. This means that the three RCs correspond to annual, semi-annual and four month cycles in SST time series and the annual and semi-annual cycles in CHL-*a* time series. Note that for CHL-*a*, in addition to two mean periods (43 and 22 8-day cycles), the spectrum indicates several shorter periods. The amplitude and phase of each harmonic were calculated using equations 6, 7, 8 and 9.

The results of the Monte Carlo test applied to two time series of SST and CHL-*a* are illustrated in Figure 6. In the case of SST (Figure 6a), the first two pairs of eigenvalues are outside of the confidence interval (CI) of AR (1). Therefore, the first and second RCs associated with these two pair eigenvalues are considered as significant at 95%. While for the CHL-*a*, only the first pair of eigenvalues associated to the first RCs lie outside the CI (Figure 6b). Therefore, the annual and semi-annual cycle for SST but only the annual cycle for CHL-*a* are statistically significant. The equations 16 and 17, represent the results of Fourier decomposition applied to the SST and CHL-*a* times series.

$$SST(t) = 28.72 + 2.5 \times \cos(\omega_0 + \frac{\pi}{7}) + 0.95 \times \cos(2\omega_0 t + \frac{\pi}{5}) \quad (16)$$

$$CHL\text{-}a(t) = 1.67 + 3.2 \times \cos(\omega_0 t + \frac{\pi}{3}) \quad (17)$$

EOF mode description

Figures 7a-7e shows the eigenvalue spectrum of the first 50 eigenvalues for SST. Following the North et al. criterion, only the two first eigenvalues exceed the 95% confidence error (Figure 7d). The first EOF mode of SST explains 93% of the variance. Its spatial pattern is in phase and positive across the whole study area (the amplitudes of the spatial EOF corresponding to the amplitudes in the real data). The noticeable feature of this spatial map is that the South-East remains warmer and the Western relatively cooler. However, as the temporal coefficient a1(t) changes signs and becomes negative from November to March, the inverse spatial pattern occurs : the product EOF1×a1(t) is negative across the gulf, the southeast remaining cooler (Figures 7a-7e). The spectral analysis of a1(t) shows a strong peak, with a period of 44 8-day (352 days), and a second one of 22 8-day (176 days). This indicates that the first EOF mode represents a combination of annual and semi-annual cycle variations of SST. The second EOF mode accounts only for 1.3% of the variance and clearly indicates a westward

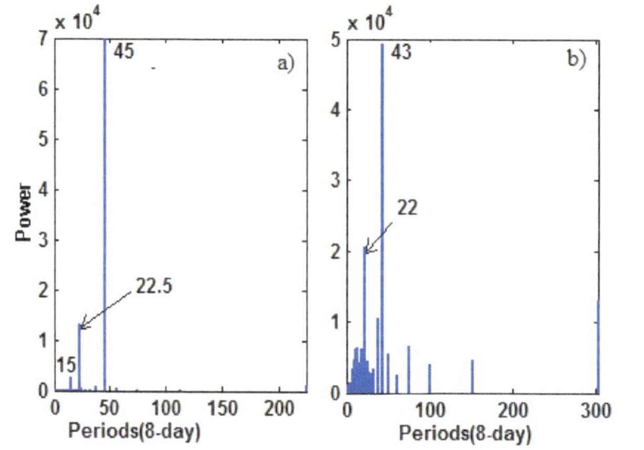

Figure 5: Periodograms of 8-day averaged SST (a) and CHL-a (b). Note that, in each case, the mean peak (period) found using the Fourier decomposition of the individual SSA temporal principal component, corresponds to one of those identified in the original time series.

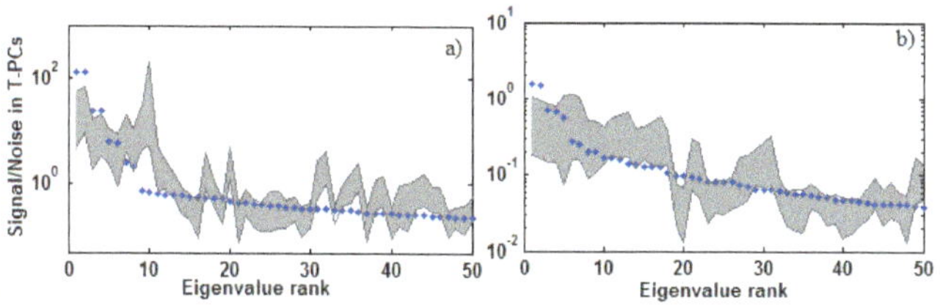

Figure 6: Results from Monte Carlo test applied to 8-day averaged SST and CHL-a data against first order autoregressive red noise AR (1). The grey area indicates the red noise confidence interval (CI) limited by the 95th (upper end) and 5th (lower end) percentiles based on 1000 Monte Carlo simulations. Blue dots show the eigenvalues from the data. The first two pairs of eigenvalues of SST are significant (outside of red noise CI) while only the first pair of eigenvalues of CHL-*a* lies outside of this CI.

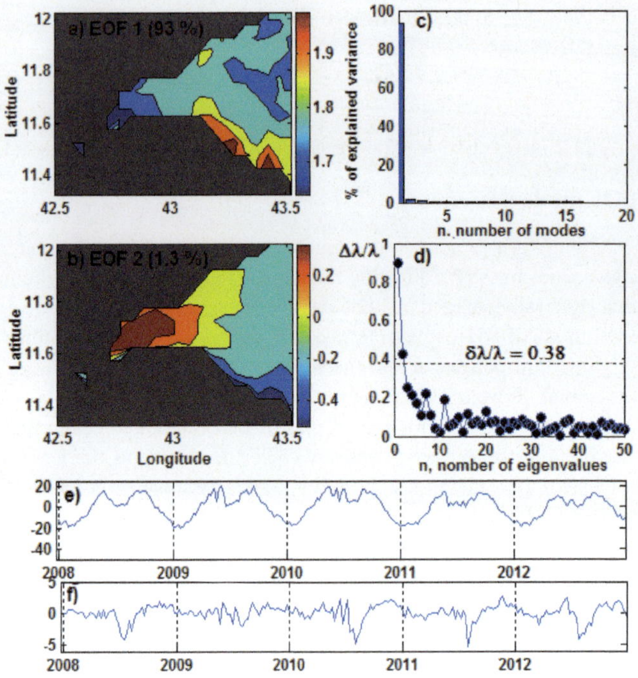

Figure 7: Results from the SST EOF analysis: spatial patterns EOF1 and EOF2 (a and b), the corresponding temporal coefficients of 8-day averaged for 2008-2012 (e and f), the percentage of variance of the first 50 eigenvectors (c), the ratio of minimum distance error (Δλ/λ), dotted line indicate the threshold (d).

gradient. The eastern part remains positive and the western negative when the temporal coefficient a2(t) is positive (November-December). This pattern is reversed when a2(t) becomes negative from May to September (Figures 7b and 7f).

For the CHL-*a* field, the first seven eigenvalues, with the exception of rank 4, exceed the confidence error (Figures 8a-8e). From these EOF modes, we selected only the first two with a higher percentage of explained variance. The first mode of CHL-*a* explains 54.1% of the variance and its spatial map shows positive amplitude in the coastal area and negative on the offshore area. This pattern occurs from January to May when the temporal coefficient is positive. From June to September, the temporal coefficient becomes negative, so the EOF1×a1(t) becomes positive in the offshore area and negative in the coastal area (Figures 8a-8e). Similarly to the SST, the spectral analysis of the temporal coefficient shows two periodic components corresponding to annual and semi-annual cycles (Figure 8e). The second mode accounts for 5.2% of the variance and shows a westward gradient in phase opposition occurring in March and August when the temporal coefficient is positive, but varying from one year to another (Figure 8f).

To check whether the spatial patterns of EOF change depending on the area, the study area was expanded eastward to 44°N and northward to 14°N, including the west of the Gulf of Aden and the southern Red Sea. The results of the EOF analysis shows that the first EOF modes are similar, while the second modes are significantly different from those identified in the GOT. In the case of the CHL-*a*, we note also that the amplitude of the spatial EOF is greater in the west of the Gulf of Aden than in the Gulf of Tadjourah. The south Red Sea appears to display very low CHL-*a* compared to the west of the Gulf of Aden.

SST and CHL-a variability related to atmospheric parameters

The M-SSA was applied to analyse the relationship of the SST and CHL-*a* with the atmospheric parameters, in particular air temperature

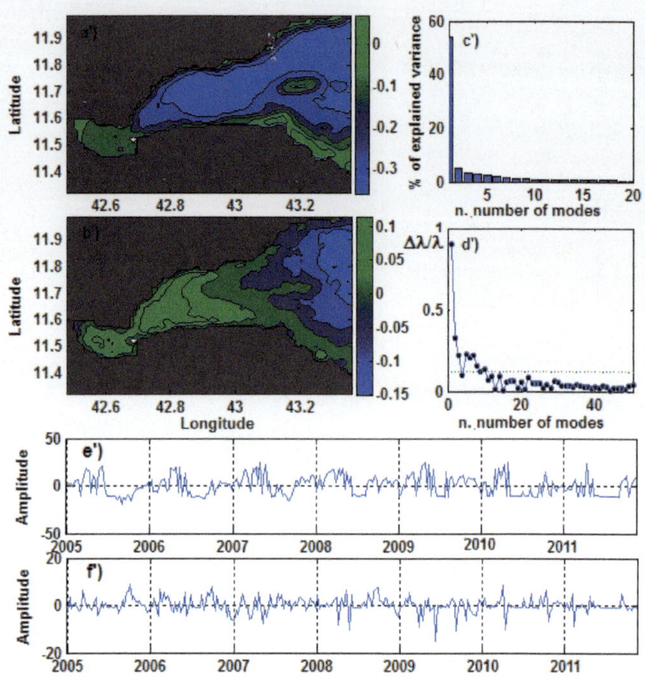

Figure 8: Results from the CHL-a EOF analysis: spatial patterns EOF1 and EOF2 (a' and b'), the corresponding temporal coefficients of 8-day averaged for 2005-2011 (e' and f'), the percentage of variance of the first 50 eigenvectors (c'), the ratio of minimum distance error (Δλ/λ), dotted line indicate the threshold (d').

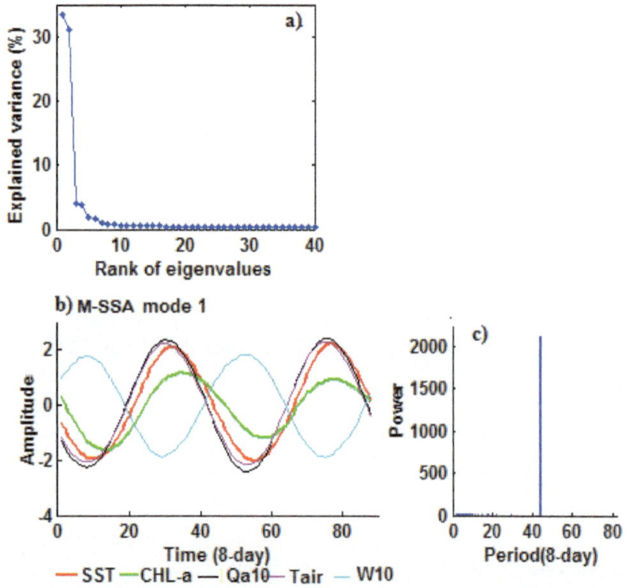

Figure 9: Results from the M-SSA analysis (window length M=88) of 8-day averaged SST, CHL-a, air temperature (Tair), air specific humidity (Qa10) and wind speed (W10) over the common period, 2008-2011: (a) eigenspectrum of the common modes, (b) First M-SSA mode, (c) the periodogram of the reconstructed components (RCs).

(Tair), specific air humidity (Qa10) and wind speed (W10). The results of the M-SSA analysis are presented in Figure 9. As shown on the eigenvalue spectrum (Figure 9a), the first pair of eigenvalues is well separated. The second pair is close to the asymptomatically linear eigenvalues. Based on the breaking of the slope in the spectrum of eigenvalues ("Screen test", see [37]), only the first M-SSA mode associated with the first pair of eigenvalue is considered significant. The associated components (SST, CHL-*a*, Tair, Qa10 and W10) of this first M-SSA mode depicts a single peak at 44 8-Day cycle, which correspond to the annual cycle (Figure 9). The annual cycle of SST is highly and positively correlated with Tair and Qa10 but negatively correlated with wind speed (Table 1). The CHL-*a* shows low correlation with all the atmospheric parameters. These results reveal that the annual cycle is the most significant common signal linking the SST to the atmospheric parameters. To examine whether this relationship varies according to the time of the year, the correlation between SST, CHL-*a* with atmospheric parameters was estimated separately for each four month period: window 1: February-May (FMAM), window 2: June-September (JJAS) and window 3: October-January (ONDJ). This indicates that SST remains highly correlated to the atmospheric parameters only during the first and the third windows, corresponding to the period extending from October to May. No significant correlation was detected for the second window which corresponds to the summer period (June-September). The CHL-a does not correlate with any atmospheric parameter during any of the three periods (Table 2).

How are SST variations linked to the atmospheric influences?

In order to analyse how the atmosphere influences the SST variation, we first computed the shortwave radiation (Qsw), longwave radiation (Qlw), latent and sensible heat flux (Ql and Qs) using daily ECMWF re-analysis data (ERA-40 interim) averaged for the study area. The net heat flux at the surface (Qsurf) was estimated by the sum of radiative flux and turbulent flux. The evolution of Qsw and Qsurf is compared to the time series of SST and wind stress time series from 2005 to 2009. As shown in Figure 10, the SST and Qsurf vary similarly; both time series show a drop during the summer period. To investigate the atmospheric influences on the SST variation more objectively, we used the Frankignoul method [38]. This method is based on the fact that the surface heat flux contributes to generate SST anomalies but

	Mode 1 (M-SSA)			
	r (0 lag)	rmax	phase	Δt(jours)
SST/Tair	0.97	0.99(-1)	-0.26	15
SST/Qa10	0.97	0.99(-1)	-0.27	16
SST/W10	-0.94	-0.99(-2)	2.79	157
CHL-a/Tair	0.79	0.96(-4)	-0.59	33
CHL-a/Qa10	0.79	0.97(-4)	-0.59	33
CHL-a/W10	-0.74	-0.96(-5)	2.46	138

Table 1: Results from the multichannel singular spectrum analysis (M-SSA) showing the correlation and phases between components.

Time series	Window 1 (FMAM)	Window 2 (JJAS)	Window 3 (ONDJ)
SST/Tair	0.85	0.08	0.88
SST/Qa10	0.85	0.56	0.94
SST/W10	-0.4	0.05	-0.8
CHL-a/Tair	-0.19	0.14	0.92
CHL-a/Qa10	-0.33	-0.33	0.95
CHL-a/W10	-0.14	-0.26	-0.84

Table 2: Results of a windowed cross-correlation using a window length of 120 days (four months) between two ensembles of oceanic and atmospheric parameters.

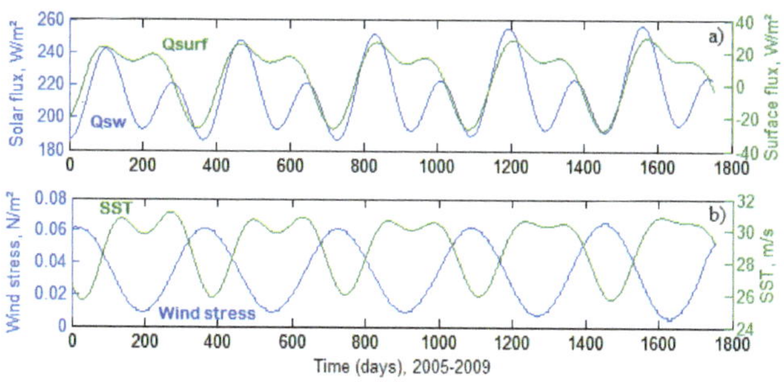

Figure 10: SST daily time series compared with the shortwave radiation (Qsw), net surface heat flux (Qsurf) and wind stress.

Feedback factor (α)	JFM	AMJ	JAS	OND
α	-3	5.6	2.1	-3.4

Table 3: Estimated atmospheric heat flux feedback factor (α) over the Gulf of Tadjoura. For each three month, α in W/m²/°C, is estimated using consecutive three months of daily data of the net heat flux (Qsurf) and the sea surface temperature (SST) time series.

it also affect their evolution after they have been generated, thereby acting as a feedback. As showed by [38], the surface heat flux anomalies can be decomposed as in Eq. (18).

$$\text{Qsurf}(\tau) = q(\tau) - \alpha\ \text{SST} \quad (18)$$

Where q is independent of the SST anomalies; -α SST is the Qsurf induced by the SST anomalies. From Eq. (18), the cross-covariance between Qsurf and SST is given by:

$$Cov_{Qsurf/SST}(\tau) = Cov_{q/SST}(\tau) - \alpha\, Cov_{SST/SST}(\tau) \quad (19)$$

$Cov_{q/SST}(\tau)$ vanishes at large negative lag (τ > 1 month). So, the feedback of the surface heat flux is estimated using the cross-covariance between SST and Qsurf, divided by the autocovariance of the SST (20).

$$\acute{a} = -\frac{Cov\left[Qsurf(-1), SSTf(0)\right]}{Cov\left[SST(-1), SST(0)\right]} \quad (20)$$

Here, α is estimated for each three consecutive monthly anomalies. For example, α during July-August-September is estimated by correlating *Qsurf* (JAS) with SST during June-July-August (JJA). The results are presented in Table 3. This indicates negative α from October to March so the surface heat flux feedback is positive (see Eq. 18). During this period, the ocean responses to the cooling by releasing heat to the atmosphere. From April to June, α remains high and positive; the feedback is negative. The atmosphere act by heating sea surface which storage energy. During summer (JAS), α is low, the atmosphere has less influence on the SST anomalies.

Discussion and Conclusions

In this paper, one-dimensional times series of SST and CHL-*a* were decomposed into a sum of temporal principal components using both singular spectrum and harmonic analysis. The results of these analyses submitted to the Monte Carlo test showed that the annual and semi-annual cycles for SST and only the annual cycle for CHL-*a* represent the most important statistically significant signals for the temporal variability of these two parameters.

The annual cycle of SST (1st harmonic) relates warming and cooling of surface waters throughout the year. These annual variations are related to the atmospheric influences. The warming of water is induced by solar heating. Indeed, the shortwave radiation flux (Qsw) increases from March to May and achieves its maximum in May-June (Figure 10a). During the same period, the SST, Tair and Qa10 increase together and achieve their maximum during the same period (May-June). However, the wind speed decreases and shows a minimum in May. This indicates that during a low wind period, solar heating induces a rise in air and sea surface temperature. The phase lag between SST and Tair could be explained by the difference in inertia between air and sea water. The increase in the specific humidity is due to the evaporation induced by the increase of SST. However, the cooling from September onward seems to be related exclusively to the strengthening of the northeast wind and the convective mixing of surface water. These prominent atmospheric influences on the annual variation of SST are consistent with the result of the surface heat flux feedback.

The spatial map of the first EOF indicates that this warming/cooling affects the entire study area. The relative difference of the southeast area, which remains warmer during summer and cooler during winter, seems to be related to the topographic features; this area is the shallowest of the Gulf (Figure 1). The remainder of the thermal variation is almost totally represented by cooling during summer, especially in July, that appears in the second SSA reconstructed component (semi-annual cycle). Both M-SSA analysis and windowed correlation analysis showed no significant correlation between the SST and atmospheric parameters at a semi-annual scale. This indicates that the drop of SST during July-August is not related to direct atmospheric influences. The low value of the retroaction coefficient (α) confirms this hypothesis.

The annual variation of the CHL-*a* is characterised by low CHL-*a* during the majority of the year and high CHL-*a* during the short period from July to September. The EOF analysis indicates that during the first period, the CHL-*a* shows a landward gradient with the CHL-*a* concentrated in the coastal waters and in the Bay of Ghoubet al-Kharab. This pattern is reversed during summer, when the CHL-*a* increases in the offshore water and diminishes on the continental shelf. The winter pattern of the CHL-*a*, limited to the coastal area seems to be related to tidal mixing. However, the summer pattern of the CHL-*a* represented by the strong peak is most likely due to the enrichment induced by upwelling. Furthermore, the singular value decomposition (not shown), applied to the SST and CHL-a fields together, indicates that the summer "offshore water enrichment" is associated with the "summer cooling" affecting the western part of the gulf, while the "winter cooling" is associated with shelf enrichment.

This process is well recognised in the west of the Gulf of Aden, where the upper ocean is mixed and the thermocline weakens during the summer period [39,40]. The shoaling of the thermocline is due to Ekman pumping, enhanced by the strong southeast monsoon wind [41-43]. This result in surface upwelling in certain areas, particularly at the southern part of the Bal-el-Mandeb strait [44]. In the Gulf of Tadjourah, as highlighted by the spatial structure of the EOF, the upwelling seems strengthening at the extreme west of the gulf.

Our study has shown that the temporal variation in SST could be described only by the first two harmonics. In the case of CHL-*a*, only the annual cycle appears as a prominent and statistically significant signal of the CHL-*a* variance. We propose a simple statistical model from theses harmonics to represent and predict seasonal variations in SST and CHL-*a*. These leading patterns were identified in for the first time in the Gulf of Tadjourah and are important for understanding how thermal structure and primary production-estimated through ocean colour-evolve in this area, but also on a larger regional scale. The EOF analysis expanded on a sub-regional scale, including the west of the Gulf of Aden, limited at 44°E and southern Red Sea below 14°N, shows the same first EOF patterns (Appendix 1). In addition, our results can be compared to some existing findings for the Gulf of Aden. For instance, the annual profile of SST found by [45] for the Gulf of Aden is comparable with our reconstructed time series. The strong summer peak highlighted by the annual cycle is also consistent with results from recent work [46] that revealed that the Gulf of Aden exhibits a prominent summer chlorophyll bloom and sustains elevated chlorophyll concentrations during autumn. This indicates that the SST and CHL-*a* patterns, isolated in the GOT, reflects what is happening regionally, particularly in the west of the Gulf of Aden.

In contrast, the second EOF modes of SST and CHL-*a*, which describe only 1.3% and 5.2% of variance respectively, seems specific to the Gulf of Tadjourah [47]. They highlight a westward gradient of both SST and CHL-*a*; these patterns seem to be related to the local wind modulated by the orography. Nevertheless, based on the satellite data of the surface water, we are limited to the basic physical process generating the isolated leading patterns [48]. Therefore, *in situ* measurement in the water column waters is required for a better understanding of the mechanisms that govern the thermal structures and primary production in this particular area of the Horn of Africa.

Acknowledgements

This work was supported financially by the "*Service de cooperation et d'action culturelle*" of the French Embassy in Djibouti and by the government of Djibouti. We are grateful to Dr Jalludin and Dr Abdulrahman for their support and encouragement during this study. Many thanks to Dr Guillaume Maze (LPO), and to the IUEM IT team (Emmanuel Tabouret, Alain Lemercier, Camille Marzeyer) and Nasser Djama for their technical assistance. We acknowledge Dr Abderrahim Bentamy and François Piollé (Ifremer/Cersat) and also Phillippe Garnesson and Olivier Sardou (Globcolour/ACRI-ST), for providing us with the satellite data.

References

1. Bosworth W, Huchon P, McClay K (2005) The Red Sea and Gulf of Aden basins. Journal of African Earth Sciences 43: 334-378.
2. Bonatti E, Cipriani A, Lupi L (2015) The Red Sea: birth of ocean. In The Red Sea: The Formation, Morphology, Oceanography and Environment of a young ocean basin.
3. Doney SC, Ruckelshaus M, Duffy JE, Barry JP, Chan F, et al. (2012) Climate Change Impacts on Marine Ecosystems. Annual Review of Marine Science 4: 11-37.
4. Levitus S, Antonov JI, Boyer TP, Baranova OK, Garcia HE, et al. (2012) World Ocean heat content and thermosteric sea level change (0-2000 m) 1955-2010. Geophys Res Lett 39: L10603.
5. Donguy JR (1970) Surface observations along the shipping lines in the part West Indian Ocean. Cah Ockznogr 22 : 353-366.
6. Donguy JR (1972) Surface data comparison with those of artificial satellites in the western part of the Indian Ocean. Cah O.R.S.T.O.M., sér.Océunogr X: 99-104.
7. Donguy JR (1974) A year of surface observations in the monsoon zone of western part of the Indian Ocean. Cah. O.R.S.T.O.M., sér. Océanogr. XII: 117-128.
8. Perez R, Braud JP (1978) Possibility of a culture of the industrial search rouge Eucheuma Spinusum in the Gulf of Tadjourah. Science et Pêche Bull Inst Pêches marit 285.
9. Robineau D, Rose JM (1982) The dugong [Dugong dugong (Müller, 1776) Sirenia , Dugongidae] in Djibouti. Biological Conservation 24: 233-238.
10. Bouhlel M (1988) Poissons de Djibouti. RDA International, Inc. Placerville, CA 95667, Etats-Unis 416.
11. Künzel T, Darar A, Vakily JM (1996) Composition, biomass and Opportunities for Fisheries Resources Exploitation Djibouti. Ministry of Agriculture and Water, Djibouti/GTZ, Germany 2: 114.
12. Awaleh MO, Hoch FB, Okieh BH, Ahmed SI, Soubaneh YD, et al. (2015) Impact

of Human Activity on Marine and Coastal Environment in the Gulf of Tadjourah. J Marine Sci Res Dev 5: 2-11.

13. Van Couwelaar M (1997) Zooplankton and micronekton biomass off Somalia and in the southern Red Sea during the SW monsoon of 1992 and the NE monsoon of 1993. Deep-Sea Research Part II-Topical Studies in Oceanography 44: 1213-1234.

14. Almogi-Labina A, Schmiedl G, Hemleben C, Siman-Tov R, Segl M, et al. (2000) The influence of the NE winter monsoon on productivity changes in the Gulf of Aden, NW Arabian Sea, during the last 530 ka as recorded by foraminifera. Marine Micropaleontology 40: 295-319.

15. Le Borgne P, Legendre G, Marsouin A, Péré S (2008) Operational SST retrieval from METOP/AVHRR, Validation Report.

16. Eastwood S, Andersen S (2007) Masking of Sea Ice for METOP SST retrieval. OSI.

17. Merchant CJ, Le Borgne P, Roquet H, Marsouin A (2009B) Sea surface temperature from a geostationary satellite by optimal estimation. Remote Sensing of Environment 113: 445-457.

18. Marsouin A, Le Borgne P, Legendre G, Péré S (2010) Homogeneous validation schemeof the OSI SAF Sea surface temperature products, proceedings of the 2010 EUMETSAT conference, Cordoba, Spain.

19. Bentamy A, Grodsky SA, Katsaros K, Mestas-Nunez AM (2013) Improvement in air-sea flux estimates derived from satellite observations. International Journal of Remote Sensing 34: 5243-5261.

20. Campbell J (1995) The lognormal distribution as a model for bio-optical variability in the sea. J Geophys Res 100: 13237-13254.

21. Hassani H (2007) Singular Spectrum Analysis: Methodology and Comparison. Journal of Data Science. 5: 239-257.

22. Broomhead DS, King GP (1986a) "Extracting qualitative dynamics from experimental data", Physica D 20: 217-236.

23. Broomhead DS, King GP (1986b) "On the qualitative analysis of experimental dynamical systems". Nonlinear Phenomena and Chaos, Sarkar S (Ed.), Adam Hilger, Bristol pp: 113-144.

24. Fraedrich K (1986) Estimating dimensions of weather and climate attractors. J Atmos Sci 43: 419-432.

25. Vautard R, Ghil M (1989) "Singular spectrum analysis in nonlinear dynamics, with applications to paleoclimatic time series". Physica D 35: 395-424.

26. Vautard R, Yiou P, M Ghil (1992) "Singular-spectrum analysis: A toolkit for short, noisy chaotic signals", Physica D 58: 95-126.

27. Golyandina N, Nekrutkin V, and Zhigljavsky A (2001) Analysis of Time Series Structure: SSA and related techniques.

28. Golyandina N, Zhigljavsky A (2013) Singular Spectrum Analysis for time series. Springer Briefs in Statistics, Springer.

29. Golyandina N, Osipov E (2007) "The 'Caterpillar'-SSA method for analysis of time series with missing values". J Stat Plan Inference 137: 2642-2653.

30. Hassani H, Mahmoudvand R (2013) Multivariate Singular Spectrum Analysis: A General View and New Vector Forecasting Approach. Int J Energy Stat 1: 55-83.

31. Allen M, Smith LA (1996) Monte Carlo SSA: Detecting irregular oscillations in the Presence of coloured noise. J Clim 93373-3404.

32. Venegas SA, Mysak LA, Straub DN (1997) Atmosphere–ocean coupled variability in the South Atlantic. J Climate 10: 2904-2920.

33. Von Storch H, Zwiers FW (1999) Statistical Analysis in Climate Research. Cambridge University Press 494.

34. North GR, Bell TL, Cahalan RF, Moeng FJ (1982) Sampling error in the estimation of empirical orthogonal functions. Mon Wea Rev 110: 699-706.

35. Thomson RE, Emery WJ (2014) Data analysis methods in physical oceanography. Newnes.

36. Zotov LV (2012) Application of multichannel singular spectrum analysis to geophysical fields and astronomical images. Advances in Astronomy and Space Physics 2: 82-84.

37. Cattell RB (1966) The screen test for the number of factors. Multivariate Behavioral Research 1: 245-276.

38. Frankignoul C, Czaja A, L Heveder B (1998) Air-sea feedback in the North Atlantic and surface boundary conditions for ocean models. J Clim 11: 2310-2324.

39. Aiki H, Takahashi K, Yamagata T (2006) The Red Sea outflow regulated by the Indian monsoon. Continental Shelf Research 26: 1448-1468.

40. Al Saafani MA, Shenoi SSC (2006) Seasonal and interannual variability of eddy field and surface circulation in the Gulf of Aden.

41. Bauer S, Hitchcock GL, Olson DB (1991) Influence of monsoonally-forced Ekmandynamics upon surface layer depth and plankton biomass distribution in the Arabian Sea. Deep-Sea Research 38: 531-553.

42. Smith SL, Codispoti LA, Morrison JM, Barber RT (1998) The 1994–1996 Arabian Sea Expedition: an integrated, interdisciplinary investigation of the response of the northwestern Indian Ocean to monsoonal forcing. Deep-Sea Research 45: 1905-1915.

43. Vecchi GA, Xie SP, Fischer AS (2004) Ocean–atmosphere covariability in the western Arabian Sea. J Climate 17: 1213-1224.

44. Baars MA, Schalk PH, Veldhuis MJW (1998) In: Sherman K, Okemwa EN, Ntiba MJ (eds.) Seasonal fluctuations in plankton biomass and productivity in the ecosystems of the Somali Current. Gulf of Aden and Southern Red Sea. Large Marine Ecosystems of the Indian Ocean – Assessment, Sustainability and Management. Blackwell Science, Cambridge, U.S. 143-174.

45. Nandkeolyar (2013) Comparative Analysis of Sea Surface Temperature Pattern in the Eastern and Western Gulfs of Arabian Sea and the Red Sea in Recent Past Using Satellite Data.

46. Yao F, Hoteit I (2015) Thermocline regulated seasonal evolution of surface chlorophyll in the Gulf of Aden. PLOS One 10: e0119951.

47. European Satellites Agency: The European Service for Ocean Colour.

48. GistaT Group: Caterpillar-SSA: Time series analysis and forecasting.

Seismic Stratigraphy and Development of Deep Fluvial Channel Incisions in the Middle Miocene Deposits, Ewan and Oloye Fields, Northwestern Niger Delta, Nigeria

Durogbitan AA*

Petrfac IES, Research and Development, Manchester, UK

Abstract

Seismic interpretation of depth converted three-dimensional seismic survey from Ewan and Oloye fields; onshore northwestern Niger Delta has helped in the identification of incised valleys (up to 350 m) deep. This study evaluated their morphologies, evolution and the local controls that influence their development. The seismic reflections of the incised valleys are characterized by low-moderate amplitude, variable internal reflections, aggradational, chaotic and progradational-sigmoid reflection patterns overlying by parallel to sub-parallel reflections configuration. The seismic reflection characteristics are probably due to variable sedimentation processes within the valleys which were affected by mass wasting. Asymmetry morphology of the valleys suggests fluvial origin while low sinuosity of the channel may indicate high gradient and high discharge. The occurrence of incised valleys landward of the shelf edge suggests and partially reflects underlying structural control. This might be caused by uplift due to shale diapirism (shelf instability). The magnitude of incision is difficult to explain in terms of sea-level fall alone because ecstatically-driven sea level variations during the Miocene are generally reported to be less than 100 m. This suggests that the character of fluvial incisions development and depositional facies preserved within the study area is locally controlled by growth faults, rapid relative sea level changes, basin physiography (shelf edge), shelf instability, variation in sediment input, slope collapse (mass flow depositional processes and downward cutting by downslope sediment flow). The incised valleys within the study area probably serve as conduits for sediment transport to the deep water. Seismic delineation of the distribution and morphology of these incised valleys may also provide critical input for reservoir modeling and volumetric analysis.

Keywords: Middle Miocene deposits; Valleys; Sediment; Seismic delineation

Introduction

Incised valleys are features of the continental margin that are characterized by erosion [1,2]. Their origin have been linked to fluvial incision, subaerial exposure, turbidity current, faulting, folding, active diapir or salt dome and slumping [3-5]. An incised valley and its depositional fill provide good evidence of lowstand to transgressive deposition in shelf-slope shallow marine depositional setting [6-10]. The study of incised valley is very important because they serve as a potential reservoir when fill with sands [9] and also preserved sedimentation and transportation history of the area. More importantly, it helps in locating the sequence boundaries and their link with the basin floor fan [1,2]. They also serve as conduits for the transportation of sediments to the slope and deeper parts of the basin in the form of slides, slumps, and debris flows.

The development of incised valleys within the onshore Niger delta and the controls on their development appears to have received little attention for the past years. However, the focus of this paper is to reveal some of the local controls that may be responsible for the development of these deep incisions as well as the influence of the depositional processes on the incised valleys.

Geological Setting

The Niger Delta is located at the southern end of Nigeria, bordering the Atlantic Ocean and extends from about longitude 30-90°E and latitude 40° 30' to 50° 20' N (Figure 1). It is situated on the Gulf of Guinea on the west coast of Central Africa. During the Tertiary it built out into the Atlantic Ocean at the mouth of the Niger-Benue river system. The total thickness of the Niger Delta sequence is not known but may be up to 12 km in the basin centre and cover area of some 75,000 km^2. The stratigraphy of the Niger Delta is described as a regressive (offlap) succession with continental coarse and gravely sandstones of deltaic top facies (Benin Formation) overlying an alternation of sandstone and shale of paralic delta front facies (Agbada Formation) which in turn grade downward into marine shales of pro-delta facies (Akata Formation) with turbidites [10-14]. Burke [12] interpreted the stratigraphy of the Niger Delta as five stratigraphic units (Figure 2). They are: a. Top continental sediment (Benin Formation) b. Transitional sand/shale unit (Agbada Formation), c. Marine shale unit (Akata shale) d. transitional shale-sand unit (newly distinguished) e. and an un-named bottom submarine fan unit. Exploration activities have concentrated in the past over these sequences (onshore) but as delta becomes better understood, exploration efforts are gradually shifting to both the offshore (Pliocene – Pleistocene section) and the flank of the delta, where Cretaceous prospects are expected (Figure 1). Although the geology, tectonics and evolution of the onshore (Eocene-Pliocene sequences) of the Niger Delta are fairly well know,

***Corresponding author:** Durogbitan A Abimbola, Principal Geoscientist, Petrfac IES, Research and Development, Manchester, UK
E-mail: adewoledurogbitan@yahoo.co.uk

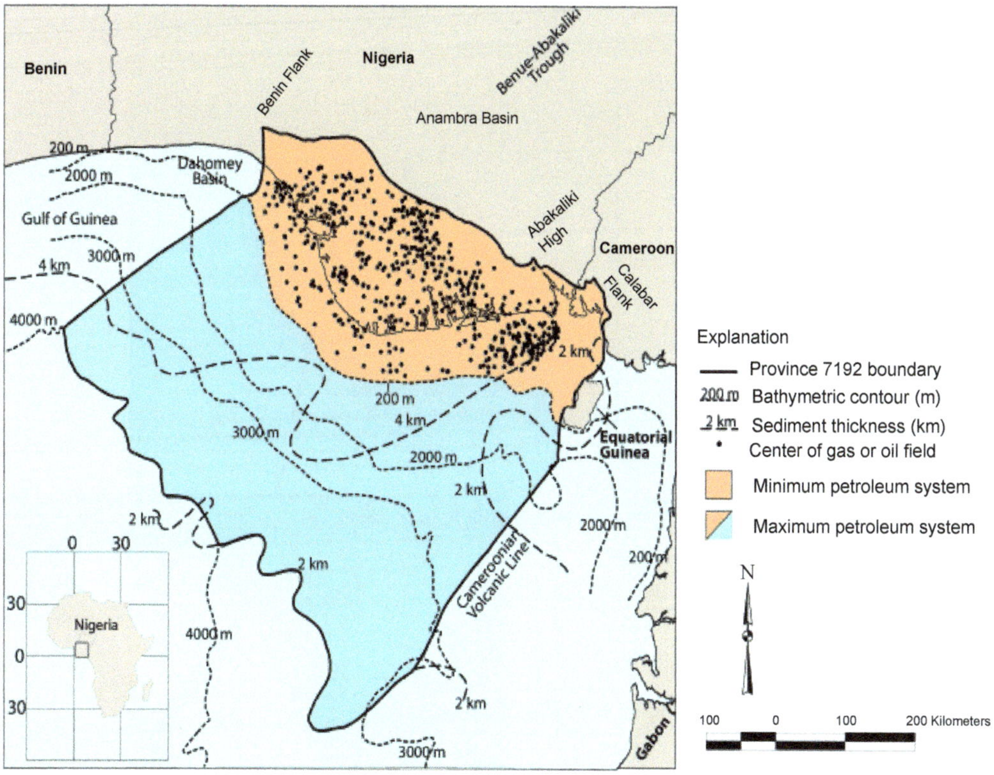

Figure 1: Location map of Nigeria showing the Niger Delta (onshore and offshore) and some of Cretaceous basins (Benue-Abakaliki Trough and Anambra basins) from Petroconsultants [28].

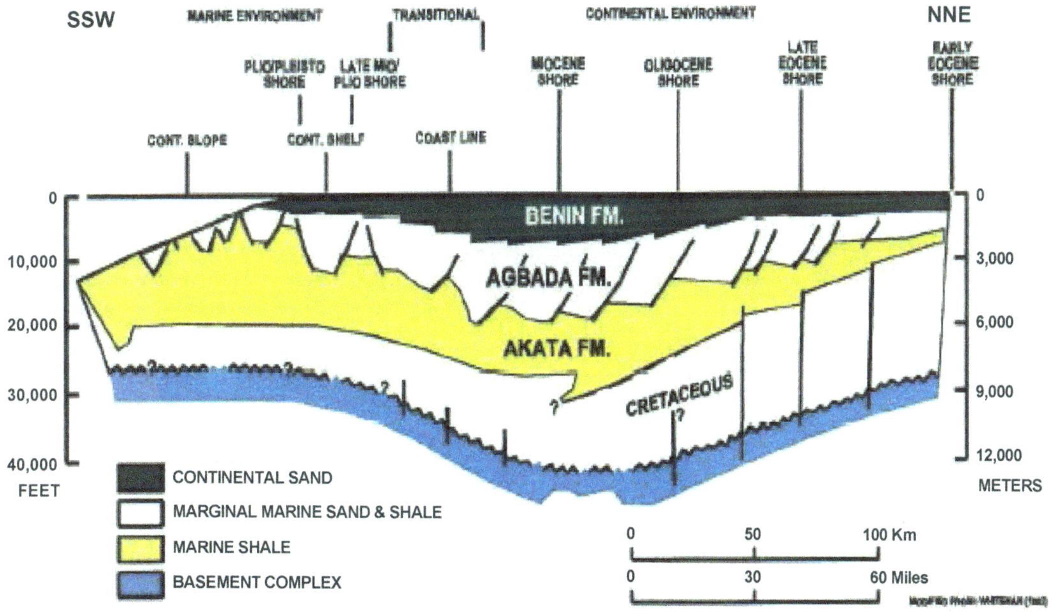

Figure 2: The stratigraphy formed by a prograding sequence of offlap cycles, with three distinct time transgressive lithologic units: the Akata, Agbada and Benin Formations [14].

these are expected to be refined as new analytical tools, concept and models evolve. This paper proposes a seismic stratigraphic analysis and evaluation of some of the local control on deep fluvial channel incisions within the northwester Niger delta using Ewan and Oloye fields as a case study.

Database and Study Location

Ewan and Oloye oil fields are located on OML 49 and 95 onshore of the north-west of the Niger Delta (Figure 3). The fields are currently operated by Chevron Nigeria. The structure that defines these fields is

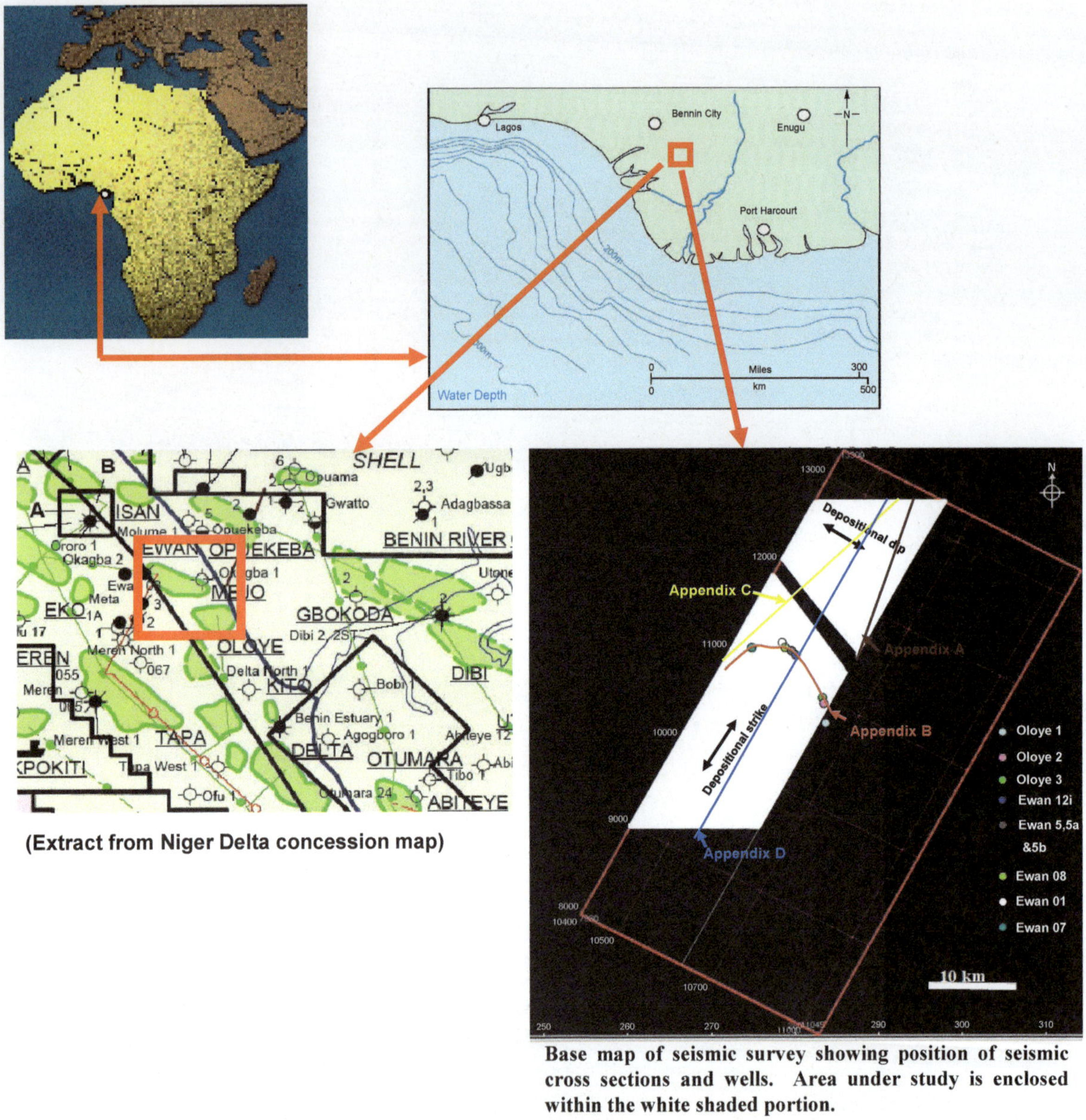

Figure 3: Detailed location and bathymetry maps of Niger Delta showing the study area - Red boxes.

a complexly faulted rollover anticline. The stratigraphy encountered in the study area are the paralic sequence of interbedded sands and shales (Agbada Formation) of Middle Miocene-Early Pliocene age as observed from the seismic survey, biostratigraphic reports and wells data.

The database includes logs of 7 wells (gamma ray, resistivity, sonic, neutron, density etc), cores and a three-dimensional seismic cube (depth converted) of the area around Ewan and Oloye fields. Biostratigraphic reports of 3 wells were also available for age determination.

Results and Discussion

Several incised valleys were identified from the seismic data having depth range between 120-350 m. This research focus on three main interpreted palaeovalleys along interpreted erosional surface shown in (Figures 4 and 5) and the deep fluvial channel incision at the upper part of the seismic section that reaches depth of about 350 m shown in figures.

Seismic facies analysis

a) Palaeovalley 1-3: The position of these palaeovalleys defined a sequence boundary characterized as an irregular unconformity that erosionally truncates dipping reflections. At the proximal area (northern part), these three WNW-ESE trending erosional lows, interpreted as palaeovalleys, cut into the underlying strata (Figure 5; valleys 1, 2 and 3). Valley 1 is the deepest of all with up to 220 m of erosional relief with valleys 2 and 3 having up to 150 m and 120 m of erosional relief respectively. Palaeovalley 2 occurred landward of the shelf break. All valleys have tributary-like features and asymmetric cross-sections, with steeper slopes along the northern margins compare to the southern margins (Figures 6-9). Truncations of the underlying stratal and onlap characterized the erosional surface.

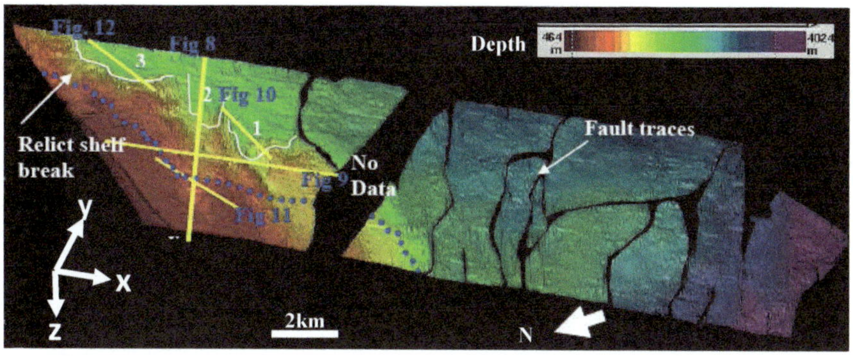

Figure 4: Depth structure map of the key surface interpreted as sequence boundary 1 (SB1).

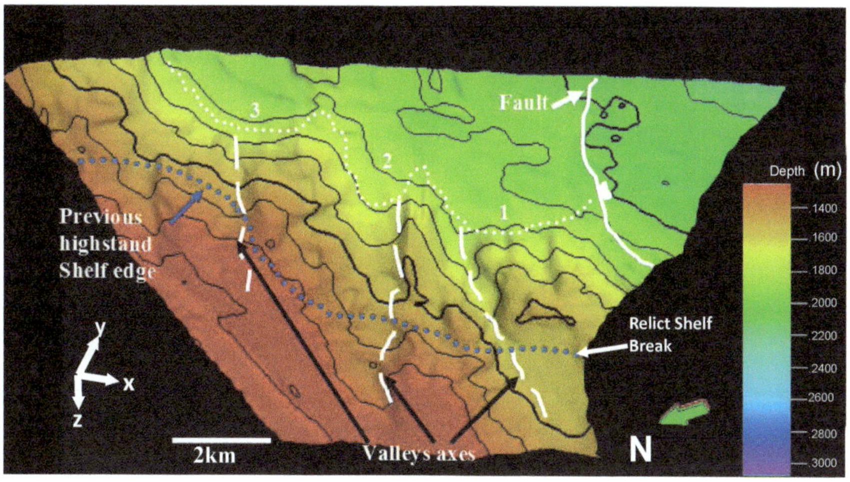

Figure 5: Depth structure map of proximal area (Northern margin) of SB1 showing position and geometry of the 3 paleovalleys fills (1-3) and their tributaries and valley axes (Vertical Exaggeration: x2).

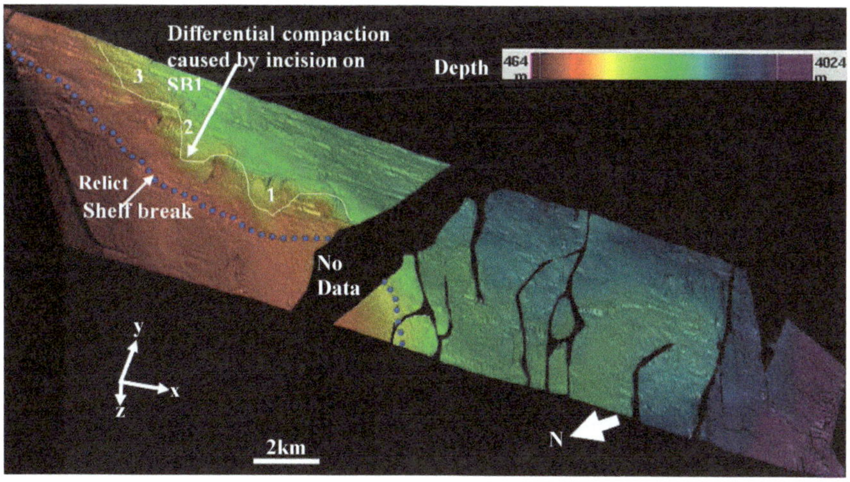

Figure 6: Depth structure map of MF1 showing differential compaction over SB1 related valleys 1-3.

Seismic reflections within the palaeovalleys shows a variable and irregular internal reflections pattern characterized by low to moderate amplitude, low frequency reflectors with prograding-aggrading clinoforms geometry that suggest gradual infills of the valleys during transgression (Figures 10-12). Thickness variation between interpreted erosional surface (sequence boundary) in which palaeovalleys occurs and the surface that caped the valleys (interpreted as maximum flooding surfaces) is average 360 m. The thickness reflects position of the incisions and sediment infill during the time of sea-level rise. Depth structure map of interpreted maximum flooding surface revealed a

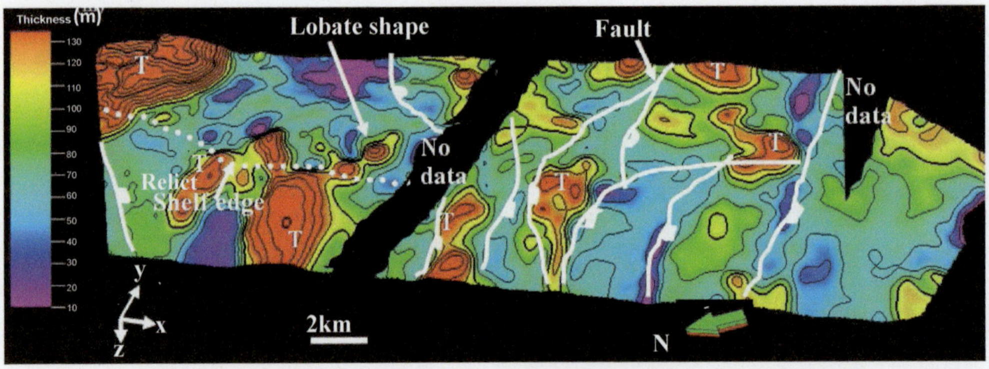

Figure 7: Thickness map between Sequence Boundary 1 and Maximum Flooding Surface 1 (about 360 m). T's are the position of thicks. The isopach map shows thicks as multi- lobate to lunate bodies but generally more lobate. White lines are the fault traces. Reds and yellows are thicks, purples and blues are lows. 20 m contour interval. Thicks reflect position of fluvial incisions.

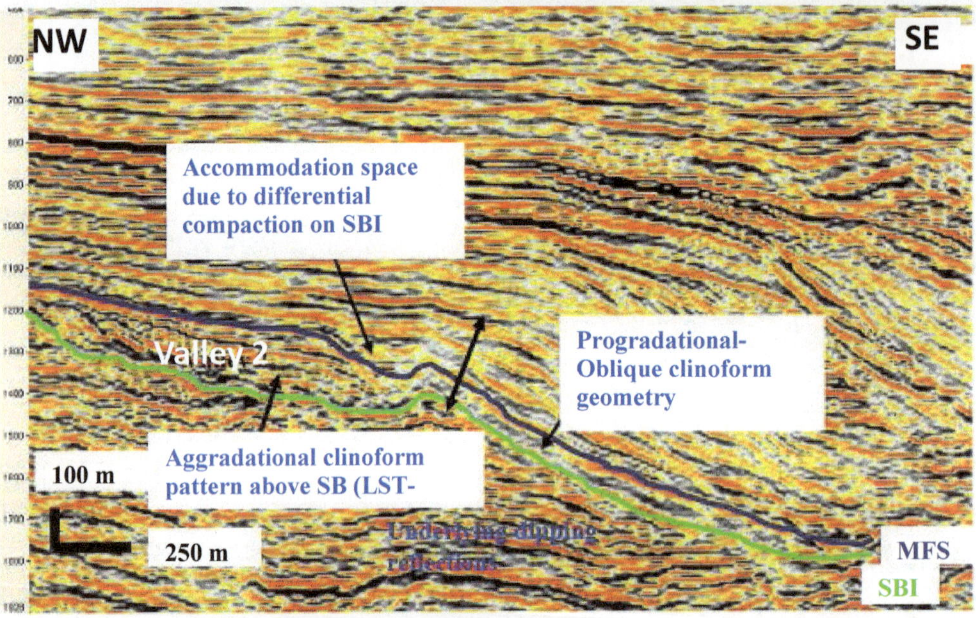

Figure 8: Seismic section of sequence 1 in the proximal area (dip line 12499). The progradation and aggradation of deltaic shelf edge due to relative sea-level rise generated by local incisions, diapir-related subsidence. See position in Figure 4.

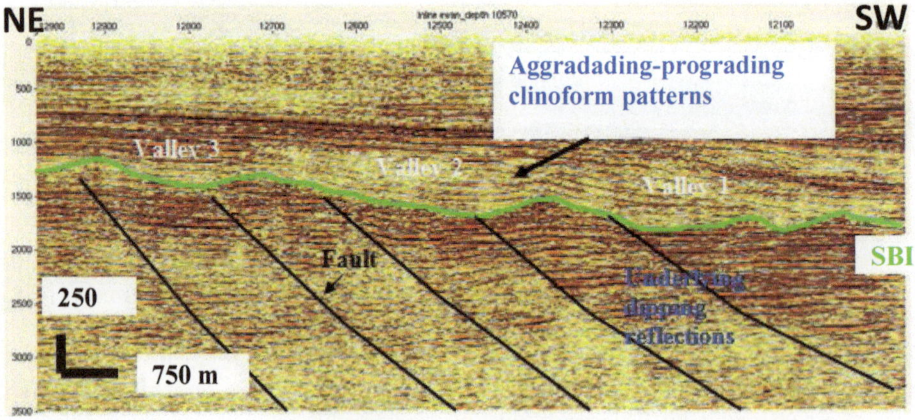

Figure 9: Seismic section in the proximal area showing the palaeovalleys. SB1 truncates the growth faults. See position in Figure 4. Reflection patterns shows arogradational - aggradational configuration.

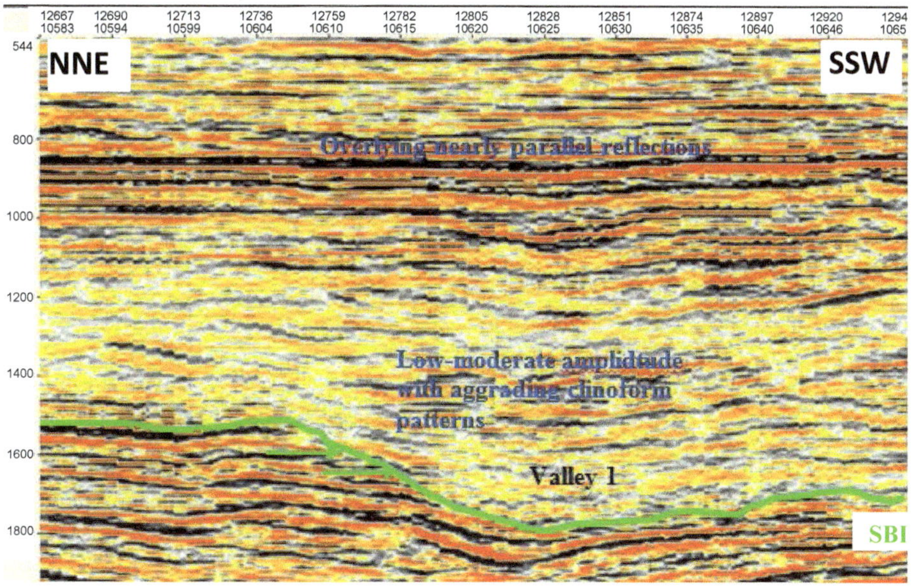

Figure 10: Random seismic section showing the palaovalley 1. Seismic reflections within the valley are characterized as low-moderate amplitude with aggradational-progradational clinoform patterns. See position in Figure 4.

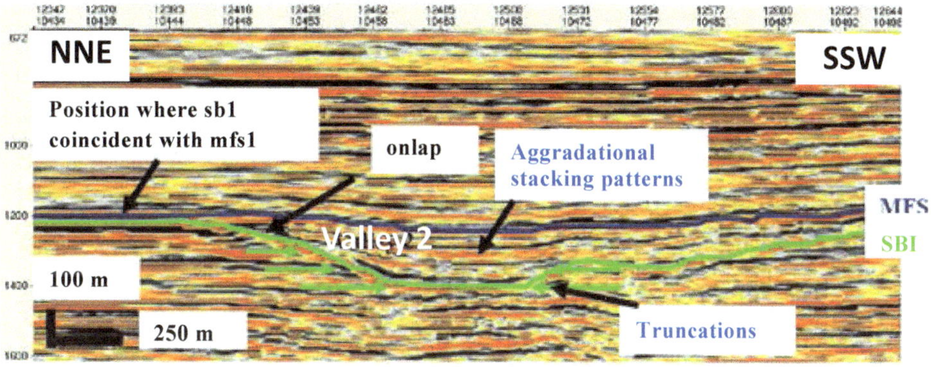

Figure 11: Random seismic line showing the palaovalley 2 and the interfluves. Seismic reflections within the valley are characterized as low-moderate amplitude with aggradational clinoform patterns. See position in Figure 4.

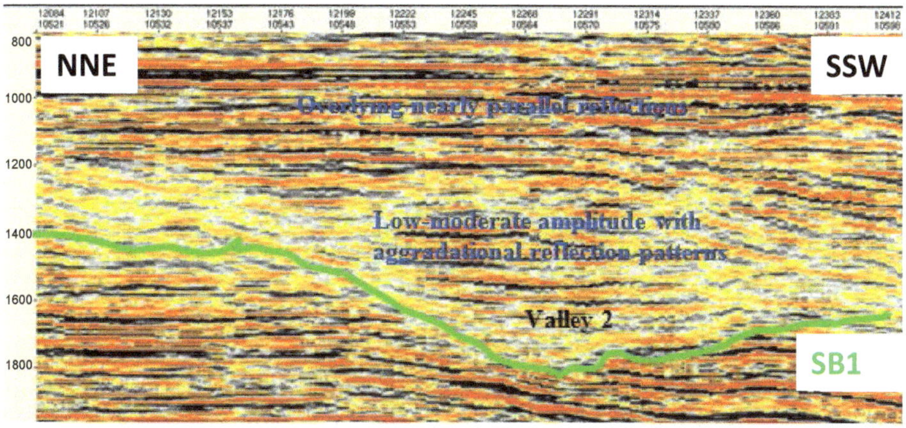

Figure 12: Random seismic line showing the palaovalley 3. Seismic reflections within the valley are characterized as low-moderate amplitude with aggradational-progradational clinoform patterns. See position in Figure 4.

differential compaction over sequence boundary related valleys (Figure 6).

Interpretation: Based on the interpreted key stratigraphic surfaces (SB and MFS) and reflection geometry, the palaevalleys infills seismic facies are interpreted to be Lowstand deposit overlying by aggradational Transgressive deposit. The Lowstand deposits are confined within the valleys, while thin and widespread transgressive deposit caped the valleys. The development of fluvial incisions (palaeovalleys) and aggrading infill reflection patterns suggest that the development of fluvial incision is characterized by fall of relative sea level below the previous Highstand System Tract (HST) shelf edge, where the sands influx to the shelf edge from the inner/middle shelf by forced regression, followed by a sea level rise (transgression). Prolonged sea-level fall have led to pronounced incisions on the shelf edge which also in turn caused instability at the shelf margin. The striking feature is the depth of palaeovalleys on the Sequence Boundary 1, which is up to 150 m (Valley 2) below the landward of previous HST shelf edge (Figure 8), partially reflects underlying structural control. This might be caused by uplift due to the shale diapirism. The fluvial incisions suggest a mechanism of lowering of relative sea level. Differential compaction over SB related valleys was observed from the depth structure map of interpreted MFS which also suggest structural control of underlying structure.

b) Deep incision at the upper seismic section: This incision has an average depth of 350 m (Figures 13 and 14). The seismic characteristics of this deep incision consist of low-moderate amplitude, variable frequency and internal reflections, chaotic and progradational-sigmoid reflection patterns (Figures 15 and 16). It is difficult to know the overall width of this fluvial channel incision due to the area extent of the survey within the study area, but the width of the fluvial channel is up to 15 km (Figure 13). Seismic reflections above the incision fill are nearly parallel, with moderate-high reflection amplitude (Figures 15 and 16). Three-Dimensional seismic slices through this incision at depth 889 m show northwest-southeast orientation of straight to low sinuous channels within a more chaotic and low amplitude background (Figure 17). The magnitude of the incision increases southward and most of the rollover anticline beds are truncated. This suggests the locus of erosion and deposition shifted basinward and has structurally controlled. This fluvial incision surfaces truncate some of the growth faults, and some of the faults cut into the channel at the top. This suggests that some of the fault were inactive during the development of this fluvial incision (Figures 15 and 16). Fault cutting into the channel may indicate tectonic

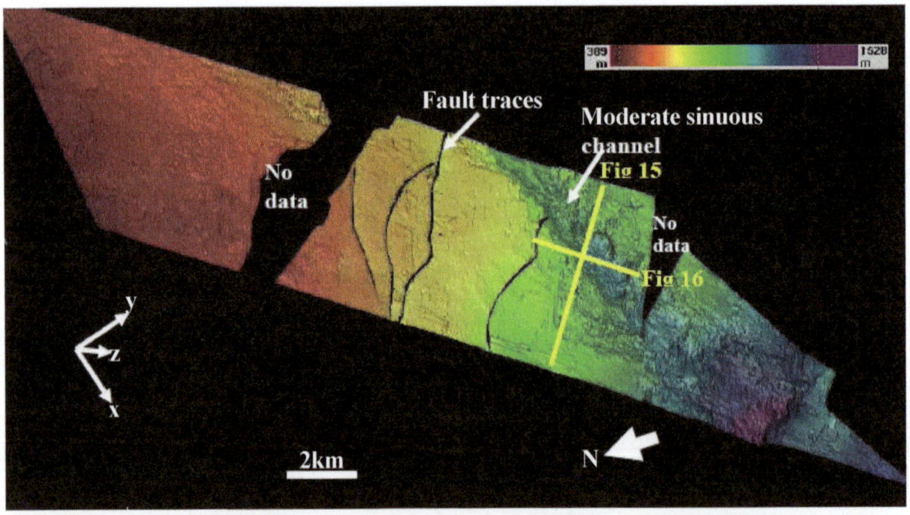

Figure 13: Depth structural map of base of incised valley trending NNE-SSW.

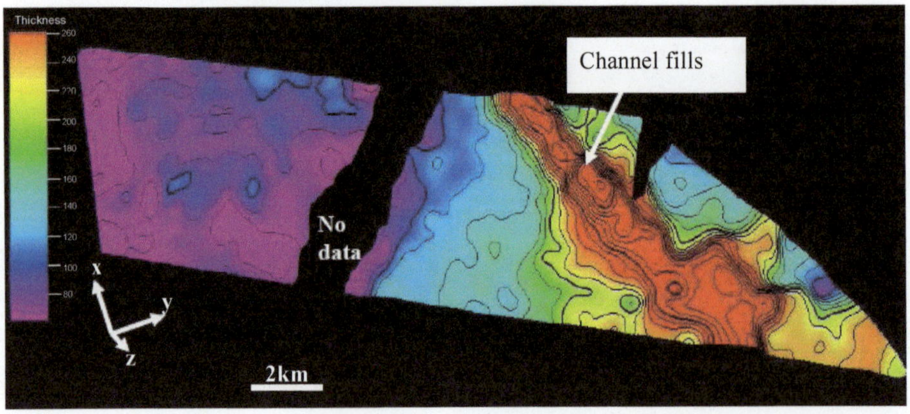

Figure 14: Isopach map between the base of incised valley and the top. Thick is within the valley which reflect infilling of accommodation space created during lowstand period.

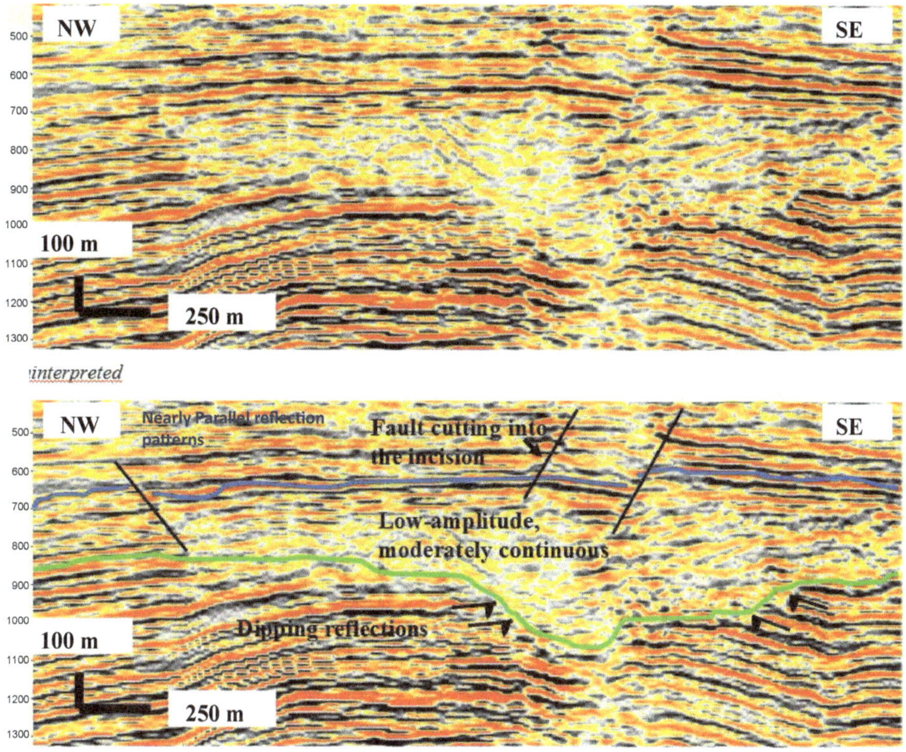

Figure 15: Seismic section of the incised valley Crossline (10367). See location in Figure 13.

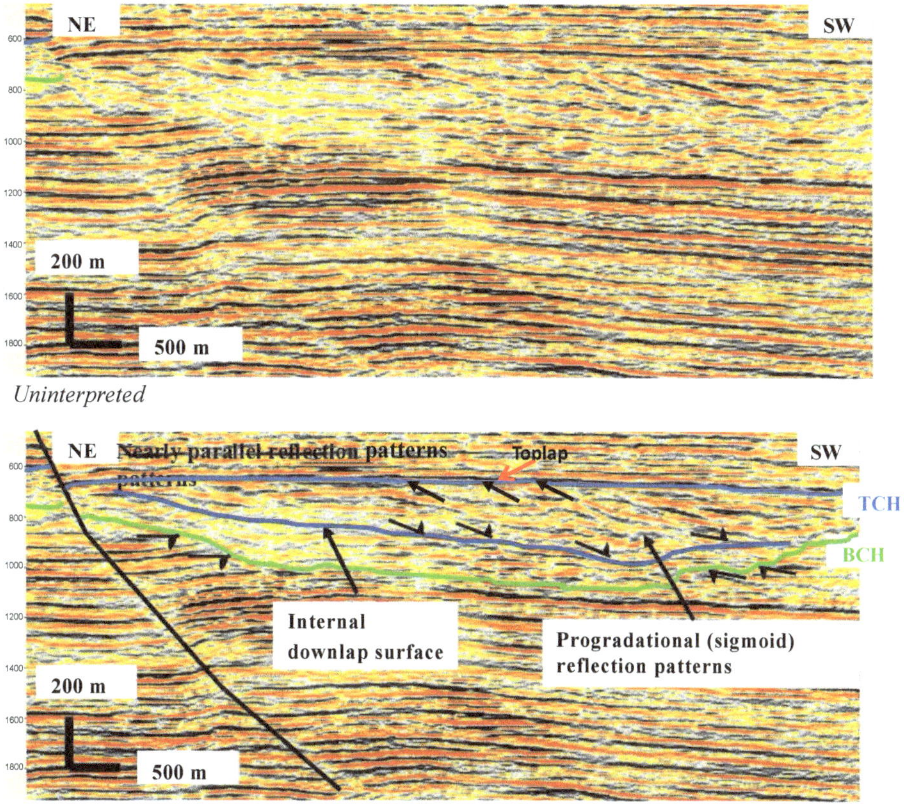

Figure 16: Seismic section of the incised valley. (In line 10547). Reflection geometry within the channel is characterized by progradational (sigmoid). See location in Figure 13.

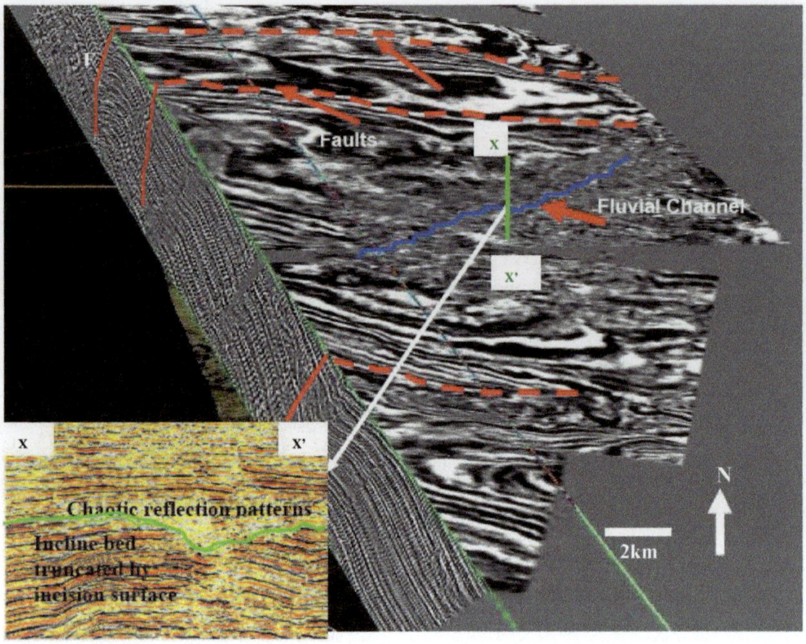

Figure 17: Seismic horizon slices showing the channel and faults orientation at depth 889 m and geometry. Channel orientation is sub parallel to the faults which suggest that, the channel orientation is structurally controlled. Inset is the seismic section of the incised valley (Crossline 10367).

response cause by the effect of differential compaction of channel fills. Thickness map between base of the channel incision (BCH) and top of the channel (TCH) is averagely, 260 m (Figure 14). Areas with thicker sediment are coincident with main axis of the channel which reflects accommodation created by relative sea level fall. This incision created significant accommodation space which is filled by aggradational-progradational of deltaic system during highstand condition. From the well logs within the study area, the uppermost section is dominantly sandy and exhibits progradational parasequences of stacked coastal barrier bar, fluvially /tidally influence channel sand (continental sand), identified from wireline stacking patterns. An increase in the sand ratio from the log motif indicates that we are at the uppermost section of the interpreted section (Agbada Formation), grading into fully continental sands deposits interval (Benin Formation).

Interpretation: It is difficult to deduce the continuity of the seismic facies and the sedimentary infill of these channel incisions based on the extent of the data. The seismic and log parameters suggest that the sedimentary infill is composed of sandstone capped by the alternating mud and sandstones. The reflections are truncated at the point of incision; this shows the effect of channeling erosion during the formation of the valley.

Progradational-sigmoid configuration suggested deposition with high sediment supply and low to medium energy regime [15,16]. Variations observed in the refection patterns can be ascribed to variable in depositional processes within the valley fills. Sigmoid facies are usually associated with rising sea level or subsidence creating an accommodation space [16]. The progradational to sigmoid reflection patterns are interpreted as alternating upbuilding and depositional bypass as sediment were fed into the incised valleys. The toplap observed represent a top discordant where the prograding reflections truncated by overlying sediment which may be as a result of low deposition and minor erosion [15]. This implies that toplap reflection termination represent period when the valley was completely filled up followed by the period of slow deposition or minor erosion. Low sinuosity of the channel indicates high gradient and high discharge which are faults controlled (Figure 17).

The depth of incision (up to 350 m) is difficult to explain in terms of sea level fall alone because ecstatically-driven sea level variations during the Miocene are generally reported to be less than 100 m, Haq [17]. This suggests that the character of fluvial incisions development and depositional facies preserved within the study area depends on relative sea level changes, relative rates of regional structural collapse and rates of sedimentation. The slope scars and shelf instability act as a locus for fluvial lowstand incision and affect sequence thickness. The occurrence of the fluvial incision cycles in the study area can be interpreted as evidence of significant relative sea level fluctuations, and the presence of Type-1 sequence boundaries may be the stratigraphic signature of major drops in relative sea level within the study area.

Evolution and Development of Incised Valleys

Seismic reflection patterns across the incised valleys and along their axes have been interpreted to evaluate the development and evolution of these valleys. The factors that were likely to be responsible for the initiation and development of these valleys include sea level changes, shelf instability, variation in sediment input and slope collapse (mass flow depositional processes and downward cutting by downslope sediment flow). These factors control the initiation and development of the valleys with strong evidence of underlying structural controls (fault and shelf instability) (Appendices A-D).

Sea level

Sea level change plays an important role in determining the position of a sequence boundary. Sea level variations, when combined with sediment supply, define the stacking pattern of each systems tract [18]. It has been established that relative sea level is the one of the major factors responsible for the formation of sequence boundaries along the continental margin [2,19-25]. A fall in sea level results in shelf exposure and subsequent erosion of the shelf results in the formation of incised valleys. The bases of these incised valleys define the sequence boundaries (Figure 18). During sea level fall, channels are generally active and re-adjust to the base level. As the sea level continues to fall, incision increases and some of the incised channels become dominant and active, and some that are inactive are abandoned and capture sediment from the channels that are inactive [26]. Progradation of sediment will continue basinward and seismic reflection patterns are always characterized by offlapping geometry (forced regressive system tracts). Rapid sea level fall has been reported to result in deeply incised valleys, which may be associated with channels and organized lobes in deep water areas [26]. No lobes were observed within the study area (away from the shelf edge), but they are expected to be present seaward of the study area.

The study area is characterized by deeply incised valleys associated with rapid basinward progradation of the sediment, with the seismic reflection patterns characterized by prograding and offlapping geometries (Appendices A-D). Therefore, it can be inferred that deep water basin floor fans / lobes may be developed basinward of the shelf edge (Figure 19).

Depositional processes

Incised valleys are normally filled during sea level rise [2]. The curved morphology exhibited by the valley wall, deep channeling, and internal reflection patterns (chaotic and moderately continuous) have

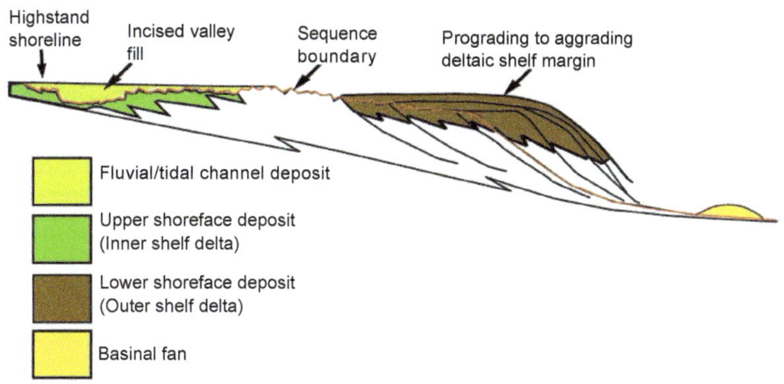

Figure 18: Schematic stratigraphic architecture of the study area (deltaic shelf margin). The occurrence of the deep incised valley cycles in the middle Miocene prograding to aggrading shelf margin succession indicated evidence of significant relative sea level fluctuations. The position of the sequence boundary corresponds to Exxon type-1 sequence boundary.

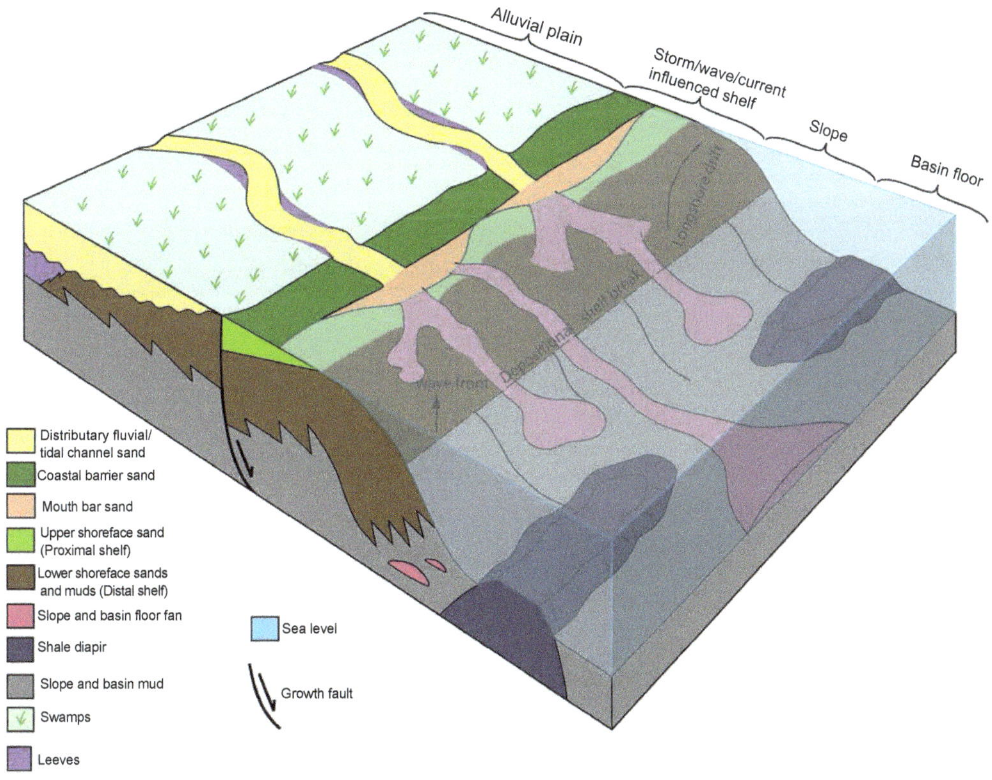

Figure 19: Schematic paleogeography and general conceptual depositional model of the study area during middle Miocene time.

been used to interpret the influence of sedimentary processes in the development of the incised valleys. The steep and curved morphology of the incision walls could have resulted from slumping and sliding of channel margins (Figure 14). The striking feature within some of the depositional sequences e.g. Depositional Sequence 1, is the depth of palaeovalleys on the Sequence Boundary 1 i.e. up to 150 m (Valley 2) below the landward part of the previous HST shelf edge (Figure 8). This is interpreted to partially reflect shelf instability related to underlying structural control or shale diapirism. The preservation of deep incision geometry below the shelf edge suggests the interplay of underlying structural controls (shelf instability) and rapid sea level fall. Over-steepening of the incision near the shelf edge results in sediment failure, a common process in the initiation and development of incised valleys/canyons [27,28].

Low to moderate amplitude, chaotic reflection patterns within the valleys are probably indicative of mass movement deposits. At the topmost part of the valley, the reflections are continuous and slightly even. They are discordant with the underlying prograding-sigmoid refection facies. This facies suggest uniform deposits formed after the valley have been filled up, probably consist of pelagic and hemi pelagic deposition formed during sea level rise.

Sediment transport

It can be inferred from this study that a large volume of sediment must have been eroded and deposited basinward based on the depth and magnitude of incisions observed from the seismic sections. The deep water sediment was probably derived from two sources (a) locally derived from where the incision occurred (endogenetic), and (b) terrigenous sediment from the landward area (exogentic). The evolution of the incised valleys was probably characterized by large episodic sediment input following rapid sea level fall during the period of sea level lowstand. These interpreted valleys within the study area have served as a pathway for transporting clastic sediment during different developmental stages to deep water.

Summary

Detailed seismic stratigraphy and fluvial incision (incised valley) study shows that the middle Miocene-Pliocene strata in the Ewan and Oloye fields of northwestern Niger delta are characterized by several deep fluvial incisions. This suggests that during the period of middle Miocene most of the areas are subaerially exposed (lowstand time).

There is local growth fault control on fluvial incisions (Paleovalleys?) development. Further local controls on fluvial incisions and facies geometry are shelf instability and slope collapse structures which have a strong geomorphological impact on the area. The occurrence of the deep fluvial incision cycles in the middle Miocene succession is interpreted as evidence of significant relative sea level fluctuations, and the presence of type-1 sequence boundaries may be the stratigraphic signature of major drops in relative sea level during middle Miocene and Pliocene. Thickness variations strongly reflect the varying depth and magnitude of incision and growth faulting.

In conclusion, seismic stratigraphic analysis has revealed some of the local controls on deep fluvial incisions within the study area to be local growth faults, rapid relative sea level changes, basin, physiography (shelf edge), shelf instability, variation in sediment input, slope collapse (mass flow depositional processes and downward cutting by downslope sediment flow).

References

1. Vail PR (1984) Jurassic unconformities, chronostratigraphy, and sea level changes from seismic stratigraphy and biostratigraphy. In: Schlee JS (Ed), Interregional Unconformities and Hydrocarbon Accumulation. AAPG Memoir. American Association of Petroleum Geologists (AAPG) 36: 129-144.
2. Van Wagoner JC, Posamentier HW, Mitchum RM, Cail PR, Sarg JF, et al. (1988) An overview of the fundamentals of sequence stratigraphy and key definitions, in CK Wilgus, Hastings BS, Posamentier HW, Van Wagoner J, Ross CA, and CC St Kendall, eds., Sea-level changes: an integrated approach: SEPM Special Publication 42: 39-45.
3. May JA, Warme JE, Slater RA (1983) Role of submarine canyons on shelf break erosion and sedimentation: modern and ancient examples. In: Stanley DJ, Moore, GT (Eds), The Shelf Break: Critical Interface on Continental Margins, SEPM 33: 315-332.
4. Twichell DC, Roberts DG (1982) Morphology, distribution and development of submarine canyons on the US Atlantic Continental Slope between Hudson and Baltimore. Geology 10: 408-412.
5. Posamentier HW (2001) Lowstand alluvial bypass systems: incised vs. unincised: AAPG Bulletin 85: 1771-1793.
6. Allen GP, Posamentier HW (1993) Sequence stratigraphy and facies models of an incised valley fill: the Gironde estuary, France, J Sedimen Petrl 63: 378-391.
7. Allen GP, Posamentiern HW (1994) Transgressive facies and sequence architecture in mixed tide and wave-dominated incised valleys: example from the Gironde estuary, France. Incised-valley systems: origin and sedimentary sequences, Dalrymple RW, Boyd RJ, Zaitlin BA (Eds), SEPM (Soc Sedient Geol) 51: 225-240.
8. Van Wagoner JC, Mitchum RM, Campion KM, Rahmanian VD (Eds) (1990) Siliciclastic sequence stratigraphy in well logs, core and outcrops: concept for high resolution correlation of time and facies. American Association of Petroleum Geologist. Methods in Exploration, Barbara H Lidz, Tulsa pp: 55.
9. Miall AD (2002) Architecture and sequence stratigraphy of Pleistocene fluvial systems in the Malay Basin, based on seismic time slices snalysis: AAPG Bulletin 86: 1201-1216.
10. Short KC, Stauble AJ (1967) Outline of Geology of Niger Delta. AAPG Bull 51: 761-779.
11. Weber KJ, Daukoru EM (1975) Petroleum geology of the Niger delta: Proceedings of the 9th World Petroleum Congress, Tokyo 2: 209-221.
12. Burke KC (1972) Longshore drift, submarine canyon and submarine fans in development of Niger Delta. AAPG Bull 56: 1975-1983.
13. Avbovbo AA (1978) Tertiary Lithostratigraphy of Niger delta: AAPG Bull 62: 295-306.
14. Doust H, Omatsola E (1990) Niger Delta in Edwards JD and Santogrossi PA (eds) Divergent/Passive Margin Basin AAPG 48: 201-238.
15. Mitchum RM, Vail PR, Sangree JB (1977) Seismic stratigraphy and global changes of sea level, part 6: stratigraphic interpretation of seismic reflection patterns in depositional sequences. In: Payton CE (Ed), Seismic Stratigraphy - Applications to Hydrocarbon Exploration. AAPG Memoir 26: 117-133.
16. Sangree JB, Widmier JM (1979) Interpretation of depositional facies from seismic data. Geophysics 44: 131-160.
17. Haq BU, Hardenbol J, Vail PR (1987) Chronology of fluctuating sea-levels since the Triassic: Science 235: 1153-1165.
18. Schlager W (1993) Accommodation and supply - a dual control on stratigraphic sequences. Sediment Geol 86: 111-136.
19. Vail PR, Mitchum Jr RM, Thompson S (1977) Seismic stratigraphy and global changes of sea level, part 4: global cycles of relative changes of sea level. In: Payton CE (Ed.), Seismic Stratigraphy - Applications to Hydrocarbon Exploration. AAPG Memoir 26: 83-97.
20. Posamentier HW, Vail PR (1988) Eustatic controls on clastic deposition: II. Sequence and systems tract models. In: Wilgus CK, Hastings BS, Kendall, CG St C, Posamentier HW, Ross CA, Van Wagoner JCŽ Eds., Sea Level Changes - An Integrated Approach. Society of Economic Paleontologists and Mineralogists 42: 125-154.
21. Posamentier HW, Jervey MT, Vail PR (1988) Eustatic controls on clastic deposition: I. Conceptual framework. In: Wilgus CK, Hastings BS, Kendall

CG St. C, Posamentier HW, Ross CA, Van Wagoner JCŽ Eds., Sea Level Changes - An Integrated Approach. Society of Economic Paleontologists and Mineralogists 42: 109-124.

22. Sarg JF (1988) Carbonate Sequence Stratigraphy: In Wilgus CK, Hastings CG, St C Kendall, Posamentier HW, Ross CA and Van Wagoner JC, eds., Sea-Level Changes: An Integrated Approach: Society of Economic Paleontologists and Mineralogists pub 42: 155-181.

23. Vail PR, Audemard F, Bowman SA, Eisner PN, Perezcruz C (1991) The stratigraphic signature of tectonics, eustacy and sedimentology – an overview. In: Einsele G, Ricken W, Seilacher A (Eds), Cycles and Events in Stratigraphy. Springer, Berlin pp: 618-659.

24. Christie-Blick N (1991) Onlap, offlap, and the origin of unconformity-bounded depositional sequences. Sedimentary Geology 70: 105-107.

25. Christie-Blick N, Driscoll NW (1995) Sequence stratigraphy. Annual Rev Earth Planet Sci 23: 451-478.

26. Ritchie BD, Gawthorpe RL, Hardy S (2004) Three-Dimensional Numerical Modelling of Deltaic Depositional sequences 2: Influence of local control. Journal of Sed Res V 74: 221-238.

27. Chuang CY, Yu HS (2002) Morphology and canyon forming processes of Upper Reach of the Penghu Submarine Canyon off southwestern Taiwan. TAO 13: 91-108.

28. Petro consultant (1996) Index map of Nigerian showing the Niger Delta and some Cretaceous Basin.

Sequence Development and Palaeogeography Evolution of Ewan and Oloye Fields (Middle Miocene), Northwestern Niger Delta

Durogbitan A Abimbola*

Petrfac IES, Research and Development, Manchester, UK

Abstract

Sequences developed within the middle Miocene in the study area are interpreted to be controlled locally by episodic shelf instabilility, basin physiography (shelf edge), fault growth and linkage. Relative sea level changes provide the major control on sequence development and deep valley incisions. The latter have a strong geomorphological impact on the area and a strong control on sequence variability. Interpretation of GDE maps shows that sands were originally deposited in river- and mouth bars with longshore drift currents redistributing the sand parallel to the shoreline. During lowstand periods, canyons developed which served as conduits for sand to the basin (deep water sand), recognized as high amplitude seismic facies on the slope and on the basin floor. This mechanism has led to the deposition of considerable amount of sandstones in the slope and offshore environments.

Idealized schematic sequence stratigraphic development models have been generated for the study area, while schematic palaeogeography and general conceptual depositional models of the study area during middle Miocene time are also presented.

Keywords: Oloye fields; Miocene; Niger delta

Introduction

The depositional setting of the study area (Figure 1) has been interpreted to range from continental to offshore environments within a shelf edge depositional setting [1]. Important depositional facies interpreted include undifferentiated continental deposits, fluvial/tidal channel sandstones, mouth bars and upper shoreface sandstones, lower shoreface heterolithics, and offshore mudstones (with discrete slope channel and basinal sand).

Seismic stratigraphic analysis of this study area consists of six depositional sequences [1] figures with base map of seismic survey showing position of seismic cross sections and wells in Figure 2. Sequences are made up aggradational to progradational HST that makes up about 50% of a sequence. TST are thin and widespread whereas LST are restricted to within WNW-ESE trending valley fills typically 4 km wide and an average 120 m deep. Outside the LST valleys, transgressive surfaces, maximum flooding surfaces and sequence boundaries are coincident on interfluves. This suggests that during LST times most of the area was periodically subaerially exposed [1]. During the time of lowstand, several incised valleys/canyons were developed. These incisions/canyons acted as a sand feeder that led to the deposition of high amplitude deposits (sandstones) on the slope and within the basin floor setting. Sequences mapped over the study area show local variation associated with growth faults. Typically, individual sequences are 10-15% thicker in the hanging wall than the footwall while LST incision is also greater in the hanging wall of the growth faults [1] (Figure 2).

Synthesis of Sequence Development and Paleogeography Evolution

Having studied and analysed the six depositional sequences [1], the spatial and temporal evolution of the depositional systems is determined using a sequence of palaeogeographies through production of Gross Depositional Environment (GDE) maps. Gross Depositional Environment maps allow synthesis of the evolution of depositional sequence palaeogeographies. Because no single map is representative for an entire stratigraphic interval, Lowstand and Highstand GDE maps were constructed for four of the depositional sequences. This has allowed the depositional fills, spatial relationships and geometry of facies to be described and changes observed between main units within a sequence as well as between sequences. Because TST in this study area are thin but widespread, Highstand GDE maps are constructed to be made up the period of transgression (TST) and progradation (HST). The position of the shoreline for each sequence (both LST and HST) are highlighted and annotated, as well position of the control wells, depocentres and faults.

Generally, shoreline geometries are lobate, and prograde basinward (south) during lowstand period, retrograde (TST) and subsequently prograde during highstand period. It is observed in all the sequences that the distal part of the study area is more tectonically active and faulted. In this area where we have syn depositional faults a more complex facies distribution/a depositional pattern is observed thereby creating some local depocentres (Figures 3-6).

Depositional sequence 1

There are few data to constrain the palaeogeography of this sequence, this due to the lack of well control/penetration and poor seismic definition at the base of Agbada Formation reflectors. Two GDE maps are constructed for this depositional sequence, for the lowstand and highstand period respectively (Figures 3a and 3b). Depositional Sequences 1 developed due to a relative sea level fall that led to the exposure of the shelf edge area. This has led to incisions within the shelf creating 3 palaeovalleys originally described (Depositional Sequence

***Corresponding author:** Durogbitan A Abimbola, Principal Geoscientist, Petrfac IES, Research and Development, Manchester, UK
E-mail: adewoledurogbitan@yahoo.co.uk

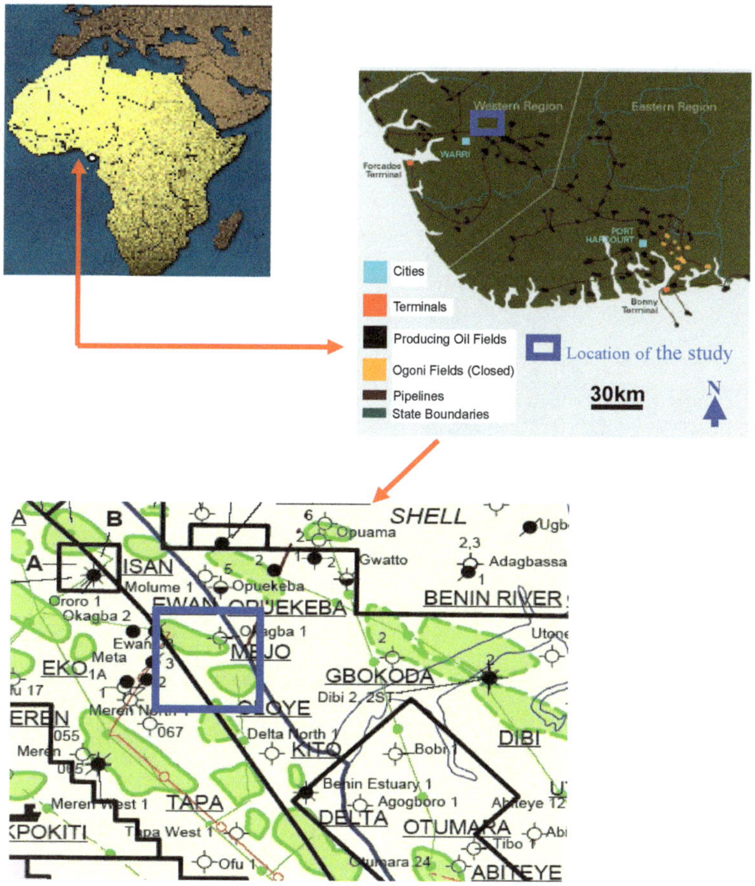

(Extract from Niger Delta Concession Map) 10km

Figure 1: Location of the study area.

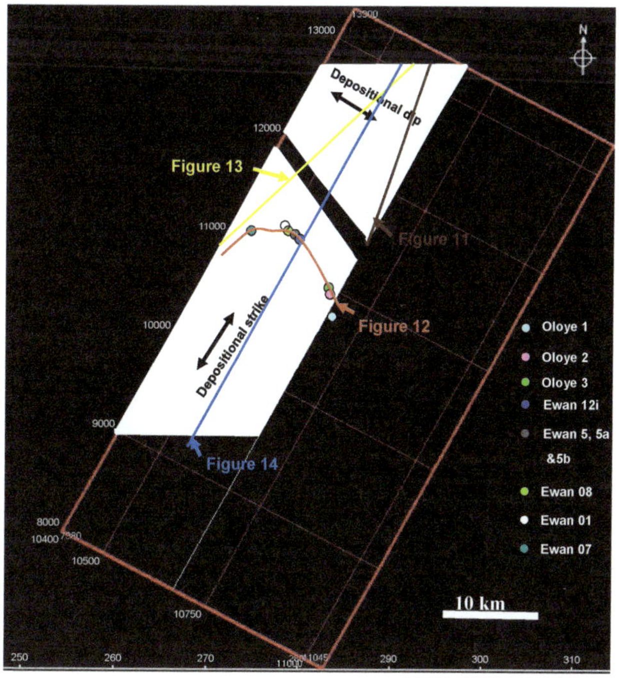

Figure 2: Base map of seismic survey showing position of seismic cross sections and wells. Area under study is enclosed within the white shaded portion [1].

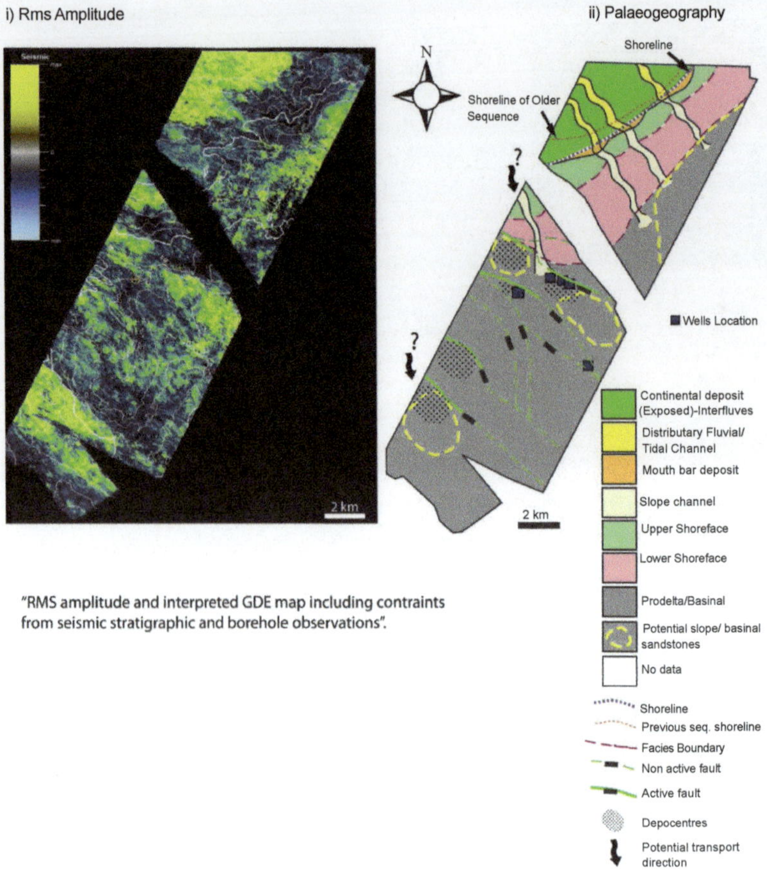

a) Sequence 1: LST.

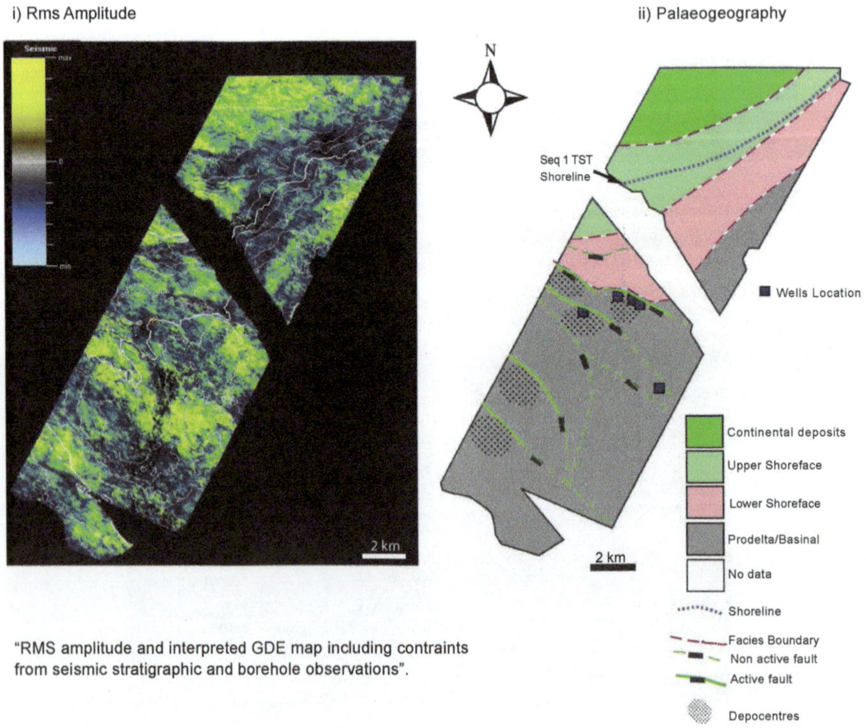

b). Sequence 1: HST.

Figure 3: Schematic Synthesis of Sequence Development and Palaeogeograhic Evolution of Sequence 1, also showing faults geometry.

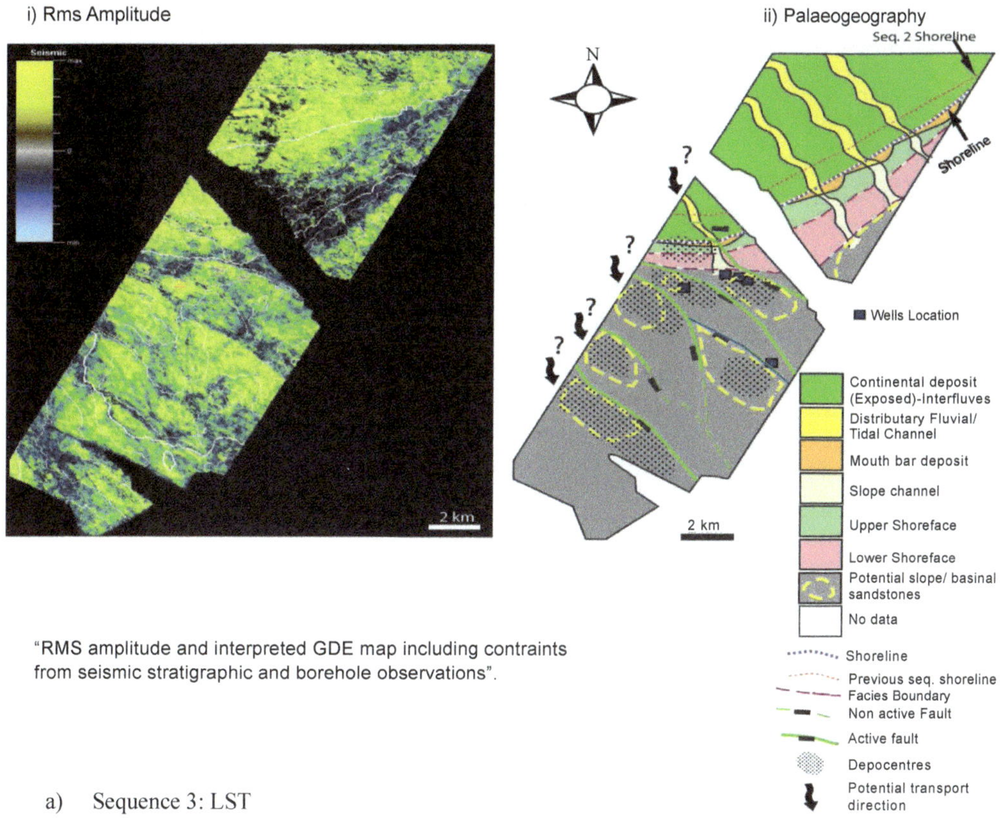

"RMS amplitude and interpreted GDE map including contraints from seismic stratigraphic and borehole observations".

a) Sequence 3: LST

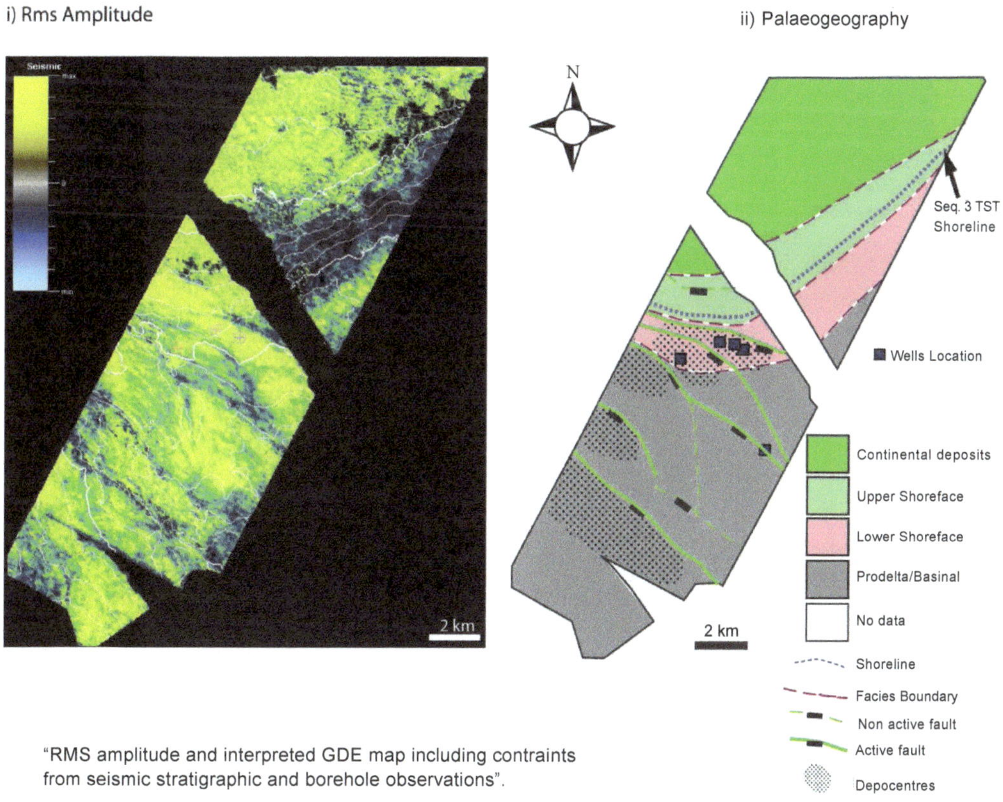

"RMS amplitude and interpreted GDE map including contraints from seismic stratigraphic and borehole observations".

b) Sequence 3: HST.

Figure 4: Schematic Synthesis of Sequence Development and Palaeogeograhic Evolution of Sequence 2, also showing faults geometry.

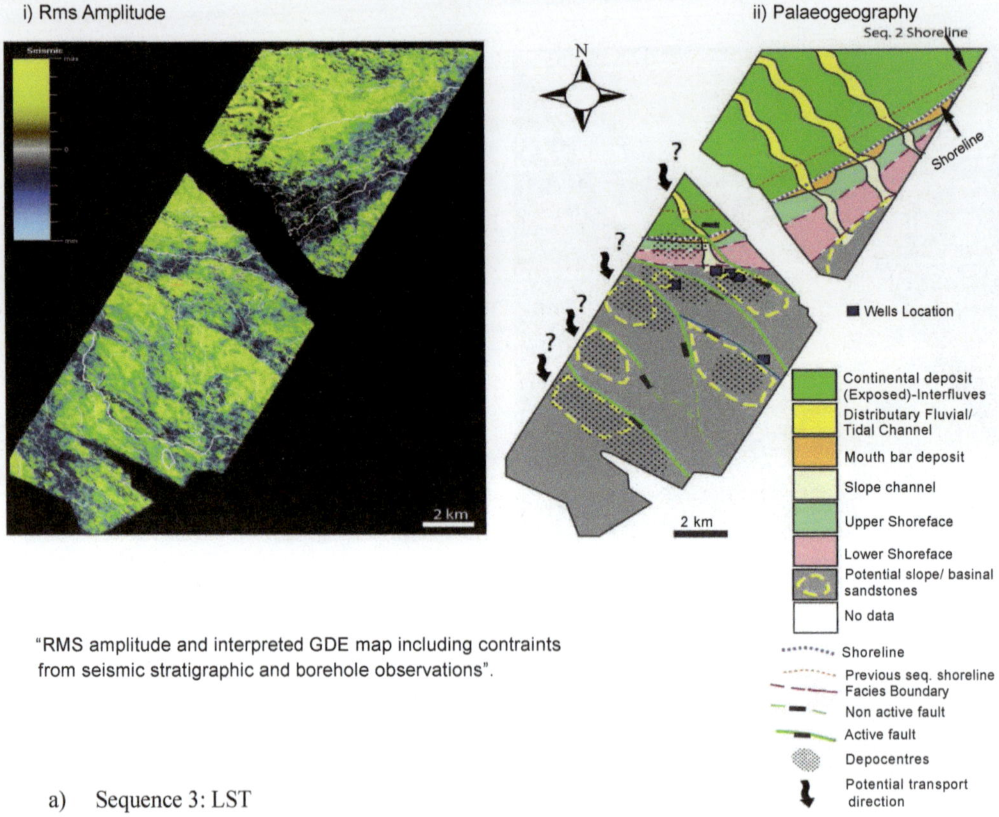

"RMS amplitude and interpreted GDE map including contraints from seismic stratigraphic and borehole observations".

a) Sequence 3: LST

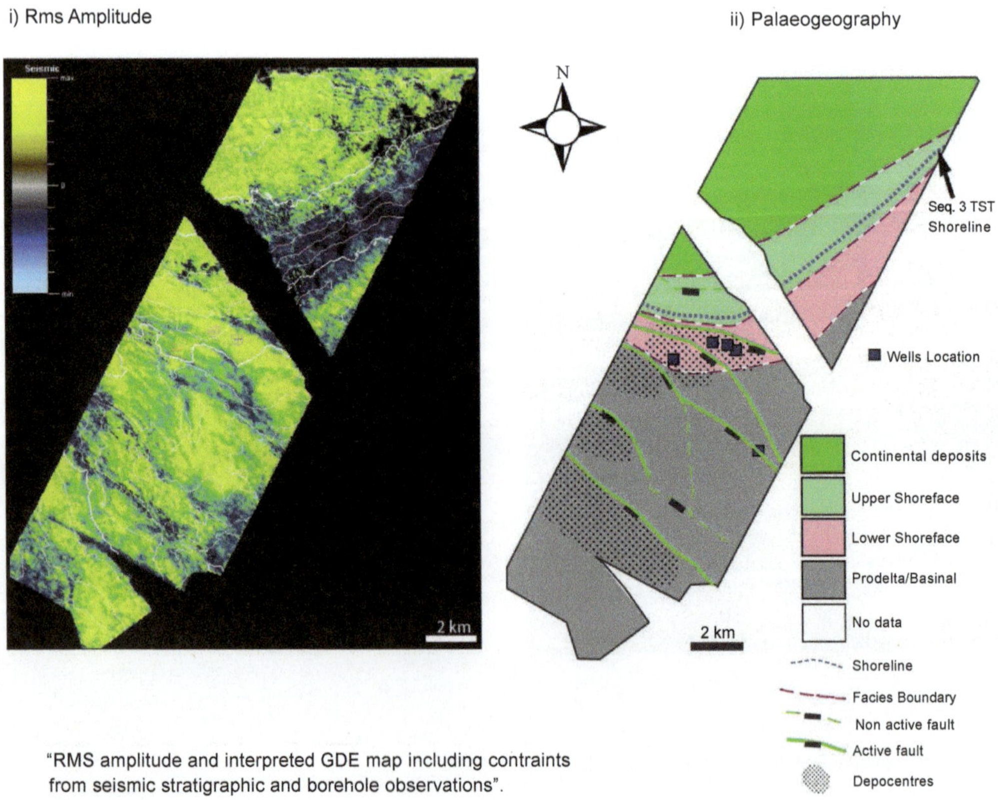

"RMS amplitude and interpreted GDE map including contraints from seismic stratigraphic and borehole observations".

b) Sequence 3: HST.

Figure 5: Schematic Synthesis of Sequence Development and Palaeogeograhic Evolution of Sequence 3, also showing faults geometry.

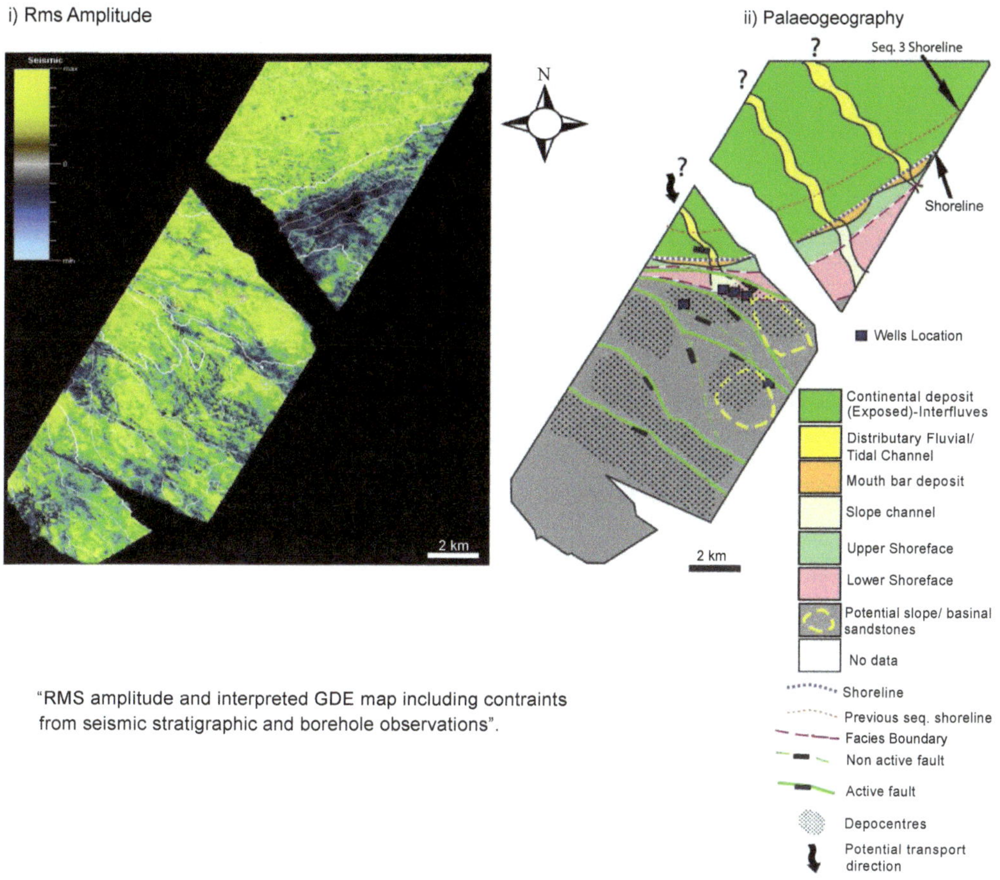

a) Sequence 4: LST

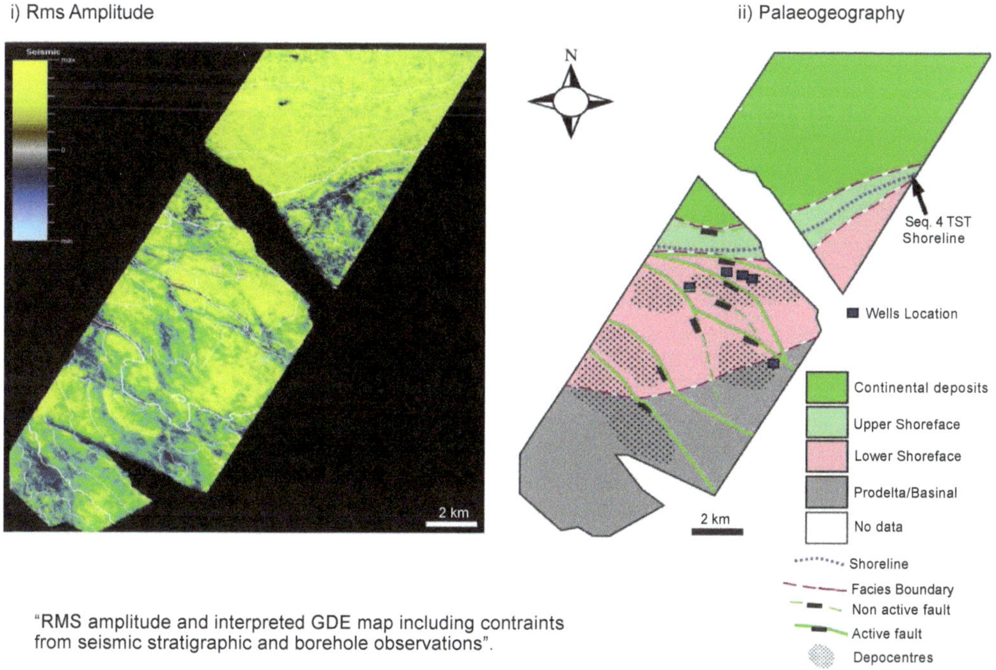

b) Sequence 4: HST

Figure 6: Schematic Synthesis of Sequence Development and Palaeogeograhic Evolution of Sequence 4 also showing faults geometry.

1- Durogbitan [1]) and subsequent deposition of sand as mouth bar deposit as a result of longshore drift and upper shoreface sandstones in front of shoreline (Figure 3a). During lowstand period, shorelines and facies belts prograde basinward (south) and canyons develop, observed from the rms amplitude map which serve as a conduit for the deposition of deep water sand on the slope and at the base of the slope (basin floor) (Figure 3a). This mechanism is interpreted to have led to transportation and deposition of potentially considerable amount of sandstones basinward during the lowstand period in all the interpreted sequences. Seismically, the deep water sandstones is characterised by mounded and prograding reflection patterns with higher amplitudes than the surrounding much-rich sediment.

From all the lowstands GDE maps, it has been inferred and interpreted that, in some places in the distal areas where high amplitude deposits were observed the deposit are similar to those observed at the base of the slope. This interpretation of the amplitudes is coupled with the observed seismic reflections patterns which are mounded and fairly continuous prograding reflection pattern. Where these potential basinal sandstones are interpreted (distal part), possible feeder channel systems and canyons are inferred to originate shelf ward of the basinal sandstone (Figures 3a, 4a, 5a and 6a). However, due to the extent of the data, it is difficult to show the direct relationship between the high amplitude bodies, interpreted to be channelized sandstones on the slope and the similar reflection characteristics on the basin floor in the distal part of the survey.

Following the lowstand period, marine flooding occurred during the TST across the shelf and most of the incisions on the shelf that have been filled by continental deposits, were transgressed and overlain by upper shoreface sandstones, lower shoreface and eventually prodelta/basinal deposits. Subsequently, the HST deposits prograde back into the basin (Figure 3b).

Depositional sequence 2

This lowstand period, characterised by relative sea level fall, resulted in incisions cutting into the underlying continental deposits of Sequence 2 (Figure 4a). This sea level fall is observed to have fallen below the shelf break and deposition of mouth bar deposit and upper shoreface sandstones are interpreted in front of the shoreline. During this lowstand period, shorelines and facies belts prograde basinward, and canyons transport potentially considerable amount of sand basinward (Figure 4a).

The following highstand system tract is characterised by rising relative sea level (flooding) and most of the incisions on the shelf that have been filled by continental deposits were also transgressed and overlain subsequently by upper shoreface sandstones, lower shoreface and eventually deeper water prodelta/basinal deposits (Figure 4b).

Depositional sequence 3

A major regression occurred across much of the area during this lowstand period. This period is characterised by several deep incisions across the shelf. This suggest there is a pronounced relative sea level fall resulting in erosion of the shelf edge These incisions/canyons acted as sand conduits and transportation of sand into deep water with sandstones deposited on the slope and within the basin floor (Figure 5a).

The highstand period follows an initial marine transgression that causes the shelf to be flooded. Subsequently, the HST deposits prograde back into the basin (Figure 5b).

Depositional sequence 4

Depositional sequence 4 evolution is similar to the sequence 3 described above. Major regression characterised the period of lowstand with several incisions cutting across shelf area (Figure 6a). This suggests there is a pronounced relative sea level fall resulting to incision into the shelf edge.

As observed from sequence 3, highstand relative sea level period is defined by marine transgressions that cause the shelf to be flooded and most of the incisions formed during lowstand period that have been filled by continental deposits, were transgressed and overlain by upper shoreface sandstones, lower shoreface and eventually prodelta/basinal deposits. Subsequently, the HST deposits prograde back into the basin (Figure 6b).

Summary of facies model of the study area is presented in Figure 7. Widespread of incisions indicate a rapid and strong progradation during the lowstand period which suggest that during LST times most of the area was periodically subaerially exposed. This has led to the development of incisions/canyons within the shelf and slope. These canyons acted as a sand feeder that led to the deposition of deep water sands on the slope and within the basin floor.

Discussion

Interpretations and descriptions of depositional sequences suggest that most sequence boundaries are eroded by sea level lowstand incision of the shelf, coupled with increased shelf instability and slope collapse structures. Patterns of sequence erosion occur both on the footwalls and hanging walls (but deeper mostly in the hanging wall) and deposits thickens across downthrown fault blocks, suggesting that most of these faults were active during the middle Miocene time of Agbada Formation deposition. The restriction of lowstands system tract deposits to within the valley fills across the entire study area, and the occurrence of deep incisions, suggests that the entire shelf region became periodically subaerially exposed during the successive relative sea level falls.

Observations from the sequence analysis show that throughout sequence development within the study area, the shelf margin has been controlled also by growth faults which have migrated basinward through time (Figures 3-7). In a case where the shelf margin is being controlled by growth faults, one of the processes of transporting sand into the basin is shelf collapse, with resultant slumping. However, there is an observed reduction in the upward propagation of faults. This may be due to changes in sediment deposition rates and fluvial discharge locations causing minor faults to become buried and truncated; fault segmentation and differential displacement on adjacent segments. Sedimentation patterns suggest that faulting, local incision and HST clinoform geometry dominate the formation of depocentres across the shelf.

Previous work showed that each depobelt is parallel to the modern coastal belt in which has been distinguished by their ages and are defined by sedimentary faults (Figure 8). Modern faults are observed to be parallel to these regional depobelts and the present day facies belts (Figures 8 and 9). Structural analysis of the study area shows a different local facies belt orientation versus the growth faults and the depobelts, which is very different from the present day orientation and relationship. Evidence from this study presented in the GDE structural maps shows that local facies belts are oblique to the regional depobelt (Figures 8-10) and often oblique to the interpreted growth faults. This

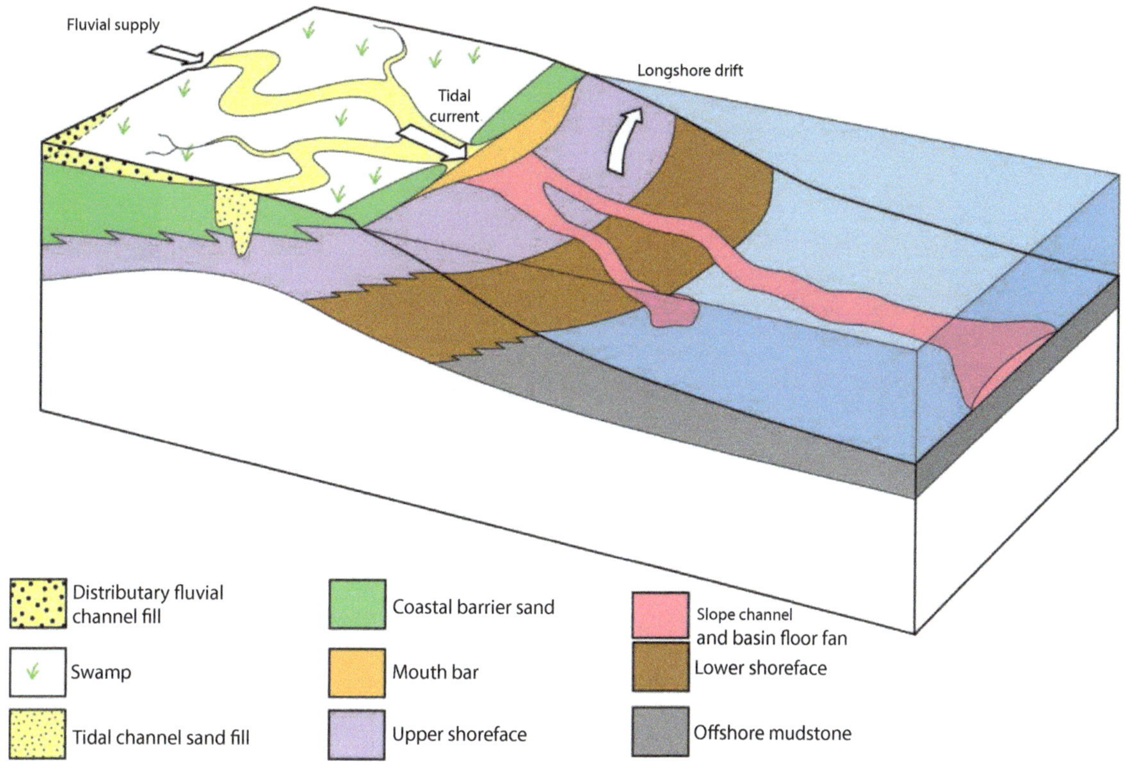

Figure 7: Block schematic diagram of the Lithofacies Relations within the Study area [2,5].

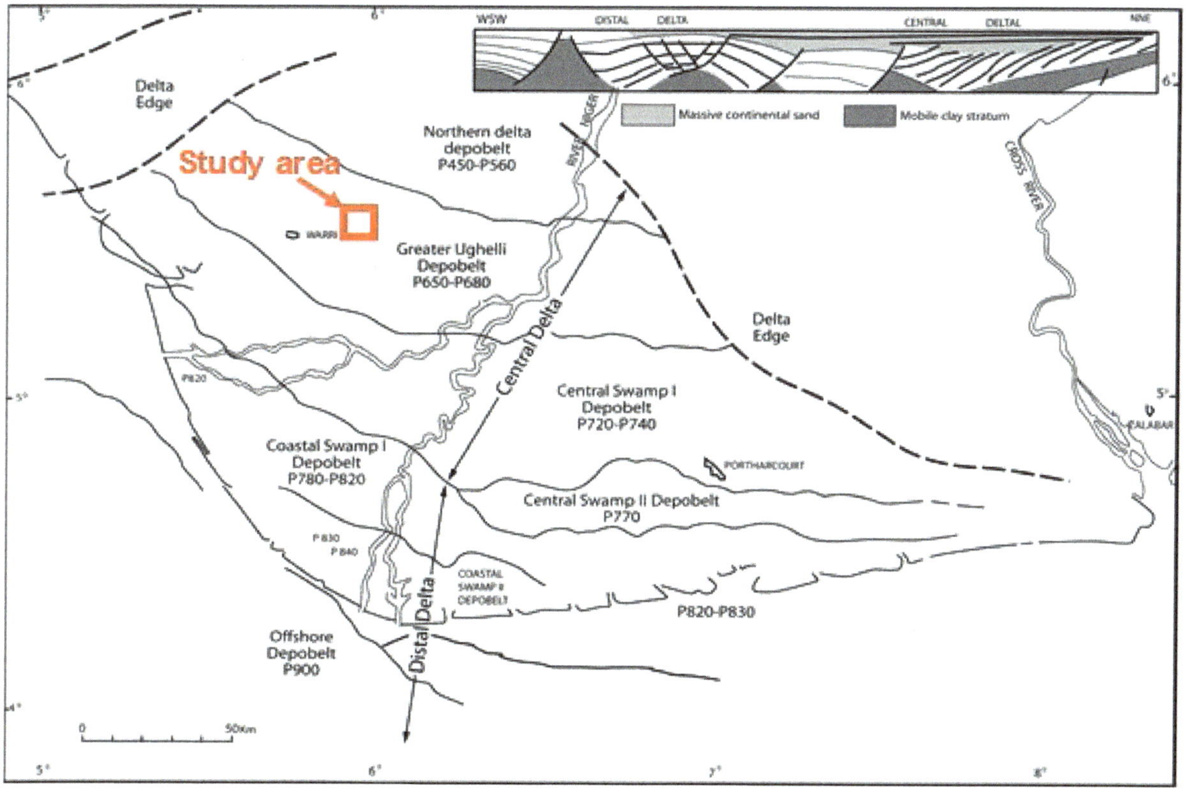

Figure 8: Age of deltaic sequences in depobelt and relationship to the broad changes in tectonic style [2,5].

observation suggests that the ancient coastal belt of the delta is more lobate /arcuate than the modern Niger delta. The delta became broadly convex to the sea during the late Miocene [2].

The overall strong progradational character of the study area as evidenced from the overall seismic geometry, prograding of shoreline break as observed from GDE maps, and dip line section at the

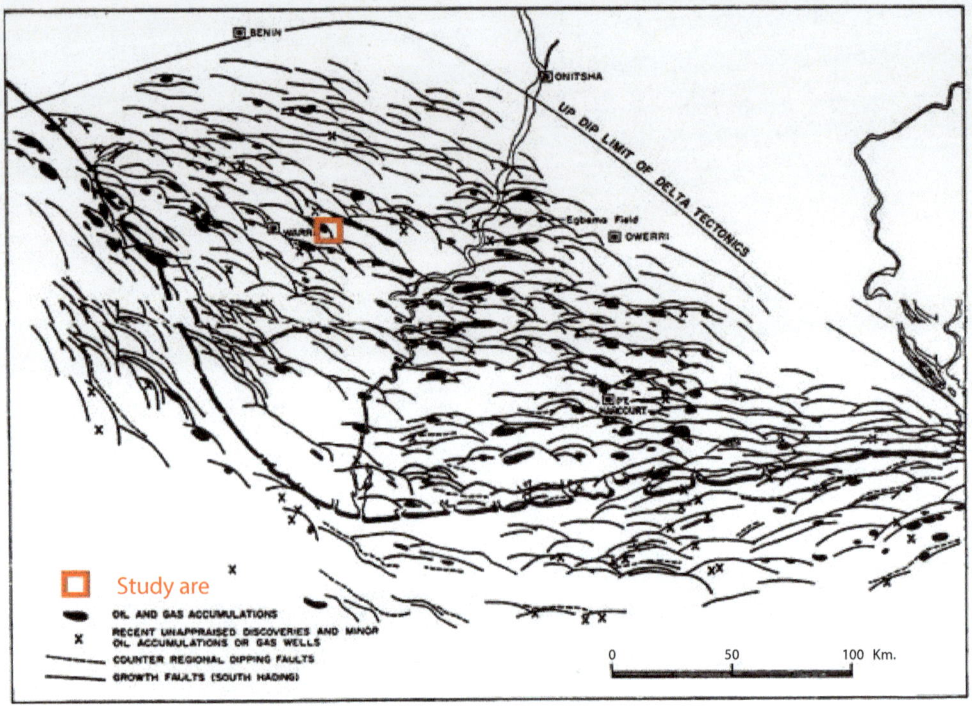

Figure 9: Growth faults and known hydrocarbon accumulation [6].

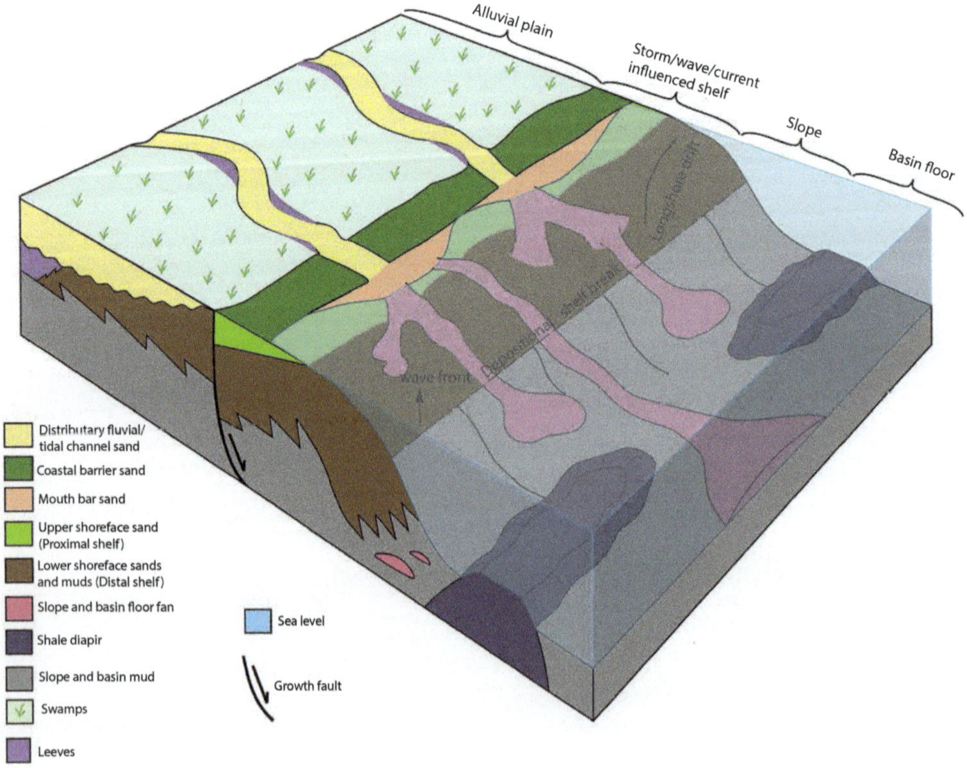

Figure 10: Schematic paleogeography and general conceptual depositional model of the study area during middle Miocene time [1].

proximal part. They are interpreted to reflect a long time decrease in accommodation triggered by a combination of tectonic uplift of the north-western Niger delta and the middle Miocene eustatic fall [2-6]. Increase in the proportion of sandstone relative to mudstone up section within the study area clearly also shows long-term progradation and graduation from the paralic Agbada formation into continental sand Benin Formation.

Conclusion

Detailed sequence stratigraphic analyses showed that the middle Miocene-early Pliocene strata in the Ewan and Oloye fields of the northwestern Niger delta consist of six depositional sequences (Figures 11-16) [1]. During lowstand period, shorelines and facies belts prograde basinward (south) and there are a development of canyons which serve as a conduit for high amplitude deposit (deep water sand) on the slope and within the basin floor. This mechanism is interpreted to have led to the deposition of considerable amount of sandstones basinward.

Sequences developed within the middle Miocene in the study area are interpreted to be controlled locally by episodic shelf instabilility, basin physiography (shelf edge), fault growth and linkage. Relative sea level changes provide the major control on sequence development and deep valley incisions. The latter have a strong geomorphological impact on the area and a strong control on sequence variability. Interpretation of GDE maps shows that sands were originally deposited in river- and mouth bars with longshore drift currents redistributing the sand parallel to the shoreline. During lowstand periods, canyons developed which served as conduits for sand to the basin (deep water sand),

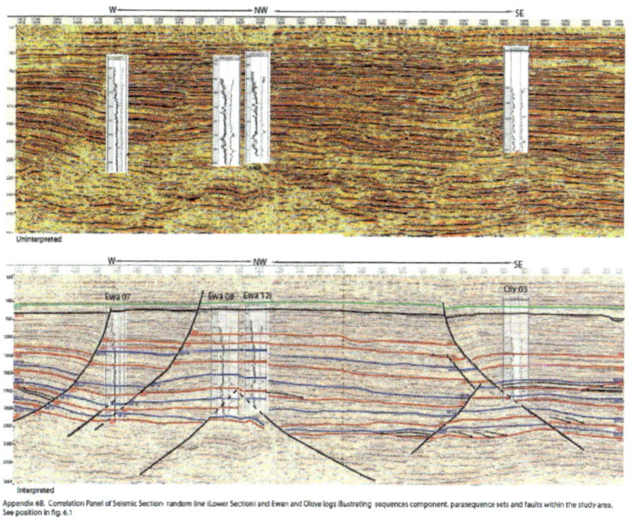

Figure 12: Correlation panel of seismic section-random line (Lower Section) and Ewan and Oloye logs illustrating sequences component, parasequence seta and faults within the study area.

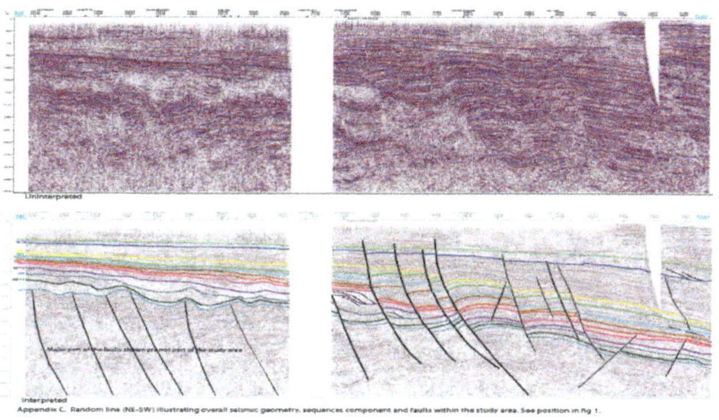

Figure 13: Random line (NE-SW) illustrating overall seismic geometry, sequences component and faults within the study area. See position Figure 2.

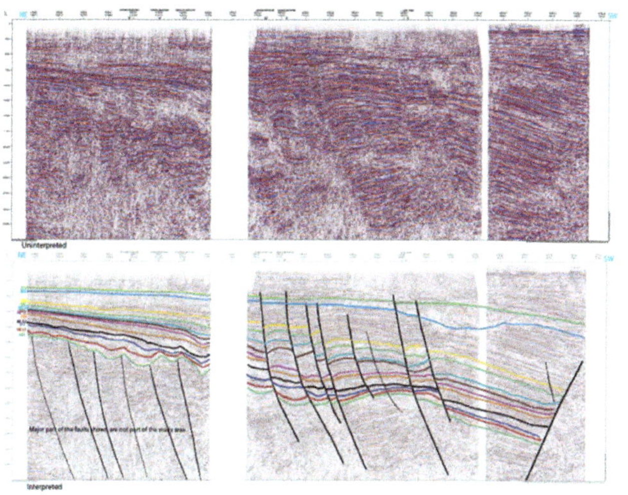

Figure 14: Seismic section (Strike line 10542) illustrating overall seismic geometry, sequences and faults within the study area. See position Figure 2.

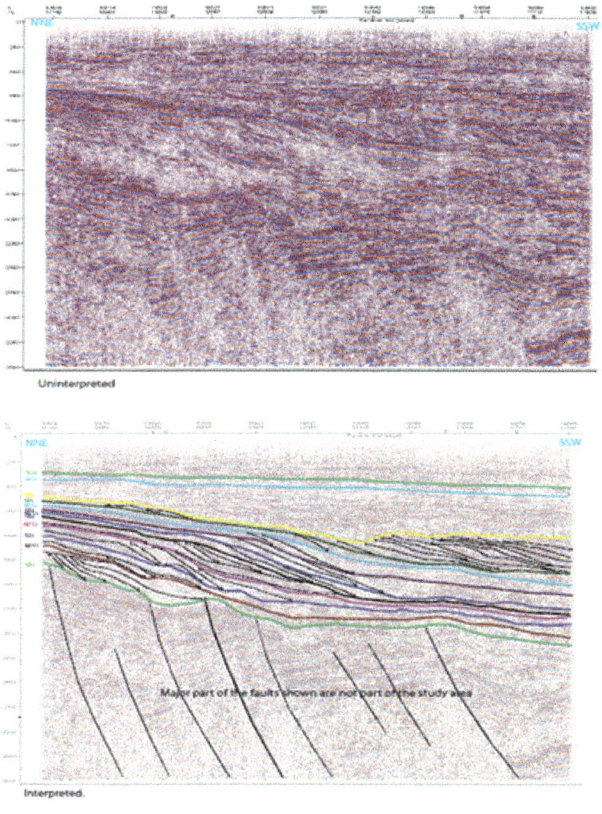

Figure 11: Random line from the study area (Upper section) showing clinoform geometry within key surfaces. See position Figure 2.

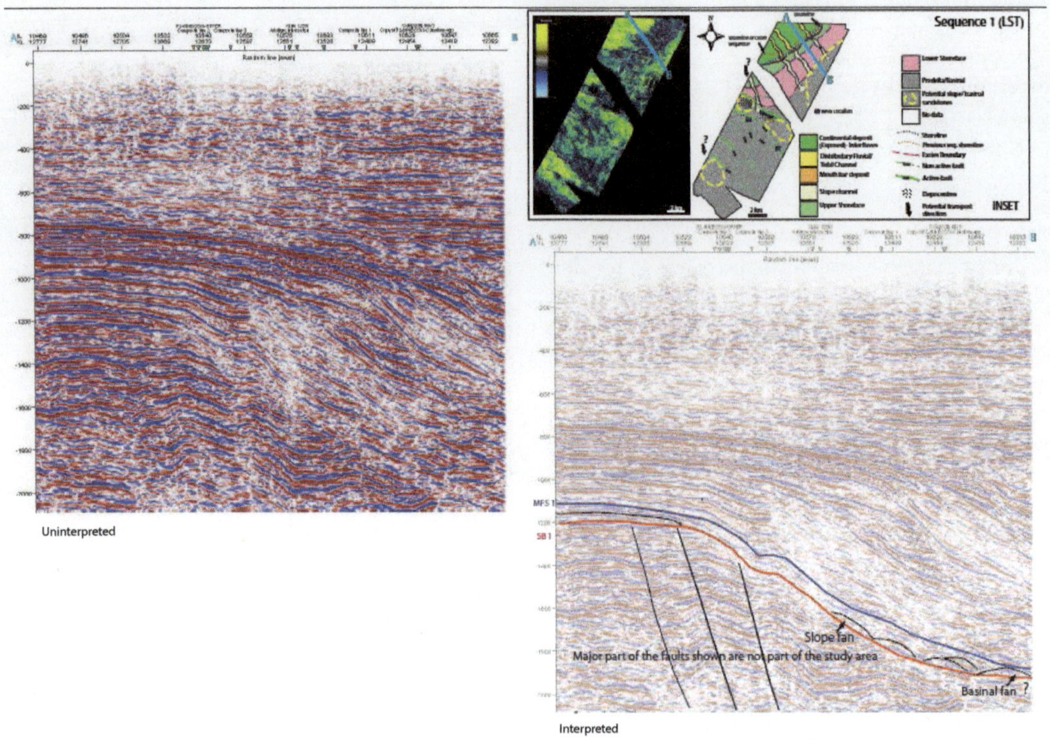

Appendix E: Random line showing the geometry, reflection patterns and position of slope and basinal fans within LST Depositional Sequence 1-(Upper Patches). See position in inset above.

Figure 15: Random line showing the geometry, reflection patterns and position of slope and basinal fans within LST Depositional Sequence 1- (Upper Patches). See position in inset above.

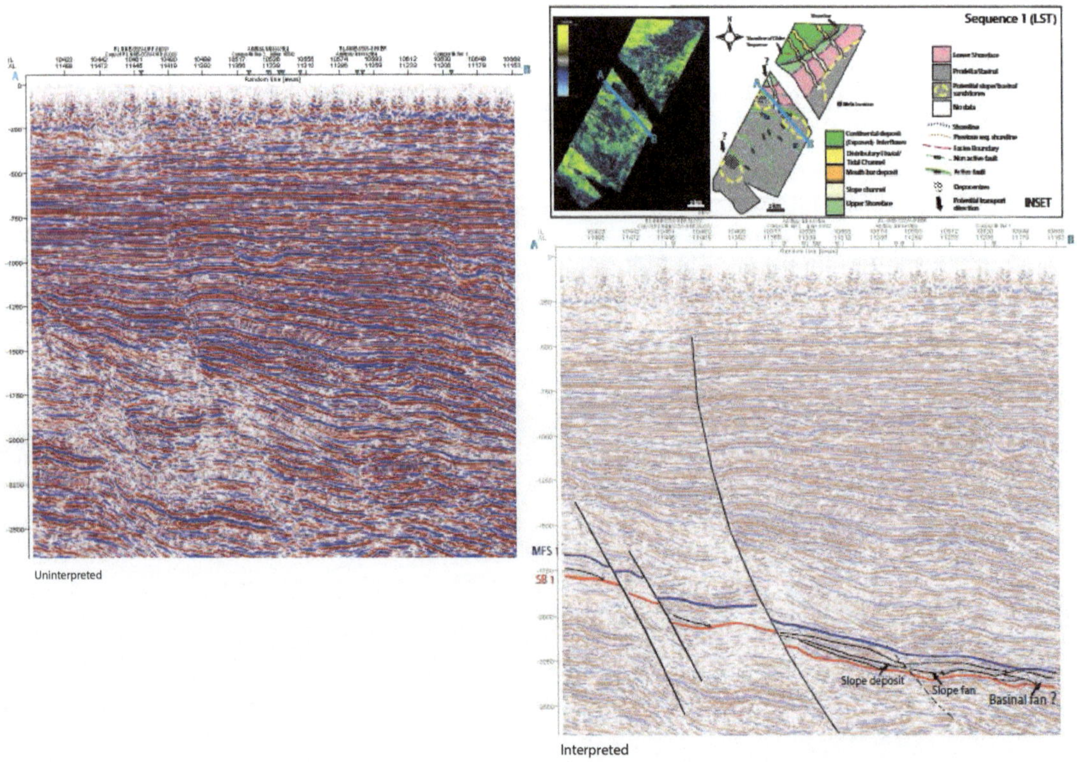

Appendix 6F: Random line showing the geometry, reflection patterns and position of slope and basinal fans within LST Depositional Sequence 1-(Lower Patches). See position in inset above.

Figure 16: Random line showing the geometry, reflection patterns and position of slope and basinal fans within LST Depositional Sequence 1-(Lower Patches). See position inset above.

recognized as high amplitude seismic facies on the slope and on the basin floor.

Idealized schematic palaeogeography and general conceptual depositional models of the study area during middle Miocene time are presented in Figures 7 and 10.

References

1. Durogbitan AA (2010) Seismic Sequence Stratigraphic and Structural Analysis of Ewan and Oloye Fields J marine Sci Res Devop 6: 197.
2. Allen JRL (1965) Late Quaternary Niger delta, and adjacent areas-sedimentary environments and lithofacies: AAPG Bull 49: 547-600.
3. Doust H, Omatsola E (1990) Niger Delta in JD Edwards and PA Santogrossi (eds) Divergent/Passive Margin Basin AAPG 48: 201-238.
4. Northwestern Niger Delta (Middle Miocene). PhD Thesis unpubl. Earth Science Department, The University of Manchester p: 337.
5. Oomkens E (1974) Lithofacies relations in the Late Quaternary Niger Delta complex. Sedimentology 21: 195-222.
6. Whiteman AJ (1982) Nigeria: Its petroleum geology, resources and potential. London, Graham & Totman 1: 166.

21

Standardization of Fishing Effort in Qatar Fisheries: Methodology and Case Studies

Stamatopoulos C[1*] and Abdallah M[2]

[1]Consultant of Fisheries Statistics, Department of Fisheries, Ministry of Environment, Doha, State of Qatar
[2]Consultant of Fisheries Resources, Department of Fisheries, Ministry of Environment, Doha, State of Qatar

Abstract

Regular statistical monitoring of fishing activities is a prerequisite for effective fisheries management. In the case of artisanal fisheries such a monitoring is often exercised by means of sample-based fisheries surveys in which catch and fishing effort (along with other basic variables) are estimated on the basis of samples relating to landings and boat-gear activity. In most cases the fishing fleet is heterogeneous and hence partitioned into boat-gear categories in each of which fishing units have similar characteristics and performance. Under this scheme catch/effort estimates are computed for each boat-gear category separately and independently of each other. It can then be assumed that in each boat-gear category fishing mortality is proportional to the total fishing effort exerted by all of its fishing units operating together. When it comes to measure the combined effect of the fishing operations of the entire fleet to the exploitation of a fish stock, it becomes apparent that adding together effort exerted by different boat-gear categories is not always meaningful without first applying effort adjustment to increase its comparability. There are various techniques for addressing such situations, the commonest of which is known as "standardization of fishing effort". In Qatar the National Fisheries Information System (NFIS) has recently incorporated effort standardization routines that combine elements of the *normalized relative effort* (used by the North Sea Round Fish Working Group, ICES, 1980) with those of *relative fishing power* developed by Robson (1966). The document presents the methodology in use by NFIS for effort standardization as well as case studies using commercial catch/effort data directly obtained from NFIS. It is envisaged that the selected approach will be further refined in order to increase the role of catch/effort data in research and stock assessment applications.

Keywords: Fisheries statistical monitoring; Sample-based surveys; Catch/effort assessment; Standardization of fishing effort

Introduction

In Qatar the fisheries resources are exploited by artisanal fishing units comprising two fishing vessel types: launches (large boats) and speedboats (or tarads). The launch is a decked vessel usually constructed of wood or fiberglass and powered by an in-board engine; the average trip duration is between 3 and 5 days. The speedboat is an open dory usually of fiberglass construction powered by one or two outboard engines. Due to its smaller size the trip duration is usually one day. It also alternates its fishing gear depending on the species sought; this however is known not to occur during the same fishing trip. All fishing units operate from the four ports of Al Shamal, Al Khor, Doha and Al Wakra. There are about 500 launches and 1000 licensed speedboats. Not all fishing units are active during a month; their operational state is variable and the number of active fishing units (a boat is considered active if it has made at least one fishing trip during the month) is enumerated on a monthly basis.

The top 11 species contributing to Qatar fish landings in 2014 were reported by the National Fisheries Information System (NFIS) to be: **Spangled emperor** *Lethrinus nebulosus* (**16.2%** of total landings) (Figure 1), **Narrow-barred Spanish mackerel** *Scomberomorus commerson* (**10.5%**) (Figure 2), **White-spot spinefoot** *Siganus canaliculatus* (**8.3%**), **Pink ear emperor** *Lethrinus lentjan* (**6.6%**), the **Orange spotted grouper** *Epinephelus coioides* (**6.0%**), **Haffara Seabream Rhabdosargus haffara** (**4.3%**), **King soldier bream** *Argyrops spinifer* (**3.7%**), **Painted sweetlips** *Diagramma pictum* (**3.5%**), **Gold toothless trevally** *Gnathanodon speciosus* (**3.4%**), **Orangespotted trevally** *Carangoides bajad* (**3.1%**) and the **Eastern little tuna** *Euthynnus affinis* (**2.2%**).

Monthly and annual landings of statistically monitored species, 57 in all, are systematically reported to and reviewed by the Fisheries Department of the Ministry of Environment, along with other relevant data that are standard part of the statistical database in use.

Until mid-2012 statistical monitoring was limited to catch information collected at the Central Auction Market in Doha. Due to its limitations in data scope and coverage a new system, the internet-supported National Fisheries Information System (NFIS), was implemented in 2012. NFIS is the first of the four principal components of the Qatar Government research project "Sustainable Management of Fisheries Resources" which is being executed by The Ministry of Environment, Department of Fisheries and in close collaboration with the Qatar University Environmental Study Centre, the Qatar Science and Technology Park and The Prince's Charities' International Sustainability Unit of the UK, which provides technical advice whenever it is required. The project's main components comprise: (i) Implementation of a web-based National Fisheries Information System; (ii) Establishment of a Marine Spatial Planning System; (iii) Development and implementation of a fishery management plan based on Maximum Sustainable Yield and, (iv) Development and implementation of a Communication Plan. The web-based NFIS has been regularly operating since September 2012 for the systematic

***Corresponding author:** Stamatopoulos C, Department of Fisheries, Ministry of Environment, State of Qatar, E-mail: cstamat@gmail.com

Figure 1: Spangled emperor (*Lethrinus nebulosus*).

Figure 2: Narrow-barred Spanish mackerel (*Scomberomorus commerson*).

recording of sample data on catch, fishing effort, catch per unit effort, fish size and prices; all such data are collected for each boat-gear category and at all four fishing ports. The data are subsequently fed into an online database accessible by user groups for statistical analyses and reporting. At present NFIS is operating at full capacity and contains data of acceptable accuracy (the threshold of accuracy is 90%) covering the period September 2012 to date. It offers a wide variety of online reports that are supplemented by easy-to-use plotting and tabulating utilities.

Owing to the heterogeneity of the artisanal fishing fleet NFIS has partitioned it into four categories of fishing units of similar characteristics and performance and in a manner that catch/effort estimates are computed separately for each boat-gear category and independently of each other. The four boat-gear categories of NFIS comprise:

1. Launches using traps;
2. Launches using kingfish nets;
3. Launches with miscellaneous gear and;
4. Speedboats (tarads) with miscellaneous gear.

It is generally accepted that when working with a specific boat-gear category (for instance launches with traps) fishing mortality is proportional to the total fishing effort exerted by its fishing units. When it comes to measure the combined effect of fishing operations of the entire fleet to the exploitation of a fish stock, it becomes apparent that adding together effort exerted by different boat-gear categories is not always meaningful without first applying effort adjustment to increase its compatibility. There are various techniques for addressing such situations, the commonest of which is known as "standardization of fishing effort". Maunder [1] gives a more general description of effort standardization as the "the ability to use catch rate data as an index of abundance by removing the impact on catch rates of changes over time of factors other than abundance".

In Qatar the National Fisheries Information System has recently incorporated effort standardization routines that combine elements of the simple (if not very recent) *normalized effort* (used by the North Sea Round Fish Working Group, ICES, 1980) and *relative fishing power* developed by Robson [2].

To be sure the existing literature offers a plethora of other more advanced methods for the standardization of catch and effort data which involve fitting statistical models to the catch and effort data. The first examples of these methods were by Gavaris [3] and Kimura in which General Linear Models (GLM) were used. Moreover the last two decades have seen a proliferation of new methods to standardizing catch and effort data, most of which extend these methods to various degrees. For instance Generalized Additive Models were used by Bigelow et al. and Rodriguez-Marin et al. Generalized Linear Mixed Models extend the GLM approach by allowing some of the parameters in the linear predictor to be treated as random variables. Several analyses of catch and effort data [4] made use of GLMM techniques.

The choice among these methods (an excellent review of which is made by Maunder [2] is based on an evaluation of the underlying assumptions of the models and the type of appropriate statistical tests and diagnostics to be employed. In addition to methodological aspects there are several operational criteria and constraints concerning the type, amount and quality of data to be used [5]. In the case of Qatar it was considered that at the first stages of effort standardization all analyses should be based exclusively on regularly collected catch/effort data from commercial fisheries and that the effort standardization routines should be part of the NFIS report generator [6]. Such being the case the concepts and approaches used by Robson [1] and ICES (1980) seemed to constitute a good and practical basis for developing the presented method.

The need for effort standardization was first pointed out by the Steering Committee of the Sustainable Management of Fisheries Resources project and was followed up by the Fisheries Department of the Ministry of Environment.

Thanks to the collective effort made by field staff and the national experts of the Fisheries Department the presented methodology was repeatedly tested using data of good quality, completeness and accuracy [7,8]. It should also be noted that the present study is only the first step in introducing effort standardization as a regular operational component of NFIS; the approach in use will be further refined when catch/effort data involving more years have been made available [9,10].

The effort standardization approach used by Qatar may be of potential interest to other neighboring countries in the Gulf region which operate similar fleets [11,12]. Effort standardization on a regional basis should not present a major problem if data protocols were set-up permitting comparability of nationally available catch/effort data. Such activities would be part of ongoing regional cooperation and considerably facilitate regional catch/effort assessment for important shared stocks.

Materials and Methods

Primary variables of the study

In this study the fishing effort exerted by a fishing unit during a fishing trip is measured by the duration of the trip and referred to as "boat-gear days". If there are **m** boat-gear categories and the statistical monitoring system produces 12 monthly catch/effort estimates per boat-gear category (as is the case with NFIS) then over a reference period of **n** years there will be (**m x 12n**) monthly effort estimates $\mathbf{E}_{i,j}$, $i=1\ldots\mathbf{m}; j=1\ldots 12\mathbf{n}$.

Along with fishing effort the system estimates monthly catch $\mathbf{C}_{i,j}$ and Catch-Per-Unit-Effort $\mathbf{CPUE}_{i,j}$.

It should be noted here that NFIS treats combined CPUE's as weighted averages and not as simple arithmetic means of their components. For instance to combine monthly CPUE values of the same boat-gear category into a an annual CPUE, the standard NFIS procedure is to re-calculate the monthly catch and effort values involved according to the standard formula $\sum(\mathbf{Catch})/\sum(\mathbf{Effort})$.

Table 1 illustrates an example of a full set of NFIS catch/effort estimates for 2014 which involves the three primary variables described above.

Computational steps in effort standardization

The objective of the presented method is to achieve effort compatibility when different boat-gear categories are combined together. Specifically, its two tasks are:

1. Producing *total standardized effort* of combined boat-gear categories;

2. Computing *standardized CPUE's* for combined boat-gear categories;

It should be noted here that the example given in this section and summarized in Tables 1 and 2 treats catch as a whole and without focusing on a specific fish stock; such a consideration is used only temporarily with the sole purpose of facilitating the presentation of the computational steps in effort standardization. In Section 3 that describes the results of the study readers will be presented with two case studies dealing with **Spangled emperor** (*Lethrinus nebulosus*) and **Narrow-barred Spanish mackerel** (*Scomberomorus commerson*) respectively; these are the two top species of the 2014 landings in Qatar.

The method starts by considering the compatibility of CPUE's of different boat-gear categories. Since these involve incompatible effort values in the denominator they cannot be combined at monthly or annual levels (notice the absent values for effort and CPUE in the totals line in Table 1). This happens since they are viewed as weighted

Table 1: NFIS catch/effort data for 2014 (all species) – Accuracy of estimates: 91.7%.

Catch in Kg	**(01)**	**(02)**	**(03)**	**(04)**	**(05)**	**(06)**	**(07)**	**(08)**	**(09)**	**(10)**	**(11)**	**(12)**	**2014**
Launches with traps	697,000	690,000	810,000	1,099,000	1,009,000	892,000	672,000	674,000	791,000	740,000	728,000	800,000	**9,602,000**
Launches with kingfish net	221,000	212,000	226,000	314,000	249,000	190,000	126,000	224,000	221,000	166,000	189,000	318,000	**2,656,000**
Launches with misc. gear	6,000	4,000	9,000	4,000	4,000	5,000	2,000	5,000	8,000	13,000	11,000	6,000	**77,000**
Speedboats with misc. gear	357,000	283,000	351,000	459,000	296,000	229,000	170,000	214,000	516,000	341,000	267,000	384,000	**3,867,000**
Combined	**1,281,000**	**1,189,000**	**1,396,000**	**1,876,000**	**1,558,000**	**1,316,000**	**970,000**	**1,117,000**	**1,536,000**	**1,260,000**	**1,195,000**	**1,508,000**	**16,202,000**
Effort in boat-gear days	**(01)**	**(02)**	**(03)**	**(04)**	**(05)**	**(06)**	**(07)**	**(08)**	**(09)**	**(10)**	**(11)**	**(12)**	**2014**
Launches with traps	3,168	3,072	3,402	3,005	3,070	3,265	2,897	3,203	3,316	2,961	3,169	3,515	**38,043**
Launches with kingfish net	1,011	1,273	1,678	1,324	1,260	1,339	1,198	1,454	1,309	1,135	1,051	1,238	**15,270**
Launches with misc. gear	159	196	193	108	114	195	108	183	213	384	333	194	**2,380**
Speedboats with misc. gear	4,775	4,544	6,181	5,941	5,580	4,450	2,584	4,276	5,784	4,112	3,082	4,468	**55,777**
Combined	...	...	...	...	...	...	...	...	...	...	...	...	...
CPUE in kg / boat-gear day	**(01)**	**(02)**	**(03)**	**(04)**	**(05)**	**(06)**	**(07)**	**(08)**	**(09)**	**(10)**	**(11)**	**(12)**	**2014**
Launches with traps	220	224.6	238.1	365.7	328.7	273.2	232	210.4	238.5	249.9	229.7	227.6	**252.4**
Launches with kingfish net	218.6	166.5	134.7	237.2	197.6	141.9	105.2	154.1	168.8	146.3	179.8	256.9	**173.9**
Launches with misc. gear	37.7	20.4	46.6	37	35.1	25.6	18.5	27.3	37.6	33.9	33	30.9	**32.4**
Speedboats with misc. gear	74.8	62.3	56.8	77.3	53	51.5	65.8	50	89.2	82.9	86.6	85.9	**69.3**
Combined	...	...	...	...	...	...	...	...	...	...	...	...	...

Table 2: Standardization of the NFIS catch/effort of Table 1.

Standardization factors	(01)	(02)	(03)	(04)	(05)	(06)	(07)	(08)	(09)	(10)	(11)	(12)	**2014**
Launches with traps	1.909	1.909	1.909	1.909	1.909	1.909	1.909	1.909	1.909	1.909	1.909	1.909	**1.909**
Launches with kingfish net	1.324	1.324	1.324	1.324	1.324	1.324	1.324	1.324	1.324	1.324	1.324	1.324	**1.324**
Launches with misc. gear	0.241	0.241	0.241	0.241	0.241	0.241	0.241	0.241	0.241	0.241	0.241	0.241	**0.241**
Speedboats with misc. gear	0.525	0.525	0.525	0.525	0.525	0.525	0.525	0.525	0.525	0.525	0.525	0.525	**0.525**

Standardized effort	(01)	(02)	(03)	(04)	(05)	(06)	(07)	(08)	(09)	(10)	(11)	(12)	**2014**
Launches with traps	6,048	5,865	6,495	5,737	5,861	6,234	5,531	6,115	6,331	5,653	6,050	6,711	72,633
Launches with kingfish net	1,339	1,686	2,222	1,753	1,669	1,773	1,586	1,925	1,733	1,503	1,392	1,639	20,221
Launches with misc. gear	38	47	47	26	27	47	26	44	51	93	80	47	574
Speedboats with misc. gear	2,509	2,387	3,247	3,121	2,932	2,338	1,358	2,247	3,039	2,160	1,619	2,347	29,305
Combined	**9,934**	**9,986**	**12,011**	**10,638**	**10,489**	**10,392**	**8,501**	**10,331**	**11,155**	**9,409**	**9,142**	**10,745**	**122,733**
Normalized	**0.971**	**0.976**	**1.174**	**1.040**	**1.026**	**1.016**	**0.831**	**1.010**	**1.091**	**0.920**	**0.894**	**1.051**	

Standardized CPUE's	(01)	(02)	(03)	(04)	(05)	(06)	(07)	(08)	(09)	(10)	(11)	(12)	**2014**
Launches with traps	115.2	117.6	124.7	191.6	172.1	143.1	121.5	110.2	124.9	130.9	120.3	119.2	**132.2**
Launches with kingfish net	165.1	125.8	101.7	179.1	149.2	107.2	79.4	116.3	127.5	110.4	135.8	194.0	**131.3**
Launches with misc. gear	156.5	84.6	193.4	153.6	145.5	106.3	76.8	113.3	155.8	140.4	137.0	128.3	**134.2**
Speedboats with misc. gear	142.3	118.5	108.1	147.1	101.0	97.9	125.2	95.3	169.8	157.8	164.9	163.6	**132.0**
Combined	**128.9**	**119.1**	**116.2**	**176.4**	**148.5**	**126.6**	**114.1**	**108.1**	**137.7**	**133.9**	**130.7**	**140.3**	**132.0**
Normalized	**0.979**	**0.904**	**0.882**	**1.339**	**1.128**	**0.961**	**0.866**	**0.821**	**1.045**	**1.017**	**0.992**	**1.065**	

averages over a period of a month or a year.

On the other hand each of these CPUE's could be temporarily viewed as the representative catch by just *one* boat from each boat-gear category during *one* day.

Using this second concept for monthly CPUE's by boat-gear category and over 12**n** periods, a 2-dimensional array of *daily* yields $\mathbf{P_{i,j}}$ can be formed where:

i=1…**m** (boat-gear categories);

j=1…12**n** (monthly estimates).

To be noted that the notation has changed from **CPUE** to **P** since a **CPUE** is expressed in Kg / boat-gear day while the newly assumed daily yields **P** are in Kg.

The method proceeds with the following notations and computations:

The sum of all daily yields is given by:

$$P = \sum_{i=1}^{m} \sum_{j=1}^{12n} P_{i,j} \tag{1}$$

The arithmetic mean of all daily yields is given by:

$$\overline{P} = \frac{1}{(mx12n)} P \tag{2}$$

Working with a boat-gear **i** it is found that its total daily yield is:

$$P_i = \sum_{j=1}^{12n} P_{i,j} \tag{3}$$

and the arithmetic mean is:

$$\overline{P_i} = \frac{1}{12n} P_i \tag{4}$$

The overall arithmetic mean $\overline{\mathbf{P}}$ shown in (2) is now assumed to represent the overall daily yield of a new (and hypothetical) boat-gear category. To compare the overall performance of each actual boat-gear to the new hypothetical one the following ratio is used:

$$f_i = \frac{\overline{P_i}}{\overline{P}} \tag{5}$$

where $\overline{\mathbf{P_i}}$ and $\overline{\mathbf{P}}$ are obtained from (4) and (2) respectively.

In the presented study this ratio is referred to as *standardization factor* since it is used for converting actual effort into a standardized one. Once calculated, the standardization factor $\mathbf{f_i}$ is considered to remain the same across all periods. Consequently each effort cell $\mathbf{E_{i,j}}$ representing effort of boat-gear **i** in period **j** can be converted to standardized effort using the expression:

$$E_{i,j}^{STD} = f_i E_{i,j} \qquad \text{i=1…}\mathbf{m}\text{; j=1…12}\mathbf{n}. \tag{6}$$

Adding up all **m** standardized (thus addable) monthly effort values for a period **j** will result in a monthly standardized effort $\mathbf{E_j^{STD}}$ which combines all boat-gear categories:

$$\mathbf{E_j^{STD}} = \sum_{i=1}^{m} \mathbf{E_{i,j}^{STD}} \qquad j=1...12\mathbf{n}. \tag{7}$$

The standardized CPUE's by boat-gear category are obtained by dividing each catch cell $\mathbf{C_{i,j}}$ by the corresponding standardized effort $\mathbf{E_{i,j}^{STD}}$ obtained from (6):

$$CPUE_{i,j}^{STD} = \frac{C_{i,j}}{E_{i,j}^{STD}} \qquad i=1...\mathbf{m};\ j=1...12\mathbf{n}. \tag{8}$$

Lastly the combined standardized catch-per-unit-effort effort in a period **j** is calculated. Here the combined monthly catch of all boat-gear categories is divided by the combined monthly standardized effort obtained from (7).

$$CPUE_j^{STD} = \frac{\sum_{i=1}^{m} C_{i,j}}{E_j^{STD}} \qquad \text{Combined standardized}$$

Catch-Per-Unit-Effort: j=1...12**n**. (9)

At this stage tasks (a) and (b) that was set-up at the beginning of this section have been achieved.

Consistency issues and need for normalization

Two points arise now regarding:

1. consistency of standardized data and
2. their numerical treatment across different periods.

It is evident that the standardization factors formulated by the presented approach depend directly on the selection of a hypothetical boat-gear category to be used as standard. According to Robson [2] the role of such a standard can also be played by any of the actual boat-gear categories, which would result in a different but equally valid set of standardization factors [13,14]. Given that in studying the fluctuation and trend of standardized variables users require consistent sets of data, it becomes apparent that the standardization effort and CPUE values so far obtained need additional treatment in order to become independent of the initial selection of a boat-gear as standard. One way of achieving this is to adopt the normalization approach that was used by the ICES North Sea Round Fish Working Group (1980). The approach consists of (i) calculating the arithmetic mean of a standardized variable across periods and, (ii) substituting each standardized value by its proportion to the mean. In such a manner the resulting normalized values are dimensionless and share a similar value scale [15,16].

It remains to be seen if such normalized values are independent of the choice of a boat-gear category as standard. This is rather easy to prove without performing tedious computations. Suffice to notice that all expressions involving standardized effort contain two factors: one which is the quotient $1/\overline{\mathbf{P}}$ and another that is independent of $\overline{\mathbf{P}}$ and depends only on the original data. Consider for instance expression (7) which computes the combined standardized effort for a given period **j**. By recalling that each $\mathbf{E_{i,j}^{STD}} = \mathbf{f_i E_{i,j}}$ and that $f_i = \frac{\overline{P_i}}{\overline{P}}$, this expression can also be written as:

$$E_j^{STD} = \sum_{i=1}^{m} E_{i,j}^{STD} = \sum_{i=1}^{m} f_i E_{i,j} = \frac{1}{\overline{P}} \sum_{i=1}^{m} \overline{P}_i E_{i,j} \tag{10}$$

When the combined monthly standardized effort is summed across periods, its arithmetic mean will also contain $\frac{1}{\overline{\mathbf{P}}}$. During normalization each standardized effort from (10) will be divided by the arithmetic mean thus canceling out $\frac{1}{\overline{\mathbf{P}}}$ and making the obtained normalized effort independent of the initial choice of a boat-gear category as standard.

Working in a similar manner with the standardized CPUE's we find that their sums and arithmetic means contain an expression of $\overline{\mathbf{P}}$ and other expressions that are independent of it. During the normalization process the expressions of $\overline{\mathbf{P}}$ cancel out thus proving that the normalized CPUE's are independent of the initial choice of a boat-gear category as standard.

Numerical example

Table 2 shows the results of the standardization approach suggested by this study after it has applied to the NFIS catch/effort data of Table 1.

Here the standardization involves **m**=4 boat-gear categories and 12 catch/effort monthly estimates resulting a total of 48 CPUE's. It is recalled that during the standardization phase the notation of these CPUE's will temporarily be changed to **P** since they will be viewed as representing daily catches. Accordingly their units will be in Kg.

Calculation of standardization factors

First the sum of all 48 daily yields (12 yields for each of the 4 boat-gear categories) is calculated.

$$P = \sum_{i=1}^{m=4} \sum_{j=1}^{12} P_{i,j} = 6{,}366 \text{ Kg}$$

The corresponding arithmetic mean $\overline{\mathbf{P}}$ in (2) will be equal to 6,366/48 = 132.6 Kg.

Next step is the calculation of average daily yields for each boat-gear category using expressions (3) and (4).

Launches with traps: $\mathbf{P_1} = 3{,}038.5$ Kg and $\overline{\mathbf{P}}_1 = 253.2$ Kg.

Launches with kingfish net: $\mathbf{P_2} = 2{,}107.5$ Kg and $\overline{\mathbf{P}}_2 = 175.6$ Kg.

Launches with misc. gear: $\mathbf{P_3} = 383.8$ Kg and $\overline{\mathbf{P}}_3 = 32.0$ Kg.

Speedboats with misc. gear: $\mathbf{P_4} = 836.2$ Kg and $\overline{\mathbf{P}}_4 = 69.7$ Kg.

Calculation of standardization factors (STD) makes use of expression (5). Each of the above averages is divided by $\overline{\mathbf{P}} = \mathbf{132.6}$ calculated earlier:

STD factor for launches with traps = 253.2/132.6 = 1.909.

STD factor for launches with kingfish net= 175.6/132.6 = 1.324.

STD factor for launches with misc. gear = 32.0/132.6 = 0.241.

STD factor for speedboats with misc. gear = 69.7/132.6 = 0.525.

These results are shown in the first block of Table 2. To be noted that once these factors have been calculated they apply to all 12 monthly columns of 2014.

Calculation of standardized effort:

The second block of Table 2 illustrates standardized effort for each of the four boat-gear categories. All standardized effort figures by boat-gear category are resulting from the application of expression (6) to all effort cells in Table 1. For instance in January 2014 the actual effort of launches with traps is 3,168 boat-gear days. The standardization factor for this boat-gear category is 1.909. By multiplying the 3,168 actual boat-gear days by this factor we obtain a standardized effort of 6,048 boat-gear days (first cell of the second block in Table 2).

To be noted that since all standardized effort values are addable it is now possible to combine them vertically across boat-gear categories and then horizontally across months, thus obtaining a total effort figure for 2014 equal to 122,733 standardized boat-gear days.

Next line shows combined standardized effort in normalized form. The arithmetic mean of the 12 effort figures is 10,228 boat-gear days. The normalized value of the first entry is 9,934/10,228 = 0.971.The rest of the normalized effort values are calculated likewise.

Calculation of standardized CPUE's:

The third block of Table 2 illustrates standardized CPUE's for each of the four boat-gear categories. All figures are resulting from the application of expression (8) to each CPUE cell in Table 1. For instance in January 2014 the standardized CPUE for launches with traps will be 697,000 Kg of catch (first cell in Table 1) divided by the corresponding standardized effort of 6,048 boat-gear days, which gives 115.2 Kg/boat-gear day.

A combined standardized CPUE is also computed using expression (9). Here the total catch for January 2014 is 1,281,000 Kg and the combined standardized effort is 9,934 boat-gear days, thus resulting a combined standardized CPUE of 128.9 Kg/boat-gear day.

Next line shows combined standardized CPUE in normalized form. The arithmetic mean of the 12 combined CPUE's figures is 131.7. The normalized value of the first entry is 128.9/131.7=0.979. The rest of the normalized effort values are calculated likewise.

To be noted that the notation for catch-per-unit-effort has returned back to **CPUE** since this variable is again calculated as a weighted average of catch divided by effort.

Figure 3 illustrates a plot of the normalized effort and CPUE contained in Table 2.

Results

Application of effort standardization to the fishery of Spangled emperor (*Lethrinus nebulosus*)

As already mentioned in Introduction **Spangled emperor** (*Lethrinus nebulosus*) was the top species in 2014 with landings representing 16.2% of the total.

This species is targeted by launches with traps and speedboats (tarads). Catches by the other two boat-gear categories are negligible and regarded as accidental (Figure 2). Consequently effort standardization focuses on the above two boat-gear categories. Launches with traps are the predominant boat-gear accounting for 76% of the species catches in 2013 and 71% in 2014.

Table 3 illustrates catch/effort data for 2013 and 2014. Since the effort exerted by the two boat-gear categories is not compatible no combined data are shown for effort and CPUE's in the last two columns.

Table 4 shows the results of the standardization process, including normalized values for effort and CPUE.

Figure 4 illustrates monthly plots of normalized effort and CPUE. There is a slight (but visible) rising trend for fishing effort and a declining one for the CPUE.

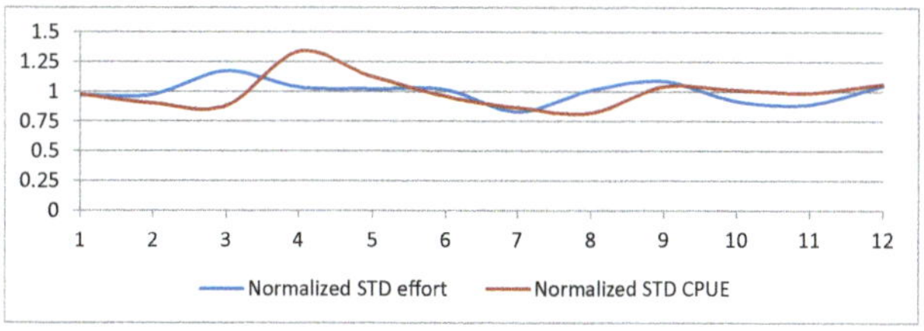

Figure 3: Plot of normalized effort and CPUE based on the 2014 NFIS catch/effort data.

Application of effort standardization to the fishery of Narrow-barred Spanish mackerel (*Scomberomorus commerson)*

This important species (second in the 2014 ranked landings and representing 10.5% of the total) is targeted by launches with kingfish net and speedboats (tarads). Catches by launches with miscellaneous gear are negligible and are not included in the case study. Launches with kingfish net are by far the predominant boat-gear accounting for 90% of the species catches in 2013 and 95% in 2014.

Table 5 illustrates catch/effort data for 2013 and 2014. Since the effort exerted by the two boat-gear categories is not compatible no combined data are shown for effort and CPUE's in the last two columns.

Table 6 shows the results of the standardization process, including normalized values for effort and CPUE.

Figure 5 illustrates monthly plots of normalized effort and CPUE. There is a slight (but visible) declining trend for both fishing effort and the CPUE.

Discussion

Comparison to other methods

As mentioned in the Introduction the National Fisheries Information System (NFIS) has recently adopted the presented approach that combines elements of the *normalized relative effort* (used by the North Sea Round Fish Working Group, ICES, 1980) and the *relative fishing power* developed by Robson [2]. It was also mentioned that although the existing literature offers a plethora of other more recent and more sophisticated methods it was nevertheless considered preferable to first try out approaches that (a) depend only on catch/effort data from commercial fisheries and, (b) are applicable to situations of limited time coverage.

The Robson basic concept of relative fishing power was adopted in formulating effort standardization factors as shown in expressions (1) - (5). The presented study uses a variation to the Robson concept; instead of arbitrarily selecting an existing CPUE to use as standard it uses for this purpose a mean daily yield of one fishing unit of a hypothetical boat-gear category. This variation does not constitute a real difference since Robson states that in choosing a CPUE standard "any boat-gear is as good as another". It is the authors' view, however, that involving all boat-gear categories in the source data makes the selection of the CPUE standard less arbitrary.

On the other hand the fact remains that users should be free to use any standard that would be appropriate or convenient for their work. This means that several standardized datasets, all equally valid but different from each other, might be resulting from the same source data. To overcome this problem the presented method further processes

Table 3: Catch/effort data for Sh'ari Lethrinus nebulosus (Spangled emperor) (2013 – 2014). Accuracy of estimates: 90.6%.

	Launches with traps			Speedboats			Combined		
Period	**Catch**	**Effort**	**CPUE**	**Catch**	**Effort**	**CPUE**	**Catch**	**Effort**	**CPUE**
1	111,130	4,561	24.37	61,683	2,594	23.78	172,813	...	...
2	98,669	3,911	25.23	69,040	2,485	27.79	167,709	...	...
3	118,539	2,625	45.17	68,689	2,313	29.70	187,228	...	...
4	382,464	3,475	110.06	57,079	3,120	18.29	439,543	...	...
5	272,046	2,790	97.50	62,590	4,086	15.32	334,636	...	...
6	190,795	2,524	75.60	39,851	1,705	23.38	230,646	...	...
7	150,694	2,812	53.59	24,752	1,143	21.65	175,446	...	...
8	129,136	3,010	42.90	12,556	816	15.39	141,692	...	...
9	89,015	2,812	31.66	33,806	2,771	12.20	122,821	...	...
10	115,355	3,267	35.31	53,946	2,769	19.48	169,301	...	...
11	119,403	3,228	36.99	47,075	6,417	7.34	166,478	...	...
12	128,556	2,907	44.22	59,225	3,733	15.87	187,781	...	...
2013	**1,905,802**	**37,922**	**50.26**	**590,294**	**33,952**	**17.39**	**2,496,096**	...	...
1	133,786	3,168	42.23	104,538	2,911	35.92	238,324	...	...
2	148,305	3,072	48.27	82,761	2,842	29.12	231,066	...	...
3	193,881	3,402	56.99	47,560	5,484	8.67	241,441	...	...
4	298,826	3,005	99.44	111,581	5,941	18.78	410,407	...	...
5	239,525	3,070	78.02	81,146	5,580	14.54	320,671	...	...
6	159,508	3,265	48.86	48,273	4,371	11.04	207,781	...	...
7	89,878	2,897	31.02	31,842	2,584	12.32	121,720	...	...
8	111,765	3,203	34.90	39,273	1,869	21.01	151,038	...	...
9	107,153	3,316	32.31	32,439	3,009	10.78	139,592	...	...
10	106,558	2,961	35.98	40,819	4,112	9.93	147,377	...	...
11	124,312	3,169	39.23	32,808	2,064	15.89	157,120	...	...
12	130,134	3,515	37.03	88,952	4,468	19.91	219,086	...	...
2014	**1,843,630**	**38,044**	**48.46**	**741,993**	**45,236**	**16.40**	**2,585,623**	...	...

Table 4: Standardized effort and CPUE for Sh'ari Lethrinus nebulosus (Spangled emperor) (2013-2014).

	Launches with traps			Speedboats			Combined			
Period	**STD factor**	**STD effort**	**STD CPUE**	**STD factor**	**STD effort**	**STD CPUE**	**STD effort**	**Norma-lized**	**STD CPUE**	**Norma-lized**
1	1.467	6,691	16.61	0.533	1,383	44.61	8,073	1.261	21.41	0.644
2	1.467	5,737	17.19	0.533	1,324	52.13	7,061	1.103	23.75	0.714
3	1.467	3,850	30.78	0.533	1,233	55.72	5,083	0.794	36.83	1.107
4	1.467	5,098	75.01	0.533	1,663	34.32	6,761	1.056	65.01	1.955
5	1.467	4,093	66.45	0.533	2,178	28.74	6,271	0.980	53.36	1.604
6	1.467	3,703	51.52	0.533	909	43.86	4,611	0.720	50.02	1.504
7	1.467	4,125	36.52	0.533	609	40.62	4,734	0.739	37.06	1.114
8	1.467	4,416	29.23	0.533	435	28.87	4,851	0.758	29.21	0.878
9	1.467	4,125	21.57	0.533	1,477	22.89	5,602	0.875	21.92	0.659
10	1.467	4,793	24.06	0.533	1,476	36.55	6,269	0.979	27.01	0.812
11	1.467	4,735	25.21	0.533	3,420	13.76	8,156	1.274	20.41	0.614
12	1.467	4,265	30.13	0.533	1,990	29.77	6,255	0.977	30.02	0.903
2013	**1.467**	**55,631**	**34.25**	**0.533**	**18,097**	**32.62**	**73,728**		**33.86**	
1	1.467	4,648	28.78	0.533	1,551	67.38	6,199	0.968	38.44	1.156
2	1.467	4,507	32.90	0.533	1,515	54.64	6,022	0.941	38.37	1.154
3	1.467	4,991	38.84	0.533	2,923	16.27	7,914	1.236	30.51	0.917
4	1.467	4,408	67.77	0.533	3,167	35.24	7,575	1.183	54.18	1.629
5	1.467	4,504	53.17	0.533	2,974	27.28	7,478	1.168	42.88	1.289
6	1.467	4,789	33.30	0.533	2,330	20.72	7,119	1.112	29.19	0.878
7	1.467	4,250	21.14	0.533	1,377	23.12	5,627	0.879	21.63	0.650
8	1.467	4,699	23.78	0.533	996	39.42	5,695	0.890	26.52	0.797
9	1.467	4,865	22.02	0.533	1,604	20.23	6,469	1.010	21.58	0.649
10	1.467	4,344	24.52	0.533	2,192	18.62	6,536	1.021	22.55	0.678
11	1.467	4,649	26.73	0.533	1,100	29.82	5,749	0.898	27.33	0.822
12	1.467	5,156	25.23	0.533	2,382	37.35	7,538	1.177	29.06	0.874
2014	**1.467**	**55,811**	**33.03**	**0.533**	**24,111**	**30.77**	**79,922**		**32.35**	

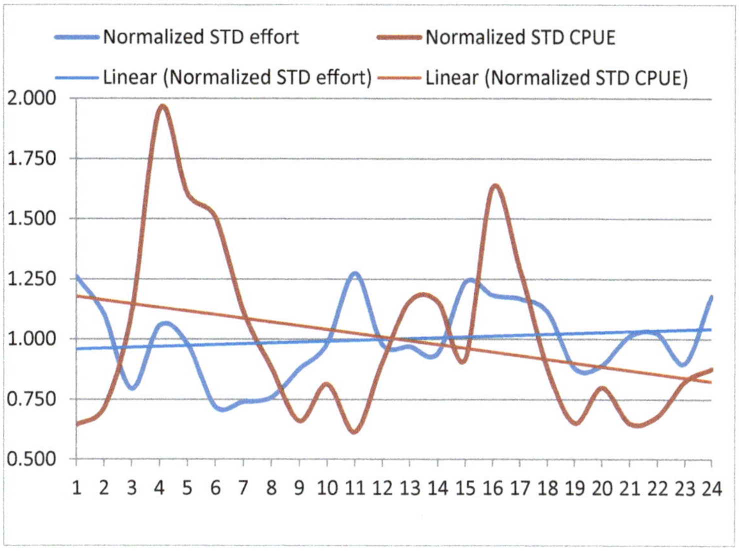

Figure 4: Monthly plots of normalized effort and CPUE for Sh'ari شعري Lethrinus nebulosus (Spangled emperor) (2013-2014). There is a slight (but visible) rising trend for fishing effort and a declining one for the CPUE.

Table 5: Catch/effort data for Narrow-barred Spanish mackerel (*Scomberomorus commerson*) (2013-2014). Accuracy of estimates: 88.3%.

Period	Launches with kingfish net			Speedboats			Combined		
1	Catch	Effort	CPUE	Catch	Effort	CPUE	Catch	Effort	CPUE
2	258,119	1,874	137.77	10,157	508	20.00	268,276	...	...
3	133,628	2,213	60.39	0	0	0.00	133,628	...	...
4	199,616	2,138	93.36	10,770	950	11.34	210,386	...	...
5	118,478	1,261	93.93	90,642	1,784	50.81	209,120	...	...
6	130,612	1,285	101.68	48,423	823	58.81	179,035	...	...
7	94,031	948	99.24	8,809	896	9.83	102,840	...	...
8	127,899	1,532	83.49	1,903	257	7.40	129,802	...	...
9	175,797	1,644	106.93	0	0	0.00	175,797	...	...
10	188,768	1,972	95.73	2,536	1,700	1.49	191,304	...	...
11	287,010	2,015	142.40	324	1,573	0.21	287,334	...	...
12	114,090	1,565	72.92	30,691	6,579	4.67	144,781	...	...
2013	112,053	1,237	90.56	22,743	2,368	9.61	134,796	...	...
1	**1,940,103**	**19,683**	**98.57**	**226,998**	**17,438**	**13.02**	**2,167,101**	...	...
2	97,925	1,011	96.85	8,168	2,911	2.81	106,093	...	...
3	117,490	1,273	92.30	0	0	0.00	117,490	...	...
4	125,140	1,678	74.56	6,281	2,250	2.79	131,421	...	...
5	255,711	1,324	193.14	11,835	5,032	2.35	267,546	...	...
6	147,988	1,260	117.44	5,351	4,786	1.12	153,339	...	...
7	102,938	1,339	76.90	1,044	1,299	0.80	103,982	...	...
8	64,857	1,198	54.16	699	68	10.25	65,556	...	...
9	112,760	1,454	77.54	1,277	1,240	1.03	114,037	...	...
10	131,961	1,309	100.82	9,974	2,285	4.37	141,935	...	...
11	113,214	1,135	99.75	20,350	2,379	8.55	133,564	...	...
12	144,751	1,051	137.69	10,800	2,064	5.23	155,551	...	...
2014	200,665	1,238	162.07	193	2,010	0.10	200,858	...	...
	1,615,400	**15,270**	**105.79**	**75,972**	**26,324**	**2.89**	**1,691,372**	...	...

the standardized data with the objective of making them consistent irrespective of the initial choice of a CPUE as standard. It was shown that such an objective can be achieved by means of a normalization process such as the one adopted by the North Sea Round Fish Working Group, ICES (1980).

Lastly the presented method follows the same concept of dynamic standardization shown in both ICES and Robson approaches. Monthly and annual standardization factors (and hence normalized effort and CPUE's) vary when the source data cover different numbers of years. For instance, launches with kingfish net have a standardization factor of 1.840 over the period January 2013-December 2014. This value

Table 6: Standardized effort and CPUE for Narrow-barred Spanish mackerel (*Scomberomorus commerson*) (2013-2014).

Period	Launches with kingfish net			Speedboats			Combined			
	STD	STD	STD	STD	STD	STD	STD	Norma-	STD	Norma-
1	factor	effort	CPUE	factor	effort	CPUE	effort	lized	CPUE	lized
2	1.840	3,448	74.86	0.160	81	125.27	3,529	1.188	76.02	1.398
3	1.840	4,072	32.81	0.160	0	0.00	4,072	1.370	32.81	0.603
4	1.840	3,935	50.73	0.160	152	70.99	4,087	1.375	51.48	0.947
5	1.840	2,321	51.04	0.160	285	318.27	2,606	0.877	80.24	1.476
6	1.840	2,364	55.25	0.160	131	368.37	2,495	0.840	71.74	1.319
7	1.840	1,744	53.93	0.160	143	61.55	1,887	0.635	54.50	1.002
8	1.840	2,819	45.37	0.160	41	46.35	2,860	0.963	45.38	0.835
9	1.840	3,026	58.10	0.160	0	0.00	3,026	1.018	58.10	1.068
10	1.840	3,629	52.02	0.160	271	9.34	3,900	1.313	49.05	0.902
11	1.840	3,709	77.38	0.160	251	1.29	3,960	1.333	72.55	1.334
12	1.840	2,879	39.62	0.160	1,050	29.22	3,930	1.323	36.84	0.677
2013	1.840	2,277	49.21	0.160	378	60.17	2,655	0.894	50.77	0.934
1	**1.840**	**36,224**	**53.56**	**0.160**	**2,784**	**81.53**	**39,008**		**55.56**	
2	1.840	1,861	52.62	0.160	465	17.57	2,326	0.783	45.62	0.839
3	1.840	2,343	50.15	0.160	0	0.00	2,343	0.788	50.15	0.922
4	1.840	3,089	40.51	0.160	359	17.49	3,448	1.160	38.12	0.701
5	1.840	2,437	104.95	0.160	803	14.73	3,240	1.090	82.58	1.518
6	1.840	2,319	63.81	0.160	764	7.00	3,083	1.038	49.73	0.915
7	1.840	2,464	41.78	0.160	207	5.04	2,671	0.899	38.93	0.716
8	1.840	2,204	29.43	0.160	11	64.20	2,215	0.745	29.60	0.544
9	1.840	2,676	42.13	0.160	198	6.45	2,874	0.967	39.67	0.730
10	1.840	2,409	54.78	0.160	365	27.34	2,774	0.933	51.17	0.941
11	1.840	2,089	54.20	0.160	380	53.57	2,469	0.831	54.10	0.995
12	1.840	1,935	74.82	0.160	330	32.77	2,264	0.762	68.70	1.263
2014	1.840	2,279	88.07	0.160	321	0.60	2,600	0.875	77.27	1.421
	1.840	**28,103**	**57.48**	**0.160**	**4,203**	**18.08**	**32,306**		**52.36**	

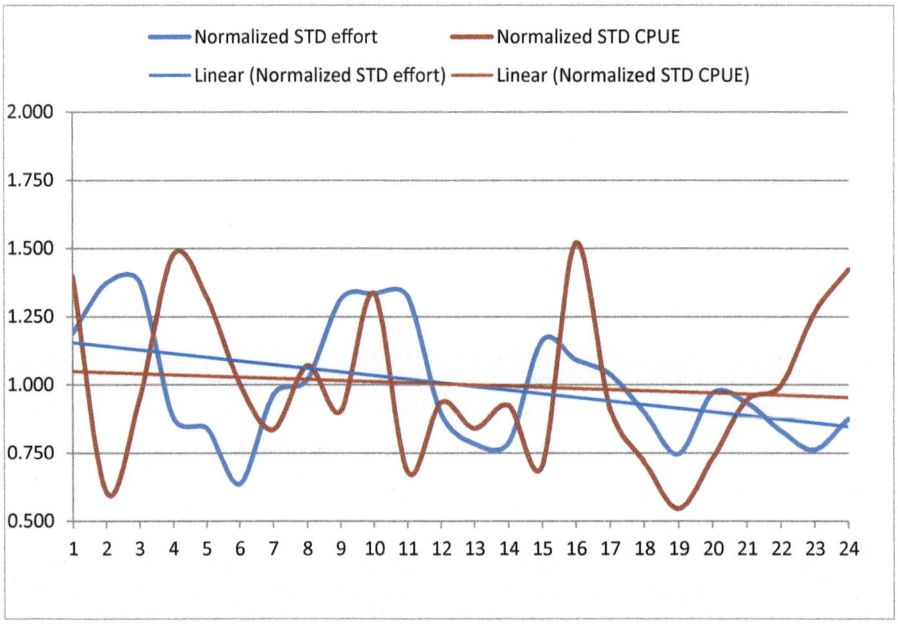

Figure 5: Monthly plots of normalized effort and CPUE for Narrow-barred Spanish mackerel (*Scomberomorus commerson*) (2013-2014). There is a slight (but visible) declining trend for both fishing effort and the CPUE

will be different when the source data will extend to December 2015, December 2016, etc. Such a consideration is essential in order for the standardized variables to be compatible across all periods, a criterion that would not be met if standardization was to apply for each year separately.

Equivalent approaches for the formulation of standardization factors

Expression (5) in Section 2 specifies that the standardization factor for a specific boat-gear category is directly defined as the ratio of its overall CPUE (viewed temporarily as the average daily yield $\overline{P_i}$ of a single fishing unit over all periods) to the average daily yield $\overline{P}$ of a hypothetical boat-gear. It is recalled that $\overline{P_i}$ is obtained from expression (4) and $\overline{P}$ from expressions (1) and (2).

The chosen approach however does not preclude the adoption of other hypotheses which can produce the same results by means of different interpretations of the CPUE's. For instance an alternative approach is to formulate standardization factors on the basis of *days needed catching the same arbitrary quantity* Q. Under such a scheme the days needed for each boat-gear to catch Q will be $Q/\overline{P_i}$. Next a hypothetical boat-gear category with catch-per-unit effort equal to $\overline{P}$ is considered. Here the number of days needed to catch Q is equal to $Q/\overline{P}$. Since the number of days needed is in reverse proportion to the relative importance of a boat-gear (i.e., higher performance implies fewer days to catch a given quantity Q) we divide the second ratio by the first, thus obtaining the same standardization factor $f_i = \frac{\overline{P_i}}{\overline{P}}$.

The problem of data gaps

Maunder [1] has stressed the importance of paying due attention to situations in which there are data gaps in the datasets. The remedies are not always simple and in some cases they become quite elaborate.

It is the authors' view that the problem of data gaps does not affect the presented method since the standardization factors are calculated on the basis of cumulative daily yields covering the entire reference period. It was shown that the standardization process applies to a matrix of source data (as shown in Tables 1,3 and 5) in which data cells may as well contain zeroes (for instance the speedboats in February and August 2013 and in February 2014). In mathematical terms the only condition for a boat-gear category to participate in the process is to have at least one non-zero entry in the matrix. In practice, however, boat-gear categories showing small and scattered quantities of accidental catch are not included in the process as was for instance the case or launches with miscellaneous gear catching kingfish.

Another point worth addressing is the reliability of catch/effort estimates that constitute the data source for the standardization process.

In Qatar the NFIS catch/effort are collected in conformance to strict norms concerning sample size and frequency of sampling. Raw data go through a gauntlet of various quality checks before they are processed and the resulting estimates are subject to quality checks relating to accuracy. The aim of such rigorous monitoring is to achieve a compound accuracy of catch/effort estimates that stays above 90%; this has been consistently achieved from 2014 onwards. In Qatar use is made of the "pessimistic" accuracy concept in which the resulting accuracy stays above a pre-set lower limit [17-19]. It is also a composite index incorporating a spatial accuracy (a function of sample size) and a temporal accuracy (that depends on sampling frequency). In addition to the above two relative indices of accuracy the Sampling Uniformity Index (SUI) monitors the uniformity of samples over the sampling days and it penalizes the temporal accuracy in cases of uneven concentrations of samples favouring certain sampling days.

References

1. Maunder MN, Langley AD (2004) Integrating the standardization of catch-per-unit-of-effort into stock assessment models: testing a population dynamics model and using multiple data types. Fish Res 70: 385-391.
2. Robson DS (1966) Estimation of the relative fishing power of individual ships. ICNAF Res Bull 3: 5-14.
3. Gavaris S (2011) Use of a multiplicative model to estimate catch rate and effort from commercial data. Can J Fish Aquat Sci 3: 2272-2275.
4. Ortiz de Zárate V, Ortiz de Urbina JM (2008) Standardized fishing effort of the Spanish baitboat fleet targeting Albacore (Thunnus alalunga) in the Northeast Atlantic, 1981-2005. Collect Vol Sci Pap ICCAT 62: 816-823.
5. Anticamaraa JA, Watson R, Gelchu A, et al. (2011) Global fishing effort (1950–2010): Trends, gaps, and implications. Fisheries Research 107: 131-136.
6. Campbell RA (2004) CPUE standardization and the construction of indices of stock abundance in a spatially varying fishery using general linear models. Fish Res 70: 209-227.
7. Francis RICC (1999) The impact of correlations in standardized CPUE indices. NZ Fisheries Association Research Document No. 99/42.
8. Gulland JA (1956) On the fishing effort in English demersal fisheries. Fishery Investigations Series II. Ministry Agriculture, Fisheries and Food, London 22: 45.
9. Hansen KE, Ferro RST, Hopper A, et al. (1998) In: Barthel KG et al. (eds) Investigation of the relative fishing effort exerted by towed demersal gears on North Sea consumption species. Proceedings of the third conference of European marine science and technology 5: 148-149.
10. Harley SJ, Myers RA, Dunn A (2001) Is catch-per-unit-effort proportional to abundance? Can J Fish Aquat Sci 58: 1760-1772.
11. Hinton MG, Maunder MN (2003) Methods for standardizing CPUE and how to select among them. ICCAT Document No. SCRS/2003/034.
12. Maunder MN, Starr PJ (2003) Fitting fisheries models to standardized CPUE abundance indices. Fish Res 63: 43-50.
13. Pearse PH (1980) Regulation of fishing effort. FAO Fisheries Technical Paper No. 197 (FIPL/T197). Food and Agriculture Organization of the United Nations, Rome 82.
14. Porter JM, Oritz M, Paul SD (2003) Updated standardized CPUE indices for Canadian bluefin tuna fisheries based on commercial catch rates. ICCAT Col Vol Sci Pap 55: 1005-1018.
15. Quinn TJ, Hoag SH, Southward GM (1982) Comparison of two methods of combining catch-per-unit-effort data from geographic regions. Can J Fish Aquat Sci 39: 837-846.
16. Tidd AN (2013) Effective fishing effort indicators and their application to spatial management of mixed demersal fisheries. Fisheries Management and Ecology.
17. Stamatopoulos C (1999) Observations on the geometrical properties of accuracy growth in sampling with finite populations. FAO Fisheries Technical Paper No. 388.
18. Stamatopoulos C (2002) Sample-based fishery surveys. A technical handbook. FAO Fisheries Technical Paper No. 425.
19. Stamatopoulos C (2003) Safety in sampling. FAO Fisheries Technical Paper No. 454.

Seismic and Sequence Stratigraphic Analysis of Ewan and Oloye Fields (Middle Miocene), Northwestern Niger Delta: Implications for Deltaic Depositional Sequences

Durogbitan A Abimbola*

Petrfac IES, Research and Development, Manchester, UK

Abstract

The main aim of this works is to define local sequence variability and its implications for deltaic depositional sequences within the studied area. This research work makes use of depth converted three dimensional seismic cube, well logs, core photography descriptions and biostratigraphic reports. Recognition and interpretation of stratal surfaces were based on facies shifts and stacking patterns and reflection terminations derived from the interpretation of seismic data and wireline logs integrated with core photography and biostratigraphic reports.

Seismic stratigraphic analysis of well and seismic data suggests that middle Miocene- early Pliocene strata within the studied area were made of six depositional sequences. Sequences are made up aggradational to progradational HST that makes up about 50% of a sequence. TST are thin and widespread while LST are restricted to within WNW-ESE trending valley fills typically 4 km wide and an average 120 m deep. Outside the LST valleys, transgressive surfaces, maximum flooding surfaces and sequence boundaries are coincident on interfluves. This suggests that during LST times most of the area was periodically subaerially exposed and several incised valleys/canyons were developed. These incisions/canyons acted as a sand feeder that led to the deposition of high amplitude deposits (sandstones) on the slope and within the basin floor setting. Sequences mapped over the study area show local variation associated with growth faults.

There is a strong local growth fault control on sequence variability. Further local controls on sequence geometry are shelf instability and slope scars which have a strong geomorphological impact on the area. The depth of incised valleys within the uppermost HST (up to 350 m) is difficult to explain in terms of sea level fall alone because eustatically-driven sea level variations during the Miocene are generally reported to be less than 100 m (330 ft). This suggests that the character of sequence development and depositional facies preserved within the study area depends on relative sea level changes, relative rates of regional structural collapse and sedimentation rate. The occurrence of several incisions in the middle Miocene succession is interpreted as evidence of significant relative sea level fluctuations, and the presence of type-1 sequence boundaries may be the stratigraphic signature of major drops in relative sea level during Miocene and Pliocene. Variability within the depositional sequences is said to be largely controlled by local growth faults, rapid relative sea level changes, basin physiography (shelf edge) and high rate of sediment influx.

Keywords: Miocene; Incisions; Depositional sequences

Introduction

Recent research work has shown that there are variabilities within sequence stratigraphy concept caused by different factors interplay. However, this study documents a seismic and sequence stratigraphic variability framework for growth-faulted delta deposits of the Agbada Formation (Middle Miocene) within the Ewan and Oloye Fields, northwestern Niger Delta (Figure 1). The application of sequence stratigraphic methods to Niger Delta sequences is complex. These complexes are due to changes in relative sea level which are strongly influenced by the interplay between shoreline advance and retreat, concomitant isostatic loading and rebound of the continental shelf growth faulting and basin physiography [1].

Aims and Objectives

The aims and objectives of this are:-

- To identify the key stratal surfaces, the geometry relationship of the surfaces and the development of sequences.
- To analyse relative sea level changes, growth faulting, sediment loading and shelf instability in the study area.
- To define local sequence variability and its implications for deltaic depositional sequences.

Data and Methods

Sequence stratigraphic correlation was based on depth converted three dimensional seismic cube, well logs, core photography descriptions and biostratigraphic reports. The model developed for the study area is based on the recognition of key stratal surfaces, facies shifts and stacking patterns derived from the interpretation of seismic data and wireline logs. The stratigraphic section was sub divided into

***Corresponding author:** Durogbitan A Abimbola, Principal Geoscientist, Petrfac IES, Research and Development, Manchester, UK
E-mail: adewoledurogbitan@yahoo.co.uk

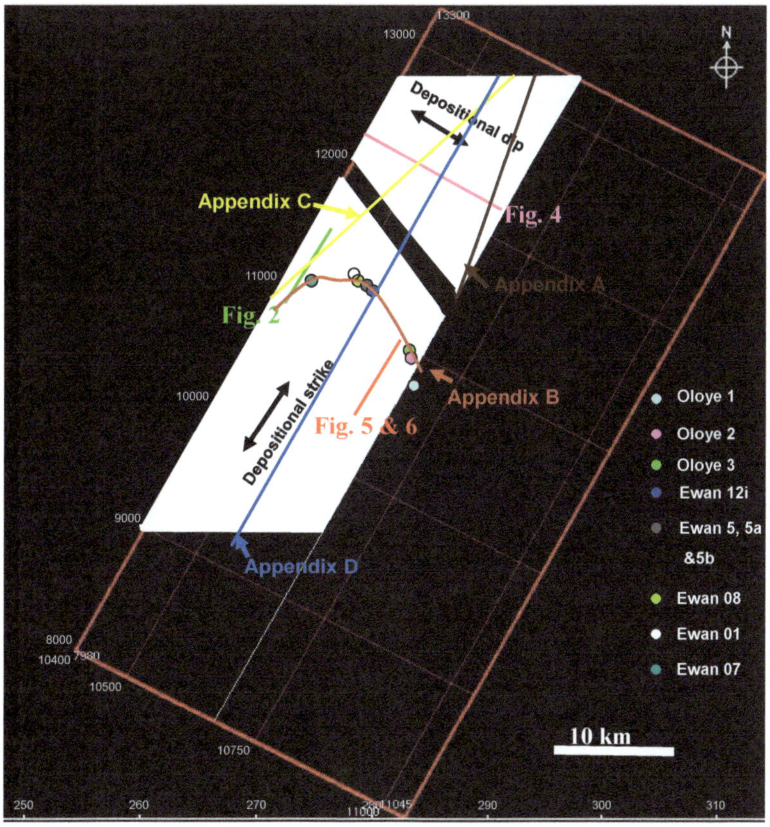

Figure 1: Base map of seismic survey showing position of seismic cross sections and wells. Area under study is enclosed within the white shaded portion [10].

packages of sediments bounded by chronostratigraphically sequence boundaries. Where possible each depositional sequence was further divided into system tracts based on reflection patterns (stratal surfaces) and well log stacking patterns.

Recognition of stratal surfaces was based on facies shifts and stacking patterns and reflection terminations derived from the interpretation of seismic data and wireline logs. In this study, the terminology developed by Exxon [2-7] has been adopted in the definition of key stratal surfaces and the sub division into lithogenetic units (system tracts).

Results

Characteristics of key stratal surfaces and packages

Four key surface types were delineated within the studied area. They are:-Sequence Boundaries, Transgressive, Flooding and Maximum Flooding Surfaces. Transgressive Systems Tracts, Highstand Systems Tracts and Lowstand Systems Tracts (in some case Lowstand Prograding Wedge) were recognised within the interpreted section.

a) **Sequence boundaries (unconformities and correlative conformities):** Sequence boundaries were identified by the geometrical relationships of strata on seismic reflections. The unconformity surfaces is characterised by truncations and toplap below the stratal surface while coastal onlap terminations characterised above the stratal surface (Figure 2) (Appendix A). In some places in the seismic data they are also recognised by the presence of incisions. On well logs, the sequence boundary is identified as a facies dislocation i.e. a sharp or occasionally gradational base overlain by fluvial/tidal channel fills and slope channel sand. The sequence boundary shows the lowest relative neutron-porosity values and the corresponding highest resistivity values as interpreted from well logs section (Figure 3); well Ewan 08-SB2-3). Six sequence boundaries have been identified and interpreted based on the seismic data (Figures 4 and 5). This has been integrated with well data (Appendix B).

b) **Transgressive surfaces:** The Transgressive Surface (TS) is referred to originally as the first significant marine flooding surface across the shelf at the top of lowstand system tract [6,7]. From the seismic data, the transgressive surface marks the tops of most of the incisions/incised valleys. On well log it is marked by the turn around surface from cleaning upward stacking pattern to dirtying up ward stacking patterns (Figure 3, well 08-SB3 underlain by slope channel sand). From this study, it is possible to define the TS based on the stratigraphic relationship with the incised valleys i.e. they commonly cap the incised valleys. Based on stratal geometry the TS is defined as a conformable surface that separates regressive strata (LST) from subsequent transgressive strata (TST) (Figure 4). With a definition based on water deepening, it represents the beginning of marine influence i.e. the first significant flooding surface when the water depth is at its shallowest [7]. On seismic data, outside the lowstand system tract valleys, the transgressive surfaces and sequence boundaries are coincident at interfluves (Figure 5). This indicates that, during the time of sea level fall, the sea level fall reaches the shelf edge before transgression begins. Within this study, transgressive surface is diachronous and reflects the landward migration of the shoreline.

c) **Maximum flooding surfaces:** On the seismic profiles, the surface is recognised as downlap surface (Figure 2 and Appendix A), i.e. the base of prograding clinoforms of the highstand system tract.

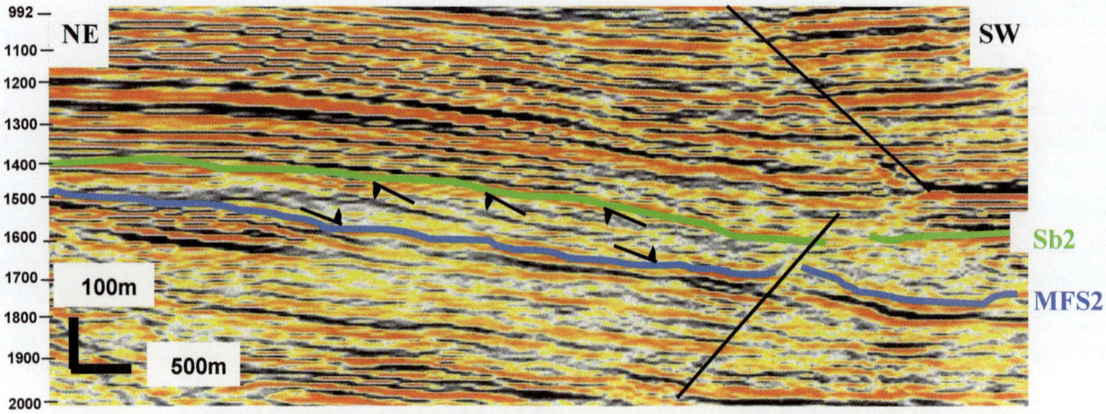

Figure 2: Seismic section (depositional strike line) showing one of the sequence boundaries (Horizon 12- green line) showing toplap and oblique progradational clinoform geometry of sedimentary package. Maximum flooding surface also showing the downlap surface. (Inline 10420). The seismic data are depth converted. See positing in Figure 1.

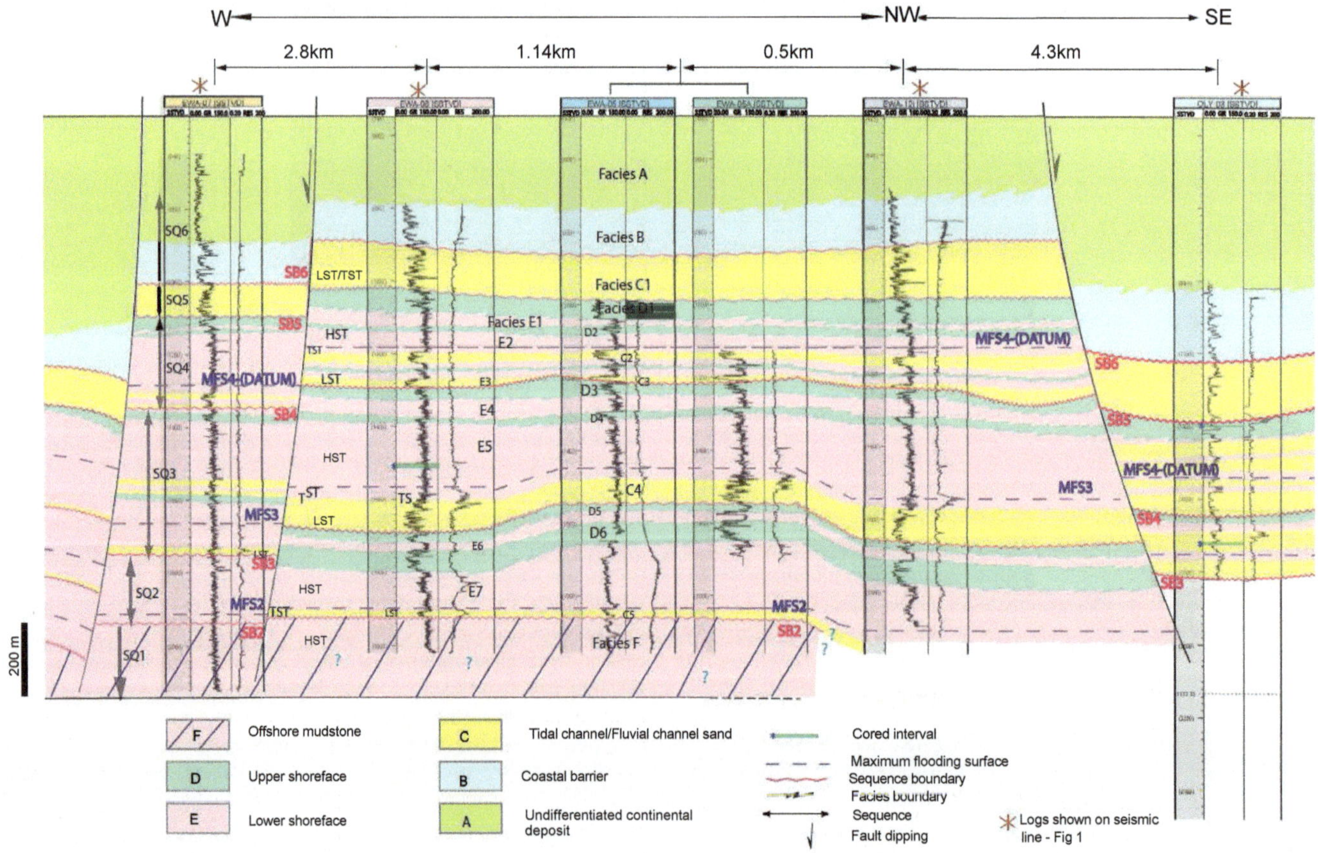

Figure 3: Correlation panel of Ewan and Oloye logs illustrating sequences component, parasequence setd and facies within the middle Miocene Agabda Formation.

Four maximum flooding surfaces have been interpreted on the seismic data capping the erosional valleys and coincide with the sequence boundaries at interfluves (Figure 6). On well logs, recognition of maximum flooding surfaces within the Niger Delta cannot be based on gamma ray pattern alone, as interbedded thin sandstone and shale of shoreface deposits normally exhibit high levels of radioactivity similar to that of marine shale. This is due to high content of mica, potassium feldspar and heavy minerals within the sandstones. On well logs, maximum flooding surface is characterised by the highest gamma ray values within the muddy interval. Using a combination of logs (gamma ray, resistivity), maximum flooding surfaces can be identified by the highest relative neutron-density values and the corresponding lowest resistivity values relative to other surrounding flooding surfaces [8] (Figure 3; well Ewan 08).

d) **Lowstand system tract (LST):** From seismic interpretation LST is characterised by prograding to aggrading reflection patterns bounded at the base by erosional surfaces (sequence boundaries)

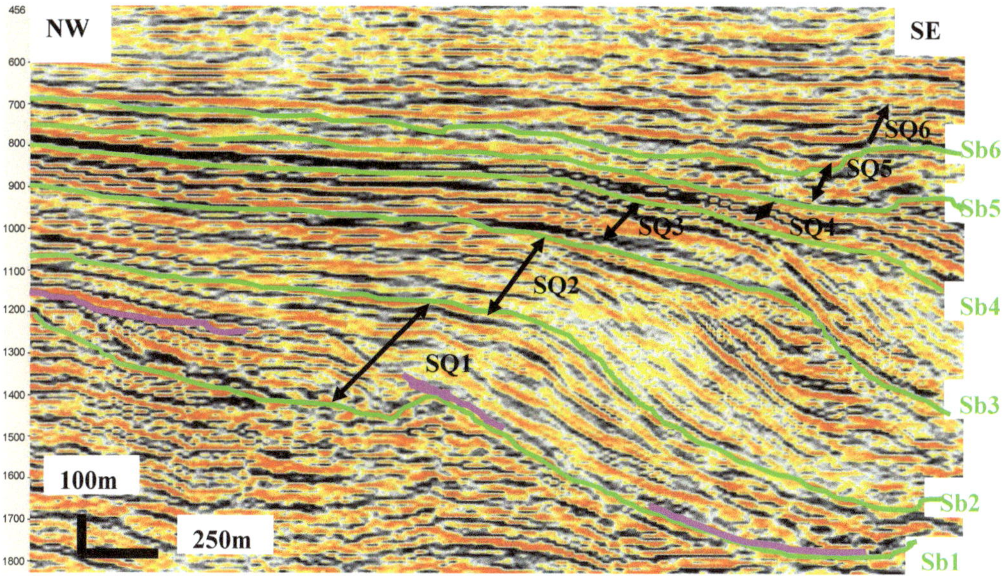

Figure 4: Seismic section showing identified sequence boundaries and transgressive surface (Crossline 12291- upper patches). Transgressive surfaces are denoted by light purple dashed line. Generally shown progradation stacking pattern, most of the incisions occurred at the shelf (s). See position in Figure 1.

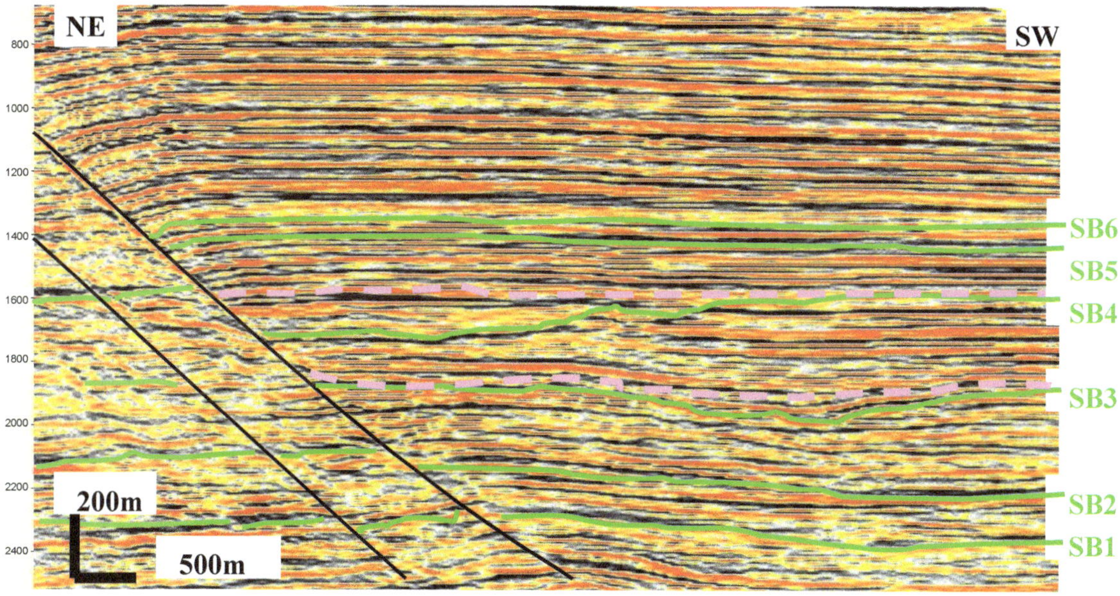

Figure 5: Seismic section showing interpreted sequence boundaries and the transgressive surfaces (inline 10620 - lower patches). Transgressive surfaces are denoted by light purple dashed line. The reflecting patterns at the lower part are more aggradational compared with the northern part. See position in Figure 1.

which is characterised by lowstand incision/ incised valley e.g. Figures 4-6. From well log correlations, it is characterised as cleaning upward stacking pattern bounded below by sequence boundaries and above the surface that marked turn around between cleaning upward and dirtying upward log motifs TS. From this study, where there is incised valley LST and Transgressive System Tract (TST) deposits are inferred with reflection pattern characterised by mostly aggradational clinoform geometry in filled the incised valleys. In this case the LST is localised within the valleys (Table 1), overlain by retrogradational TST. The localisation of lowstand system tract (LST) favour the position of the study area to be shelf-edge, where most of the LST are deposited basinward and the incised valleys within the sub aerial are filled with LST deposit.

e) **Transgressive systems tract:** This systems tract within the study area is characterised by aggradational reflection patterns with little retrogradational bounded at the top by downlap of the toes of overlying clinoforms. This unit is thin, widespread and traceable within the study area (Table 1). Where there is incision, it forms part of the incised fills, generally overlie the LST, in such places the upper bounding surface coincided with the transgressive surfaces and sequence boundary. Generally, reflections of clinoform sets within the transgressive system tracts are discontinuous and they are of low amplitude and variable

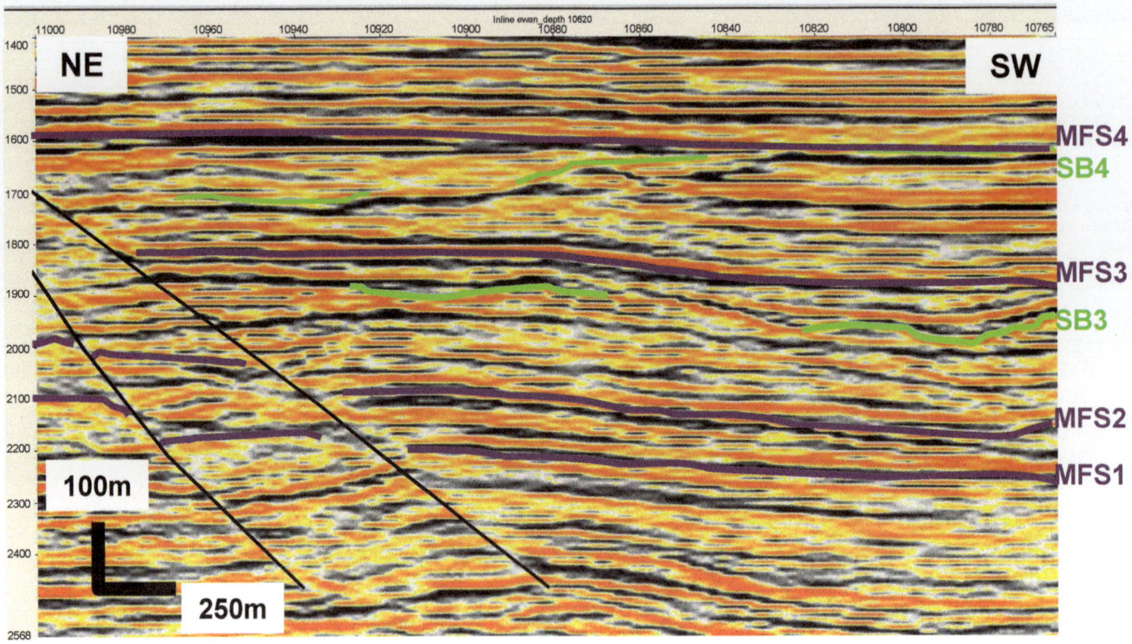

Figure 6: Seismic section showing interpreted maximum flooding surfaces and sequence boundaries (SB 3 and 4) denoted by green dashed line. (Inline 10620- lower patches). See position in Figure 1.

Depositional Units	External Form	Internal Form	Amplitude	Continuity	Lateral Distribution	Lower Boundary
LST	Sheet	Variable reflection patterns	Low	Low, more discontinuous	localised	Concordant
TST	Sheet	Variable reflection patterns/aggrad-ation	Moderate	Moderate, more discontinuous	Thin and widespread	Concordant
HST	Sheet/	Progradational–Oblique offlap	Moderate	Moderate	Widespread	Downlap

Table 1: Characteristics seismic facies of depositional sequences recognised in the study area.

reflection patterns. On well logs, it is characterised by high serrated increasing gamma ray motif bounded by a turn around from cleaning upward to dirtying upward and at the top maximum flooding surface. TST is assumed to be deposited at a time during which there is a rise in the relative sea level, i.e. when topset accommodation volume is increasing faster than the rate of sediment supply [4] (Figure 7).

f) **Highstand systems tract:** The Highstand System Tract within the study area is characterised by strong aggradational to progradational clinoform geometry and its downlap on the maximum flooding surface (Figure 2 and Appendix A). It is bounded at the base by downlap surface (MFS) and at the top by the sequence boundaries which always truncate the HST . Generally, reflections of the clinoform set within the HST are continuous and they are of moderate amplitude and oblique progradational clinoform geometry (Table 1). HST generally prograde over the underlying LST/TST (Figures 2 and 4). On a well log, highstand system tract is recognised by deposit form after maximum flooding as the relative sea level rise slows down during the late part of a relative sea level rise. At this time, the rate of creation of accommodation is typically less than the rate of sediment supply (Figure 7).

Depositional sequences

Seismic and sequence stratigraphic analysis has helped in defining six depositional sequences that can be recognised within the Ewan and Oloye fields, north-western Niger Delta. These sequences are illustrated in Appendices A-D and Figures 3 and 4; described below in order of the sequence development. The boundaries of these sequences are marked by reflections that represent unconformities.

Depositional sequence 1: Depositional Sequence 1 is bounded by Sequence Boundary 1 (Figure 8) at its base and at the top by Sequence Boundary 2. The sequence has maximum thickness of about 480 m (Figure 9). This sequence was partly penetrated by wells Ewan 07 and 08 and the gamma ray log signature within this sequence reflect a monotonous mudstone-dominated interval. The sequence is characterised by major valley incisions at its base and contains aggrading and prograding clinoforms (Figure 10). The aggrading valley fills is up to 100 m thick immediately overlying Sequence Boundary 1, located landward of the shelf break. Maximum Flooding Surface 1 overlies the prograding-aggrading clinoforms that downlap onto the sequence boundary. Towards the top of sequence 1 prograding clinoform downlap on top of Maximum Flooding Surface 1 (Figure 10). These clinoform are up 120 m thick and have a sigmoidal geometry. In the southern part of the study area Sequence 1 is less incised but more complexly faulted (Appendices C and D).

The Sequence Boundary 1 is an irregular unconformity that erosionally truncates dipping reflections. At the proximal area (northern part), three WNW-ESE trending erosional lows, interpreted as palaovalleys, are cut into the underlying strata (Figures 8b and 11; valleys 1, 2 and 3). Valley 1 is the deepest of all with up to 220 m of erosional relief with valleys 2 and 3 having up to 150 m and 120 m of erosional relief respectively. All valleys have tributary-like features and asymmetric cross-sections, with steeper slopes along the northern margins compared to the southern margins (Figures 8 and 11). Truncations of the underlying stratal and onlap characterised Sequence Boundary 1. This sequence boundary is traceable across

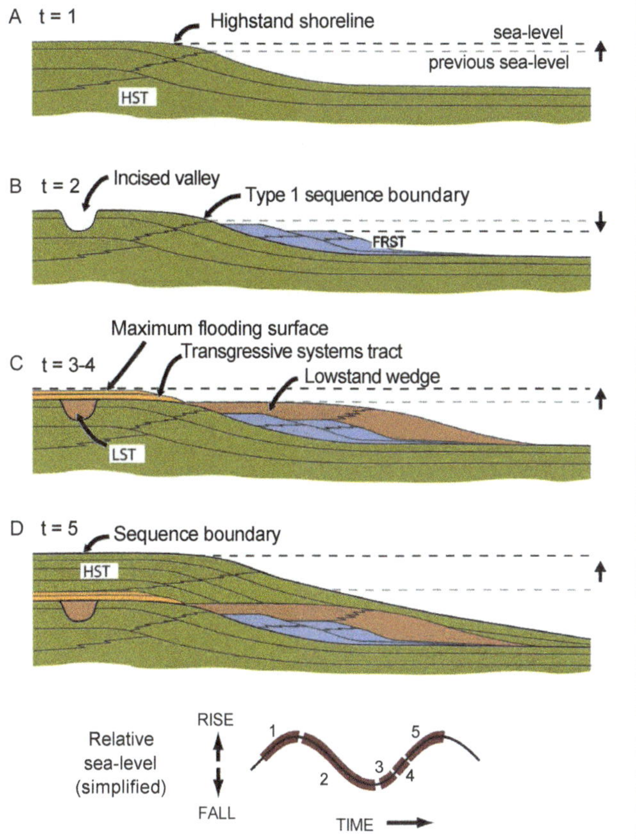

HST, TST, LST & FRST are Highstand, Transgressive, Lowstand and Forced Regressive systems tracts respectively.

Figure 7: Sequence architecture, sea level, systems tracts and inferred deposition during a complete relative sea level cycle on passive continental margins [14].

the study area and truncates underlain growth faults in the proximal area (Appendices A-D), which indicates there is rapid progradation of sediment basinward, where the sedimentation is controlled by faults

The Maximum Flooding Surface (MFS1) is defined by a prominent downlap surface within Depositional Sequence 1, separates the sequence into two seismic packages (Figure 10). The thickness between Sequence Boundary 1 (SB1) and Maximum Flooding Surface 1 (MFS1) is on average approximately 360 m and thins towards the SE. The Maximum Flooding Surface 1 coincides with sequence boundary 1 between the palaeovalley 2.

Depositional Sequence 1 shows a variable and irregular internal reflections pattern characterised by low to moderate amplitude, low frequency reflectors with prograding clinoforms geometry that suggest southward progradation (Appendices A, C and D) (Figure 10). Between SB1 and MFS1 seismic facies consist of variable and progradational-aggaradational reflection pattern locally infilling palaeovalleys in the proximal-northern part of the survey with the maximum thickness of about 360 m. In the distal southern part of the survey, seismic facies also consist of variable and aggradational-progradational reflections and they are highly faulted (Appendices A- D) (Figure 10). Above MFS1, the seismic facies consist of progradational clinoforms, with moderate amplitude and reflection continuity (Appendix. A) (Figures 10a and 10c). The maximum thickness of the unit MFS1-SB2 is 270 m (Figure 9c). Thickness variations within the Depositional Sequence 1 (Figure 9a) mostly occur along shelf break of highstand prograding clinoforms with height of above 200 m (Figure 10a). This thickness variation within this Depositional Sequence is as a result of the incisions/incised valleys, HST prograding clinoforms geometry and syn-depositional faulting. Thickness map between SB1 and MFS1 also show evidence of thicks which account for the erosional relief and the unfilling of

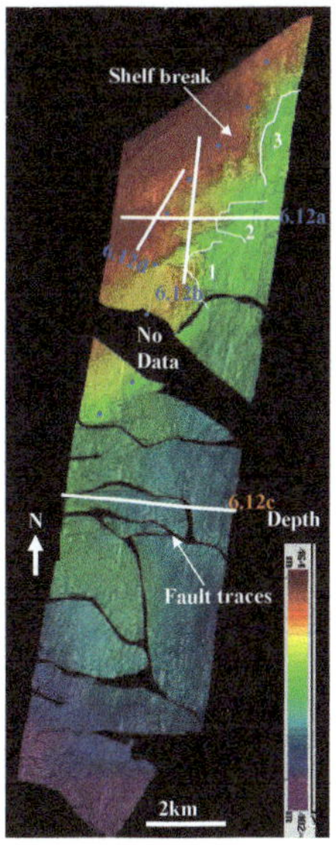

A . Depth structure map of Sequence Boundary 1 in a map view. 1,2 and 3 denoted the position of the palaeovalleys

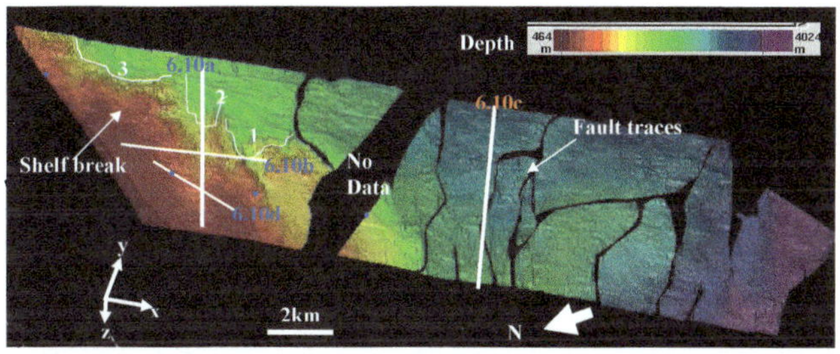

B . Depth structure map of Sequence Boundary 1.

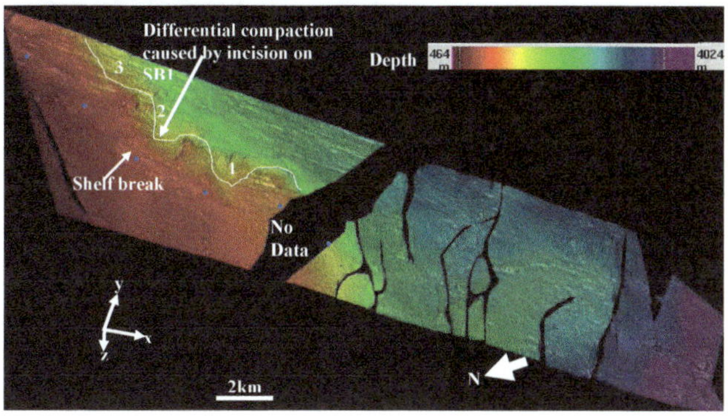

C. Depth structure map of Maximum Flooding 1

Figure 8: (A) Map view of depth structure map of Sequence Boundary 1, showing position and geometry of the erosional surface, relict shelf break and the position of seismic cross sections. (B) Oblique view of depth structure map of Sequence Boundary 1, showing clearly position and geometry of the erosional surface, relict shelf break, valley axes and the position of seismic cross sections. (C) Depth structure map of Maximum Flooding 1 showing differential compaction caused by incised valleys over SB1. Gamma ray log signature above this surface show an overall aggrading stacking pattern (Figure 3).

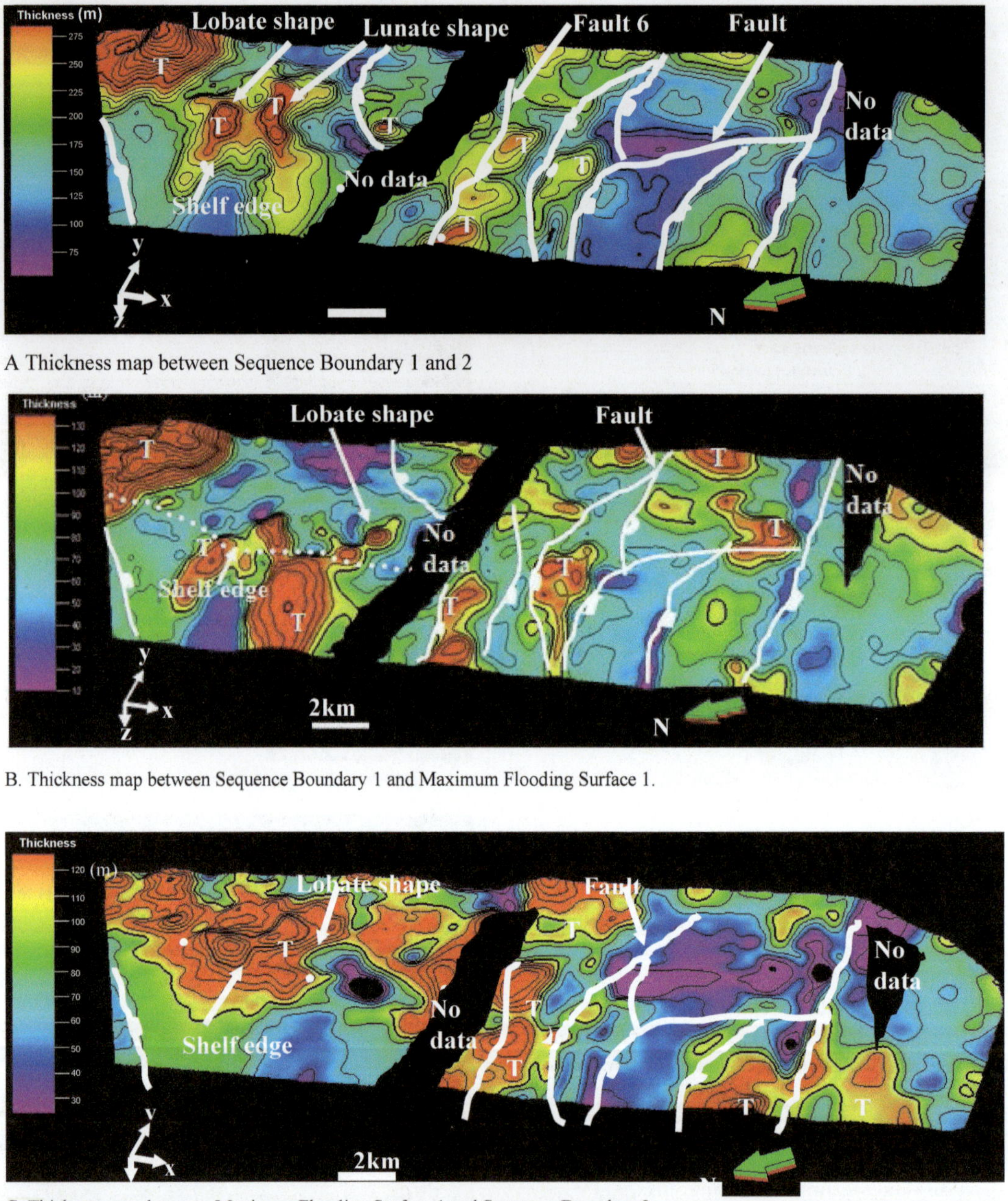

A Thickness map between Sequence Boundary 1 and 2

B. Thickness map between Sequence Boundary 1 and Maximum Flooding Surface 1.

C. Thickness map between Maximum Flooding Surface 1 and Sequence Boundary 2.

Figure 9: A. Thickness map between Sequence Boundary 1 and Sequence Boundary 2. The isopach map shows thicks as multi- lobate to lunate bodies but generally more lobate. B. Thickness map between Sequence Boundary 1 and Maximum Flooding Surface 1. The isopach map shows thicks as multi- lobate to lunate bodies but generally more lobate. C. Thickness map between Maximum Flooding Surface 1 and Sequence Boundary 2. The isopach map shows thicks as lobate bodies. Generally, T's are the position of thicks, white lines are the fault traces and contour interval is 20 m.

these incisions resulting in elongate thicks along the palaeovalley axes on SB1. The ESE-WNW trending linear thicks in the northern part of the study are coincident with main palaeovalley axes. Thickness map between MFS1 and SB2 (Figure 9c) also show evidence of thicks which reflects HST prograding clinoform geometry, differential compaction caused by the incision on SB1 and the fault throw (thicks at the hanging wall). The height of HST prograding (about 200 m) thick determined from prograding clinoforms geometry on the shelf break suggesting an increase in accommodation space caused by increase in relative sea level and sediment influx. The isopach maps of sequence 1 show thick bodies as multi-lobate to lunate bodies, but generally more lobate in geometry (Figure 9). The thick bodies show an elongation along the HST 1, and are about 1 km to 4 km wide.

Based on their stratigraphic position within the sequence and reflection geometry, the palaeovalleys fill seismic facies are interpreted to be Lowstand System Tract (LST) and Transgressive System Tract (TST). The unit above MFS1 is interpreted as a Highstand System Tract (HST) that is truncated by the overlying Sequence Boundary 2. The HST prograde over the underlying LST in the down dip.

Interpretation shows that, during the time of lowstand, several incised valleys/canyons were developed. These incised valleys/canyons

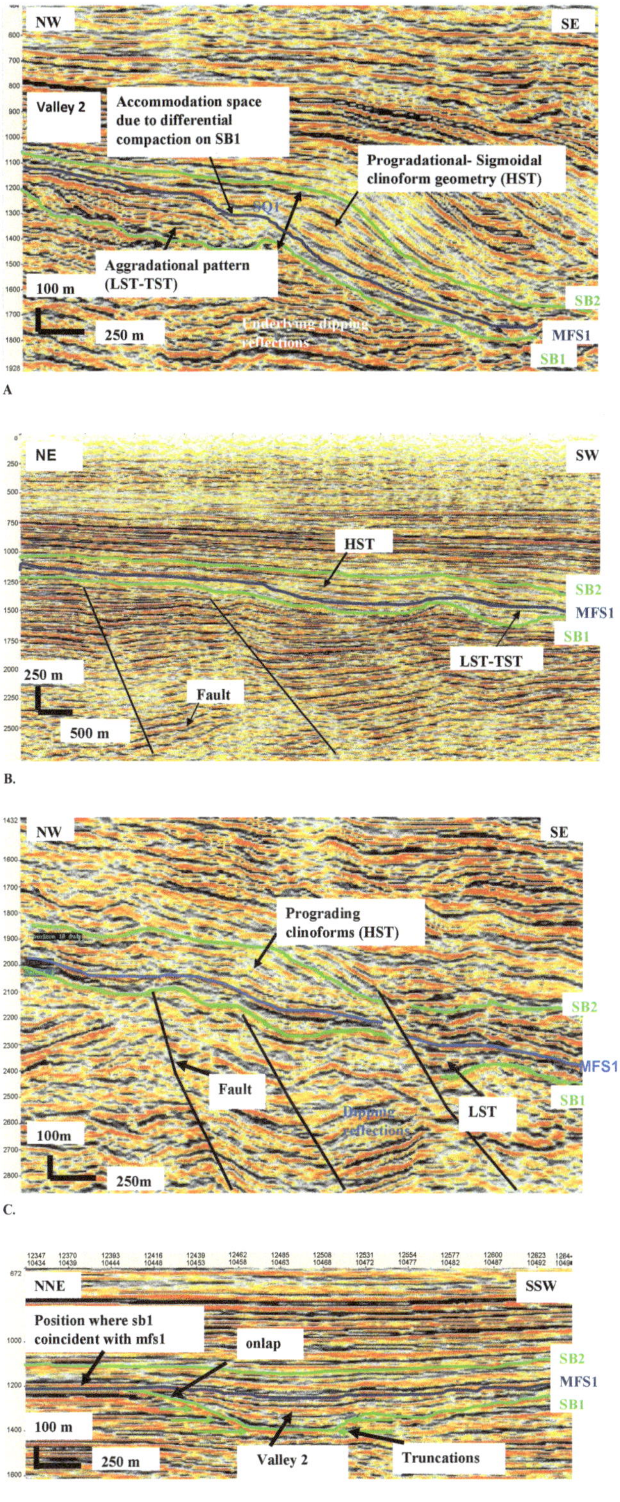

Figure 10: (A) Seismic section of sequence 1 in the proximal area (crossline 12499). The aggradation of deltaic shelf edge due to relative sea-level rise generated by local incisions, diapir-related subsidence. See position in Figure 8a. (B) Seismic section of sequence 1 in the proximal area (Inline line 10533). SB1 truncates the faults. See position in Figures 8a and 8b. (C) Seismic section of sequence 1 in the distal area (crossline 11268). See position in Figures 8a and 8b. (D) Random seismic section showing the palaovalley 2 and the interfluves. See position in Figures 8a and 8b.

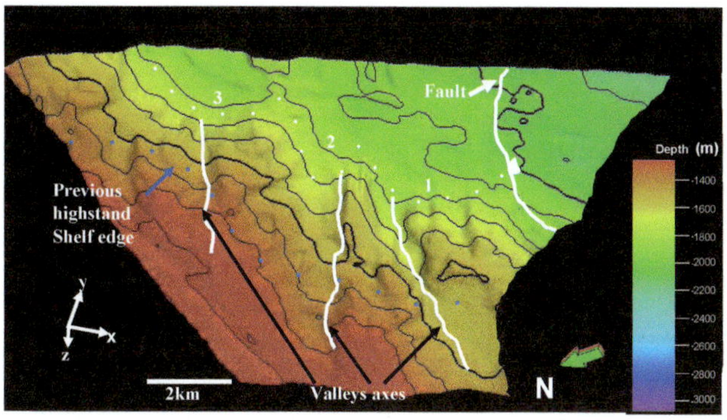

Figure 11: Depth structure map of proximal area (Northern margin) of Sequence Boundary 1 showing position and the geometry of the 3 palaeovalleys (1-3) and their valley axes.

acted as a sand feeder that led to the deposition sandstones on the slope and within the basin floor setting. The above observations show that the development of this Depositional Sequence is characterised by fall of relative sea level below the previous HST shelf edge, where the sands influx to the slope and basin floor by forced regression, followed by a sea level rise. Sea level fall have led to pronounced incisions on the shelf edge which also in turn caused instability at the shelf margin. The striking feature within this Depositional Sequence is the depth of palaeovalleys cuts on the Sequence Boundary 1 i.e. up to 220 m; partially reflect underlying structural control (Figure 10a). This might be caused by uplift due to the shale diapir.

Depositional sequence 2: Depositional Sequence 2 has a maximum thickness approximately 410 m and is bounded at the base by Sequence Boundary 2 and at the top by Sequence Boundary 3. This sequence is penetrated by all the Ewan wells that show high gamma ray serrated and cleaning upward motifs (Appendix B) (Figure 3). Internally, Depositional Sequence 2 is characterised by aggrading and prograding clinoforms geometry on dip direction. Maximum Flooding Surface 2 overlies the prograding-aggrading clinoforms that downlap onto the sequence boundary while the highstand prograding clinoforms downlap onto the Maximum Flooding Surface 2 on dip direction (Appendix A). At the shelf break of the previous HST, clinoforms geometry is up to 200 m thick. At the proximal part of the study area (northern area) SB2 is characterised by an irregular erosional surface (Figure 12). This irregular erosional surface is similar to Sequence Boundary 1 (Figure 8b). This can be viewed in respect of the position and geometry of palaovalleys. These subsidences do not prove they are valley as described in SB1 but they are effect differential compaction over SB1 related valleys. These related valleys also have asymmetric cross-sections with a steeper geometry at the northern margin compared to southern margin. Truncations of the underlying Sequence 1 strata and onlap of younger Sequence 2 strata characterise Sequence Boundary 2.

Maximum Flooding Surface (MFS2) lies within Depositional Sequence 2 and is a defined by downlap surface at the down dip (Appendices A and C) (Figures 2 and 12b). The stratal unit between SB2 and MFS2 with maximum thickness of 250 m (Figure 13). The second seismic unit is between MFS2 and SB3 with thickness up to 300 m. From well logs, this Maximum Flooding Surface was encountered at approximately 1925 m in Ewan 07 and at a progressively shallower depth in Ewan 08, 05, 05b and 12i at about 1894 m, 1870 m, 1854 m and 1810 m respectively. Gamma ray log signature within this surface is defined by high serrated gamma ray log motif.

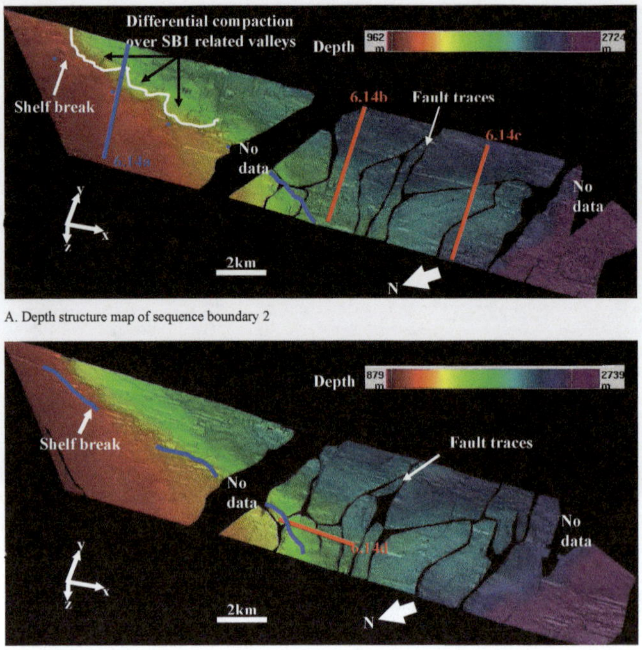

A. Depth structure map of sequence boundary 2

B. Depth structure map of MFS 2.

Figure 12: (A). Depth structure map of sequence boundary 2, showing position and geometry of the erosional surface, relict shelf break and the position of seismic sections. Distal part is more tectonically faulted. (B). Depth structure map of MFS 2. Gamma ray logs patterns above MFS 2 showing an overall aggrading to prograding pattern.

Depositional Sequence 2 is characterised by three distinct seismic facies (i) moderate amplitude, variable and irregular internal reflections, (ii) low frequency reflectors and (iii) internal aggrading and prograding clinoforms (Figure 14). Between SB2 and MFS2, facies consist of progradational-offlap clinoforms onlapping onto SB2 at the shelf edge and downlapping seaward (Appendices A and C). Gamma ray logs within this interval show a cleaning up motif (up to 35 m thick) terminated by a flooding surface in the Ewan 07, 08, 05 and 12i wells. The facies within this interval is interpreted to be slope channel sands overlying by lower shoreface muddy sandstones within the Ewan Field. In the Ewan 07 well, the facies is characterised by monotonous fine sand with no deposition of slope channel sand. This indicates that the slope channel sand has thinned out or faulted out.

Between MFS2 and SB3, the seismic unit comprises aggrading to prograding clinoforms, characterised by moderate amplitude that downlap onto MFS2. This package is truncated by Sequence Boundary 3 and more complexly faulted in the distal area compared to proximal area, which indicates that they are more tectonically active in the down dip (Appendices A, C, and D) (Figure 14). Well logs within Ewan field, indicate that this interval is characterised by high serrated gamma ray motifs intercalated with cleaning upward stacking patterns. Sand development is prominent in the other Ewan wells (Ewan 08, 05, 05a, 05b and 12i) except Ewan 07. From the well log correlation, gamma ray logs through the up thrown blocks within this succession (Ewan 08), suggest a mix of upward cleaning, blocky and upward fining successions. At the down-thrown side of the same fault block, the gamma ray log motif through this sequence shows monotonous mudstone and fine grained sandstone as observed from Ewan 07 (Figure 3 and Appendix B). This reflects local variability in type of sediment influx and grain size distribution. Thickness variations within Depositional Sequence 2 (SB2-SB3; Figure 13) mostly occurred along the shelf break of highstand prograding clinoforms. This thickness variation is as a result of HST clinoform geometry, accommodation space created by differential compaction and syn-deposition faulting. The HST prograding thick is determined by the height of prograding clinoforms geometry on the shelf break. Thickness map between SB2 and MFS2 show evidence of thickening mostly along the previous HST shelf break, where there is evidence of incision/incised valley and in the hanging wall of the growth faults at the distal part. This thickness variation reflects effect of differential compaction coupled, with relative sea level fall and fault movement (thickening within the hanging wall block) (Figure 13b). Thickness map between MFS2 and SB3 show evidence of thickening along Sequence 1 HST shelf break. This reflects Sequence 2 HST aggrading-prograding clinoforms geometry and also syn-depositional fault movement during Sequence 2 development (Figure 13c). Thicks bodies show lobate bodies along shelf break. Thick are about 6-7 km wide.

Depositional sequence 3: Depositional Sequence 3 has a maximum thickness of about 380 m and is bounded at its base by Sequence Boundary 3 (Figure 15) and at the top by Sequence Boundary 4. These two bounding surfaces show a clearly defined incision along their interpreted surfaces. Seismically, this sequence consists of aggrading and prograding clinoform patterns (Appendices A, C and D). MFS 3 (Figure 16) overlies the prograding-aggrading clinoforms that downlap onto the sequence boundary 3 while the highstand prograding

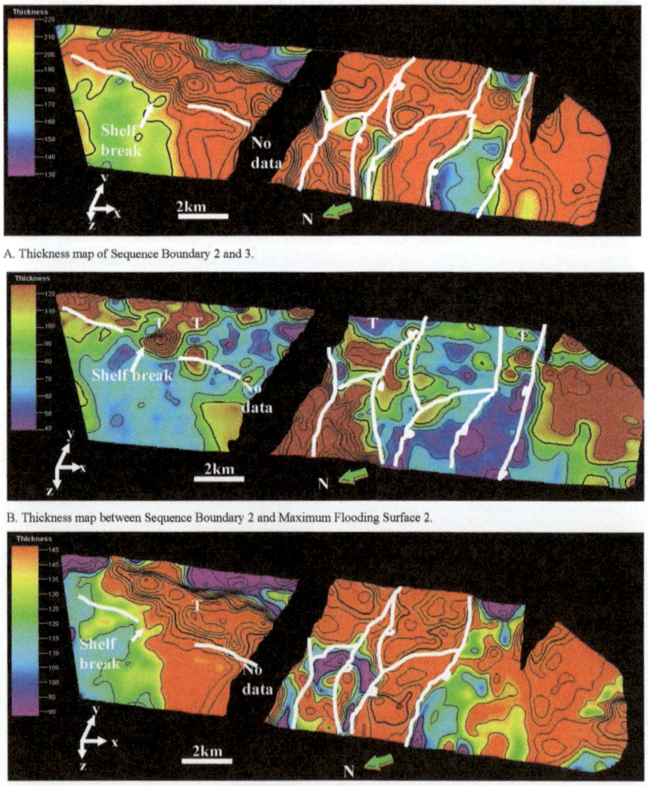

A. Thickness map of Sequence Boundary 2 and 3.

B. Thickness map between Sequence Boundary 2 and Maximum Flooding Surface 2.

C. Thickness map between Maximum Flooding Surface 2 and Sequence Boundary 3.

Figure 13: A. Thickness map of Sequence Boundary 2 and 3. The isopach map shows thicks along shelf break and basinward. B. Thickness map between Sequence Boundary 2 and Maximum Flooding Surface 2. The isopach map shows thicks as lobate bodies. C. Thickness map between Maximum Flooding Surface 2 and Sequence Boundary 3. The isopach map shows thicks as lobate bodies. Generally, T's are the position of thicks, white lines are the fault traces and contour interval is 20 m.

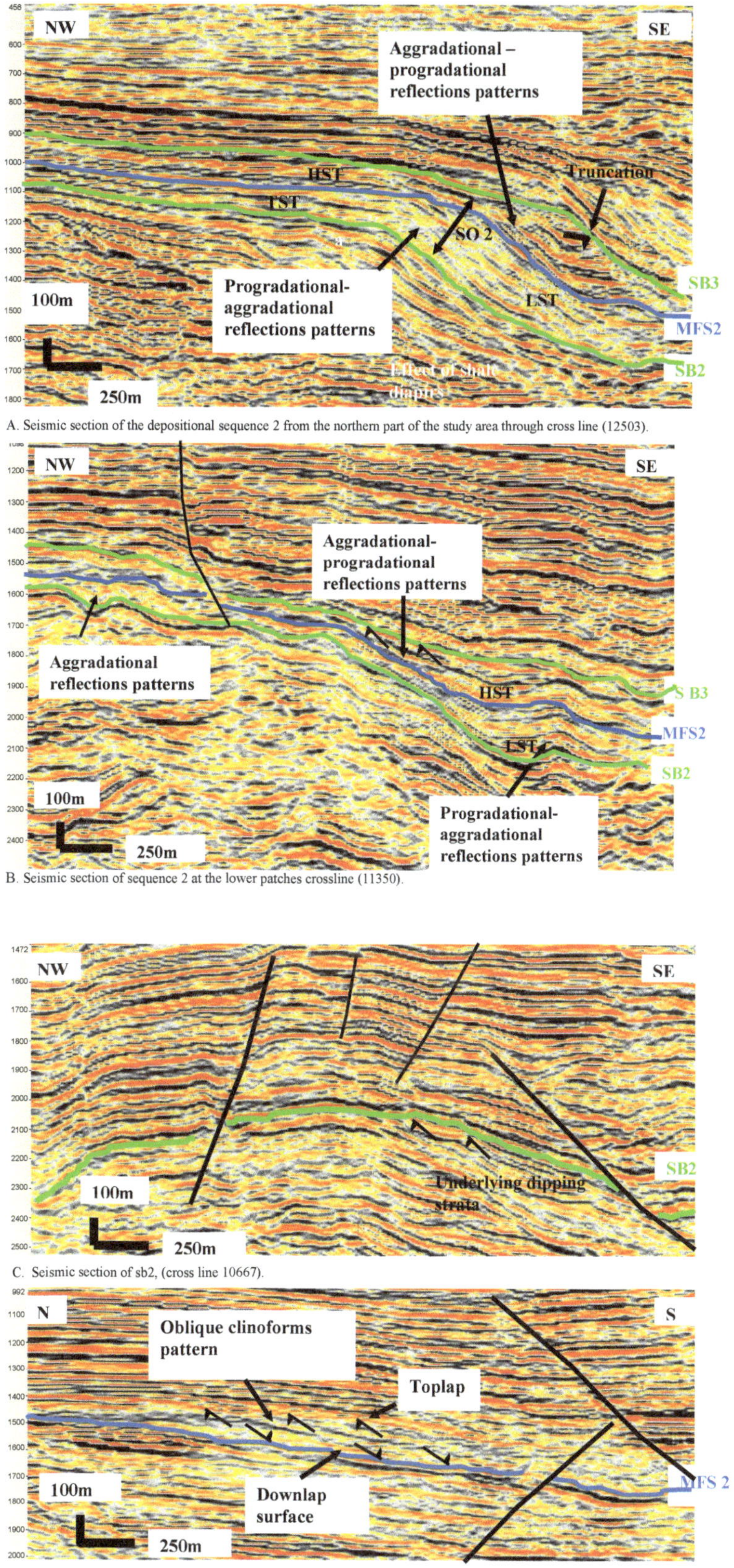

A. Seismic section of the depositional sequence 2 from the northern part of the study area through cross line (12503).

B. Seismic section of sequence 2 at the lower patches crossline (11350).

C. Seismic section of sb2, (cross line 10667).

D. Seismic section of MFS 2 Inline (10420).

Figure 14: (A) Seismic section of the depositional sequence 2 from the northern part of the study area through cross line (12503). See position in Figure 12a. (B) Seismic section of sequence 2 at the lower patches dip line (11350) showing the key surfaces and internal reflections patterns. See position in Figure 12a. (C) Seismic section of sb2, showing truncation and underlying dipping reflections terminated below sb2 (cross line 10667). See position in Figure 12a.

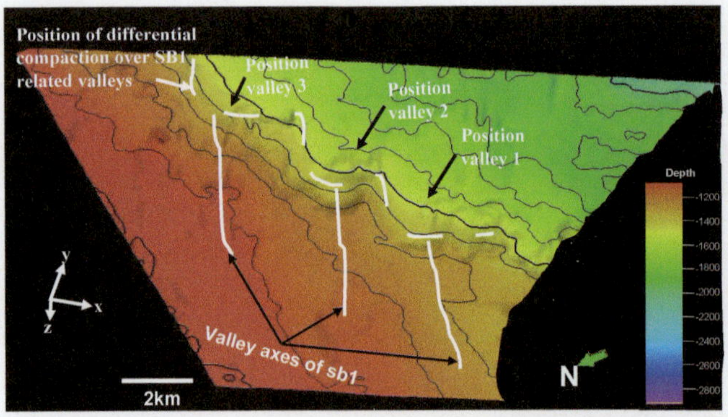

Figure 15: Depth structure map of sequence boundary 2 at the proximal area (Northern part) showing the effects of differential compaction over sequence boundary 1 related valleys.

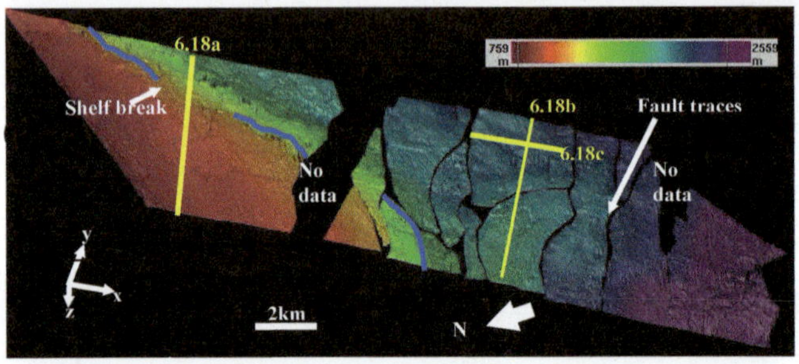

A. Depth structure map of sequence boundary 3.

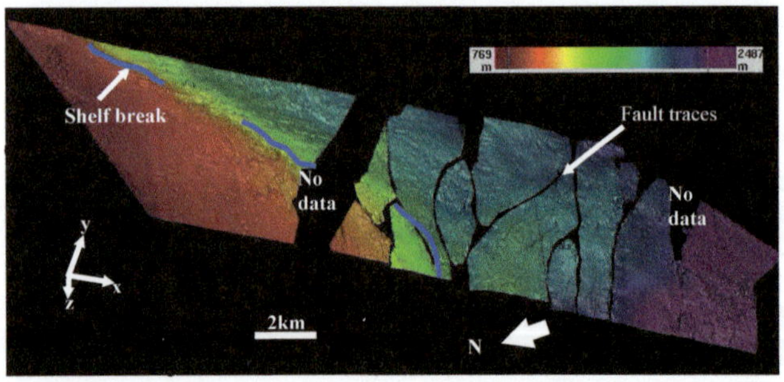

B. Depth structure map of mfs 3.

Figure 16: A) Depth structure map of sequence boundary 3 showing the over all geometry and the position of seismic sections. Distal area is more tectonically faulted. B) Depth structure map of MFS 3.

clinoforms downlap onto the Maximum Flooding Surface 3 (Appendix A). The highstand prograding clinoforms reflection patterns is up to 200 m thick and it thickens basinward. This sequence is penetrated by all the Ewan wells and partially penetrated by Oloye 03 well. Gamma ray log motifs shows high serrated gamma, cleaning upward and dirtying upward motifs (Figure 3 and Appendix B).

Sequence Boundary 3 is characterised as an irregular erosional surface (Figures 6 and 16). This sequence boundary is characterised by incision up to 120 m. This incision is widening in the seaward direction. Truncations of the underlying stratal and onlap characterised Sequence Boundary 3 (Appendix A). Seismic reflection above the sequence boundary is characterised by progradational-aggradational reflection patterns. Gamma ray log signatures above this surface within Ewan and Oloye fields shows a turn around from cleaning upward and dirtying upward motifs (Figures 17 and 18).

The Maximum Flooding Surface within this sequence is defined by a downlap surface. At the distal area incision directly overlies the SB3 on interfluves. Gamma ray log signature within this surface is defined by high serrated gamma ray log motif.

Depositional Sequence 3 is characterised by moderate to high amplitude, variable and irregular reflection patterns with internal prograding and aggrading clinoforms geometry that suggest north-south progradation (Appendices A, C and D) (Figures 18a and 18b). Seismic facies between SB3 and MFS3 consist of variable, progradational-aggradational reflection patterns overlying by MFS3 in the proximal area (Figure 18). At the distal part where there is incision the package is characterised by aggrading reflections patterns infilling the incised valleys (Figure 18c). Gamma ray log signature within this unit is characterised by dirtying upward, which has been interpreted to be slope channel sand overlain by a transgressive (Figure 3). Above MFS3 the seismic facies consists progradational clinoforms of moderate amplitude that downlap directly on MFS3. This unit is truncated by the overlying erosional surface of SB4 (Appendix A) (Figure 18c). Gamma ray log motifs within this interval generally show high serrated gamma ray which are intercalated with gradual cleaning upward motifs in the upper part of the unit (Figure 3). The high gamma ray serrated motifs have been interpreted to be lower shoreface deposit while the gradual

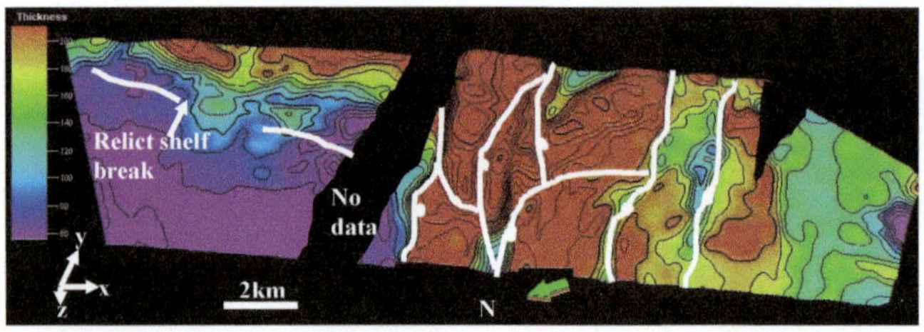

A. Thickness map of SB 3 and 4.

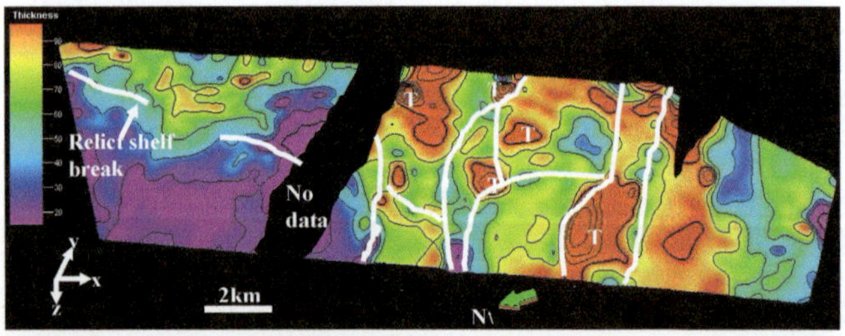

B. Thickness map between Sequence Boundary 3 and Maximum Flooding Surface 3.

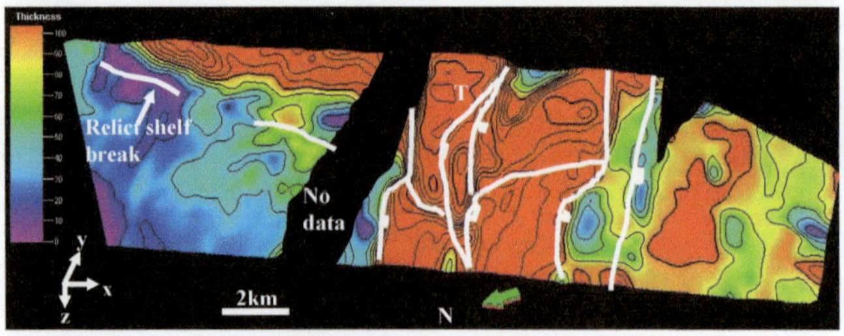

C. Thickness map between Maximum Flooding Surface 3 and Sequence Boundary 4.

Figure 17: A. Thickness map of SB 3 and 4. The isopach map shows thicks basinward and at the hanging wall of faults. This reflects aggrading and prograding geometry, incisions and faulting. B. Thickness map between Sequence Boundary 3 and Maximum Flooding Surface 3. T's are the position of thicks caused by I incision clinoforms geometry. Contour interval is 20 m C. Thickness map between Maximum Flooding Surface 3 and Sequence Boundary 4. The isopach map shows thicks basinward in response to HST clinoforms geometry and growth faulting. Generally, T's are the position of thicks, white lines are the fault traces and contour interval is 20 m.

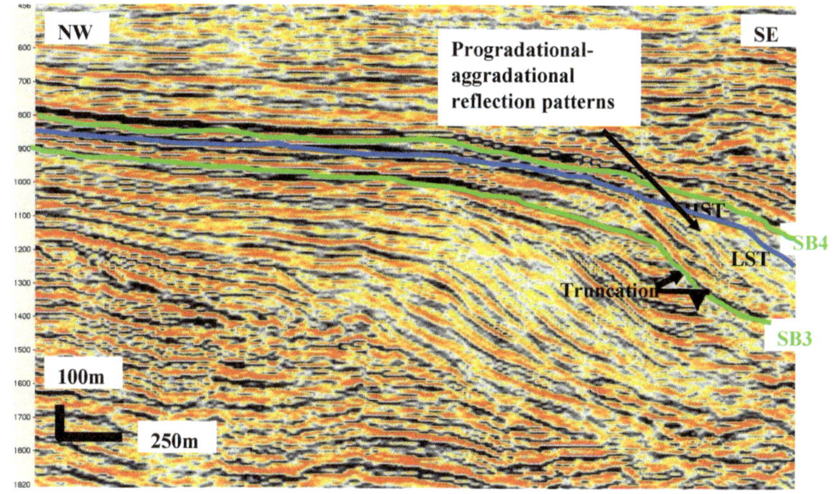

A. Seismic section of sequence 3 at the proximal part crossline (12503)

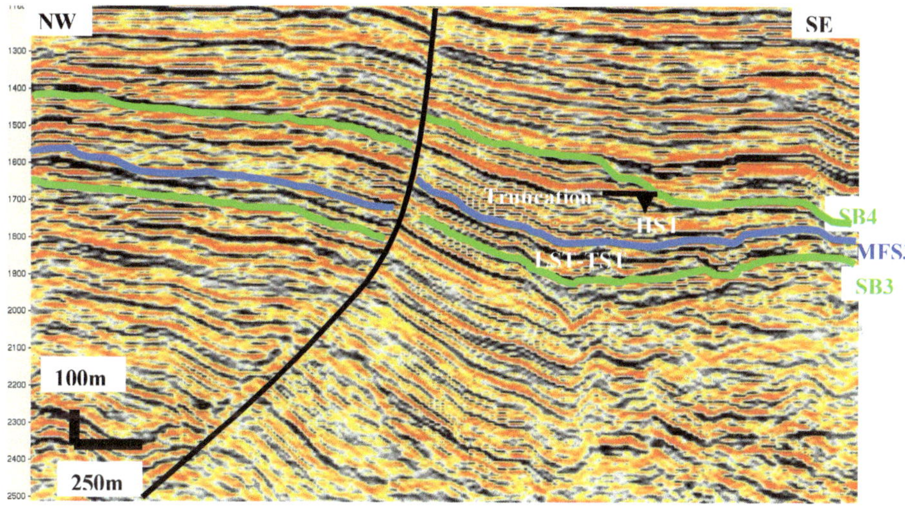

B. Seismic section of sequence 3 at the distal part crossline (10939)

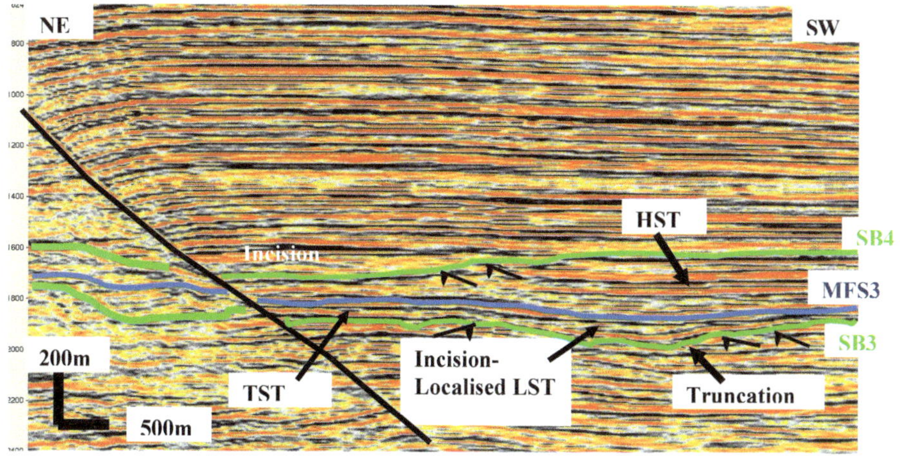

C. Seismic section of depositional sequence 3 inline (10620).

Figure 18: (A). Seismic section of sequence 3 at the proximal part crossline (12503) showing the key surfaces, reflection termination, system tracts and internal reflections patterns. See positing in Figure 16a. (B). Seismic section of sequence 3 at the distal part dip line (10939) showing the key surfaces and reflection termination. See position in Figure 16a. (C). Seismic section of depositional sequence 3 inline (10620). Showing incision at the distal part, reflections termination and system tracts. See position in Figure 6.

cleaning upward motifs have been interpreted to be upper shoreface deposits. Thickness variations within Depositional Sequence 3 (Figures 17a-17c) mostly occurred at the basinward part and the thicks reflect clinoform geometry, incision and syn-depositional faulting. The maximum thickness between the SB3 and MFS3 is about 170 m (Figure 17b). Thick occurs as lobate bodies mostly where there incision and at the down thrown side of the growth faults (Figure 17b). The maximum thickness between the MFS3 and SB4 is about 260 m (17c) thicks are located seaward which reflects prograding clinoforms geometry of HST and accommodation space created by tectonic activities (growth faulting).

The development of lowstand prograding wedge within this sequence (Figure 18a) indicates reestablish of delta at the shelf edge during rise of relative sea level, after a period of lowstand below the shelf edge, [2]. Area where the surface is deeply incised, (Figure 18c) LST deposits are restricted to the valley fills and the maximum flooding surface (MFS3) marks the top of the incision. Outside the LST valleys, MFS3 and SB3 are coincident on interfluves. This suggests that during LST times most of the area was periodically subaerially exposed. The incision along the bounding sequence boundaries at the distal area suggests the locus of erosion and deposition were shifted basinward during the development of the sequence. From seismic data, this sequence thickens basinward; this may also reflect basinward shifts in depocentres coupled with the effect of accommodation created by incision/canyons basinward. From the well log (Figure 3) (Appendix 6B) the slope channel sand unit overlying the SB3 is also thin on the hanging wall of the fault block as observed from Ewan 07; this might be faulted out or shale out.

Contour interval is 20 m C. Thickness map between Maximum Flooding Surface 3 and Sequence Boundary 4. The isopach map shows thicks basinward in response to HST clinoforms geometry and growth faulting. Generally, T's are the position of thicks, white lines are the fault traces and contour interval is 20 m.

Depositional sequence 4: Depositional Sequence 4 is bounded by Sequence Boundary 4 at the base and Sequence Boundary 5 at the top. This sequence is fully penetrated by all the Ewan wells and Oloye 03 well (Appendix B) (Figure 3). Seismically, this sequence consisting of two units characterised by aggrading and prograding clinoforms (Appendices A, C and D) and gamma ray log signatures show a variety of blocky, high serrated gamma and cleaning upward motifs. Maximum Flooding Surface 4 overlies prograding-aggrading clinoforms that downlap onto the sequence boundary 4 while the highstand prograding clinoforms downlap onto the Maximum Flooding Surface 4 (Appendix A) Sequence Boundary 4 is characterized by an irregular erosional surface with evidence of truncation of inclined reflection on the shelf edge break. In the distal area of seismic data, the sequence boundary is characterized by d incision which is up to 110 m of localized truncation in the underlying HST (Figures 19-21). Above SB4 in some places i.e., where there is incision Maximum Flooding Surface 4 coincide with SB4 Figures 21b and 21c) and gamma ray log signatures above this surface within Ewan and Oloye fields show blocky stacking patterns. The Maximum Flooding Surface (MFS 4; Figure 19b) within this sequence is defined by downlap surface and where there is an incision, the maximum flooding surface caps the top of the incision (Figures 21b and 21c). This surface separates this sequence into two seismic packages (i) aggrading unit and (ii) prograding unit. Gamma ray log signature above this surface is defined by high serrated gamma ray log motif (Figure 3 and Appendix B).

Seismically, Depositional Sequence 4 is characterised by low to moderate amplitude, overall prograding clinoforms that suggest southward progradation (Appendices A, C and D) (Figure 21). Between SB4 and MFS4 seismic facies consist of variable and progradational-aggaradational clinoforms. Within incised valley fill, seismic character is defined as low to moderate amplitude and variable reflections (Figure 21). Gamma ray logs motifs within this interval show an upward cleaning to blocky log motif trend overlain by high gamma serrated log motif. This seismic facies have been interpreted to be slope channel sand and lower shoreface deposits respectively [9,10].

Above MFS4, seismic facies consist of aggradational-progradational clinoforms, moderate amplitude and reflections continuity both at the proximal and distal areas. Gamma ray logs signatures within this interval show high serrated trace, overlain by a gradual cleaning upward trend which has been interpreted to be lower shoreface muddy sandstone and upper shoreface sand respectively. Thickness variations with this sequence (SB4-SB5)) reflect highstand aggrading-prograding clinoforms geometry, incisions and growth faulting (Figure 20a). Between SB4 and MFS4 (Figure 20b) show thicks reflect incision and accommodation space created within the subsidence in the hanging wall of growth faults (Figure 20b). Thickness map between MFS4 and the overlying sequence boundary SB5, (Figure 20c) show thicks which reflects aggrading-prograding clinoforms of HST and accommodation space created by tectonic activities (growth faulting).

Depositional sequence 5: Depositional Sequence 5 is bounded at the base by Sequence Boundary 5 (Figures 22 and 23) and at the top by Sequence Boundary 6. This sequence is characterized by progradational geometry (Appendices A, C and D) (Figure 24a). In seismic cross section, this sequence is also characterized by low to moderate amplitude, moderate continuity reflection and widespread lateral distribution (Figure 24). This sequence is fully penetrated by all

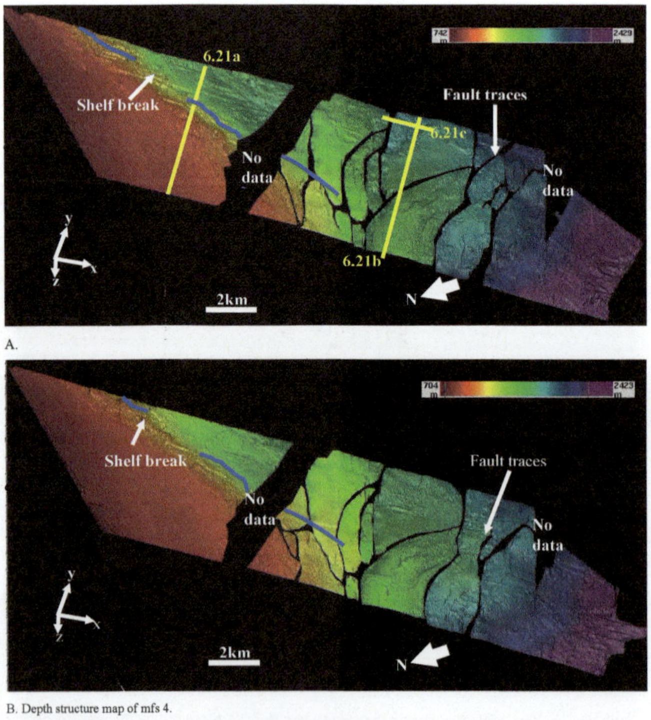

Figure 19: (A) Depth structure map of sequence boundary 4 showing its overall geometry and the position of seismic sections. Distal area is more tectonically faulted. B. Depth structure map of MFS 4.

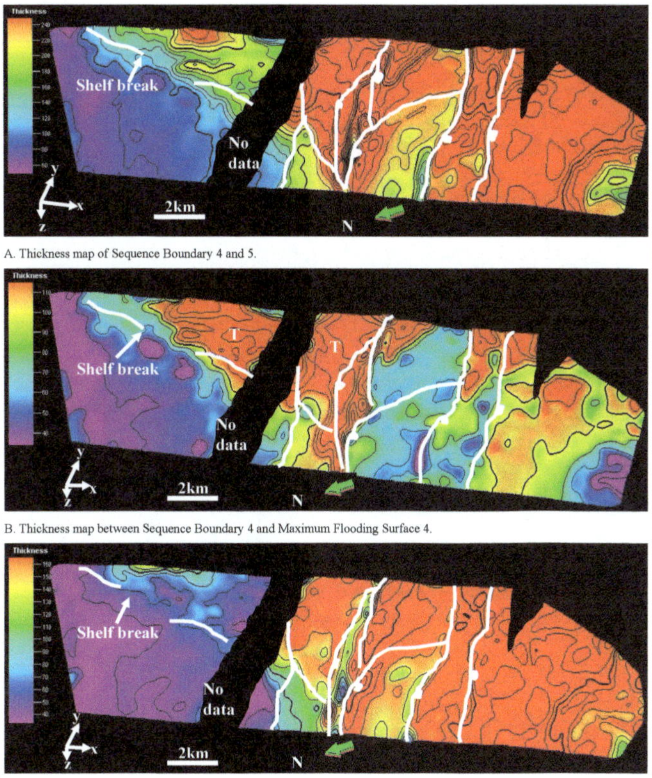

A. Thickness map of Sequence Boundary 4 and 5.

B. Thickness map between Sequence Boundary 4 and Maximum Flooding Surface 4.

C. Thickness map between Maximum Flooding Surface 4 and Sequence Boundary 5.

Figure 20: (A). Thickness map of Sequence Boundary 4 and 5. The isopach map shows thicks basinward and within the hanging wall of growth faults. This reflects aggrading and prograding geometry, incisions and faulting. (B). Thickness map between Sequence Boundary 4 and Maximum Flooding Surface 4. T's are the position of thicks caused by incision. White lines are the fault traces. (C). Thickness map between Maximum Flooding Surface 4 and Sequence Boundary 5. The isopach map shows thicks basinward in response to HST clinoforms geometry and growth faulting. Generally, white lines are the fault traces and contour interval is 20 m.

the Ewan wells and Oloye 03 well. Gamma ray log signatures within this sequence show dirtying and cleaning upward signature and it is described as bow shape geometry (Figure 3 and Appendix B).

Sequence Boundary 5 (Figure 22) is an irregular unconformity with some evidence of incisions (Figures 24a and 24b). At the southern part of the survey, further evidence of erosion along this surface is provided by the truncation of moderately inclined reflections (Figure 24b). Within this sequence, no evidence for the presence of a maximum flooding surface was observed, may be, the mfs is likely to be outside of the data or cut by a major candidate sequence boundary (SB6), and probably the transgressive surface has been eroded at this location. Gamma ray log signature above this surface marked the base of cleaning upward motif (Figure 3) (Appendix B) (Figures 22-25).

Thickness map between SB5 and SB6 is average 320 m maximum (Figure 23). Thickening mostly occurred at the shelf break and extends basinward, which reflect LST incision and HST aggrading-prograding clinoforms geometry (Appendix A). These thicks occurred as lobate bodies and are about 1-6 km wide. The dirtying upward followed by cleaning upward gamma ray log signatures (bow-shape log motif) within this sequence suggests waxing and waning of sand sedimentation rate (Emery and Myers, 1996) Figure 3, Appendix B). The sequence is sandy than all the lower sequences, and is interpreted as tidal/fluvial channel sand [10]. The increase in sand content upwardly, (Figure 3) indicates that we are at the uppermost section of the interpreted section (Agbada Formation), grading into the fully continental sand interval of the Benin Formation.

Depositional sequence 6: This Depositional Sequence is bounded at its base by Sequence Boundary 6 overlying by coastal barrier sands while the upper part of the sequence is made up of undifferentiated continental deposit. This sequence is penetrated by all the wells except Ewan 05, 5a and 5b (Figure 3) (Appendix B). Reflection patterns within this sequence are defined by the progradational-aggradational stacking patterns (Appendices A, C and D). Their gamma ray logs reflect a stacked of cleaning up stacking patterns (Figure 3).

Sequence Boundary 6 is characterised by an irregular erosional surface, truncations of the underlying stratal and onlap (Figure 25). There is evidence of incisions up to 150 m deep and up to 16 km wide that cut into the underlying LST/TST of sequence 5 (Figure 26) (Appendix A). In more distal areas, this incision are less pronounced but further evidence of erosion along this surface is provided by the

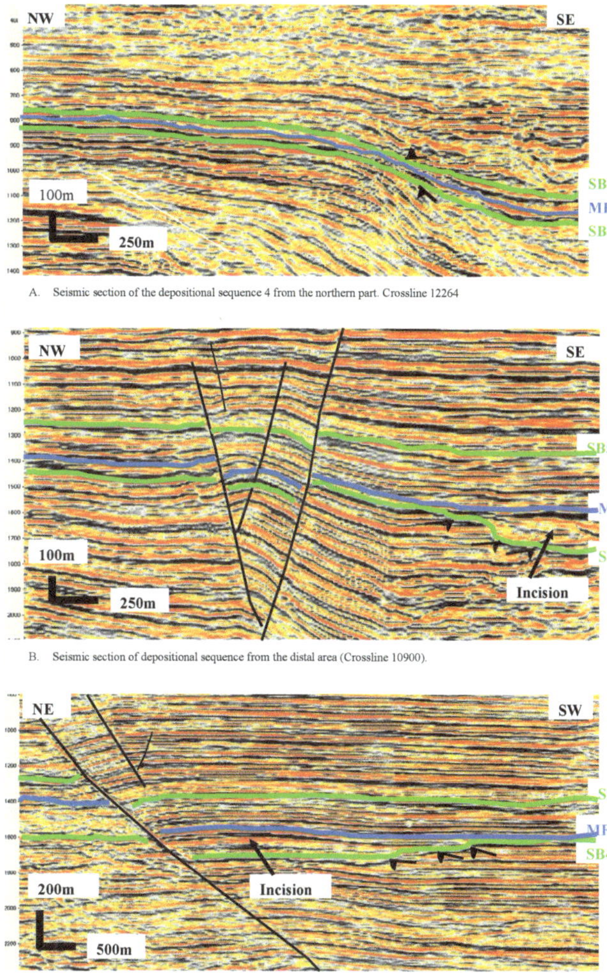

A. Seismic section of the depositional sequence 4 from the northern part. Crossline 12264

B. Seismic section of depositional sequence from the distal area (Crossline 10900).

C. Seismic sections of depositional sequence from the distal area (Crossline10638).

Figure 21: (A). Seismic section of the depositional sequence 4 from the northern part of the study area through. Crossline 12264, characterised by prograding clinoform geometry. See position in Figure 19a. (B). Seismic section of depositional sequence form the distal area showing the major incision and the reflection patterns and termination within the sequence (Crossline 10900). See position in Figure 19a. (C). Seismic sections of depositional sequence from the distal area showing the major incision and the reflection patterns and termination within the sequence (Crossline10638). See position in Figure 19a.

truncation and incision of moderately inclined reflections (Figure 26). Gamma ray log signature above this surface marked the base of cleaning upward motif. Within this sequence, no evidence for the presence of a maximum flooding surface was observed, may be, the mfs is likely to be outside of the data.

Seismically, this sequence is characterised by moderate amplitude,

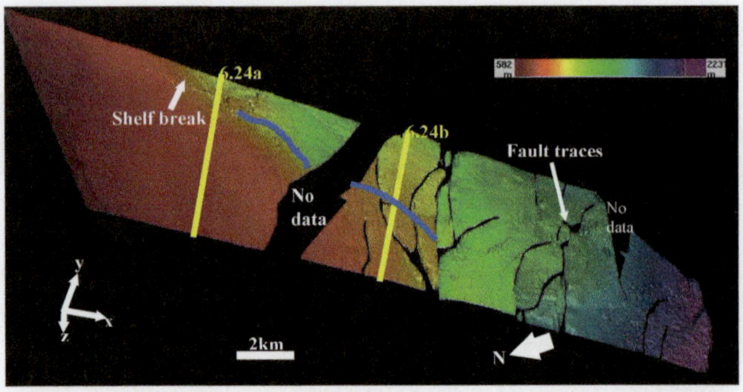

Figure 22: Depth structure map of sequence boundary 5. Gamma-ray logs show typical log pattern above the surface.

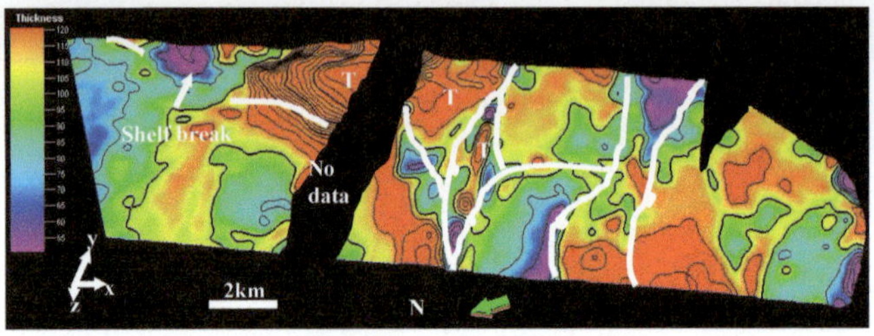

Figure 23: Thickness map between Sequence Boundary 5 and 6. T's are the position of thicks caused by incision and HST clinoforms geometry. The isopach map shows thicks as lobate bodies. White lines are the fault traces.

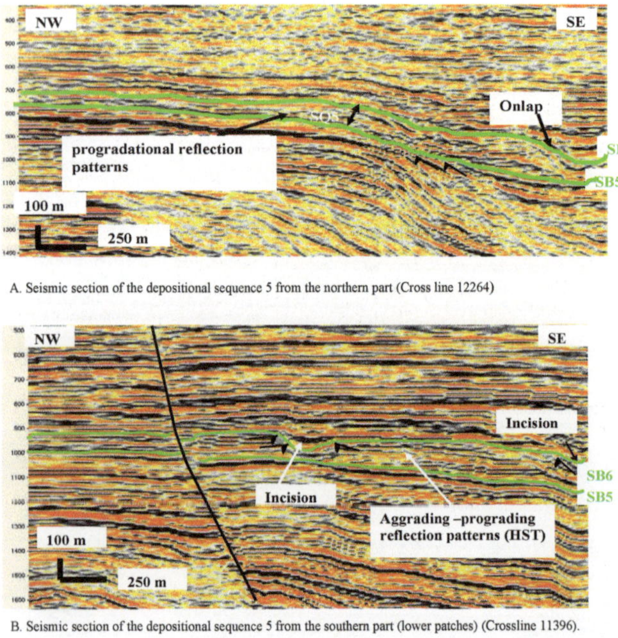

Figure 24: (A). Seismic section of the depositional sequence 5 from the northern part of the study area (Cross line 12264), characterised by progradational geometry. See position in Figure 22. (B). Seismic section of the depositional sequence 5 from the southern part (lower patches) of the study area (Crossline 11396). The sedimentary package within the sequence (LST/TST) show progradational-aggradational stacking patterns with the SB 6 incised/removed the MFS and HST. See position in Figure 22.

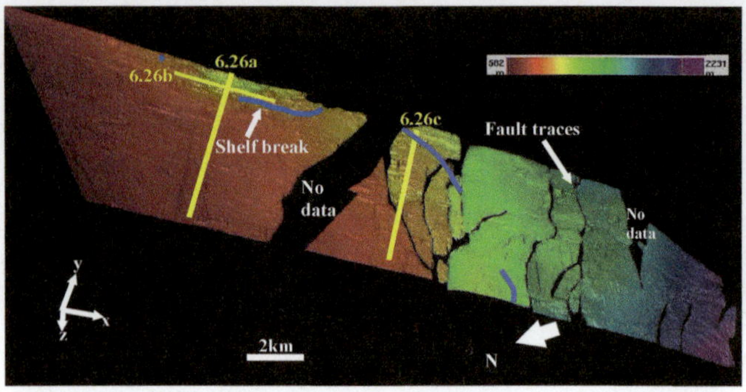

Figure 25: Depth structure map of sequence boundary 6. Gamma-ray log pattern above the surface ranges from blocky to coarsening upward trend. F's are the fault traces.

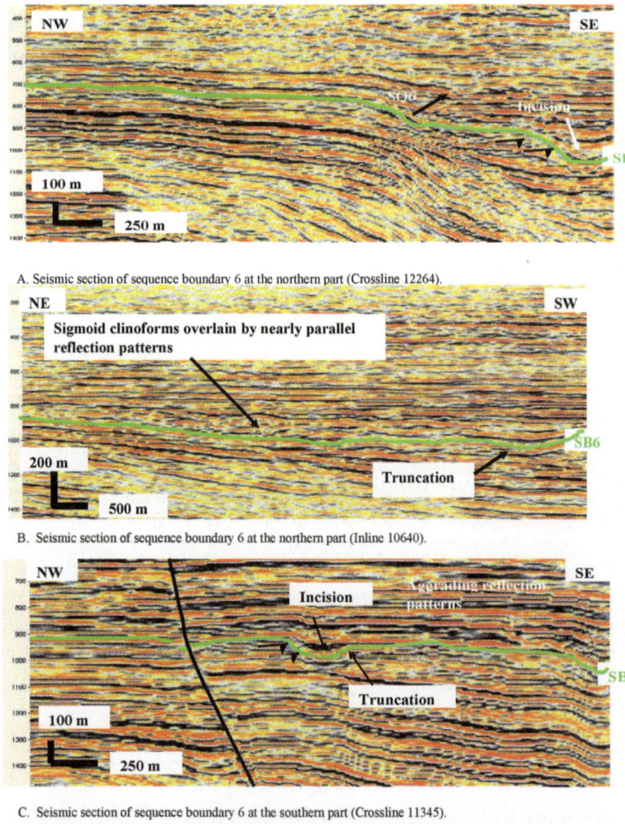

Figure 26: (A). Seismic section showing sequence boundary 6 at the northern part (Crossline 12264). See position in Figure 25. (B). Seismic section of sequence boundary 6 at the northern part (Inline 10640). See position in Figure 25. (C). Seismic section of sequence boundary 6 at the southern part (Crossline 11345). See position in Figure 25.

reflection continuity. There is a major erosional incision (incised valley) cuts into this sequence in the distal area located at the upper part of the sequence. The depth of this incised valley is up to 350 m deep and greater than 15 km wide, but it is difficult to know the overall width due to the incision extending beyond the area of the survey (Appendices C and D) (Figures 27 and 28). Seismic reflections within this incision are of low amplitude, moderately continuous-chaotic and progradational (sigmoid) while the reflections above the valley fill are nearly parallel, with moderate-high reflection amplitude (Figure 29). Seismic horizon slices through this incised valley at depth 889 m show northwest-southeast orientation of straight to sinuous channels within a more chaotic and low amplitude background (Figure 30). The magnitude of the incised valley increases southward truncating the underlying rollover anticline strata. This suggests the locus of erosion and deposition shifted basinward and it is structurally controlled. This major incised valley surface truncates some of the growth faults, and some faults are cut into the valley at the top. This suggests that some of the faults were inactive during the creation of sequence boundary 6. Fault cutting into the valley fills. Thickness map between base of the valley (BCH) and top of the valley (TCH) is 260 m maximum (Figure 28). Thicks are coincident with main axis of the incised valley which reflects accommodation space created by relative sea level fall which caused the valley to be incised couple with shelf instability. This incised valley created significant accommodation space which is filled by a progradational-aggradational of deltaic system during the early sea level rise. From well logs, this uppermost sequence is dominantly sandy and exhibits progradational parasequences coastal barrier and undifferentiated continental deposit, identified from wireline stacking patterns. An increase in the sand content upwardly within the sequence indicates that we are at the uppermost section of the interpreted section

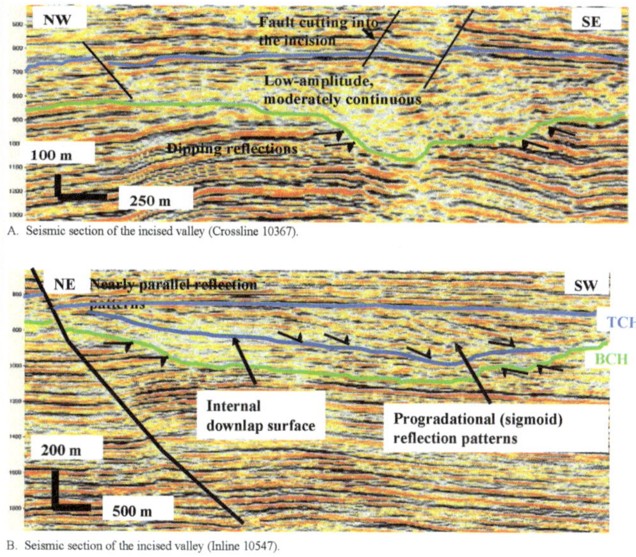

Figure 29: (A). Seismic section of the major incised valley (Crossline 10367). See position in Figure 27. (B). Seismic section of the major incised valley. (Inline 10547). Reflection geometry within the valley is characterized by progradational (sigmoid) with internal downlap surface. See position in Figure 27.

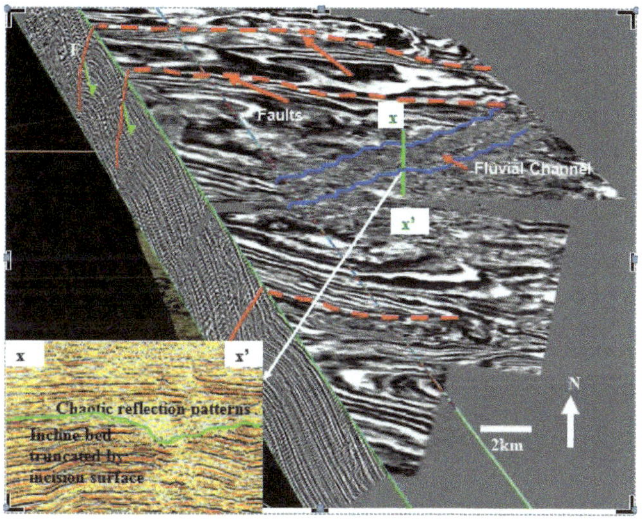

Figure 30: Seismic horizon slices showing the channel and faults orientation at depth 889m and geometry. Channel orientation is sub parallel to the faults which suggest that, the channel orientation is structurally controlled. Inset is the seismic section of the incised valley (Crossline 10367).

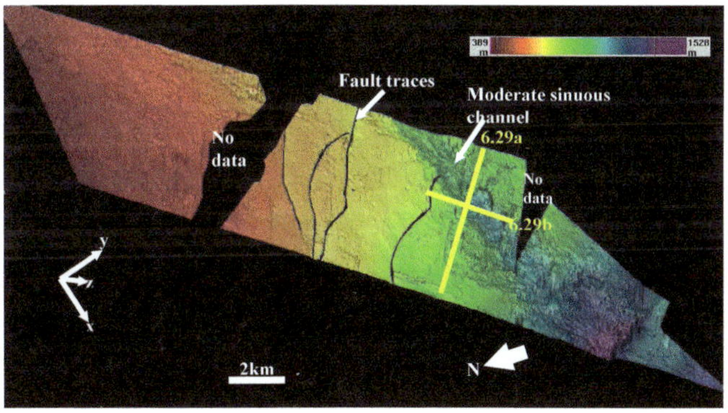

Figure 27: Depth structure map of base of incised valley trending NNE-SSW.

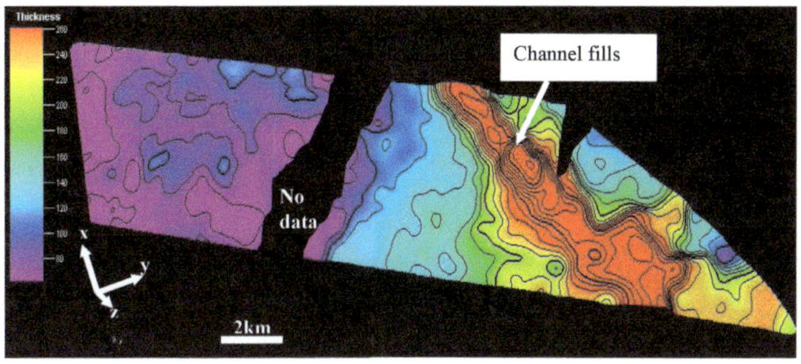

Figure 28: Isopach map between base of incised valley and the top. Thick is within the valley which reflect infilling of accommodation space created during the lowstand period.

(Agbada Formation), grading into fully continental sands deposits interval (Benin Formation).

Idealized sequence stratigraphic development model within the study area are summarized in Figures 31-33.

Discussion

Specific ages are assigned to the key surfaces based on the integration of biostratigraphic report with documentation by Doust and Omatsola, 1990 and the regional biozonal ages of Niger delta complex proposed by Shell Nigeria. The difficulty with a subsurface study in the Niger Delta is that most of the subsurface data especially the biostratigraphic data, is not publicly available. Therefore, the age constraint within

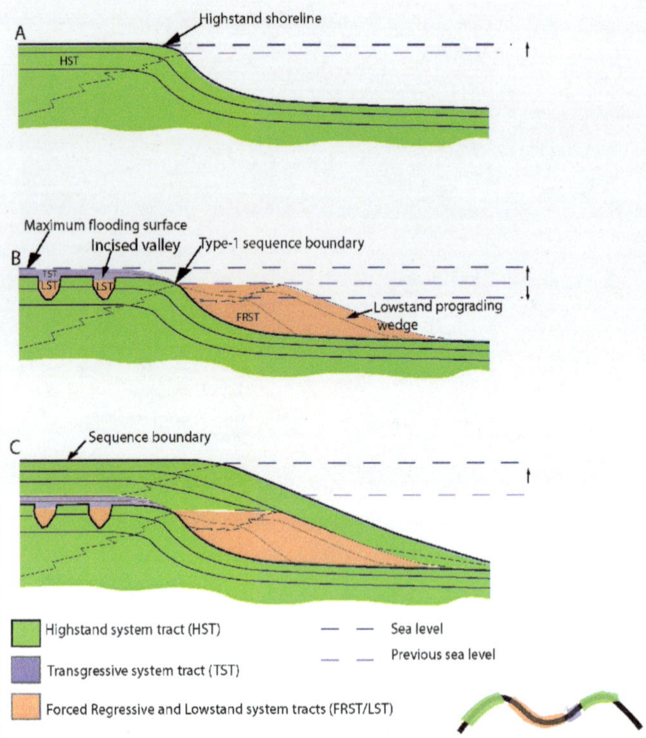

Figure 31: Idealised schematic sequence stratigraphic model in dip-section showing system tracts and key stratal surfaces developed during a complete depositional sequence within the study area. Based on [14] model.

this studies relies upon integration of the biostratigraphic report from three wells (Ewan 1x, Oloye 1 and 2). The age dating is based on palynostratigraphic zonation of Niger Delta documented by Evamy [9]. It should be noted that Ewan 1x is at the proximal end and is higher in the stratigraphic section compared to Oloye 1 and 2 (Figure 1).

The age of the MFS4 surface that defines the downlap surface of depositional sequence 4 is constrained by the palynostratigraphic zonation of [9] within Ewan 1x at a depth 1326 m as the top of regular occurrence of *Belskipollis elegans* which is 11.5 Ma (middle Miocene).

The age of MFS3 is constrained to be the top rich occurrence of *Belskipollis elegans* at depth 1655 m. However, based on the above constrains, the base of the sequence 4 will be the unconformity below MFS4 which is constrained to be 12.1 Ma while the immediate sequence boundary is dated as 10.6 Ma (Figures 32 and 33). The base of sequence boundary 3 will be older in age compared with the age of MFS3 and it is interpreted to correspond to 13.1 Ma. The ages of the key surfaces mapped below this sequence will be older in age. Based on the biostratigraphic zonation report above, the age of the studied section is middle Miocene to early Pliocene (Figure 33).

Based on above constraints, the four depositional sequences recognised from the well logs and the maximum flooding surfaces identified are;-11.5 Ma MFS, 12.8 Ma MFS, and 15.0 Ma MFS at depth c1260 m, c1650 m and c1925 m respectively on Ewan 07 (Figure 3). The MFS of other wells was extrapolated and correlated using well log stacking patterns.

Interpretations and descriptions of depositional sequences suggest that most sequence boundaries are eroded by sea level lowstand

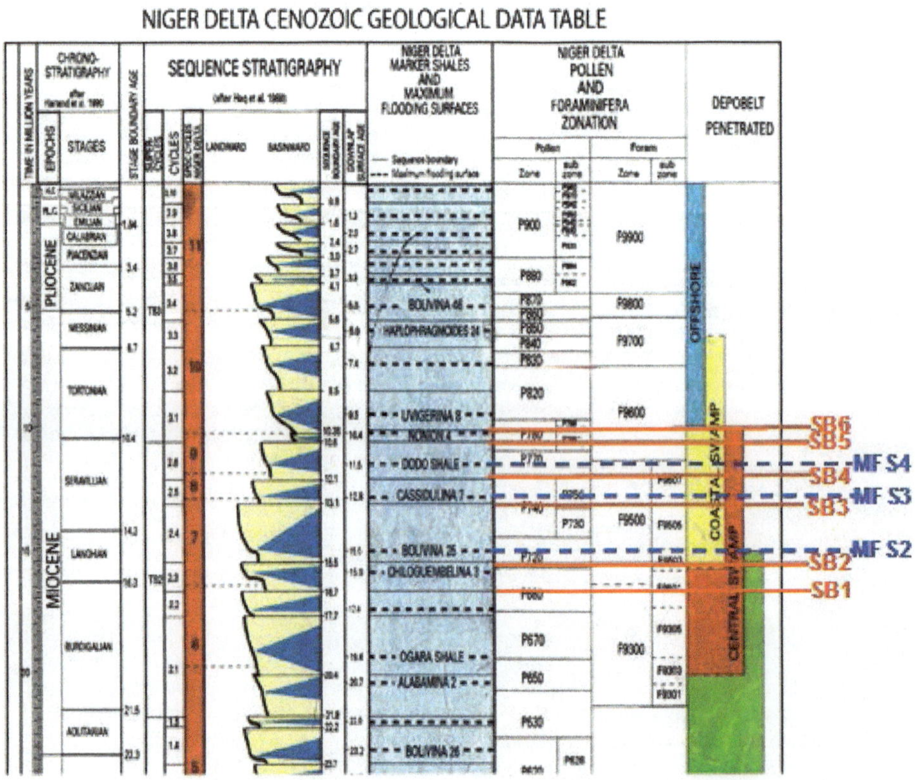

Figure 32: Regional Biozonal Ages for the Niger delta Complex from Miocene to Recent. The figure is updated with the key surfaces interpreted within the study area. (Modified from Shell Petroleum Development Company, Nigeria).

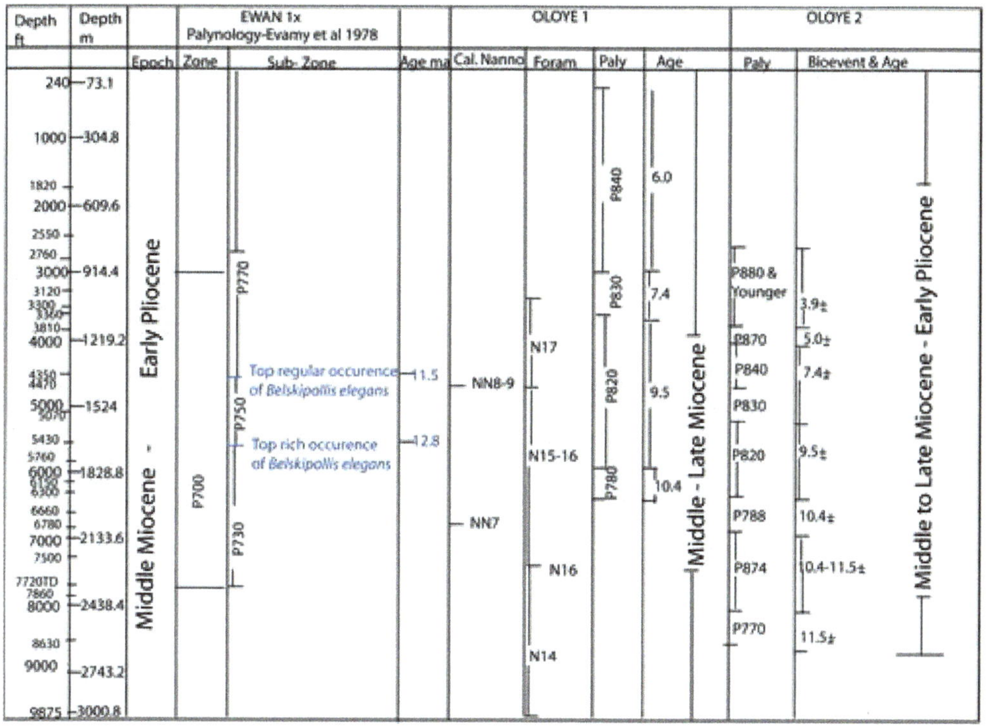

Figure 33: Biostratigraphic Zonation of Ewan and Oloye Fields Integrated from the Biostratigraphic Reports of Ewan 1x, Oloye 1 and 2 obtained form Chevron, Nigeria. Note- age is based on the Palynostratigraphic zonation of [9] using biozonal ages of Niger Delta complex. Palynology (P)-[9], Foraminifera (N)- Blow and Calcareous Nannofossil (NN)- Martini.

incision of the shelf, coupled with increased shelf instability and slope collapse structures. Patterns of sequence erosion occur both on the footwalls and hanging walls (but deeper mostly in the hanging wall) and deposits thickens across downthrown fault blocks, suggesting that most of these faults were active during the middle Miocene time of Agbada Formation deposition. The restriction of lowstands system tract deposits to within the valley fills across the entire study area, and the occurrence of deep incisions, suggests that the entire shelf region became periodically subaerially exposed during the successive relative sea level falls.

Within the study area, a general thinning of the sequences upward and basinward were observed. The thinning of the sequences basinward from the shelf edge occurred over a few kilometres. This supports a shelf-edge deltaic depositional setting. Thickness changes between stratigraphic surfaces within the formation are strongly influenced by structural deformation, with sediment thinning away from faults and over the crest of rollover anticlines. These thickness variations also strongly reflect the varying depth and magnitude of incision along the bounding sequence boundaries and the HST progradational clinoforms geometry. An upward decrease in thickness in the hanging wall of the fault blocks is observed between the isopach of successive sequence boundaries and maximum flooding surfaces within the studied area. This may be due to upward decrease in movement along some faults tips which in turn reduced the accommodation space. Progressive increase in local incision at the successive sequences and decrease in the thickness of successive sequences reflects rapid accommodation infilling, increased rates of sediment bypass, and decreased local rates of sediment loading. As a result of this accommodation infilling, the locus of sediment deposition and location of rapid sediment loading would also have shifted seaward. This is revealed by strong erosion and rapid progradation of sediments over the inactive faults updip, and this has led to rapid sedimentation and initiation of growth faults basinward (Appendices A, C and D). This reflects larger-scale deltaic progradation into the basin. Observation from seismic cross sections shows that areas of deepest incision follow onshore-offshore elongate trends and they do occur at the shelf break. Interpretations of these sequences cut by incised valley fills suggest relative sea level fall below the shelf edge which have resulted in shelf instability and structural collapse.

The depth of incision within the uppermost sequence (Depositional Sequence 6; up to 350 m) (Figures 27- 30) is difficult to explain in terms of sea level fall alone because eustatically-driven sea level variations during the Miocene are generally reported to be less than 100 m (330 ft) [11]. This suggests that local controls on incision, sequence development and the depositional facies preserved depend on relative sea level fall controlled by a complex interaction of eustacy, regional shelf/slope subsidence and structural collapse, shelf instability and rates of sedimentation. Sea level falls may facilitate delivery of large volumes of sediment that can destabilise the shelf and slope and eventually lead to slumping [12]. As a result of instability within the shelf, the deposits within the incised valleys in turn become unstable causing the infills to be eroded and transported to deeper water. This headward erosion caused the valleys to be deeply incised as a result of shelf instability. The edge-ward erosion assumed to have ceased incising when the shelf became stable and the gradient within the shelf area become gentle (landward of shelf). These developments of deep incision at the shelf edge have some implication for deltaic depositional sequences variability.

It is not possible to extract an analysis of longer-term fault behaviour and evolution using the isopachs generated in this study, due to the length of the observation period (about 16.7 – 10.35 myr) and length of the faults, because most of the faults extend beyond the study area and the full length of the faults is not imaged. Observation from the sequence analysis shows that throughout sequence development within the study area, the shelf margin has been controlled also by growth faults which have migrated basinward through time (Appendices A, C and D). In a case where the shelf margin is being controlled by growth faults, one of the processes of transporting sand into the basin is shelf collapse, with resultant slumping. However, there is an observed reduction in the upward propagation of faults. This may be due to changes in sediment deposition rates and fluvial discharge locations causing minor faults to become buried and truncated; fault segmentation and differential displacement on adjacent segments. Sedimentation patterns suggest that faulting, local incision and HST clinoform geometry dominate the formation of depocentres across the shelf.

The overall strong progradational character of the study area as evidenced from the overall seismic geometry, prograding of shoreline break as observed from GDE maps, structural maps of the sequence boundaries and dip line section at the proximal part. They are interpreted to reflect a long time decrease in accommodation triggered by a combination of tectonic uplift of the north-western Niger delta and the middle Miocene eustatic fall [1,13]. Increase in the proportion of sandstone relative to mudstone up section within the study area clearly also shows long-term progradation and graduation from the paralic Agbada Formation into continental sand Benin Formation.

Conclusion

Detailed sequence stratigraphic analyses showed that the middle Miocene-early Pliocene strata in the Ewan and Oloye fields of the northwestern Niger delta consist of six depositional sequences (Appendices A-D) (Figures 4 and 5). In general, sequences are widespread. They are made up of progradational HST, TST are thin but widespread, and LST are restricted to within valley fills. Outside the LST valleys, transgressive surfaces and sequence boundaries are coincident on interfluves. This suggests that during LST times most of the area was periodically subaerially exposed. During lowstand period, shorelines and facies belts prograde basinward (south) and there are a development of canyons which serve as a conduit for high amplitude deposit (deep water sand) on the slope and within the basin floor. This mechanism is interpreted to have led to the deposition of considerable amount of sandstones basinward [14,15].

Individual sequences can be mapped across the study area and biostratigraphic data allow a correlation with regional biozonal ages for the Niger delta complex. However, there is strong local growth fault control on sequence variability. Further local controls on sequence geometry are shelf instability and slope collapse structures which have a strong geomorphological impact on the area. The slope scars act as a locus for lowstand incision and affect overall sequence thickness. The incisions are narrower and deeper in lower sequences whereas those at the higher sequences are broader and less deeply incised. The additional accommodation space above slope scars allows local development of thick HST with seismically resolvable clinoforms. The occurrence of the several incisions in the middle Miocene succession is interpreted as evidence of significant relative sea level fluctuations, and the presence of type-1 sequence boundaries may be the stratigraphic signature of major drops in relative sea level during Miocene and Pliocene.

Sequences developed within the middle Miocene in the study area are interpreted to be controlled locally by episodic shelf instability, basin physiography (shelf edge), fault growth and linkage. Relative sea level changes provide the major control on sequence development and deep valley incisions. The latter have a strong geomorphological impact on the area and a strong control on sequence variability. Idealized schematic sequence stratigraphic development models have been generated for the study area and are summarized in Figure 31.

References

1. Doust H, Omatsola E (1990) Niger Delta in JD Edwards and PA Santogrossi (eds) Divergent/Passive Margin Basin AAPG 48: 201-238.
2. Vail PR, Mitchum Jr RM, Thompson S (1977) Seismic stratigraphy and global changes of sea level, part 4: global cycles of relative changes of sea level. In: Payton CE (Ed), Seismic Stratigraphy – Applications to Hydrocarbon Exploration. AAPG Memoir 26: 83-97.
3. Posamentier HW, Jervey MT, Vail PR (1988) Eustatic Controls on Clastic Deposition I: Conceptual Framework: In Wilgus CK, Hastings, CG, St C Kendall, HW Posamentier, CA Ross and Van Wagoner, JC eds. Sea-Level Changes: An Integrated Approach: Society of Economic Paleontologists and Mineralogists Special Publication No 42 pp: 109-124.
4. Posamentier HW, Vail PR (1988) Eustatic Controls on Clastic Deposition II: Sequence and Systems Tract Models: In Wilgus CK, Hastings, CG, StC. Kendall, HWPosamentier, CA Ross and Van Wagoner, JC eds Sea-Level Changes: An Integrated Approach: Society of Economic Paleontologists and Mineralogists Special Publication No 42: 124-154.
5. Van-Wagoner JC, Posamentier HW, Mitchum RM, Vail PR, Sarg JF, et al. (1988) An Overview of the Fundamentals of Sequence Stratigraphy and Key Definition: In Wilgus, C.K, Hastings, CG, St C Kendall, HW Posamentier, CA Ross and Van Wagoner, JC eds. Sea-Level Changes: An Integrated Approach: Society of Economic Paleontologists and Mineralogists Special Publication No 42: 39-45.
6. Van-Wagoner JC, Mitchum RM, Campion KM, Rahmanian VD (1990) Siliciclastic Sequence Stratigraphy in Well Logs, Cores, and Outcrops: Concepts for High-Resolution Correlation of Time and Facies. AAPG Methods in Exploration series No 7.
7. Posamentier HW, Allen GP (1999) Siliciclastic Sequence Stratigraphy-Concepts and Application. SEPM Concepts in Sedimentology and Paleontology no pp: 7-210.
8. Williams H, Berge C, Hartung M, De Ruiter R, Ugueto GA (1997) Prediction of Nigerian reservoir sands through integration of seismic imaging and sequence stratigraphy. In: Shanley, K.W and Perkins, BE (Eds) Shallow Marine and Non-marine Reservoirs. GCSSEPM pp: 401-408.
9. Evamy BD, Haremboure J, Kimberling P, Knap WA, Molloy FA, et al. (1978) Hydrocarbon habitat of Tertiary Niger delta. AAPG Bull 62: 1-39.
10. Durogbitan AA (2010) Seismic, Sequence Stratigraphic and Structural Analysis of Ewan and Oloye Fields (Middle Miocene), Northwestern Niger Delta. PhD Thesis unpubl. Earth Science Department, The University of Manchester pp: 337.
11. Haq BU, Hardenbol J, Vail PR (1987) Chronology of fluctuating sea levels since the Triassic. Science 235: 1156-1167.
12. Colleman JM, Prior DB, Lindsay JF (1983) Deltaic influences n shelf-edge instability processes. In Stanley, DJ Moore GT (Eds). The Shelf Break: Critical interface on Continental Margins. Spec Publ Soc Econ Paleontol. Mineral 33: 121-137.
13. Weber K.J, Daukoru EM (1975) Petroleum Geology of the Niger Delta Proc. Ninth World Petrol. Congress 2: 209-221.
14. Gawthorpe RL, Hall M, Sharp I, Dreyer T (2000) Tectonically enhanced forced regressions: example from growth fold in extensional and compressional settings, the Miocene of the Suez rift and the Eocene of the Pyrenees. In Hunt D, Gawthorpe, RL (Eds.). Sedimentary Response to Forced Regression, vol 172 Geol Soc London Special Publ pp: 177-191.
15. Oomkens E (1974) Lithofacies relations in the Late Quaternary Niger Delta complex. Sedimentology 21: 195-222.

The Control of Vibration Transmission from an Engine to its Resonant Base Structure

MWS Lau[1]*, SC Fok[2], GL Seet[3] and E Low[4]

[1]*School of Mechanical and Systems Engineering, Faculty of Science, Agriculture and Engineering, University of Newcastle, Newcastle (UK), International Singapore*
[2]*Department of Mechanical Engineering, The Petroleum Institute, P.O. Box 2533, Abu Dhabi, UAE*
[3]*School of Mechanical and Aerospace Engineering, Nanyang Technological University, 50 Nanyang Avenue, Singapore*
[4]*Engineering Accreditation Board, Institution of Engineers, 70 Bukit Tinggi Road, Singapore*

Abstract

Engines and motors on board marine vessels are often mounted on flexible base supports. When the levels of vibration transmitted from such machines to the base structure are intolerable, vibration reduction techniques have to be used. In instances where machines are mounted via very stiff mounts to maintain shaft alignment, the transmission paths of the vibratory force are through the mounts. In such cases, active force cancellation at the mount locations using opposing controlled forces can be used. This paper discusses the use of frequency domain techniques to reduce the vibration with the H_∞ norm as a performance measure. The model of the system, identified using frequency response method is used in the design of the H_∞ controllers for active force cancellation. Two approaches were presented. Experimental results show a reduction of between 15 ~ 65% of the measured transmitted force at different mount locations when the machine is operated at a speed of about 1320 rpm. This frequency matches that of the machine-base system principal mode. Where the system response is not significant, reduction in transmitted force level is minimal. This limitation is the result of the control goal used in the design.

Keywords: Machine on flexible support; Frequency domain; H_∞ controller design; Vibration transmission control

Introduction

Vibration is a common issue in marine vessels [1]. In particular, engine and drive line induced vibration is a major concern. These disturbances, which vary with propulsion inertia and sea conditions, not only can affect passenger comfort [2], but also would lead to deterioration of systems' performances. Excessive vibrations can ultimately lead to systems failure. To overcome such problems, minimizing vibration in marine vessels is critical [3].

One of the most common approaches to isolate engine vibrations is to use passive mounts [4-6]. In this approach, light damping is needed to absorb the vibrations. However, the use of "soft" (relatively low stiffness) engine mounts may not be appropriate in certain situations where it is necessary to hold the engine firmly in place. Under such circumstances, "hard" engine mounts with high stiffness and heavy damping are normally needed to prevent excessive movements. These engine mounts often require active anti-vibration control [7].

The basic principle of the active anti-vibration is to counter the disturbance with an equal and opposite oscillating signal. This requires real-time monitoring of the vibration and a control strategy to generate the compensating forces. A typical active anti-vibration system would consist of sufficiently stiff engine mounts, force actuators, sensors and a controller. The sensors measured the vibration and feedback the information to the controller, which determines the compensating forces. The information is send to the actuators to generate the required forces at the mounts to cancel the vibration. Force actuators for active vibration isolation include pneumatic, electro-hydraulic, electro-magnetic as well as piezo-electric actuators.

There are many available techniques for the design of the active vibration control strategy [8-14]. These techniques range from classical to modern multiple variable feedback control theories to intelligent control. Although active anti-vibration controllers had been widely investigated, the practical application of simultaneous control of multiple engine mounting points on a flexible base has not been very wide spread, although engines on flexible structures are common in marine vessels. Such structures can be excited to near resonance state by sea conditions, for example, and can result in excessive engine movements.

The complexity of a vibrating system of very stiff mounts on a flexible base was discussed in the work of Scribner et al. [15]. For active control within a narrow frequency band, they highlighted the desirability to have a high modal overlap at disturbance frequencies to achieve a stable active vibration isolation system. With low modal overlaps, the base structure should have sufficient damping [16]. One way to satisfy these constraints is to design the active controller using the robust H_∞ control strategy [17].

This work investigates via modeling and experiments various methods of control using the H_∞ approach for the active control of a near resonant system associated with an engine "hard" mounted onto a flexible base structure. Two approaches to the design of the H_∞ type controllers were carried out and implemented to generate sets of counteracting forces at each mount location to reduce the root mean square of the transmitted vibration to the base. Their effectiveness are examined and compared.

***Corresponding author:** MWS Lau, School of Mechanical and Systems Engineering, Faculty of Science, Agriculture and Engineering, University of Newcastle, Newcastle (UK), International Singapore, E-mail: michael.lau@ncl.ac.uk

Materials and Methods

The approach taken involves a system identification of the model of a motor system mounted on a flexible support and the design of dynamic controllers using the H_∞ approach. The excitation frequency range used in determining the model covers the range of operation of the motor in operation.

Figure 1 shows the experimental setup. It consists of a motor, the mounts, actuators and the supporting platform. A Digital Signal Processor (DSP) is used as a multiple input-multiple-output controller.

The 5.5 kW, 3-phase, 50 Hz AC industrial motor simulates the engine hard mounted at four points onto an aluminium flexible plate of dimensions 1.5×0.5×0.01 m supported on a rigid square-section mild steel frame. This represents the mounting platform. During normal operation of the motor, the flexible structure can be excited to near resonance.

The vertical forces at the interfaces between the four mounts and the aluminium plate are measured by four in-line PCB 201B01 Quartz force ring sensors. Each sensor has a dynamic range of 0.01Hz to 90kHz and can measure a compressive force of about 44.48 N.

The set of controlled forces $\mathbf{F}_c = \{f_{c1}, f_{c2}, f_{c3}, f_{c4}\}^T$ are provided by actuators on top of each of the hard (aluminum ring) mounts. These actuators are Gearing Watson GWV4 Electro-dynamic shakers each reacting against a small inertia mass of 0.25kg. The actuator has a sinusoidal peak force of 17.8N corresponding to a peak acceleration of 892 m/s^2.

Two main component models have to be separately identified via experiments: the actuator model $\mathbf{G}_a$ and the machine hard-mounted on the base system model $\mathbf{G}_p$ using an external shaker to excite each of the four mount locations separately. The input excitation was a multi-sine signal given by:

$$x(t) = \sum_{k=1}^{N} A_k \cos\left(2\pi f_k t + \varphi_k\right) \; ; \; f_k = \frac{n_k}{T_r} \tag{1}$$

In equation (1), T_r is the measurement period and n_k is an integer. This time signal will be transformed by Discrete Fourier Transform into the frequency domain without spectral leakage errors when the sampling frequency is an integer multiple of the fundamental frequency $1/T_r$.

During the machine-base system model identification, the motor and the actuators were not operated.

Machine-base System Model G_p

The frequency range used for the machine-base system model identification has to have sufficient frequency resolution to resolve the frequency spectral peaks within the motor operating speed range, which is between 1320 rpm (22Hz) and 1500 rpm (25Hz). A suitable range of 0.125~50 Hz with resolution of 0.125 Hz (400 harmonics) was used.

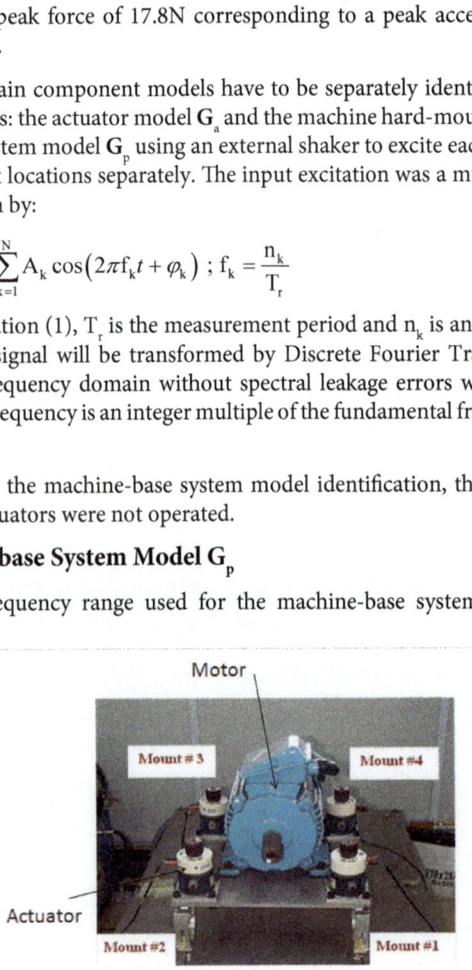

Figure 1: Experimental setup.

A frequency response system identification method [18] was used to identify all sixteen transfer functions providing a transfer function matrix $\mathbf{G}_p(s)$. A direct transformation of the transfer function matrix into a state space model $\mathbf{G}_{p154}$ will result in 154 states. A two-step balanced residualisation model reduction method is used [19] to reduce the number of states. The states are first arranged in descending Hankel singular values (HSV). Using the method of residualisation [20], states with very small HSV are removed. The process results in a minimal state space model $\mathbf{G}_{p12}$ of 12 states (corresponding to 6 modes) that are both observable and controllable. The identified model frequency responses are shown in Figures 2,3,4, and 5. There is a good match between the respective measurements and that obtained from the identified transfer functions. Figure 6 shows a good match in the maximum singular value $\overline{\sigma}$ ($\mathbf{G}_{p154}$) of the frequency plot between the G_{p154} model and $\overline{\sigma}$ ($\mathbf{G}_{p12}$) of the minimal state model $\mathbf{G}_{p12}$. In addition, in Figure 7, it is observed that $\overline{\sigma}$ ($\mathbf{G}_{p12}$) compares quite well with the sum (point-wise) of all measured frequency response plots. This illustrates that the identified model is a reasonable representation of the system.

Based on the input frequency resolution of 0.125Hz, the frequencies of the six modes are determined as: {12.875, 16, 21.5, 21.625, 33.875, 38.35} Hz. This set matches 6 of the 8 frequencies identified using ICATS (software for modal analysis and identification by Imperial College, London): {12.9, 15.9, 21.5, 21.6, 24.3, 33.8, 38.46, 46.7} Hz. The two modes that are not matched are underlined and are circled in Figure 7. The one at 24.3Hz was removed during model reduction. The higher frequency mode at 46.7 Hz was omitted during the identification process to reduce the order of the transfer function.

Actuator Model G_a

The frequency response of the actuator attached with a 0.25kg mass is shown in Figure 8. The input voltage was 1V peak. The amplitude at the 1st peak frequency around 20 Hz is about 8 mV and with a sensitivity of 1N (mV)$^{-1}$, the corresponding force is about 8 N. The actuator force magnitude tapered off to about 5N until the frequency approaches the 3rd resonance at about 2000Hz. The shaker has mass of 0.02 kg and the suspension axial stiffness is 4414.5 N/m. The actuator first resonant frequency is calculated to be:

$$f_a = \frac{1}{2\pi}\sqrt{\frac{4414.5}{0.25+0.02}} = 20.35Hz \tag{2}$$

The damping is approximated to be 0.7 and the model of the force actuator can be approximated by a 2nd order system:

$$G_a(s) = \frac{kV_f}{V_{in}} = k\frac{s^2}{s^2 + 2\left(0.7\right)\left(\omega_a\right)s + \omega_a^2} \tag{3}$$

where k is the amplifier variable. With this transfer function, a 4×4 diagonal matrix $\mathbf{G}_a = G_a \cdot \mathbf{I}$ is used to model all the four actuators dynamics.

Controller Design and Implementation

As the rotating machine-base system model $\mathbf{G}_p$ and actuators dynamics $\mathbf{G}_a$ are obtained from frequency domain identification method, it is natural to attempt to design the controller based on frequency domain concept. A performance (weighting) function $\mathbf{W}_p(s)$ based on the frequency plots of $\mathbf{G}_p(s)$ can be specified to design a set of controllers that will produce the specified system outputs. As H_∞ is a performance index norm measuring the maximum value of a transfer matrix (which is a function of a complex variable) in the frequency domain, the use of H_∞ design methodology will be appropriate. In addition, the method is suitable for multi-variable systems with cross couplings.

In the discussion that follows, the generic symbol $\mathbf{G}_p$ shall be used to represent the motor-based dynamics. Disturbance $\mathbf{F}_e$ arising from engine operations will result in forces $\mathbf{F}_b$ being transmitted to

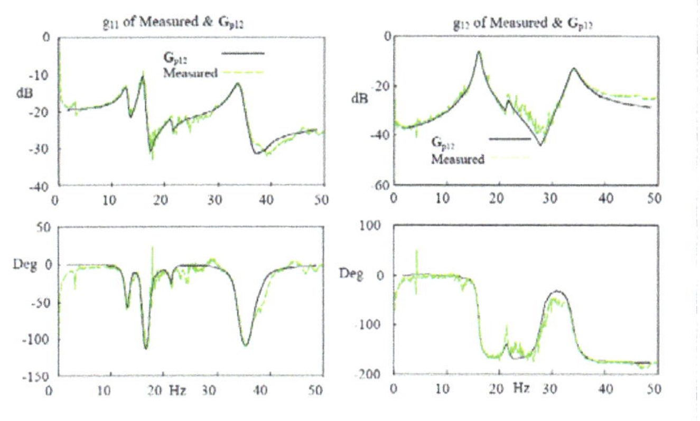

Figure 2: Sample Bode plots of g_{1i} of measured and G_{p12}.

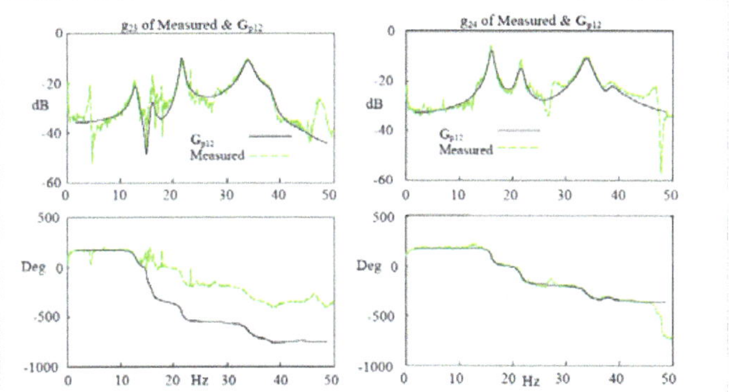

Figure 3: Sample Bode plots of g_{2i} of measured and G_{p12}.

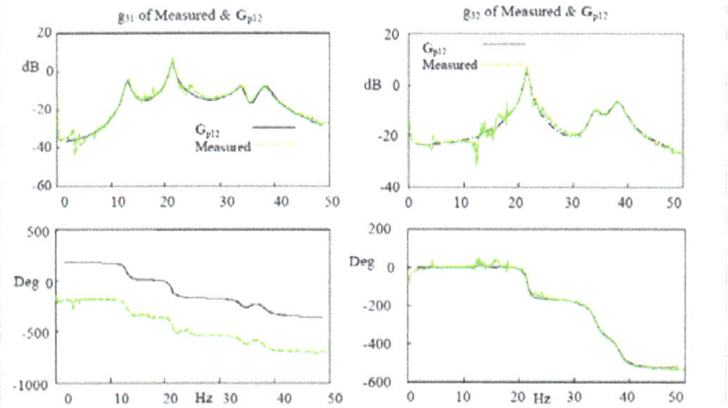

Figure 4: Sample Bode plots of g_{3i} of measured and G_{p12}.

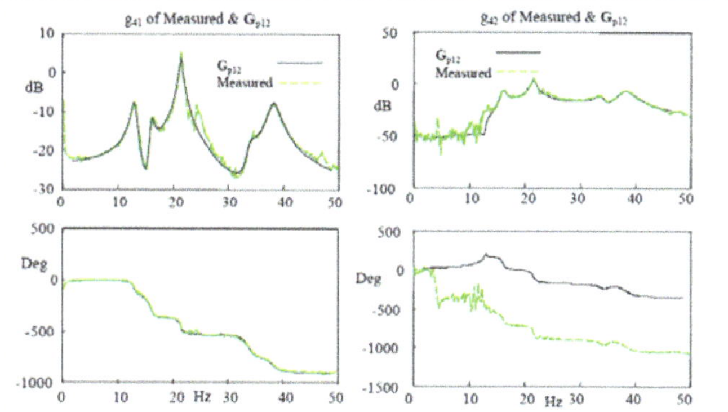

Figure 5: Sample Bode plots of g_{4i} of measured and G_{p12}.

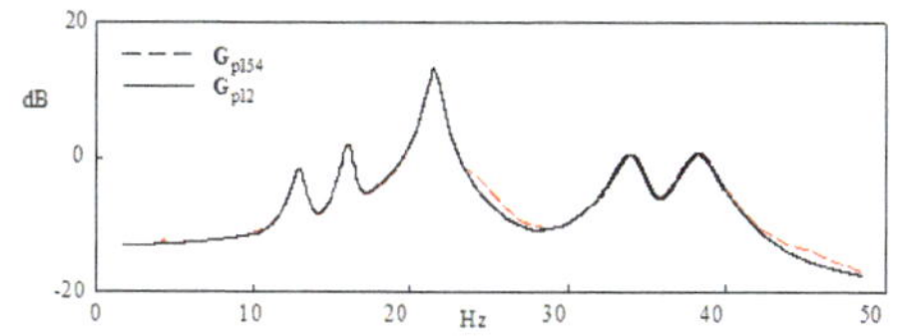

Figure 6: Plots of $\bar{\sigma}$ ($\mathbf{G}_{p12}$) and $\bar{\sigma}$ ($\mathbf{G}_{p154}$).

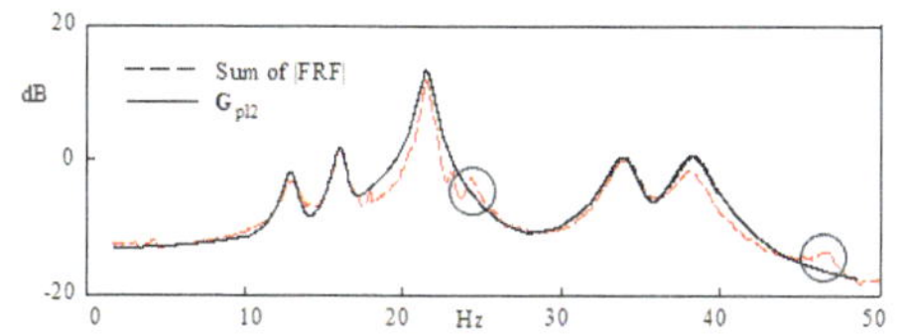

Figure 7: Plots of $\bar{\sigma}$ ($\mathbf{G}_{p12}$) and (factor $\sum_{i,j=1}^{4}$ |measured g_{ij}|).

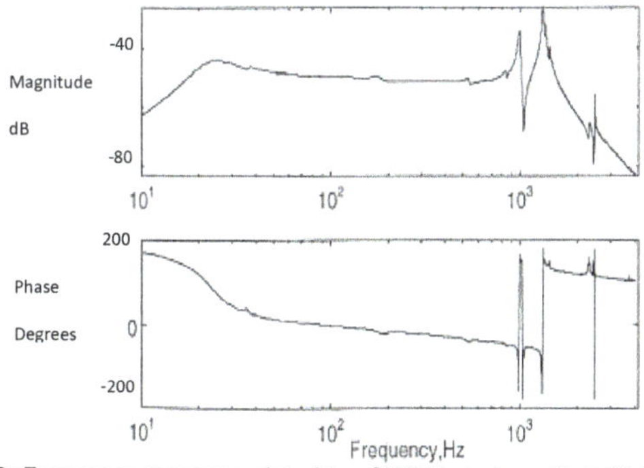

Figure 8: Frequency response plot of the GW V4 shaker with 0.25 kg mass.

the flexible base via the feet and the hard mount. A multiple-input multiple-output (MIMO) feedback controller $\mathbf{K}$ is designed to output signals to the actuators to generate a set of control forces, $\mathbf{F}_c$ to reduce $\mathbf{F}_b$. The term $\mathbf{F}_b$' will refer to the measured transmitted force after control, whereas $\mathbf{F}_b$ is the transmitted force due only to $\mathbf{F}_e$ (i.e. without force compensation).

The block diagram of the control system is illustrated in Figure 9. The closed loop transfer matrix is given by

$$\mathbf{T}: \mathbf{F}_e \rightarrow \mathbf{F}_b' = \mathbf{S} \cdot \mathbf{G}_p \tag{4}$$

In equation (4), $\mathbf{S}$ is the sensitivity matrix, $\mathbf{T} = \mathbf{I} - \mathbf{S}$ is the complimentary sensitivity matrix and **I** is the identity matrix.

Since the H_∞ norm of **T**, $||\mathbf{T}||_\infty = ||\mathbf{S}\cdot\mathbf{G}_p||_\infty \le ||\mathbf{S}||_\infty \cdot ||\mathbf{G}_p||_\infty$, designing $||\mathbf{S}||_\infty << 1$ would result in $||\mathbf{F}_b'|| < ||\mathbf{F}_b||$. However, for physical systems, $||\mathbf{S}||_\infty$ can only be less than 1 for some $\omega < \omega_h$. Under this condition, a controller **K** is designed such that $||\mathbf{S}||_\infty << 1$ for $\omega \in [\omega_l, \omega_h]$, resulting in

$$\bar{\sigma}\ (\mathbf{T}(j\omega)) < \bar{\sigma}\ (\mathbf{G}_p(j\omega)) \tag{5}$$

where $\bar{\sigma}$ (·) is the maximum singular value of the function. A systematic way to design a **K** is to specify a performance function $\mathbf{W}_p(s)$ comprising of simple low order transfer functions such that for a selected frequency range ω:

$$\gamma\cdot\bar{\sigma}\ (\mathbf{W}_p^{-1}(j\omega)) < \bar{\sigma}\ (\mathbf{G}_p(j\omega)) \tag{6}$$

The scalar parameter $\gamma > 0$ is associated with the design of **K** and $\mathbf{W}_p(s)$ has to be non-singular, with non-minimum-phase zeros, and $\mathbf{W}_p^{-1}(s)$ has to be stable. The performance weighting function $\mathbf{W}_p$ is designed to "shape" the output of the resulting closed system $\mathbf{W}_p\cdot\mathbf{T}$ using the controller **K** such that the resulting weighted closed loop system is stable i.e. the real part of the poles of $\mathbf{W}_p(s)\cdot\mathbf{T}(s) < 0$ and $||\mathbf{W}_p\cdot\mathbf{T}||_\infty$ is minimized or

$$||\mathbf{W}_p(s)\cdot\mathbf{T}(s)||_\infty \le \gamma \tag{7}$$

Equation (7) can be regarded as minimization problem which can be solved using the standard γ-iteration algorithm of Doyle et al. [17] to find **K** for a suitable γ value. With a specified $\mathbf{W}_p$, the design process starts with an initial γ value. This value is reduced until equation (6) is satisfied. Otherwise, another $\mathbf{W}_p$ has to be specified and the process is iterated.

Since $\underline{\sigma}\ (\mathbf{W}_p(j\omega))\cdot\bar{\sigma}\ (\mathbf{T}(j\omega)) \le \bar{\sigma}\ (\mathbf{W}_p(j\omega)\cdot\mathbf{T}(j\omega)) \le \gamma$ at each ω, (where $\underline{\sigma}$ (·) is the minimum singular value of the function),

$$\bar{\sigma}\ (\mathbf{T}(j\omega)) \le \frac{\gamma}{\underline{\sigma}\left(\mathbf{W}_p\left(j\grave{u}\right)\right)} = \gamma\cdot\bar{\sigma}\left(\mathbf{W}_p^{-1}\left(j\omega\right)\right) \tag{8}$$

Satisfying equation (8) implies that the desired condition given by (5) is achieved. The constraint on the frequency range $[\omega_l, {}_h]$ is essential for system model identification and for real time implementation. It is not unduly restrictive since the forces existing at the interface outside certain frequencies beyond the system bandwidth would be sufficiently small.

The actuator weighting function $\mathbf{W}_a$, which can be used to prevent actuators saturation, was set to **I**. The output weighting function $\mathbf{W}_p$ was now selected based on the measured $\bar{\sigma}\ (\mathbf{G}_p(j\omega))$ for $\omega \in [0.125, 50]$ Hz or $\omega \in [0.8, 310]$ rad/s. $\mathbf{W}_p$ is a 4×4 matrix with a maximum of sixteen transfer functions to be selected. To limit the number of states for the overall system, $\mathbf{W}_p$ was selected to be diagonal

$$\mathbf{W}_p = \begin{bmatrix} w_{11} & 0 & 0 & 0 \\ 0 & w_{22} & 0 & 0 \\ 0 & 0 & w_{33} & 0 \\ 0 & 0 & 0 & w_{44} \end{bmatrix} \tag{9}$$

where

$$w_{ii} = \frac{\left(s\tau_{i2}+1\right)^2}{\alpha_i^2\left(s\tau_{i1}+1\right)^2}; i = 1,2,3,4 \tag{10}$$

The simplest design is to make all elements w_{ii} the same. This requires the selection of α, τ_1 and τ_2. Typically w_{ii} are 2nd order lag-lead filters (i.e. low pass follow by high pass) so that weighted closed function $\mathbf{W}_p\cdot\mathbf{T}$ will have sufficient roll off at high frequency. The low pass cut off frequency was set to be above 50 Hz or 314 rad/s (the maximum frequency used for model identification) and $\bar{\sigma}\ (\mathbf{W}_p^{-1}(j\omega))$ should be small in the range where $\bar{\sigma}\ (\mathbf{G}_p(j\omega))$ was large. The high frequency cut-off can be adjusted to get a suitable design. A w_{ii} with $\alpha^2 = 0.3$, $\tau_1 = 1/400$ and $\tau_2 = 1/3000$ had been selected and the frequency response plot is shown in Figure 10.

The H_∞ controller was designed using $\mathbf{G}_a$ and the identified $\mathbf{G}_p(s)$ or its reduced version $\mathbf{G}_{p12}$. Two ways of designing the controller **K** had been attempted.

The first set of $\mathbf{K}_8$ controllers was designed using $\mathbf{G}_{p154}$ derived from the identification of the 16 transfer functions matrix $\mathbf{G}_p(s)$. The goal was to have a controller for all the possible modes of response and that could give a closed loop stable system. The controller with the smallest γ value was used for the implementation. With the $\mathbf{H}_\infty$ method, the controller **K** will have as many states as the total number of system states. With 154 states from $\mathbf{G}_{p154}$, 8 from $\mathbf{G}_a$, 8 from $\mathbf{W}_p$ and 0 from $\mathbf{W}_a$ being a constant matrix, the designed controller had 170 states, which would be difficult to implement. Using the two-step balanced residualisation model reduction method discussed, a $\mathbf{K}_8$ controller, with 8 observable and controllable states, was obtained.

The other $\mathbf{K}_{28}$ (with 28 states in the state space form) controller was designed using the $\mathbf{G}_{p12}$ model. The $\mathbf{K}_{28}$ picked up 12 states from the motor-mount model $\mathbf{G}_{p12}$, 8 from $\mathbf{G}_a$, and 8 from $\mathbf{W}_p$. Since the $\mathbf{K}_{28}$ controller was to be implemented on the actual system, it had to be tested for closed loop stability on the original plant model $\mathbf{G}_p(s)$, which though contains excess poles nevertheless, had some poles neglected in the reduced model $\mathbf{G}_{p12}$. In brief, the design procedure involved:

a) Finding a suitable $\mathbf{W}_p$

b) Performing a sub-optimisation to find a stable $\mathbf{K}_{28}$

c) Checking that the $\mathbf{K}_{28}$ stabilises the closed loop system using $\mathbf{G}_p(s)$

d) Verifying that the resulting closed loop system, $||\mathbf{T}||_\infty < ||\mathbf{G}_p(s)||_\infty$

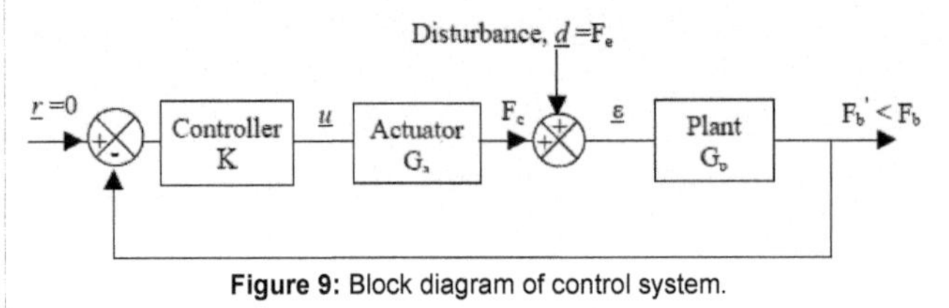

Figure 9: Block diagram of control system.

Figure 12a shows the maximum singular values of $\mathbf{G}_{p12}$ and the resulting **T** using the $\mathbf{K}_{28}$ controller. Similarly, figure 12b shows the maximum singular values of $\mathbf{G}_{p154}$ and the resulting **T** using the $\mathbf{K}_8$ controller. Both figures show that $\bar{\sigma}\ (\mathbf{T}(j\omega)) < \bar{\sigma}\ (\mathbf{G}_p(j\omega))$ for $\omega \in [70,280]$ rad/s. Maximum reduction occurred at $\omega \approx 135$ rad/s, which met the design requirement.

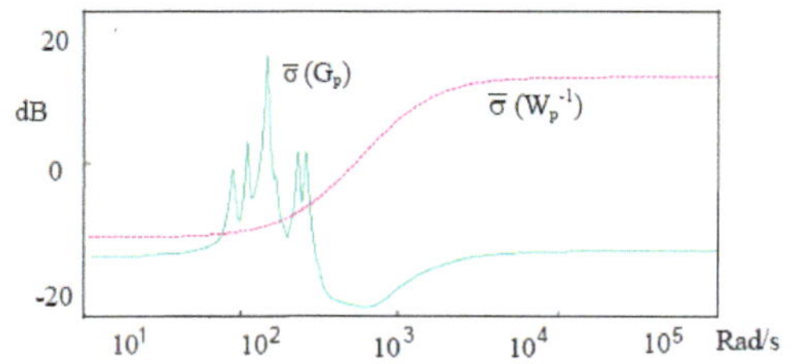

Figure 10: The selected $\mathbf{W}_p$ as a disturbance rejection performance specification.

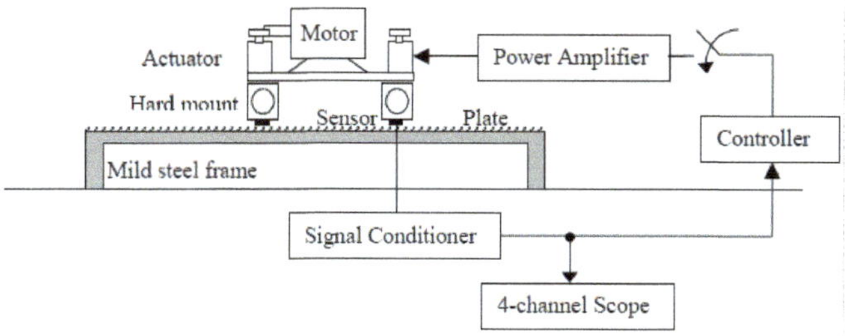

Figure 11: The DSP based control system.

The two controllers $\mathbf{K}_8$ and $\mathbf{K}_{28}$ were implemented in C-language on a PC based Digital Signal Processor (DSP) controller which has 4 channels of Analogue-to-Digital and Digital-to-Analogue converters. The control sampling time was selected to be 0.000125 msec and 1 msec respectively. Figure 11 shows the layout of the DSP based control system with only a one-channel loop shown for clarity. A set of switches was used to turn on and off the controller **K** to demonstrate their effectiveness.

Results and Discussion

For the experiment, the motor was operated between 1300 and 1500 rpm. This operating frequency range covered the dominant system resonant mode of about 21.5 Hz (1290 rpm). Figure 13 showed the force sensor output at each mount location when the machine was operated at 1320 rpm (22 Hz), which was quite close to the resonant mode.

Between 1300 and 1380 rpm, both controllers had been shown to be quite effective. The $\mathbf{K}_{28}$ controller worked very well for the machine operating at 1320 rpm and vibration reduction was better compared with the $\mathbf{K}_8$ controller. Results for both controllers indicated that attenuation at mount #2 was not as good as the other mount locations, in particular at mount #3. This could be due to the center of gravity of the motor being located closer to rear of the motor. In general, the measured transmitted forces at motor rear mount locations #3 and #4 were typically higher than those at its front, and a higher reduction of vibration was correspondingly achieved. Reductions in RMS values for both the controllers were summarised in Table 1.

Figure 14 showed force sensor outputs when the motor was operated at 1440 rpm. No significant attenuation was observed for both controllers.

The reduction in RMS values (shown in Table 1) was expected as the gain of the system can be defined as:

$$\|\mathbf{G}_p\|_{RMS\_gain} = \sup_{|\mathbf{F}_e| \neq 0} \frac{\|\mathbf{G}_p \cdot \mathbf{F}_e\|_{RMS}}{\|\mathbf{F}_e\|_{RMS}} = \max_{\omega} \bar{\sigma}(\mathbf{G}_p(jw)) = \|\mathbf{G}_p\|_\infty \quad (11)$$

If $\mathbf{g}_p(t)$ is the time domain model of $\mathbf{G}_p(s)$ then the gain $||\mathbf{g}_p||_2$ of the system is an induced 2-norm, which is also related to the infinity norm of $\mathbf{G}_p(s)$ i.e.

$$\|\mathbf{G}_p\|_\infty = \max_{\omega} \bar{\sigma}(\mathbf{G}_p(j\omega)) = \|\mathbf{g}_p\|_2 .$$

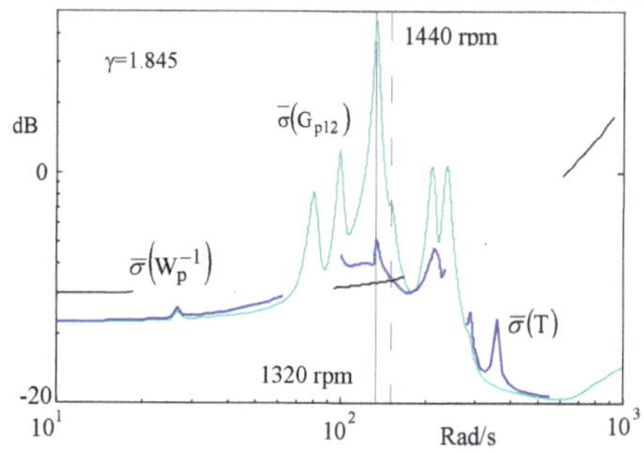

Figure 12a: $\mathbf{K}_{28}$ design with maximum singular values of $\mathbf{G}_{p12}$ and **T**.

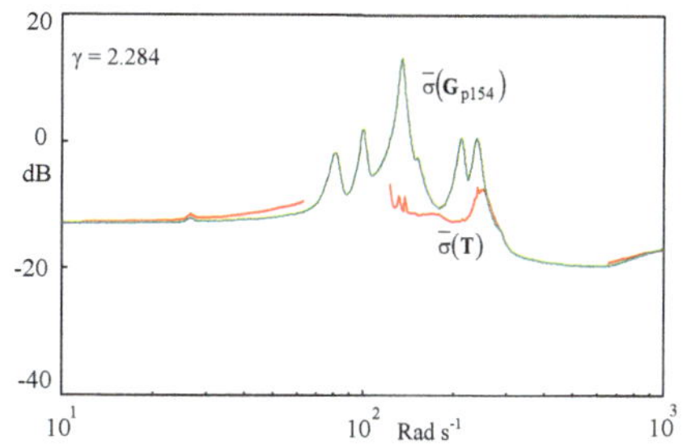

Figure 12b: $\mathbf{K}_8$ design with maximum singular values of $\mathbf{G}_{p154}$ and **T**.

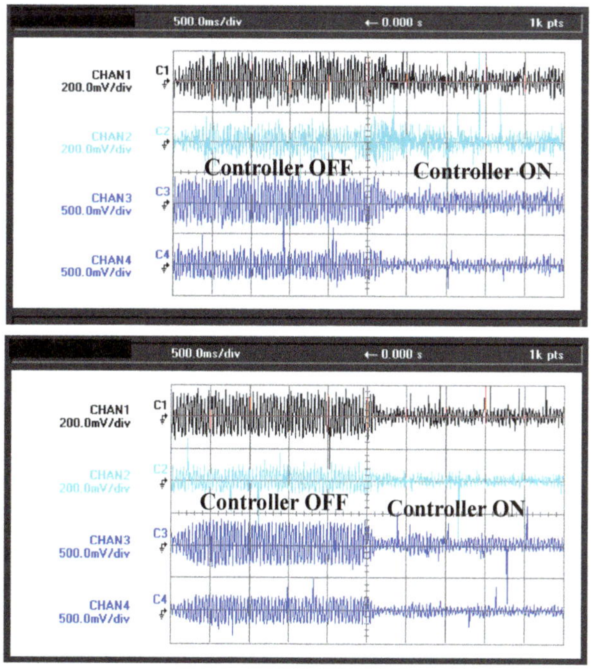

Figure 13: Results of $\mathbf{K}_8$ (top) and $\mathbf{K}_{28}$ controller for motor operated at about 1320 rpm.

	Mount #1 RMS values		Mount #2 RMS values		Mount #3 RMS values		Mount #4 RMS values	
	Control OFF	Control ON	Control OFF	Control ON	Control OFF	Control ON	Control OFF	Control ON
K_{28}	0.065	0.033	0.046	0.038	0.179	0.058	0.101	0.050
	49.7%		17.1%		67.8%		50.2%	
K_8	0.034	0.026	0.039	0.036	0.144	0.068	0.074	0.048
	21.0%		7.7%		52.9%		35.3%	

Table 1: RMS reduction in responses with and without controllers for motor at 1320 rpm.

As shown in Figures 12a and 12b the designed controllers resulted in lower values of $\overline{\sigma}$ $(\mathbf{T}(j\omega))$ at all the peaks frequencies of $\overline{\sigma}$ $(\mathbf{G}_p(j\omega))$. These frequencies were with the modal frequencies of the system. The most significant reduction was around the highest peak at about 21.5 Hz (135 rad/s or 1290 rpm). This indeed was the case in the experiment when the motor was operated at 1320 rpm (22 Hz).

When the motor was operated at 1440 rpm (24 Hz), for instance, the system had no resonant mode there and reduction was insignificant at all locations both in the design and in the experiment. Through the selection of $\mathbf{W}_p$ and the minimisation of $||\mathbf{W}_p{\cdot}\mathbf{T}||_\infty$, the $\mathbf{H}_\infty$ controller design method targeted and "pushed down" the worst case gain.

As shown in the experiment, reducing the controller order was not necessarily beneficial since faster processors are currently available. Moreover, the reduction in the order of the controller below the plant minimal order could reduce the controller effectiveness.

One advantage of using the $\mathbf{H}_\infty$ control design method is that worst case peak can be "pushed down". However, this has a "waterbed" effect. As shown in Figure 12a, there would be peaks of $\overline{\sigma}$ $(\mathbf{T}(j\omega))$ appearing outside the range of interest. For example, peaks appeared beyond 50Hz frequency (3000 rpm), but it was beyond the motor rated operation. In the experiment, the pushed down effect is significant only at the frequency of largest peak and at the mount locations where the resonance is more pronounced. At mount #2, for example, the reductions are much less than at mount #3. This was also a consequence of the design goal and method i.e. a reduction in overall system gain, and not for example an equal reduction at all mounts. The disadvantage of the push down effect is that one cannot specifically design to reduce vibration at any particular speed, for example in this case the rated speed of the motor at 24Hz or 1440rpm. At this speed as the system is not in resonance, the reduction is not significant.

In many instances, the use of active soft mounts had been shown to be quite effective in providing high stiffness at relatively low frequency and compliance (larger than the base structure) at frequencies where isolation in needed. Alternatively active vibration control of very stiff mount system that naturally provided the high stiffness at low frequencies and active force cancellation at higher frequencies can also be found in the literature. This method has the advantage in that during actuation, the motor-base system is not vibrated by the disturbances.

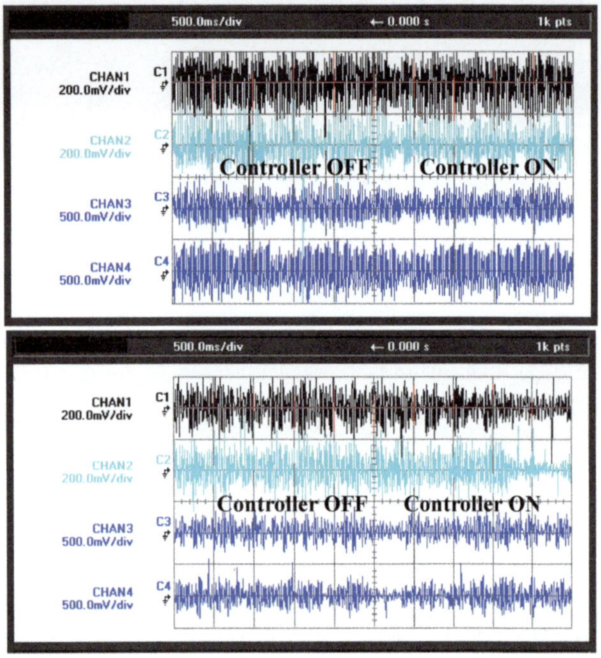

Figure 14: Results of K_8 (top) and K_{28} controller for motor operated at about 1440 rpm.

Concluding Remarks

In this paper, a feedback compensation method for an MIMO active vibration isolation system has been discussed and its ability to reduce the vibration near machine-base resonance (ranging from 15% to 65% at the four mounts) has been demonstrated experimentally. The method uses the uncompensated system frequency response as a design parameter and H_∞ norm as the design objective in a $\mathbf{H}_\infty$ controller design method. This method is quite appropriate when there is a large peak given by the system $\mathbf{H}_\infty$-norm or $\overline{\sigma}$ $(\mathbf{G}_p(j\omega))$ and is effective at this frequency. Controller order reduction is not necessarily beneficial and the reduction in the order of the controller below the plant minimal order could reduce the controller effectiveness.

References

1. Bambill DV, Escanes SJ, Rossit CA (2003) Forced vibrations of a clamped-free beam with a mass at the free end with an external periodic disturbance acting on the mass with applications in ships' structures. Ocean Eng 30: 1065-1077.

2. Coe TE, Xing JT, Shenoi RA, Taunton D (2009) A simplified 3-D human body–seat interaction model and its applications to the vibration isolation design of high-speed marine craft. Ocean Eng 36: 732-746.

3. Dylejko PG, Kessissoglou NJ, Tso Y, Norwood CJ (2007) Optimisation of a resonance changer to minimise the vibration transmission in marine vessels. J Sound Vib 300: 101-116.

4. Tao JS, Liu GR, Lam KY (2000) Design optimization of marine engine–mount system. J Sound Vib 235: 477-494.

5. Sirafi M, Chang YP, Qatu MS (2006) Robustness of mount systems for idle NVH, Part I: Centre of Gravity (CG) mounts. Int J Vehicle Noise and Vibration 2: 317-333.

6. Sirafi M, Chang YP, Qatu MS (2006) Robustness of mount systems for idle NVH, Part II: pendulum mounts. Int J Vehicle Noise and Vibration 2: 334-340.

7. Buchholz K (2000) Good vibrations. Automotive Eng 108: 85-89.

8. Gradonio P, Elliot SJ, Pinnington RJ (1997) Active isolation of structural vibration on a multiple-degree-of-freedom system Part 2: effectiveness of active control strategies. J Sound Vib 207: 95-121.

9. Nakaji Y, Satoh S, Kimura T, Hamabe T, Akatsu, et al. (1999) Development of an active control engine mount system. Vehicle Syst Dyn 32: 185-193.

10. Kim SM, Elliot SG, Brennan MJ (2001) Decentralized control for multichannel active vibration isolation. IEEE Trans Instrum Meas 9: 93-100.

11. Lee YW, Lee CW (2002) Dynamic analysis and control of active engine mount system. Journal of Automobile Engineering 216: 921-931.

12. JWu JD, Su FC, Tseng WK (2005) Vibration isolation for engine mount systems using an active hybrid robust controller. Int J Vehicle Noise and Vibration 1: 251-264.

13. Darsivan FJ, Faris WF, Martono W (2008) Active engine mounting controller using extended minimal resource allocating networks. Int J Vehicle Noise and Vibration 4: 150-168.

14. Hasbullah F, Faris WF (2010) An evaluation of LQR and fuzzy logic controllers for active suspension using half car model. Int J Vehicle Noise and Vibration 6: 200-214.

15. Scribner KB, Sievers LA, von Flotow AH (1990) Active Narrow-band Vibration Isolation of Machinery Noise from Resonant Substructures. J Sound Vib 167: 17-40.

16. von Flotow AH, Vos DW (1991) The Need for Passive Damping in Feedback Controlled Flexible Structures. Proceedings of the Recent Advances in Active Control of Sound and Vibration.

17. Doyle JC, Glover K, Khargonekar PP, Francis B (1989) State-space solutions to the standard H_2 and H_∞ control problems. IEEE Trans Automat Contr 34: 831-847.

18. Schoukens J, Pintelon R, Renneboog J (1988) A Maximum Likelihood Estimator for Linear and Non-Linear Systems - A Practical Application of Estimation Techniques in Measurement Problems. IEEE Trans Instrum Meas 37: 10-17.

19. Glover K (1984) All optimal Hankel-norm approximations of linear multi-variable systems and their L_∞ error bounds. International Journal of Control 39: 1115-1193.

20. Skogestad S, Postlethwaite I (1996) Multivariable feedback control – analysis and design, John Wiley & Sons.

24

Spatial and Temporal Variations in Phytoplankton Abundance and Species Diversity in the Sundarbans Mangrove Forest of Bangladesh

Rahaman SMB[1]*, Golder J[2], Rahaman MS[3], Hasanuzzaman AFM[1], Huq KA[1], Begum S[2], Islam SS[1] and Bir J[1]

[1]*Fisheries and Marine Research Technology Discipline, Khulna University, Khulna 9208, Bangladesh*
[2]*Environmental Science Discipline, Khulna University, Khulna 9208, Bangladesh*
[3]*Department of Chemistry, Comilla University, Comilla 3500, Bangladesh*

Abstract

The study examined taxonomic composition, abundance and spatial distribution of phytoplankton, and water quality of three major river systems of the Sundarbans. A total of 134 phytoplankton species were identified, and diversity and abundance were found to fluctuate with time and space. A total of 97 species were enumerated in Rupsha-Pashur while 122 and 110 in Khalpatua-Arpangachia and Bhola-Baleswar river system respectively. Abundance was lowest ($3.709\times10^3 \pm 4.257\times10^2$ cellsL^{-1}) in monsoon and highest ($2.174\times10^5 \pm 1.723\times10^5$cellsL^{-1}) in summer in Bhola-Baleswar. Species composition was dominated by Bacillariophyta over the area except in summer in Bhola-Baleswar, where Cyanophyta become dominated. Species diversity, richness and evenness index varied between 2.03-4.64, 1.2-2.44, 0.77-1.5 in Rupsha-Pashur; 2.47-3.85, 1.8-5.84, 0.78-0.94 in Khalpatua-Arpangachia; and 0.66-4.27, 1.19-5.12, 0.59-1.29 in Bhola-Baleswar. Water temperature, pH, DO, Transparency and Salinity were determined between 19.92°C-31°C; 6.7-7.87; 3.93 mgL^{-1}-7.37 mgL^{-1}; 7.5 cm-60 cm; 2-23 ppt, respectively. Nutrient elements i.e. NO_3^-, PO_4^{3-}, NH_4^+, SiO_4^{4-} fluctuated seasonally from 0.0062 to 1.633 mgL^{-1}, 0.005 to 0.772 mgL^{-1}, 0.038 to 2.467 mgL^{-1}, 3.124 to 27.234 mgL^{-1}, respectively. Chlorophyll-*a* concentrations were fluctuated seasonally within 0.24 to 5.94 µgL^{-1} and highest phytoplankton biomass was observed in Bhola-Baleswar in summer. Chlorophyll concentration was found to be correlated positively with transparency, salinity and nutrients.

Keywords: Water quality; Phytoplankton; Biomass production; Monsoon; River system; Estuary; Sundarbans

Introduction

The Sundarbans; the largest single chunk of tidal halophytic mangrove forest in the world; lies in the vast delta on the Bay of Bengal formed by the greater confluence of the Ganges; Brahmaputra and Meghna rivers. It covers 10,000 sq.km of which about 6000 sq.km area is within the political boundary of Bangladesh; and the remaining in India. The area experiences a subtropical monsoonal climate with an annual rainfall of 1600-1800 mm and severe cyclonic storms. Enormous amounts of sediments carried by the rivers contribute to its expansion and dynamics. The Hydro-geochemical environment of this area is highly dynamic in nature with numerous drainage channels and significant coastal processes [1]. The biodiversity includes about 350 species of vascular plants; 250 fishes and 300 birds; besides numerous species of phytoplankton; fungi; bacteria; zooplankton; benthic invertebrates; molluscs; reptiles; amphibians and mammals. Species composition and community structure vary east to west; and along the hydrological and salinity gradients. The Sundarbans with diverse gene pool for flora and fauna provides livelihoods for about 2.5 million people of Bangladesh [2]. Being in the coastal waters; the abiotic and biotic resources of the Sundarbans are highly variable in response to the coastal dynamic processes. The regulation of river flows by a series of dams; barrages and embankments for diverting water upstream for various human needs and for flood control has caused large reduction in freshwater inflow and seriously affected the biodiversity because of an increase in salinity and changes in sedimentation. Ecological characteristics; particularly intermixing of saline and fresh water; seasonally fluctuating salinity and the silt brought down by the rivers greatly influence the distribution and abundance of algae in the Sundarbans. The observed distribution of a given species is a function of differential adaptations (differences in reproduction and growth) in response to environmental factors such as climate; freshwater inflow; salinity; tidal coverage; sediment type; nutrients; etc.

The algal flora of the Sundarbans is very poorly known; but the available information suggests that the Sundarbans has a highly diverse algal flora comprised of both benthic and planktonic forms ranging from the freshwater to marine environments. There was no previous record of algal flora of the Sundarbans before studies by Islam [3]. A few works on the phytoplankton community structure and its relation to abiotic variables in the Sundarbans river systems were hitherto studied in Bangladesh [4-8]. Various published reports on the algal flora provide only a patchy picture as they are based on short-term surveys of small isolated areas. Since algal flora play very important role in ecological context; the study of phytoplankton community structure is utmost important. Phytoplankton community structures change be influenced by their surrounding environment; physically; chemically; and biologically; in ways which favor or disfavor their continued persistence. The study of phytoplankton community response to these variables is considered useful for interpreting hydro-chemical variations in coastal areas [9]. So the temporal and spatial composition of the phytoplankton population may act as an indicator of the water quality fluctuation in response to changing environment.

The present study was conducted to understand the seasonal variation of phytoplankton population with water chemistry of the major three river systems (Rupsha-Pashur; Khalpatua-Arpangachia;

***Corresponding author:** Dr. SM Bazlur Rahaman, Professor, Fisheries and Marine Resource Technology Discipline, Khulna University, Khulna-9208, Bangladesh
E-mail: riti_rahaman@yahoo.com

Bhola-Baleswar) in the Sundarbans. Thus it can provide clues to do further research in order to get the factors responsible for disturbance in ecological imbalance of the Sundarbans; one of the unique aquatic habitats of the globe.

Materials and Methods

Sampling strategy

Fifteen stations were selected for the examination of water quality indices. Rupsha; Passur; Shibsha; Arpangachia; Khalpatua; Malancha; Raymangal; Baleswar; and Bhola are major rivers of the Sundarbans. Sampling stations were designed to cover the three major river systems of the Sundarbans such as Rupsha-Pashur; Khalpatua-Arpangachia; and Bhola-Baleswar (Figure 1). Samples were collected temporally namely in monsoon (June-October); winter (November-February); and summer (March-May) from 5 stations of Rupsha-Pashur river system namely (1) Karamjol (89°35.962"E; 22°25.97'N) (2) Karamjol Canal (89°35.445'E; 22°25.737'N); (3) Joymoni (89°37.757'E; 22°21.053'N); (4) Harbaria (89°36.563'E; 22°17.974'N); and (5) Harbaria Canal (89°36.985'E; 22°17.998'N); and from 6 stations of Khalpatua-Arpangachia namely (1) Pashurtala (89°11'934"E; 22°14'038"N); (2) Pashurtala Canal (89°12'043"E; 22°14'077"N); (3) Kalagashi (89°14'541"E 22°12'685"N); (4) Kalagashi Canal (89°14'638"E; 22°12'392"N); (5) Nildumur (89°14'708"E; 22°14'828"N); and (6) Arpangachia (89°18'581"E; 22°12'406"N); and also 4 stations of Bhola-Baleswar namely (1) Bogi (89°50'21.9"E; 22°12'54.4"N); (2) Sharankhola (89°48'42.4"E; 22°12'35.6"N); (3) Supati (89°49.07'E; 22°03.442'N); and (4) Supati Canal (89°48.93'E 22°3.128' N).

Sample collection

For phytoplankton community structure study; 20 L of water collected from surface water of the study area was passed through a plankton net (20 μm mesh sized; silk bolting cloth or nylon monofilament screen cloth). Then the concentrated samples were preserved with Lugol's Solution (20 g potassium iodide and 10 g iodine crystals dissolved in 200 ml distilled water containing 20 ml glacial acetic acid). For chlorophyll *a* estimation 200 m^3 volume samples were collected in the same way using plankton net and samples for nutrient analysis were collected using a grab sampler. For nutrient analysis water samples were collected from surface; middle bottom of the particular location which was filtered immediately through pre-cleaned 0.45 μm pore-size cellulose filters. The samples were preserved in deep frozen in the dark before analysis.

Sample analysis

Temperature; pH; transparency; DO; and salinity were measured

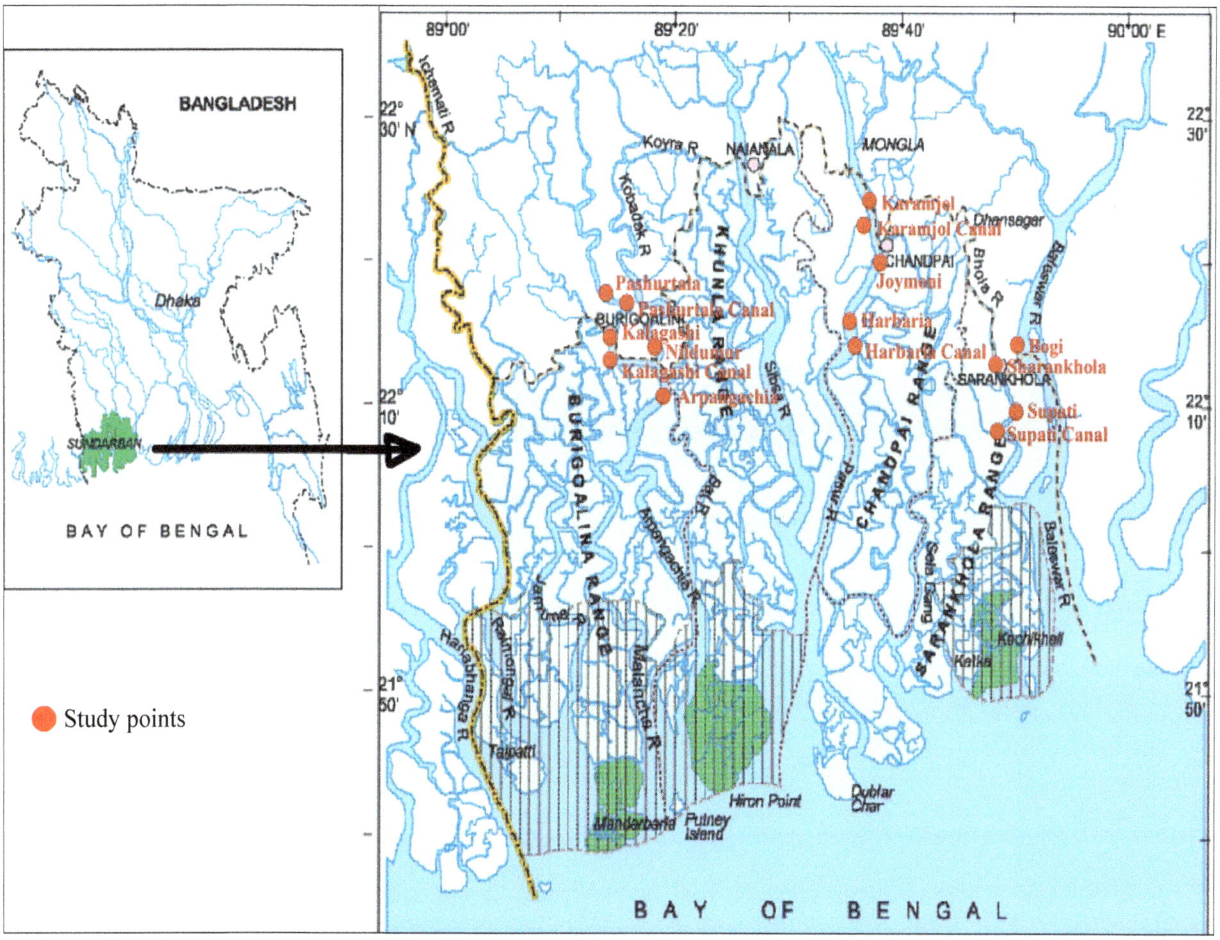

Figure 1: Location of the study points in Rupsha-Pashur, Khalpatua-Arpangachia, Bhola-Baleswar river systems of the Sundarbans.

in-situ and water samples collected for analyzing nutrients (NO_3^-; PO_4^{3-}; NH_4^+; SiO_4^{4-}); chlorophyll a and phytoplankton community composition were measured later in the laboratory. Water temperature was measured with a Centigrade Mercury thermometer; pH with a Microprocessor pH meter (HANNA instruments; pH 211); salinity with a TDS meter (HI 9635; portable multirange conductivity/TDS meter; HANNA); transparency with a secchi disc; and DO in Winkler's method (APHA 1992).

Quantitative estimation of phytoplankton was done by Sedgewick-Rafter counting chamber (S-R cell) method [6] using Labomed Imaging Device (ivu 15000 microscope). Phytoplankton genera and species were identified [10-18]. Chlorophyll a was determined by spectrophotometric method (APHA 1992); and nutrients were measured by colorimetric methods described in Yin et al. [19].

Species richness; diversity and evenness index calculation

Species richness index (d); species diversity index (H); and evenness index were calculated according to following equations

- Species richness index (d) [20]

  $d=(S-1)/ \text{Log } N$

  Where:

  d=Species richness index

  S=Number of species in a population

  N=Total number of individuals in S species.

- Species diversity index (H) [21]

  $Hs=\Sigma P_i \text{ log} P_i$

  Where

  Hs=Diversity Index

  i=Counts denoting the ith species ranging from 1–n

  P_i=Proportion that the ith species represents in terms of numbers of individuals with respect to the total number of individuals in the sampling space as whole.

- Evenness index (j) [22]

  $j=H_s / \text{Log } S$

  Where

  J=Equitability index

  H_s=Shannon and weaver index

  S=Number of species in a population

Data analysis

To establish differences of the phytoplankton community descriptors among the fifteen sites; one-way ANOVA was carried out. Two tailed Pearson correlation was performed to identify relation among various physico-chemical and biological parameters. Analyses were performed using the software package SPSS Statistics 17.0.

Results

Water quality

Water quality parameters namely temperature; pH; transparency; dissolved oxygen; salinity; nitrate; phosphate; ammonium; silicate; and chlorophyll *a* were determined temporally in 15 stations situated in three major river systems of the Sundarbans presented in Table 1.

Temperature: Mean water temperature fluctuated seasonally. In Rupsha-Pashur river system temperature ranged from 20.02°C-31°C; highest in summer and lowest in winter. In Khalpatua-Arpangachia and Bhola-Baleswar River systems it was within the range 22.17°C-30.45°C; and 19.92°C-30.25°C respectively and also showed the same kind of seasonal variability of higher values in summer and lower values in winter.

pH: In Rupsha-Pashur; Khalpatua-Arpangachia and Bhola-Baleswar river system pH varied from 7.29-7.87; 7.31-7.82; and 6.7-7.53 respectively. In Rupsha-Pashur pH values showed a seasonal trend of variation with higher values in monsoon then showed a slight gradual reduction through winter and summer. But in Khalpatua-Arpangachia and Bhola-Baleswar river system no clear seasonal trend was observed.

Transparency: In all three river systems; transparency values were very low throughout the year; but also showed a seasonal trend of variation. In both Rupsha-Pashur and Bhola-Baleswar river system; transparency was the lowest in monsoon but was found to rise gradually in winter and summer. It ranged from 8cm-36cm in Rupsha-Pashur with highest in Harbaria during summer and lowest in Joymoni during monsoon. The transparency ranged between 18 cm and 36 cm in Bhola-Baleswar highest in Sharankhola during summer and lowest in Bogi during monsoon. But in Khalpatua-Arpangachia River system the transparency was lower in monsoon but rose in winter and again fell in summer. It was found within the range of 16 cm-66 cm; highest in Pashurtala canal during winter and lowest in Kalagashi during monsoon. Annual mean transparency also was highest (34.66 ± 18.84 cm in Pashurtala) in Khalpatua-Arpangachia.

Dissolved oxygen: Dissolved Oxygen concentration (DO) fluctuated spatially and temporally. In Rupsha-Pashur; Khalpatua-Arpangachia; and Bhola-Baleswar river systems; DO varied from 4.37 mgL^{-1} to 7.37 mgL^{-1} (highest at Harbaria Canal in monsoon & lowest at Karamjol Canal in summer); 3.93 mgL^{-1} to 6.4 mgL^{-1} (highest at Kalagashi Canal in monsoon & lowest at Kalagashi in summer); and 4.88 mgL^{-1} to 5.65 mgL^{-1} (highest at Sharankhola in summer & lowest at Supati Canal in monsoon) respectively. In Bhola-Baleswar DO fluctuated very little and not showed any seasonal trend of variation. But in Rupsha-Pashur and Khalpatua-Arpangachia a little fluctuation of seasonal trend was observed with higher values in monsoon and gradually decreased over the winter and summer period. The highest annual mean value was found 6.25 ± 0.80 mgL^{-1} at Harbaria Canal (Rupsha-Pashur).

Salinity: The highest salinity was observed (23‰) at Nildumur in winter. In Rupsha-Pashur; Khalpatua-Arpangachia; and Bhola-Baleswar river systems salinity ranged from 5.17‰ to 16‰ (highest at Harbaria in summer and lowest at Harbaria in monsoon); 9‰ to 23‰ (highest at Nildumur in winter and lowest at Arpangachia in summer); and 2‰ to 16‰ (highest at Supati in summer and lowest at Bogi in monsoon) respectively. In both Rupsha-Pashur and Bhola-Baleswar river system salinity showed similar seasonal variation of lower values in monsoon then rose through the period of winter and summer. But in Khalpatua-Arpangachia salinity was low in monsoon; and then rose in winter and again dropped in summer. Lowest annual mean salinity was 8.29% ± 4.50‰ (Sharankhola) in Bhola-Baleswar river system.

Nitrate: In maximum sampling stations annual mean nitrate (NO_3^-) concentrations were lower than 1 mgL^{-1}. The highest concentration (>1 mgL^{-1}) was observed in Bhola-Baleswar in summer. In Rupsha-

	Bhola-Baleswar River System					
	Bogi	Sharankhola	Supati	Supati Canal		
Geographical Location	89°50'21.9"E, 22°12'54.4"N	89°48'42.4"E, 22°12'35.6"N	89°49.07'E, 22°03.442'N	89°48.93'E, 22°3.128'N		
Temperature(°C)	25.39 ± 3.72	25.19 ± 3.76	25.95 ± 4.09	25.71 ± 4.05		
pH	7.04 ± 0.21	6.87 ± 0.14	7.19 ± 0.23	7.23 ± 0.30		
Transparency(cm)	25 ± 5.72	28.33 ± 6.55	25.33 ± 5.25	24.66 ± 4.11		
DO(mgL^{-1})	5.43 ± 0.03	5.53 ± 0.08	5.29 ± 0.16	5.16 ± 0.21		
Salinity	9.17 ± 5.24	8.29 ± 4.50	10.33 ± 5.72	9.72 ± 5.40		
Nitrate(mgL^{-1})	0.603 ± 0.73	0.425 ± 0.48	0.470 ± 0.54	0.433 ± 0.49		
Phosphate(mgL^{-1})	0.015 ± 0.005	0.029 ± 0.011	0.011 ± 0.005	0.023 ± 0.012		
Ammonium(mgL^{-1})	0.041 ± 0.037	0.038 ± 0.018	0.064 ± 0.039	0.066 ± 0.041		
Silicate(mgL^{-1})	16.106 ± 6.81	17.56 ± 9.22	16 ± 8.27	17.20 ± 9.56		
Chlorophyll-a(μgL^{-1})	2.82 ± 2.28	1.71 ± 0.87	2 ± 0.87	2.33 ± 1.45		
	Rupsha-Pashur River System					
	Karamjol	Karamjol Canal	Joymoni	Harbaria	Harbaria Canal	
Geographical Location	89°35.962'E,22°25.97'N	89°35.445'E,22°25.737'N	89°37.757'E,22°21.053'N	89°36.563'E,22°17.974'N	89°36.985'E,22°17.998'N	
Temperature(°C)	27.15 ± 4.48	27.32 ± 4.49	27.57 ± 4.30	27.05 ± 4.63	26.78 ± 4.79	
pH	7.55 ± 0.18	7.55 ± 0.24	7.51 ± 0.18	7.64 ± 0.14	7.67 ± 0.10	
Transparency(cm)	13.33 ± 2.36	20 ± 7.48	16 ± 5.66	23.33 ± 11.09	22.66 ± 8.99	
DO(mgL^{-1})	6.1 ± 0.81	5.91 ± 1.09	5.93 ± 0.94	6.06 ± 0.76	6.25 ± 0.80	
Salinity	9.96 ± 3.13	9.86 ± 3.64	9.22 ± 2.52	11.22 ± 4.51	10.54 ± 3.72	
Nitrate(mgL^{-1})	0.322 ± 0.26	0.627 ± 0.42	0.375 ± 0.27	0.404 ± 0.30	0.317 ± 0.21	
Phosphate(mgL^{-1})	0.449 ± 0.26	0.215 ± 0.17	0.291 ± 0.14	0.168 ± 0.12	0.256 ± 0.10	
Ammonium(mgL^{-1})	0.050 ± 0.025	0.052 ± 0.026	0.049 ± 0.022	0.051 ± 0.023	0.047 ± 0.025	
Silicate(mgL^{-1})	12.19 ± 5.61	14.04 ± 4.50	11.45 ± 6.22	13.11 ± 7.36	12.19 ± 6.29	
Chlorophyll-a (μgL^{-1})	1.38 ± 0.81	1.12 ± 0.57	1.44 ± 0.86	1.53 ± 1.04	1.63 ± 1.07	
	Khalpatua-Arpangachia River System					
	Pashurtala	Pashurtala Canal	Kalagashi	Kalagashi Canal	Nildumur	Arpangachia
Geographical Location	89°11'934"E, 22°14'038"N	89°12'043"E, 22°14'077"N	89°14'541"E, 22°12'685"N	89°14'638"E, 22°12'392"N	89°14'708"E, 22°14'828"N	89°18'581"E, 22°12'406"N
Temperature(°C)	27.02 ± 3.40	27.17 ± 3.25	27.12 ± 3.19	27.26 ± 3.49	27.15 ± 3.57	27.21 ± 3.56
pH	7.48 ± 0.19	7.64 ± 0.09	7.63 ± 0.11	7.65 ± 0.13	7.60 ± 0.02	7.65 ± 0.16
Transparency(cm)	34.66 ± 18.84	34 ± 22.69	24.33 ± 7.93	30.66 ± 13.70	31.33 ± 10.66	29.66 ± 10.34
DO(mgL^{-1})	4.96 ± 0.50	5.02 ± 0.79	4.94 ± 0.72	5.08 ± 0.99	5.18 ± 0.84	5.04 ± 0.54
Salinity	14.91 ± 4.69	15.39 ± 4.27	14.66 ± 5.46	14.72 ± 5.90	14.72 ± 5.94	14.22 ± 6.03
Nitrate(mgL^{-1})	0.269 ± 0.26	0.247 ± 0.13	0.083 ± 0.03	0.123 ± 0.04	0.091 ± 0.003	0.145 ± 0.02
Phosphate(mgL^{-1})	0.108 ± 0.03	0.109 ± 0.03	0.099 ± 0.01	0.119 ± 0.008	0.118 ± 0.04	0.108 ± 0.03
Ammonium(mgL^{-1})	0.153 ± 0.19	0.117 ± 0.13	0.176 ± 0.16	0.155 ± 0.17	0.109 ± 0.12	0.121 ± 0.14
Silicate(mgL^{-1})	9.404 ± 6.32	10.689 ± 8.30	9.98 ± 7.26	9.98 ± 7.06	8.238 ± 5.17	8.311 ± 4.26
Chlorophyll-a(μgL^{-1})	0.813 ± 0.24	1.14 ± 0.91	0.87 ± 0.35	0.806 ± 0.39	1.17 ± 0.84	0.89 ± 0.52

Table 1: Annual mean values of water quality variables in Rupsha-Pashur, Khalpatua- Arpangachia and Bhola-Baleswar river systems.

Pashur river system it ranged from 0.021 mgL^{-1} to 1 mgL^{-1} (highest at Karamjol Canal in winter & lowest at Karamjol in monsoon); and in Bhola-Baleswar it varied from 0.0062 mgL^{-1} to 1.633 mgL^{-1} (highest at Bogi in summer & lowest at Supati in monsoon). Both of these river systems showed a seasonal trend of lower NO_3^- values in monsoon and higher values in winter and summer. But in Khalpatua-Arpangachia NO_3^- values were found within the range of 0.05 mgL^{-1}-0.65mgL^{-1} and no clear seasonal trend was observed.

Phosphate: Throughout the study areas; phosphate (PO_4^{3-}) concentrations were found lower than 1 mgL^{-1}. In Rupsha-Pashur; Khalpatua-Arpangachia; and Bhola-Baleswar river systems it was observed within the range of 0.04 mgL^{-1}-0.772 mgL^{-1} (highest at Karamjol in summer & lowest at Harbaria in monsoon); 0.063 mgL^{-1}-0.161 mgL^{-1} (highest at Kalagashi Canal in monsoon & lowest at Pashurtala Canal in winter); and 0.005 mgL^{-1}-0.045 mgL^{-1} (highest at Sharankhola in monsoon & lowest at Supati in winter); respectively. In Rupsha-Pashur; and Bhola-Baleswar river systems PO_4^{3-}concentrations showed no significant seasonal trend of variation. But in Khalpatua-Arpangachia river system a seasonal trend with higher concentrations in monsoon and summer and lower concentrations in winter were observed. Annual mean PO_4^{3-} value was the lowest (0.011 ± 0.005 mgL^{-1} in Supati) in Bhola-Baleswar river system.

Ammonium: Mean Annual ammonium (NH_4^+) value was found the lowest (0.038 ± 0.018 mgL^{-1} in Sharankhola) in Bhola-Baleswar river system. NH_4^+ values varied from 0.015 mgL^{-1}-0.087 mgL^{-1} (highest at Karamjol Canal in winter & lowest at Karamjol in monsoon) in Rupsha-Pashur river system. But in Khalpatua-Arpangachia and Bhola-Baleswar river systems NH_4^+ values ranged from 0.008 mgL^{-1}-0.434 mgL^{-1} (highest at Pashurtala in monsoon & lowest at Pashurtala in winter); and 0.008 mgL^{-1}-0.102 mgL^{-1} (highest at Supati Canal in winter & lowest at Supati both in monsoon ans winter) respectively. In Rupsha-Pashur and Bhola-Baleswar NH_4^+ values were lower in monsoon and higher in winter and summer. But in Khalpatua-Arpangachia trends were found different here NH_4^+ values were much higher in monsoon and lower in winter and summer season.

Silicate: Silicate (SiO_4^{4-}) concentration was observed extremely high throughout the study area. Annual mean SiO_4^{4-} concentration was found within the range of 8.238 ± 5.17 mgL^{-1}-17.56 ± 9.22 mgL^{-1} (highest in Sharankhola; Bhola-Baleswar and lowest in Nildumur; Khalpatua-Arpangachia). Both Rupsha-Pashur and Bhola-Baleswar showed similar seasonal trend with lower values in monsoon and higher values in winter and summer. But in Khalpatua-Arpangachia SiO_4^{4-} concentrations were recorded lower in monsoon; and it rose in winter and again fell in summer.

Chlorophyll: Concentration of chlorophyll *a* fluctuated within the range of 0.24 μgL^{-1}-3.11 μgL^{-1} in Rupsha-Pashur; 0.42 μgL^{-1}-2.43 mgL^{-1} in Khalpatua-Arpangachia; and 0.57 μg/L-5.94 μg/L in Bhola-Baleswar. In Rupsha-Pashur and Bhola-Baleswar river systems; phytoplankton biomass showed a seasonal trend with lower values in monsoon and gradually raised levels in winter and summer. But in Khalpatua-Arpangachia it was lower in monsoon; rose in winter and again dropped in summer. Annual mean biomass production was the highest (2.82 ± 2.28 μgL^{-1} in Bogi) in Bhola-Baleswar river system.

There were significant differences in water quality data *($p<0.05$)* among river systems and also sampling sites.

Phytoplankton community composition

Throughout the study areas a total of 134 phytoplankton species dominated by diatoms were identified. 99 species from 41 genera of Bacillariophyta; 18 species from 6 genera of Pyrophyta; 12 species from 9 genera of Chlorophyta; 4 species from 4 genera of Cyanobacteria; and 1 Species of Ochrophyta were present. There were significant differences *($p<0.05$)* in abundance and diversity of phytoplankton communities. In Rupsha-Pashur; Khalpatua-Arpangachia; and Bhola-Baleswar river systems; abundance varied in the range of 3.755×10^3 cellsL^{-1}-1.015×10^5 cellsL^{-1} (highest at Harbaria Canal in summer & lowest at Karamjol in monsoon); 2.951×10^3 cellsL^{-1} to 4.197×10^4 cellsL^{-1} (highest at Kalagashi in winter & lowest at Kalagashi Canal in summer); and 3.246×10^3 cellsL^{-1}-5.031×10^5 cellsL^{-1} (highest at Bogi in summer & lowest at Bogi in monsoon) respectively. In Rupsha-Pashur and Bhola-Baleswar similar seasonal trend was observed with lower density in monsoon and then rose rapidly through winter and summer. But in Khalpatua-Arpangachia phytoplankton density was low in monsoon; rose in winter and again dropped in summer. Annual average density was the highest (8.347×10^4 cellsL^{-1}) in Bhola-Baleswar river system (Table 2).

Species diversity: Most diverse genera were *Coscinodiscus* with 12 species; *Protoperidinium* with 11 species; *Thalassiosira* with 10 species; *Chaetoceros* and *Paleurosigma* with 7 species. Maximun species diversity was observed in Khalpatua-Arpangachia during winter (Table 3).

Species diversity index; richness index and evenness index values showed spatial and temporal variation. Annual mean diversity index was highest (2.63 ± 0.49) in Khalpatua-Arpangachia. It ranged from 1.31to 4.64 (highest at Karamjol in winter & lowest at Karamjol in monsoon) in Rupsha-Pashur; 1.58to 3.85 (highest at Arpangachia in winter & lowest at Nildumur in winter) in Khalpatua-Arpangachia; and 0.66 to 4.27 (highest at Bogi in winter & lowest at Supati in summer) in Bhola-Baleswar. Richness values were found in the range of 1.21-4.99 in Rupsha-Pashur; 1.18-11.24 in Khalpatua-Arpangachia; and 1.19-5.12 in Bhola-Baleswar; and evenness index varied from 0.77 to 1.5 in Rupsha-Pashur; 0.82 to 0.97 in Khalpatua-Arpangachia; and 0.59 to1.21 in Bhola-Baleswar. All these index values showed no clear seasonal trend of variation (Table 2).

Throughout the study area over all three river system Bacillariophyta dominated the community composition. It was 81.82%-100% in Rupsha-Pashur; 64.63%-96.57% in Khalpatua-Arpangachia river system. But in Bhola-Baleswar river system Bacillariophyta dominated the composition in monsoon and winter (81.81%-98.48%); but during summer Cyanophyta dominated the composition (54.42%-78.76%) (Figure 2).

Discussion

Phytoplankton community in the Sundarbans river systems showed with noted abundance and diversity with spatial and temporal variation to some extent. The observed differences are likely related to local forcing functions such as transparency; salinity and in some cases with nutrient concentrations variations in addition with some unknown reasons in this study.

Phytoplankton community in relation to water quality condition

Concentration of nitrate (NO_3^-); phosphate (PO_4^{3-}); ammonium (NH_4^+); and silicate (SiO_4^{4-}) play important role as nutrient both

River System	Item	Quantity	Range of Variation
Rupsha-Pashur	Total Species	93 species	
	Abundance	2.28×10^4 cellsL^{-1} (Average)	3.755×10^3-1.015×10^5 cellsL^{-1}
	Diversity Index (H)	2.39 (Average)	1.31-4.64
	Richness Index (d)	2.32 (Average)	1.21-4.99
	Evenness Index(j)	0.94 (Average)	0.77-1.5
Khalpatua-Arpangachia	Total Species	122 species	
	Abundance	1.052×10^4 cellsL^{-1} (Average)	2.951×10^3-4.197×10^5 cellsL^{-1}
	Diversity Index (H)	2.63 (Average)	1.58-3.85
	Richness Index (d)	3.36 (Average)	1.8-11.24
	Evenness Index(j)	0.90 (Average)	0.82-0.97
Bhola-Baleswar	Total Species	110 species	
	Abundance	8.347×10^4 cellsL^{-1} (Average)	3.246×10^3-5.03×10^5 cellsL^{-1}
	Diversity Index (H)	2.80 (Average)	2.47-3.85
	Richness Index (d)	2.75 (Average)	1.8-5.84
	Evenness Index(j)	0.89 (Average)	0.78-0.94

Table 2: Annual mean values of phytoplankton species abundance in three major river systems.

River systems	Seasons	Species
Rupsha-Pashur	Monsoon	*Thalassiosira oestrupii, Thalassiosira ecentrica, Thalassiosira decipens, Coscinodiscus centralis, Coscinodiscus concinus, Coscinodiscus spiniferus, Cyclotella striata, Roperia tesselata, Nitzschia lorenziana*
	Winter	*Thalassiosira wongii, Coscinodiscus granii, Cyclotella stylorum, Chaetoceros affinis, Chaetoceros debile, Skeletonema costatum, Thalassionema nitzschioides, Synedra ulna, Pleurosigma angulatum, Cylindrotheca fusiformis, Ankistrodesmus falcatus*
	Summer	*Ditylum brightwelli, Anabaena cf. flos-aquae* and *Spirulina platens*
Khalpatua-Arpangachia	Monsoon	*Thalassiosira ecentrica, Thalassiosira oestrupii, Thalassiosira decipens, Thalassiosira angulata, Coscinodiscus centralis, Coscinodiscus spiniferus, Coscinodiscus angsti, Coscinodiscus concinus, Cyclotella striata, Cyclotella striata, Pleurosigma angulatum, Pleurosigma directum, Pleurosigma cf. elongatum, Cerataulina bicornis, Thalassionema nitzschioides, Ceratium fusus*
	winter	*Thalassiosira wongii, Thalassiosira punctigera, Thalassiosira pseudonona, Thalassiosira anguste-lineata, Coscinodiscus marginatus, Coscinodiscus wailesii, Coscinodiscus radiatus, Coscinodiscus pavillardi, Actinocyclus anulatus, Actinocyclus Pruniosus , Cylindrotheca closterium, Pleurosigma normani, Navicula meninscus, Ditylum brightwelli, Odontella sinensis, Odontella mobiliensis, Bacillaria paxillifera, Surirella gemma, Entomoneis sulcata, Entomoneis paludosa, Leptocylindrus minimus, Cladopyxis hemibrachiata, Protoperidinium biconicum, Protoperidinium subinerme, Protoperidinium claudicans, Protoperidinium leonis, Ceratium furca, Eudorina elegans, Netrium oblongum*
	Summer	Not available
Bhola-Baleswar	Monsoon	*Thalassiosira ecentrica, Thalassiosira oestrupii, Thalassiosira decipens, Coscinodiscus centralis, Coscinodiscus spiniferus, Nitzschia behrei, Roperia tesselata, Cerataulina dentate, Eudorina elegans, Hydroductyon etc.*
	Winter	*Thalassiosira lundiana, Thalassiosira wongii, Coscinodiscus angsti, Skeletonema costatum, Cylindrotheca fusiformis, Fragilaria sp., Entomoneis sulcata, Entomoneis paludosa, Chaetoceros affinis, Actinocyclus anulatus, Pleurosigma estuarii, Odontella mobiliensis, Cladopyxis hemibrachiata, Ceratium furca, Protoperidinium punctulatum, Protoperidinium subinerme, Netrium oblongum, Pediastrum simplex, Pediastrum duplex* etc.
	Summer	*Anabaena cf. flos-aquae, Microcystis sp, Spirulina platens, Anacystis aeruginosa* etc

Table 3: List of the most abundant phytoplankton genera found in the Sundarbans river-systems at different seasons.

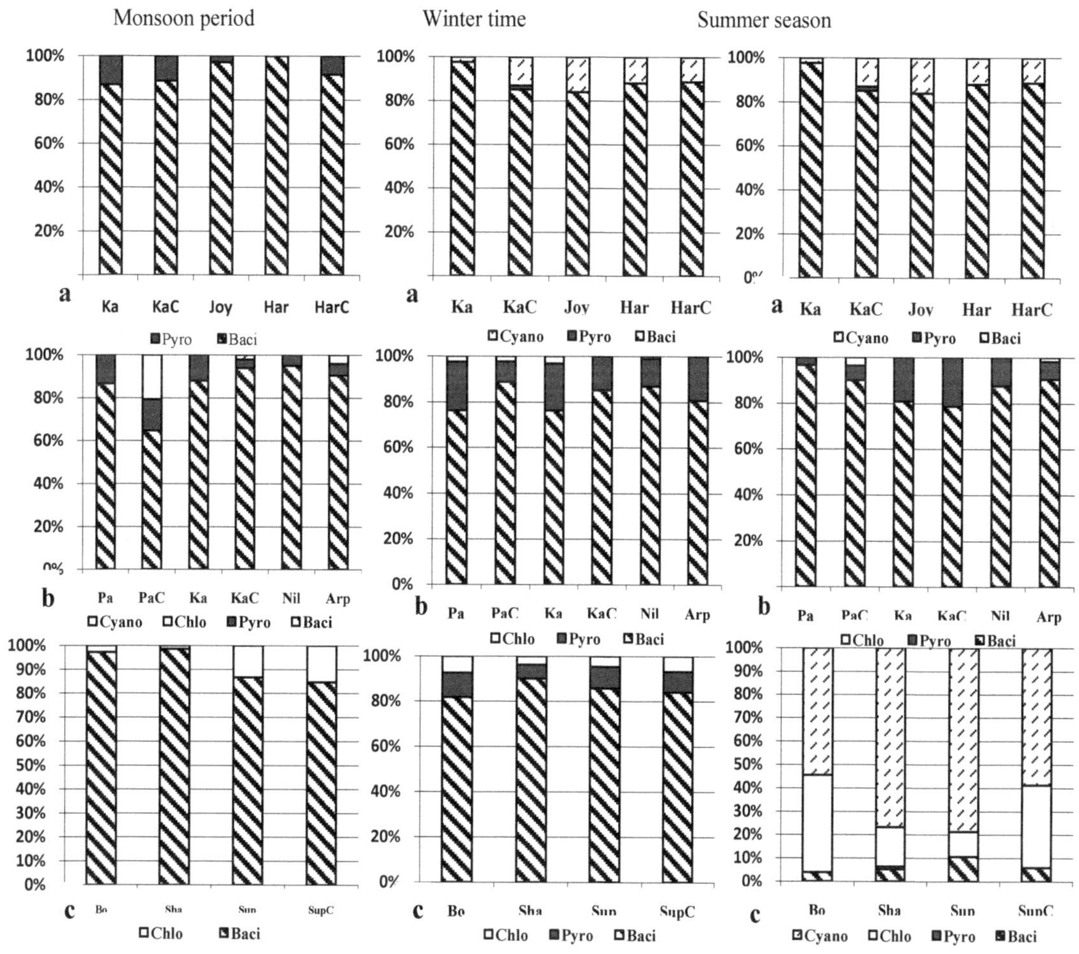

Figure 2: Spatial distribution of relative abundance of phytoplankton at different measuring stations of (a) Rupsha-Pashur, (b) Khalpatua-Arpangachia and (c) Bhola-Baleswar river system during monsoon, winter, and summer respectively.

for phytoplankton growth and production. These were estimated on a seasonal basis during the study. Results showed no uniform seasonal trend of all these nutrient elements. Nitrate (NO_3^-) followed a seasonal trend of lower concentration in monsoon and increased through winter and summer in Rupsha-Pashur; and Bhola-Baleswar. But in Khalpatua-Arpangagchia; it showed no clear seasonal trend. Phosphate didn't show any seasonal pattern in Rupsha-Pashur; and Bhola-Baleswar; but in Khalpatua-Arpangachia it showed a trend of lower amount in monsoon then rose in winter and again dropped in summer. Only in Bhola-Baleswar a seasonal pattern of Ammonium (NH_4^+) was observed with lower values in monsoon; and gradual higher values through winter and summer. In Rupsha-Pashur silicate (SiO_4^{4-}) showed a pattern of lower concentrations in monsoon and higher in winter and summer. But in Khalpatua-Arpangachia it was different; higher concentrations were found in winter and lower in monsoon and summer. This significant variation suggests the coexistence of different processes in water quality conditions. In marine coastal systems; there can be many sources of nutrients like upwelling; river input; sediment resuspension or remineralization; aquaculture effluents; and urban; agricultural; and industrial wastewater; thereby making it difficult to determine the relative contribution of nutrient sources to coastal water quality [23,24].

Concentration of Chlorophyll *a* (algal biomass) is normally used as an index of the productivity [25] and trophic condition of estuaries; coastal and oceanic waters [26-28]. It reflects the net result (standing stock) of both growth and loss processes. There is generally a good agreement between planktonic primary production and algal biomass. Algal biomass is associated with the visible symptoms of eutrophication. It is considered the principal variable to use as a trophic state indicator. Present study analyzed algal biomass production in terms of chlorophyll a concentration on seasonal basis. Throughout the study area highest biomass production (5.94 μgL^{-1}) was observed in the Bhola-Baleswar river system in summer. Rupsha-Pashur and Bhola-Baleswar river systems showed the similar trend of seasonal variation with lower concentrations in monsoon and gradually increased value through winter and summer. But in Khalpatua-Arpangachia river system concentrations were lower in monsoon; rose in winter; and again fell in summer. Growth and production of phytoplankton in an area is a function of environmental factors prevailed in that area [29]. By pearson correlation analysis (Figure 3) we found biomass production i.e. chlorophyll a was positively correlated with transparency; salinity; and nitrate (NO_3^-) in Rupsha-Pashur river system. In Khalpatua-Arpangachia it was positively correlated with transparency; salinity and silicate (SiO_4^{4-}); and negatively correlated with temperature and phosphate (PO_4^{3-}). But in Bhola-Baleswar positive significant correlation was observed with transparency; salinity; nitrate (NO_3^-); ammonium (NH_4^+); and silicate (SiO_4^{4-}). That is; in all three river systems chlorophyll a didn't show strong and uniform correlation with temperature and all nutrient elements.

Water temperature and transparency are most important among various physical factors affecting the distribution and seasonal variation of phytoplankton growth [30]. In the present study we have found significant positive correlation with transparency but no significant positive correlation with temperature. It may be due to the cross interaction between temperature and transparency. In summer temperature and transparency both are high and biomass production was also high; but in monsoon temperature was high but low transparency limited the light penetration and as a result limited the growth [31].

Phytoplankton abundance and biomass production both showed similar trend of seasonal variation in Rupsha-Pashur and Bhola-Baleswar river systems with highest production in summer but differed in Khalpatua-Arpangachia with lower production in summer. This noncompliance may be due to the late sampling time for the summer season in Khalpatua-Arpangachia (26th May-2011); when early monsoonal impact (144mm rainfall in May-2011) has already started.

Silicate concentrations varied a great deal among the sites and seasons. It was found within the range of 3.201 mgL^{-1}-26.122 mgL^{-1}. These higher concentrations of SiO_4^{4-} may result from the shrimp-farm effluents and nutrient export from the harbor that enriched the coastal water with nutrient [32]. Often; SiO_4^{4-} acts as a limiting nutrient for diatom growth; and it could therefore control replacement of diatoms by dinoflagellates in conditions of Si deficiency; which means that SiO_4^{4-}

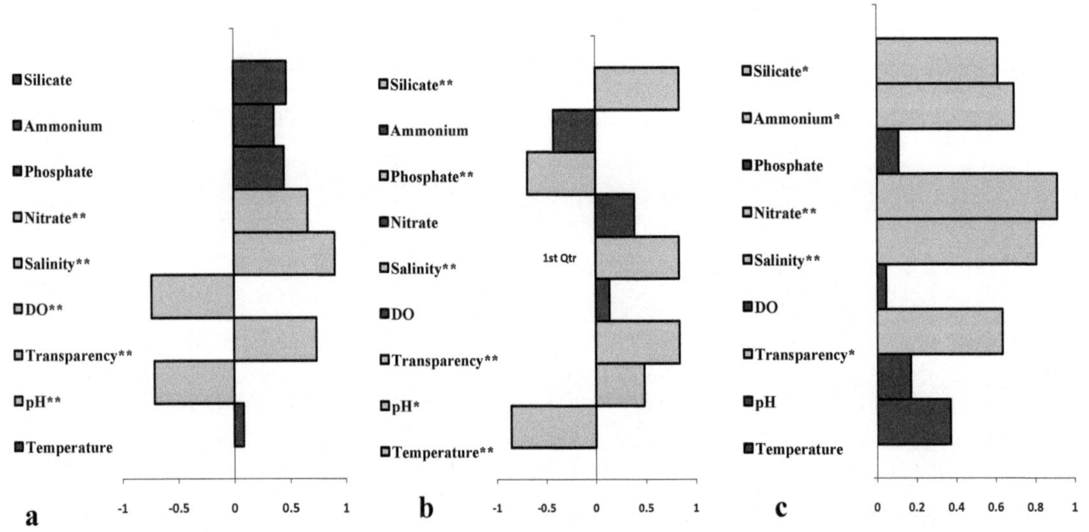

Figure 3: Pearson correlation between phytoplankton biomass and different water quality parameters in (a) Rupsha-Pashur, (b) Khalpatua-Arpangachia, and (c) Bhola-Baleswar river systems (* indicate significant correlation at 0.05 level (2 tailed) & ** indicate significant correlation (2 tailed) at the 0.01 level).

can play an important role in phytoplankton community-structure changes [33]. High inputs of SiO_4^{4-} could also cause an imbalance of the normal phytoplanktonic communities by stimulating diatom growth; some of which may have harmful effects.

The salinity is also one of the main parameters that can be attributed to the phytoplankton diversity and acts as a limiting factor which influences the distribution of planktonic community [34-37]. Generally; changes in the salinity of the brackish water habitats such as estuaries; backwaters and mangrove are due to the influx of freshwater from land run off; caused by monsoon or by tidal variations. Presently recorded higher values in dry season could be attributed to the higher degree neritic water dominance from sea [38,39].

Differences among sites in dissolved oxygen concentrations; as well as seasonal changes within sites; result from the balance between physical (e.g. turbulence; diffusion and solubility of oxygen) and biogeochemical (e.g.; consumption; production; remineralization) processes [40]. In the study area comparatively higher values were observed in monsoon and lower in summer. In monsoon it may be due to precipitation and higher turbulence and in summer which could be a function of lower turbulence and warmer water temperatures that increase respiration and decrease solubility of oxygen.

Phytoplanktonic community composition as bioindicator

The phytoplankton community structures in coastal Sundarbans were determined as significantly dynamic with respect to spatial and temporal level. These differences in phytoplankton production may be related to a variety of environmental factors in aquatic environment [29].

The number of phytoplankton species identified in major three river systems of Sundarbans; reflecting the high species richness characteristics of tropical coastal areas [41]. In general diatoms dominate the whole area. But in Bhola-Baleswar Cyanophyta dominate the composition in summer. Dinoflagelates are common only in Khalpatua-Arpangachia. Chlorophytes are abundant in Bhola-Baleswar. These variations in species composition suggests; however hydrological connectivity exists among these river systems; it was local hydrological conditions that determined phytoplankton community structure at each site.

Silicate concentrations were much higher in the study area. It probably acted as a factor for stimulating diatom growth [33]. In Bhola-Baleswar community composition was dominated by Cyanophyta and in summer; when highest NO_3^- concentrations were observed among all sites. This NO_3^- enriched condition probably stimulated the growth of Cyanophyta [42]. This community composition indicated high nutrient status and the presence of toxic contaminants [43-45]. It also reflects the possibility of anthropogenic activities for the enrichment of nutrient [42].

Phytoplankton community structural changes are a good indicator of water quality or aquatic ecological status as they show complex and rapid responses to fluctuations of environmental conditions [46]. Skeletonema costatum is an abundant species in the study areas. It is a eurytharmal and euryhaline species. So its presence is an indicator of stress environmental condition. Thalassionema nitzschioides was also abundant throughout the area; and its presence indicates resuspension and remineralization processes happening there. Amphora sp. and Nitzschia sp. etc. are benthic diatoms. These were mostly abundant in Khalpatua-Arpangachia river system. Their presence indicates higher concentrations of SiO_4^{4-} and resuspension of sediments. Nitzschia longissima and Protoperidinium were also abundant throughout the area. These are opportunistic and indicator species of shrimp farm effluents [24,47].

Conclusion

The Sundarbans which environmental conditions are highly dynamic and variable in nature is rich in phytoplankton community. Variation both in abundance and diversity is due to rainfall; transparency; salinity and nutrient concentartions. Variable water quality conditions at spatial and temporal level indicates the influence of various climatic and local forcing functions. A relationship among local water quality characteristics with phytoplankton community is noticed. Favorable climatic condition and nutrient status of water led to a blooming condition in summer. The study will provide a number of information regarding water chemistry of the Sundarbans river systems and phytoplankton community structure in order to find factors responsible for biodiversity loss.

Acknowledgements

The study was supported by the Grants for Advanced Research in Science from the Ministry of Education (MoE) of Bangladesh. The authors are thankful to the laboratory staff of Fisheries & Marine Resource Technology Discipline and Environmental Science Discipline of Khulna University for their cooperation during sample analysis. Thanks are due to the graduate and undergraduate students for their generous cooperation during field visit and laboratory analysis. The writers acknowledge the support given by the authorities of Forest Department of Bangladesh for providing necessary support to carry out field observation and sampling in the Sundarbans Mangrove Forest. The writers would also like to thank the many others who have been involved in the field observations.

References

1. Wahid SM, Babel MS, Bhuiyan AR (2002) Hydrologic monitoring and analysis in the Sundarbans mangrove ecosystem, Bangladesh. Journal of Hydrology 332: 381-395.
2. Hasanuzzaman AFM, Sayeed MAB, Rahaman SMB, Huq KA (2010) Implications of climate change for fisheries and aquaculture in the Sundarbans region of Bangladesh, Khulna University Studies Special Issue 73-82.
3. Islam AKMN (1973) The algal flora of Sundarbans Mangrove forest, Bangladesh. J Bot.
4. Nazrul-Islam AM (1995) Ecological conditions and species diversity in Sundarban mangrove forest community, Bangladesh.
5. Hossain MZ, Chowdhury AH (2008) Phytoplankton abundance in relation to phyco-chemical conditions of the Sundarbans estuary. J Asiat Soc Bangladesh 34: 103-112.
6. Shah MMR, Hossain MY, Begum M, Ahmed ZF, Ohtomi J, et al. (2008) Seasonal Variations of Phytoplanktonic Community Structure and Production in Relation to Environmental Factors of the Southwest Coastal Waters of Bangladesh. Journal of Fisheries and Aquatic Science 3: 102-113.
7. Mamun MM, Golam MG, Ali MA, Rahaman SMB, Huq KA (2009) Abundance and Distribution of Plankton in the Sundarbans Mangrove Forest. J Innov Dev Strategy 3: 43-54.
8. Aziz A, Rahman M, Ahmed A (2012) Diversity, Distribution and Density of Estuarine Phytoplankton in the Sundarban Mangrove Forests, Bangladesh. J Bot 41: 87-95.
9. Troccoli GL, Herrera-Silveira JA, Comı´n FA (2004) Structural variations of phytoplankton in the coastal seas of Yucatan, Mexico. Hydrobiologia 519: 85-102.
10. Round FE, Crawford RM, Mann DG (1990) The Diatoms: Biology and Morphology of the Genera. Cambridge University Press, Cambridge, 747.
11. Hendey NI (1973) The plankton diatoms of the Southern Seas. 16: 151-364.
12. Lebour MV (1925) The dinoflagellates of northern seas. The Marine Biological Ass. UK.
13. Hasle GR, Syvertsen EE, Steidinger KA, Karl Tangen, Tomas CR (1996) Identifying Marine diatoms and Dinoflagellates 598.

14. Blaze KL (1984) Morphology and taxonomy of Diplomenora gen:nov. (Bacillariophyta). Br Phycol Journal 19: 217-225.

15. Dodge JD (1985) Atlas of Dinoflagellates: A Scanning Electron Microscope Survey. London: Farrand Press.

16. Subrahmanyan R (1971) The Dinophyceae of the Indian Seas. Part 2: Family Peridiniaceae Schutt emend Lindemann. Marine Biological Association of India, Memoir II, Cochin, India.

17. Taylor FJR (1987) The Biology of Dinoflagellates. Bot Monogr 21 Blackwell & Univ Calif Press, 786.

18. APHA (2012) Standard methods for the examination of water and waste water. American Public Health Association, Washington, USA.

19. Yin K, Qian PY, Wu MCS, Chen JC, Huang L, et al. (2001) Shift from P to N limitation of phytoplankton biomass across the Pearl River estuarine plume during summer. Marine Ecological Progress Series 221: 17-28.

20. Margalef R (1951) Species diversity in natural comunidales. Publ Inst Biol Apl 9: 5-27.

21. Shannon CE, Weaver W (1949) The mathematical theory of communicatio. Univ. Illinois Press. Illinois, 125.

22. Pielou EC (1969) An Introduction to Mathematical Ecology. John Wiley, New York, 286.

23. Gobler C, Boneillo GE (2003) Impacts of anthropogenically influenced groundwater seepage on water chemistry and phytoplankton dynamics within a coastal marine system. Marine Ecology Progress Series 255: 101-114.

24. Alvarez-Góngora C, Herrera-Silveira JA (2005) Variations of phytoplankton community structure related to water quality trends in a tropical karstic coastal zone. Mar Pollut Bull 52: 48-60.

25. Yao P, Yu Z, Deng C, Liu S, Zhen Y (2010) Spatial–temporal distribution of phytoplankton pigments in relation to nutrient status in Jiaozhou Bay, China. Estuarine, Coastal and Shelf Science 89: 234-244.

26. Steele JH (1962) Environmental control of photosynthesis in the sea. Limnology and Oceanography 7: 137-150.

27. Cullen JJ (1982) The deep chlorophyll maximum Comparing vertical profiles of chlorophyll a. Canadian Journal of Fisheries and Aquatic Science 39: 791-803.

28. Boyer JN, Kelble CR, Ortner PB, Rudnick DT (2008) Phytoplankton bloom status: Chlorophyll a biomass as an indicator of water quality condition in the southern estuaries of Florida, USA. Ecological Idicators 9: S56-S67.

29. Cetin AK, Sen B (2004) Seasonal distribution of phytoplankton in Orduzu Dam Lake (Malatya, Turkey). Turk J Bot 28: 279-285.

30. Mosisch TD, Bunn SE, Davies PM, Marshall CJ (1999) Effects of shade and nutrient manipulation on periphyton growth in a subtropical stream. Aquat Bot 64: 167-177.

31. Bouman HA, Nakane T, Oka K, Nakata K, Kurita K, et al. (2010) Environmental controls on phytoplankton production in coastal ecosystems: A case study from Tokyo Bay. Estuarine Coastal and Shelf Science 87: 63-72.

32. Reyes-Rodrı´guez MA (2001) Hidrologı´a y estado tro´fico de la columna de agua de la zona marina costera de Sisal. Instituto Tecnolo´gico de Me´rida MsC Thesis 217.

33. Tilstone GH, Figueiras FG, Lorenzo LM, Arbones B (2003) Phytoplankton composition, photosynthesis and primary production during different hydrographic conditions at the Northwest Iberian upwelling system. Marine Ecology Progress Series 252: 89-104.

34. Kouwenberg JHM (1994) Copepod distribution in relation to seasonal hydrographic and spatial structure in the north-western Mediterranean (Gulf du Lion). Estuar. Coastal Shelf Sci 38: 69-90.

35. Ramaiah N, Nair V (1997) Distribution and abundance of copepods in the pollution gradient zones of Bombay Harbour-Thana Creek-Bassein Creek, west coast of India. Indian J Mar Sci 26: 20-25.

36. Chandramohan P, Sreenivas N (1998) Diel variations in zooplankton population in mangrove ecosystem at Gaderu canal, South-east coast of India. Indian J Mar Sci 27: 486-488.

37. Sridhar R, Thangaradjou T, Kumar SS, Kannan L (2006) Water quality and phytoplankton characteristics in the Palk bay, southeast coast of India. J Environ Biol 27: 561-566.

38. Subramanian B, Mahadevan A (1999) Seasonal and diurnal variations of hydro biological characters of coastal waters of Chennai (Madras) Bay of Bengal. Indian J Mar Sci 28: 429-433.

39. Senthilkumar S, Santhanam P, Perumal P (2002) Diversity of phytoplankton in Vellar estuary, south-east coast of India. AFSIB, Mangalore and AeA, Bhubanewar, India, 245-248

40. Corbett DR, Kump L, Dillon K, Burnett W, Chanton J (2000) Fate of wastewater-borne nutrients in the subsurface of the Florida Keys, USA. Marine Chemistry 69: 99-115.

41. Philippart C, Cade´e G, Raaphorst W, Riegman R (2000) Long-term phytoplankton nutrient interactions in a shallow sea: algal community structure, nutrient budgets, a denitrification potential. Limnology & Oceanography 45: 131-144.

42. Domingues RB, Anselmo TP, Barbosa AB, Sommer U, Galvão HM (2011) Nutrient limitation of phytoplankton growth in the freshwater tidal zone of a turbid, Mediterranean estuary. Estuarine, Coastal and Shelf Science 91: 282-297.

43. Nwankwo DI (1998) Seasonal changes in phytoplankton population and diversity in Epe Lagoon, Nigeria. Acta Hydrobiologia 40: 82-92.

44. Vanlandingham SL (1982) Guide to the identification and environmental requirements and pollution tolerance of freshwater blue-green algae (cyanophyta). US Environmental Protection Agency, EPA-60.

45. Nwankwo DI, Akinsoji A (1992) Epiphyte community of water hyacinth, Eichhornia crassipes (MART) Solms in coastal waters of South Western Nigeria. Archiv fur Hydrobiologie 124: 501-511.

46. Pa´ez-Osuna F, Guerrero-Galvan SR, Ruiz-Fernandez AC (1998) Discharge of nutrients from shrimp farming to coastal waters of the Gulf of California. Marine Pollution Bulletin 38: 585-592.

47. Jones AB, O'Donohue MJ, Udy J, Denison WC (2001) Assessing ecological impacts of shrimp and sewage effluent: biological indicators with standard water quality analyses. Estuarine, Coastal and Shelf Science 52: 91-109.

The Summer Distribution of Dissolved Inorganic Iodine along 18°N in the South China Sea

Long A[1,*], Dang A[1,2], Xiao H[1] and Yu X[1,3]

[1]*State Key Laboratory of Tropic Oceanography, South China Sea Institute of Oceanology, Chinese Academy of Sciences, 164 West Xingang Road, Guangzhou 510301, P.R China*

[2]*South China Sea Branch of the State Oceanic Administration, 353 Middle Xingang Road, Guangzhou, 510301, P.R China*

[3]*University of Chinese Academy of Sciences, Beijing, 100039, P.R China*

Abstract

Dissolved inorganic iodine (iodate and iodide) in seawater samples collected in the northern South China Sea (SCS) were investigated using differential pulse polarography and cathodic stripping square wave voltammetry, respectively. High concentration of iodide was observed in the surface layer in agreement with those previously reported for tropical waters. Iodine speciation and macro-nutrients were strongly coupled as observed in other tropical oceans. The vertical distributions of iodine and macro-nutrients can be classified into three biogeochemical regimes: (1) in the upper 50 m of the water column, low iodate was observed as nutrients were depleted and photosynthesis was low in summer. In contrast, substantial amount of iodide produced by high phytoplankton productivity and intense bacterial action in winter and spring was saved till summer; (2) in the water column between 50 m and the Tropical Water at around 150 m, the amount of iodide oxidation varied proportionally with nitrification, leading to rapid increase of nitrate and iodate with depth; (3) below the Tropical Water, iodate and nitrate are regenerated mainly due to re-mineralization of organic matter sinking from the euphotic zone. The distribution of dissolved inorganic iodine and hydrographic parameters (salinity, temperature, dissolved oxygen, and macro-nutrients) were influenced by the downwelling of the mesoscale warm eddy.

Keywords: Iodine; Nutrients; Photosynthesis; Nitrification; South china sea

Introduction

There has been interest in the biogeochemical cycling of iodine in recent decades for its own sake and due to the importance of this element in human health, hydrologic processes and radioecology [1]. Compared to low levels of iodine found in soils, rivers, lakes and terrestrial plants, oceans are the major source of iodine in the global geochemical cycle with an average oceanic iodine concentration of approximately 0.45 μM. Although thermodynamic calculation indicates that iodate should be more stable than iodide in oxic seawater [2-4], field studies have observed pronounced conversion of iodate to iodide in permanently stratified surface waters not only in tropical and sub-tropical oceans [5-9], but also in temperate continental shelf seas where iodide concentration increase shoreward [10-14]. Thus, the overall longitudinal distribution of iodate in the surface layer shows that iodate concentration increases polewards whereas iodide concentration maxima is observed near the equator [5,15]; meanwhile, the general vertical trend shows that iodide decreases with depth to around the detection limit below the euphotic zone while iodate increases with depth to an approximately constant level [16]. Furthermore, iodide can be enriched in anoxic basins and oxygen minimum zones.

Understanding the controls on dissolved iodine speciation is important because of its role in the global biogeochemical cycle of iodine and indirect effect on the supply of iodine to the atmosphere from the oceans [17]. Once iodine enters the atmosphere, it can promote particle formation reactions in the marine boundary layer and very effectively destroy ozone, thus influencing climate [18].

In a redox cycle, iodate and iodide are interconverted, and this is the most pronounced effect for iodine speciation in oxic seawater [19]. However, the evidence for the actual mechanisms of iodate reduction to iodide in oxic oceanic waters remains unclear [20-22]and our knowledge on iodide oxidation to iodate has been limited [23], making it difficult to adequately model iodine cycling in the ocean.

The South China Sea (SCS) is the largest marginal sea of the world [24] and is situated in the tropical-subtropical zone where active interconversion between iodate and iodide is expected. The study of iodine in the SCS may provide insight into mechanisms controlling global ocean iodine cycle. However, few data are available on the spatial distribution of iodine in the SCS. In this work, the spatial distribution of dissolved inorganic iodine in the SCS and the implications for the interconversion between iodate and iodide are studied and the implication on the biogeochemical dynamics of the dissolved iodine system in the marginal seas is discussed.

Study area

As a result of a unique combination of environmental conditions, the characteristics of its upper water are significantly different from those in other low-latitude areas and these differences have led to patterns in the intra-seasonal variations in phytoplankton biomass that are atypical of low-latitude waters [25-27].

The SCS is the largest semi-enclosed tropical marginal sea in the northwest Pacific Ocean with a total area of 3.5 million km2 often

***Corresponding author:** Long A, State Key Laboratory of Tropic Oceanography, South China Sea Institute of Oceanology, Chinese Academy of Sciences, 164 West Xingang Road, Guangzhou 510301, P.R China, E-mail: longam@scsio.ac.cn

referred to as the Asian Mediterranean, e.g., Morton and Blackmore [28]. The maximum depth of the SCS exceeds 5000 m, while its mean depth is 1350 m [24]. Thus, the South China Sea combines characteristics of open oceans and confined marginal seas.

The northern part of the SCS has a broad continental shelf (150-250 km offshore) which has an average depth of shallower than 200 m and has isobaths that are parallel to the continental coastline. Along the northern boundary, the Pearl River Estuary (PRE) is located between the Taiwan Shoal and the Hainan Island. The annual average river discharge through the PRE is 10 524 m3s-1, with 80% occurring between April and September during the period of the summer monsoon. Under the influence of the summer monsoon, the less saline river discharge of over 16,000 m-3 s-1 influences the broad region over the shelf [29]. In addition, the southwesterly summer monsoon can induce upwelling at many places over the continental shelf, which is a seasonal common phenomenon with the spatial extent of a basin-scale in the NSCS [30,31].

In this study, we focused on the 18°N transect including 13 stations stretching from 110°E to 119.5°E (Figure 1). In the west section of this transect stations 1 and 2 are close to the Hainan Island, with depths of 94 m and 156 m respectively. The station 13 locates at about 100 km from the south of Luzon Strait.

Sampling and analyses

Sampling was conducted aboard the R/V "Shiyan 3" during KJ cruise in June 2014. Seawater samples were collected using a conductivity-temperature-depth (CTD) rosette equipped with 12 modified Hydro-Bios 2 L water bottles.

Sub-samples for dissolved oxygen (DO) were measured by the standard manual Winkler's method [32]. Samples for the determination of nutrients were filtered and frozen until analysis within 3 weeks after sample collection. The concentrations of nitrate, nitrite and phosphate were determined by standard methods [32].

Water samples for dissolved inorganic iodine species (iodate and iodide) analysis were collected were filtered and stored frozen in the dark. All samples were analyzed within 2 months. Both iodate and iodide were determined by using Metrohm Model 797 VA Computrace (Switzerland) in conjunction with a hanging mercury drop electrode, a platinum auxiliary electrode and a reference Ag/AgCl/KCl (3 mol/L) electrode. Iodate was determined directly by differential pulse polarography according to the method of Wong and Zhang [33] adapted from protocol of Herring and Liss [34]. Iodide was determined by Cathodic Stripping Square Wave Voltammetry according to the original method of Luther et al. [35]. The analyses were carried out using standard addition with calibration plot method. Three additions were added to the original sample and each was scanned in duplicate. Each sample was measured in duplicate or triplicate. The coefficient of variation was typically less than ± 5% for iodate and ± 9% for iodide. Detection limits of the iodate and iodide analyses were around 20 nM and 5 nM, respectively.

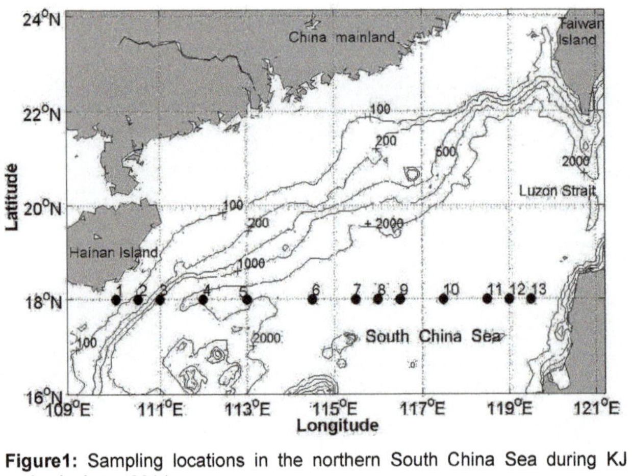

Figure1: Sampling locations in the northern South China Sea during KJ cruise in June, 2014.

Results

Hydrography

The vertical distributions of salinity and temperature above 1500 m are in accordance with those reported previously in the northern SCS [36]. Salinity increases with depth to a maximum at around 150 m, which represents the Tropical Water. Then it decreases to a minimum at around 500 m, which represents the North Pacific Intermediate Water (NPIW). Below the NPIW, salinity increases slightly from 34.4 to 34.6 at 1500 m. In contrast, temperature decreases linearly with depth, with a strong gradient in the top 200 m layer. The T-S plot (Figure 2) shows that salinity and temperature were uniform below 200 m and variable in the top 200 m as shown in Figure 3.

These variations were caused by typical hydrological events of the South China Sea. A mesoscale eddy converging warm waters was progressing between 116°E and 118°E during the sampling period. The downwelling movement of the water within this warm eddy was clearly demonstrated by the depressed isopleths of salinity and temperature. While stations 9 and 10 were located at the center of this eddy, stations 6, 7 and 11, 12 recorded the peripheral vertical convections. On the western side of this latitudinal transect, stations 1 and 2 were located on the continental shelf where upwelling prevails in summer [31]. The upwelling centers of QDU (in the east of the Hainan Island) were located at (19°45'N, 111°10'E) and (18°25'N, 110°15'E), and stations 1 and 2 were close to the latter one. The higher salinity at these two stations, relative to the more offshore water, was attributed to the influence of upwelling at that time.

High insolation causes relatively high sea-surface temperature in the SCS, exceeding 29°C (Figure 3a). Meanwhile, large inputs of fresh water from terrestrial runoffs and precipitation during this season reduced

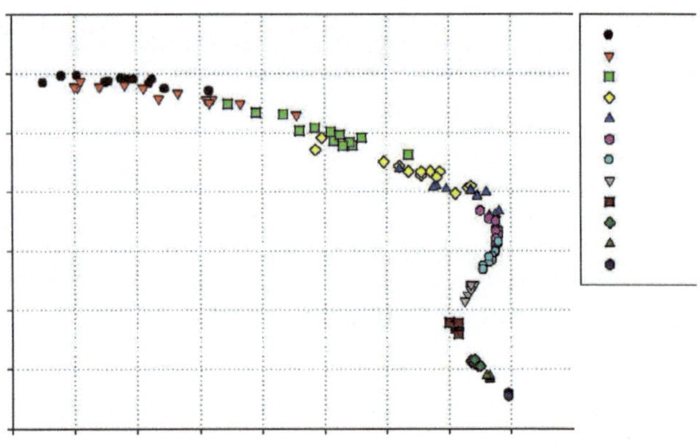

Figure 2: The T-S plot of the 18°N transect (110-119.5°E) in June 2014.

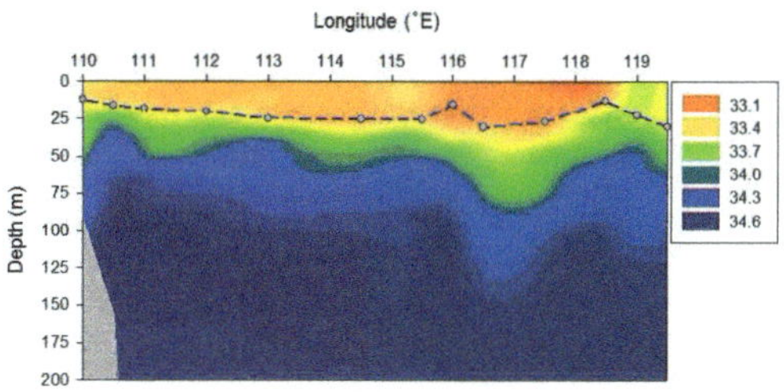

(3b) Temperature.

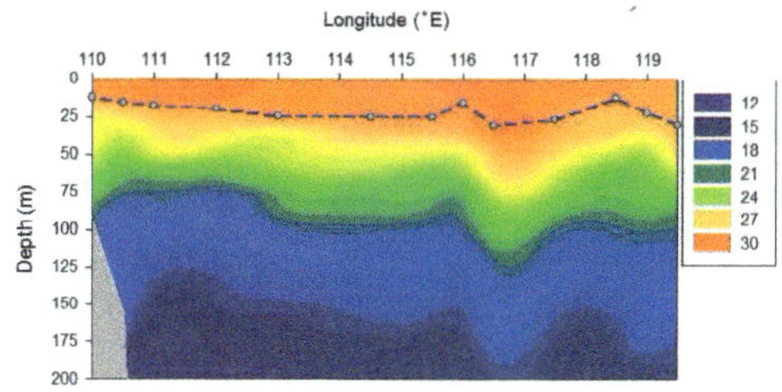

(3c) Density.

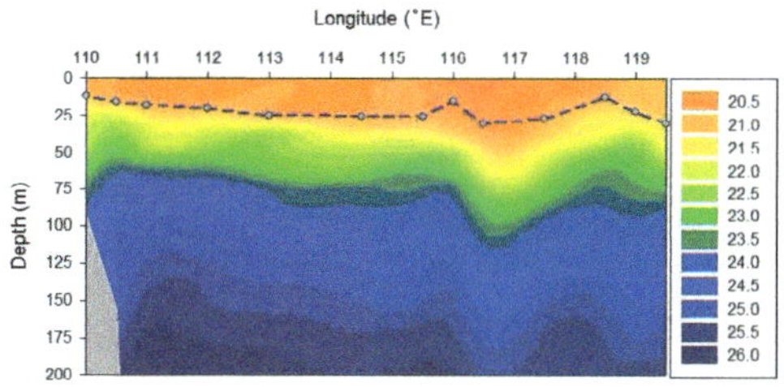

Figure 3: Spatial variations of (3a) salinity, (3b) temperature and (3c) density in the top 200m along the 18°N transect (110-119.5°E) in June 2014. The values of mixed layer depth (MLD) were derived from a drop of 0.5°C from the sea-surface temperature (Karl and Lukas, 1996) and shown with blue short-dash line. The gray circles represent sampling stations. (3a) Salinity.

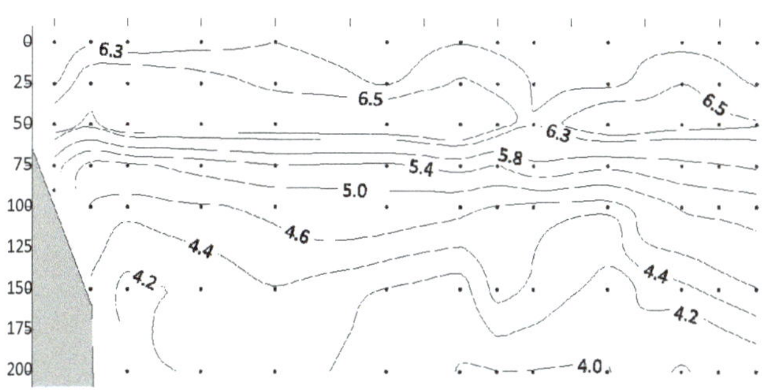

Figure 4: Spatial variations of DO in the top 200m along the 18°N transect (110-119.5°E) in June 2014. The unit is mg O_2/L. Dark dots represent the sampling sites.

the salinity to below 33.7 in the upper 30 m of the water column (Figure 3b). Therefore, the upper layer was stratified and the mixed layer depths (MLD) were less than 30 m along this transect (Figure 3c). This result is consistent with Wong et al. [36], which shows a mixed layer depth of 20 ± 6 m during summer.

Dissolved oxygen (DO)

The vertical distribution of DO above 200 m (Figure 4) exhibits a subsurface maximum between 25 and 50 m, and a rapid decrease with depth below 50 m, in particular between 50 m and 100 m. This DO profile appears to be related to biological activities and re-mineralization.

Nutrients

The vertical distributions of dissolved inorganic nutrients in the upper 1500 m water column and the distributions of these parameters in the top 200 m are shown in Figures 5a and 6. Within the upper 50 m of the water column, nitrate was below detection limit, except the low values (0.1~0.2 μM) at stations 1 and 2 on the continental shelf. The

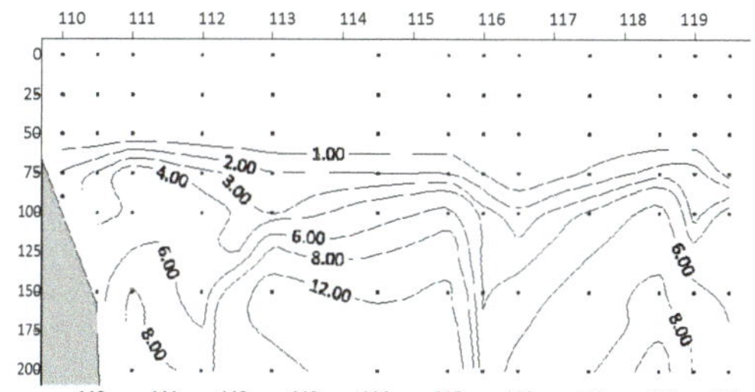

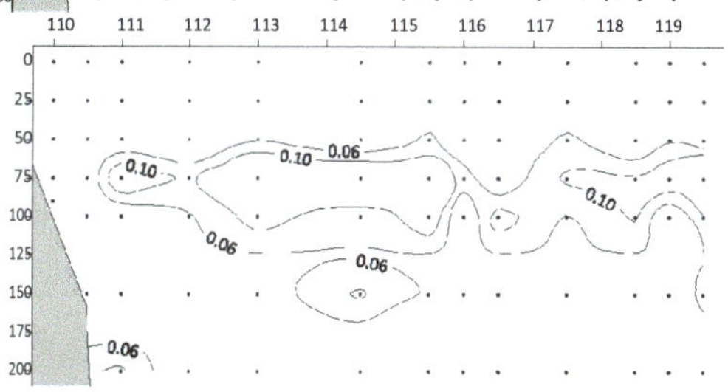

(5c) Phosphate.

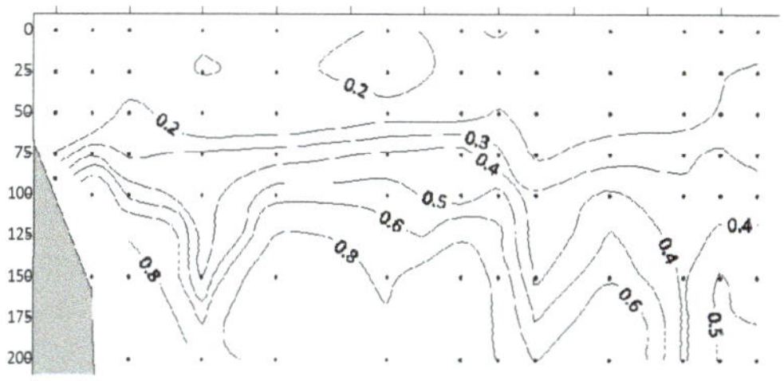

Figure 5: Spatial variations of (5a) nitrate, (5b) nitrite, (5c) phosphate in the top 200m along the 18°N transect (110-119.5°E) in June 2014. Dark dots represent the sampling sites. The unit is μM.(5a) Nitrate. (5b) Nitrite. (5c) Phosphate.

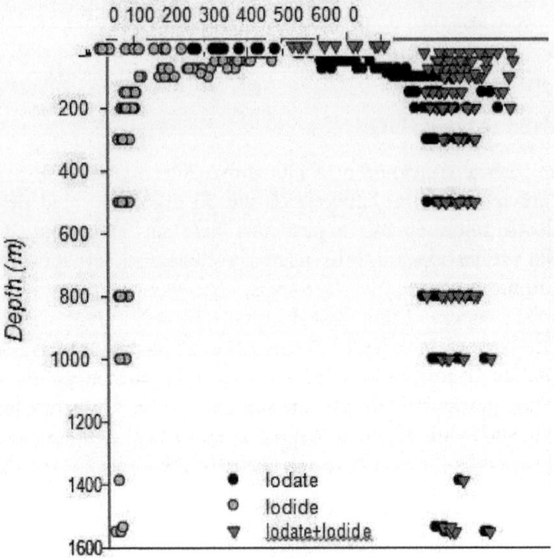

Figure 6: Vertical distributions of iodine above 1500m along the 18°N transect (110-119.5°E) in June 2014.

presence of nitrate there may be due to terrestrial input or upwelling input. Furthermore, phosphate in this water was quite low (~0.1 μM), even undetectable in the warm eddy and on the continental shelf. Below 50 m, both nitrate and phosphate increased with depth (Figure 5b). A strong gradient was observed between 50 m and 500 m probably due to intensive re-mineralization of sinking organic matter. The primary nitrite maximum (>0.1 μM) was located between 50 and 100 m where nitrification was extensive [37,38]. Although the vertical spacing of samples was inadequate to precisely depict and could introduce distortions into the details in the distribution of nitrite in this layer of water, the general feature was clear. Because of the converging and downwelling effect of the warm core eddy, nitrate, nitrite and phosphate isopycnals were apparently depressed, just like those of temperature and salinity.

Dissolved inorganic iodine

The vertical distributions of iodine above 1500 m and the spatial variations of iodate and iodide in the top 200 m layer along this transect were plotted in Figures 5c and 7a,7b. All 13 stations can be grouped into two different patterns according to iodine distribution. At stations 2, 6, 7, 10, 11 and 12, iodine was quite constant in the water column without obvious surface depletion. In contrast, significant iodine surface enrichments were observed in stations 1, 3, 4, 5, 8, 9 and 13. Iodide and iodate concentrations at the surface of these stations were higher than 200 nM and 300 nM, respectively. Despite the different surface patterns, DII was similar below 200 m at all these stations. Nonetheless, vertical distributions of iodate and iodide had different patterns. From the surface to the Tropical Water at around 150 m, a decrease in iodide concentration was accompanied by an increase of iodate concentration, with a strong gradient between 50 m and 100 m. Below the Tropical Water, both iodide and iodate concentrations are constant. The high concentration of iodide in the surface layer was consistent with that previously reported for the tropical waters. The isopycnals of iodate and iodide in the warm eddy were also depressed in a way similar to salinity, temperature and nutrients.

Discussion

Hydrography and nutrient condition

In the north of SCS, the annual excess precipitation (total precipitation minus evaporation) reaches 1039 mm and most of this precipitation occurs between April and September. The input of fresh water from terrestrial runoffs also occurs mostly in these months. A combination of a generally intense radiative flux together with high rainfall in the north of the SCS induced stratification during the sampling period. The combination of these two fresh water sources reduces the surface salinity in most regions of the central SCS to below 34 [39-41]. Furthermore, the intensive irradiation in tropics induces persistent high SST (Sea-Surface Temperature) in the South China Sea. Thus, the upper water column is always well stratified [41] with the mixed layer shallower than 30 m during most of the year. Meanwhile, the photic zone depth stays constantly at around 85 m throughout the year [26]. The photic zone depth exceeds the mixed layer depth and the top of the nutricline, whereas the mixed layer does not reach the top of the nutricline. Hence, the nutrients in the upper nutricline cannot be made available to fuel photosynthetic activities via vertical mixing [36]. Therefore, the northern South China Sea is oligotrophic during summer and primary production is limited primarily by the availability of the nutrients rather than the availability of light.

Nitrate-salinity and phosphate-salinity graphs are useful in classifying the water column into three distinct biogeochemical regimes, exemplified here by that for nitrate (Figure 8a). The relationship between phosphate and salinity followed the same pattern qualitatively. As shown in Figure 8a, nitrate was almost undetectable in the first 50 m layer where salinity varied between 33.1 and 34.1 (representing by the horizontal limb). Nutrients are imported into the surface layer due to surface cooling and wind-induced vertical mixing in winter, and then depleted by high biological productivities during winter and

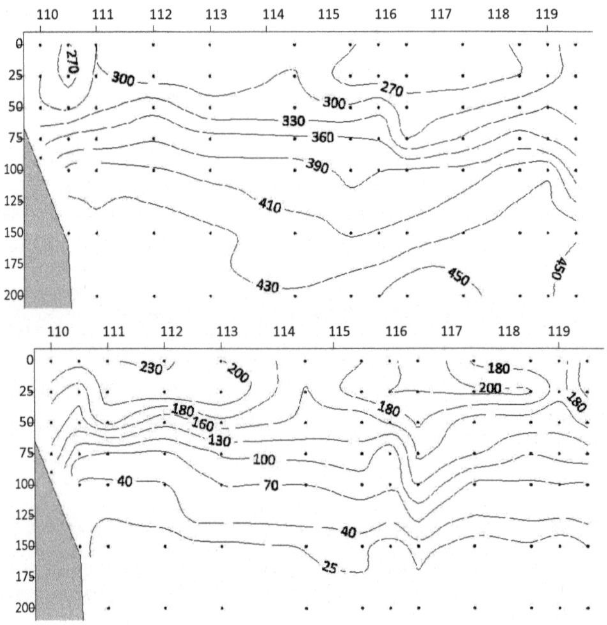

Figure 7: Spatial variations of (7a) iodate, (7b) iodide in the top 200m along the 18°N transect (110-119.5°E) in June 2014.The unit is nM. (7a) Iodate. (7b) Iodide.

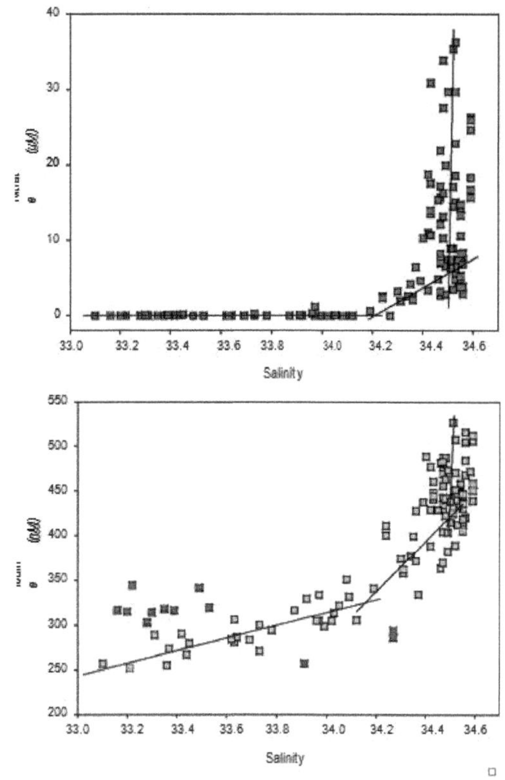

Figure 8: The relationship between nitrate and salinity (8a) and iodate and salinity (8b) of the 18°N transect in the South China Sea.Overimus more

spring. Before summer, nutrients were exhausted and photosynthesis was subdued by limited abundance of nutrients, in particular by N. In the layer between the Tropical Water and 1500 m (representing by the nearly vertical line), nitrate increased with depth gradually. The main biogeochemical process here was the re-mineralization of sinking organic matter produced by surface productivity. The gradient of nitrate concentration was higher above 500 m because most of the biologically produced organic matter within the euphotic zone was degraded before falling below 400 m. From 50 m to the layer of the Tropical Water with a salinity maximum at around 150 m (representing by the line connecting the horizontal line and the nearly vertical line), nitrate increased sharply. This layer of rapid nitrate increase layer coincides approximately with the primary nitrite maximum layer. Since nitrate is generated through biologically mediated process of nitrification and nitrite is an intermedium of this process, the nitrite maximum and rapid increase of nitrate between 50 m and 150 m suggest a high activity of nitrifers. Because of intensive nitrification, DO decreases rapidly in this layer.

Dissolved inorganic iodine

As mentioned above, the vertical profiles of Dissolved Inorganic Iodine (DII) can be categorized into two patterns. While stations 2, 6, 7, 10, 11 and 12 had essentially the same concentrations of DII, stations 1, 3, 4, 5, 8, 9 and 13 showed surface enrichments of DII. Apparently, the surface enrichment of DII was interrupted by eddy-induced convection at stations 10, 11, 12 and 6, 7, and by upwelling at station 2. It appears that the surface enrichment of DII is a common phenomenon in the northern South China Sea. The enrichment of total iodine has been reported in many different environments due to diversified reasons. e.g., in brines such as the Orca Basin [42] and the Mediterranean Sea [43], simultaneous mobilization of chloride and iodide during the dissolution of evaporate beds results in extraordinarily high iodide and total iodine; thin sub-surface maxima of iodide and total iodine found in the pycnocline at 150 m in the Orca Basin [42] and at 100 m in the Venezuela Basin [2] may be resulted from advection or re-mineralization of organic matter forming iodide as the initial product; advection from margin sediments off of India may be the most plausible source of iodine and contributed ~96% of the total excess iodine to the OMZ in the Arabian Sea. Anyhow, these iodine enrichments possessed extraordinarily high iodide concentrations and none of these happened in surface waters. There is only one reported iodine enrichment happened at surface water, i.e., Truesdale et al. [44] observed unusually high iodide concentrations (320-850 nM) in surface waters in the Skagerrak, particularly during the summer cruise. Unfortunately, the cause for such iodine enrichment remained unclear although a surface phenomenon has been speculated. Moreover, this enrichment was also caused by extraordinarily high iodide. In contrast to all these instances enumerated above, the enrichment of iodine within the mixed layer in the northern South China Sea is not so exaggerated and the concentration of iodide still falls within the range of concentrations for surface tropical waters from previous oceanographic measurements. Furthermore, iodate is also slightly higher within the mixed layer at most of these stations showing surface enrichment than the others. Thus, the iodate/iodide ratios of these stations are still similar to those of the other stations.

The surface waters in the SCS exchange freely with those in the East China Sea, the Java Sea, the Sulu Sea and the western Philippine Sea. In addition, a multitude of rivers drain into the South China Sea after they have passed through densely populated areas. External import of sources with elevated iodine from these areas and relatively longer residence time of the mixed layer may cause enrichment of DII in the northern South China Sea. Atmospheric deposition may be another potential source of excess total iodine to the mixed layer.

Interconversion of iodide and iodate

A close examination of data in Figure 5a and 5b reveals that the vertical distribution of iodate resembles that of nitrate (also phosphate, not shown). Relationships between iodate and nitrate/phosphate show a strong positive correlation, yielding 0.9 and 0.8 respectively. As for nitrate, we use the iodate-salinity graph to classify different biogeochemical regimes of iodine (Figure 8b). The iodide-salinity graph is symmetrically reversed to iodate-salinity graph.

The similarity between the iodate-salinity graph and the nitrate-salinity graph indicates strong coupling of iodine speciation and nutrients. In the upper 50 m layer (salinity between 33.10 and 34.12), iodate is proportionally depleted and nitrate is almost undetectable respectively. In this photic layer, iodate reduction dominates and iodide should be the main product of iodate reduction since iodide accounts for ~40% of DII here. Although we did not measure organic iodine, the organic iodine may also be produced. Iodate reduction might depend upon biotic processes including phytoplankton [7,16,45,46] and bacterial action during respiration [9,47], and abiotic processes including photochemical reduction initiated by organic matter [48] and reaction with sulfide [14]. Since the photochemical reduction of iodate by iodide at natural concentrations in seawater cannot account for the extent of iodate reduction in surface seawaters [49] and chemical redox reactions are not possible to occur in open oceanic environment, the formation of iodide here must be caused mainly by biological activities in surface waters of the northern South China Sea. While nutrients were already exhausted by biological activities and photosynthesis was low in the summer, the large percent of iodide may be produced by high phytoplanktonic productivity during winter and spring and intense bacterial action. The red points above the line indicated relatively high iodate concentrations and enrichment of DII in the mixed layer. The blue points indicated lower iodate concentrations on the continental shelf. Meanwhile, the DII of these three blue points were lower than the ambient waters, possibly indicating the formation of organic iodine.

In the layer between the Tropical Water and 1500 m (representing by the vertical line in Figure 8b), iodate increases with depth slightly from 150 m to 200 m and then becomes almost invariant with several higher concentration points. While a large percent of iodate was reduced to iodide at the surface, it became the predominant species of iodine below the Tropical Water. Meanwhile, nitrate increases gradually in this thick layer due to re-mineralization of sinking particles. As compared with nitrate, very little iodine is turned into particulate form through biological activities. Meanwhile, most of the biologically produced organic matter within the euphotic zone was degraded before it reached 400 m. Thus iodate stayed quite constant while nitrate increased gradually through the water column to 1500 m.

From 50 m to the depth of the Tropical Water, iodate increases while iodide decreases. Relative to the surface 50 m layer, this layer of water has high concentration of macro-nutrients. Furthermore there is a subsurface chlorophyll maximum [50].

But iodide formation was lower in despite of higher phytoplanktonic activities. Thus, the opposite process should dominate. Song et al. [51] found that bacteria played an important role in the oligotrophic waters of the South China Sea and bacterial production was closely correlated to the phytoplankton biomass and primary production, and maximum bacterial production occurs near bottom layers of the euphotic zone. Combining with their findings, the nitrite maximum and rapid increase of nitrate between 50 m and the Tropical Water at around 150 m suggest high activity of nitrifers, i.e., strong nitrification. On the other hand, iodide oxidation may occur in parallel with bacterial nitrification at the base of the photic zone in the northern South China Sea. The "nitrification hypothesis" was proposed by Truesdale et al. [52] who apply a box model to the Black Sea and the hypothesis is supported by direct observations of iodide oxidation and iodate formation in anchialine systems [23,47,53]. In the global biogeochemical model of nitrification proposed by Yool et al. [54], for much of the world ocean a substantial fraction of the nitrate that is taken up biologically is generated through recent nitrification near the surface [55]. At the global scale, nitrification accounts for about half of the nitrate consumed by growing phytoplankton [56]. Based on strong nitrification, the extent of iodide oxidation in parallel with nitrification may be presumed to be sufficient to compete with iodide formation at the base of the photic zone and thus iodide was reduced quickly.

Conclusions

Vertical distributions of iodate and iodide have different patterns. Above the Tropical Water at around 150 m, iodide concentration decreased while iodate concentration increased. Below the Tropical Water, both iodide and iodate are constant. Iodide concentrations were high at the surface mixed layer but fall within the range of concentrations previously reported for surface tropical waters. The speciation of iodine was strongly influenced by biological activities such as photosynthesis and bacterial action in the South China Sea. Thus, iodine speciation is coupled with nutrients. In the mesoscale warm eddy, the isopycnals of hydrographic factors (salinity, temperature), dissolved oxygen, nutrients and iodine species were depressed due to downwelling. This work is a preliminary survey of the dissolved inorganic iodine in the northern South China Sea. Further investigations are needed to better understand the biogeochemistry of iodine in this typical marginal sea.

Acknowledgements

The authors would like to thank the captain and crew of R/V "Shiyan 3" for their help during field sampling. We thank Mr. Cheng for his professional help with the instrument and thank Wang Guifen for her drawing of the station map. This work was financially supported by the National Natural Science Foundation of China (No. 40876039), the Science and Technology Program of China (No. 2008FY110100), and Global change scientific research program of China (No. 2010CB951201).

References

1. Leblanc C, Colin C, Cosse A, Delage L, La Barre S, et al. (2006) Iodine transfers in the coastal marine environment: the key role of brown algae and of their vanadium-dependent haloperoxidases. Biochimie 88: 1773-1785.
2. Wong GTF, Brewer PG (1977) The marine chemistry of iodine in anoxic basins. Geochimica et Cosmochimica Acta 41: 151-159.
3. Wong GTF (1980) The stability of dissolved inorganic species of iodine in seawater. Marine Chemistry 9: 13-24.
4. Wong GTF (1991) The marine geochemistry of iodine. Reviews in Aquatic Sciences 4: 45-73.
5. Tsunogai S, Henmi T (1971) Iodine in the surface water of the ocean. Journal of Oceanographical Society of Japan 27: 67-72.
6. Elderfield H, Truesdale VW (1980) On the biophilic nature of iodine in seawater. Earth and Planetary Science Letters 50: 105-114.
7. Campos M, Farrenkopf AM, Jickells TD, Luther GW (1996) A comparison of dissolved iodine cycling at the Bermuda Atlantic Time-Series station and Hawaii Ocean Time-Series Station. Deep-Sea Research Part II: Topical Studies in Oceanography 43: 455-466.
8. Farrenkopf AM, Luther Ш GW (2002) Iodine chemistry reflects productivity and denitrification in the Arabian Sea: evidence for flux of dissolved species from sediments of western India into the OMZ. Deep Sea Research Part II: Topical Studies in Oceanography 49: 2303-2318.
9. Truesdale VW, Bailey GW (2002) Iodine distribution in the Southern Benguela system during an upwelling episode. Continental Shelf Research 22: 39-49.
10. Wong GTF, Zhang LS (1992a) Changes in iodine speciation across coastal hydrographic fronts in southeastern United States continental shelf waters. Continental Shelf Research 12: 717-733.
11. Truesdale VW (1994) A re-assessment of redfield correlations between dissolved iodine and nutrients in oceanic waters and a strategy for further investigations of iodine. Marine Chemistry 48: 43-56.
12. Truesdale VW, Bailey GW (2000) Dissolved Iodate and Total Iodine during an Extreme Hypoxic Event in the Southern Benguela System. Estuarine, Coastal and Shelf Science 50: 751-760.
13. Truesdale VW, Jones K (2000) Steady-state mixing of iodine in shelf seas off the British Isles. Continental Shelf Research 20: 1889-1905.
14. Truesdale VW, Upstill-Goddard R (2003) Dissolved iodate and total iodine along the British east coast. Estuarine, Coastal and Shelf Science 56: 261-270.
15. Truesdale VW, Bale AJ, Woodward EMS (2000) The meridional distribution of dissolved iodine in near-surface waters of the Atlantic Ocean. Progress In Oceanography 45: 387-400.
16. Wong GTF, Zhang LS (2003) Seasonal variations in the speciation of dissolved iodine in the Chesapeake Bay. Estuarine, Coastal and Shelf Science 56: 1093-1106.
17. Chance R, Malin G, Jickells TD, Baker AR (2007) Reduction of iodate to iodide by cold water diatom cultures. Marine Chemistry 105: 169-180.
18. Von Glasow R. (2005) Seaweed, Iodine, New Particles and Atmospheric Chemistry-The Current State of Play. Environmental Chemistry 2: 243.
19. Bluhm K, Croot P, Huhn O, Rohardt G, Lochte K (2011) Distribution of iodide and iodate in the Atlantic sector of the southern ocean during austral summer.

Deep Sea Research II: Topical Studies in Oceanography 58: 2733-2748.

20. Truesdale VW, Nausch G, Baker A (2001a) The distribution of iodine in the Baltic Sea during summer. Marine Chemistry 74: 87-98.

21. Waite TJ, Truesdale VW, Olafsson J (2006) The distribution of dissolved inorganic iodine in the seas around Iceland. Marine Chemistry 101: 54-67.

22. Chance R, Weston K, Baker AR, Hughes C, Malin G, et al. (2010) Seasonal and interannual variation of dissolved iodine speciation at a coastal Antarctic site. Marine Chemistry 118: 171-181.

23. Zic V, Caric M, Viollier E, Ciglenecki I (2010) Intensive sampling of iodine and nutrient speciation in naturally eutrophicated anchialine pond (Rogoznica Lake) during spring and summer seasons. Estuarine, Coastal and Shelf Science 87: 265-274.

24. Chen CTA, Wang SL, Wang BJ, Pai SC (2001) Nutrient budgets for the South China Sea basin. Marine Chemistry 75: 281-300.

25. Cai SQ, Su JL, Gan ZJ, Liu Q (2002) The numerical study of the South China Sea upper circulation characteristics and its dynamic mechanism, in winter. Continental Shelf Research 22: 2247-2264.

26. Tseng CM, Wong GTF, Lin I.I., Wu CR, Liu KK (2005) A unique seasonal pattern in phytoplankton biomass in low-latitude waters in the South China Sea. Geophysical Research Letters 32: 4.

27. Yuan YC, Liao GH, Yang CH (2009) A diagnostic calculation of the circulation in the upper and middle layers of the Luzon Strait and the northern South China Sea during March 1992. Dynamics of Atmospheres and Oceans 47: 86-113.

28. Morton B, Blackmore G (2001) South China Sea. Marine Pollution Bulletin 42: 1236-1263.

29. Gan JP, Li L, Wang DX, Guo XG (2009) Interaction of a river plume with coastal upwelling in the northeastern South China Sea. Continental Shelf Research 29: 728-740.

30. Su JL (2004) Overview of the South China Sea circulation and its influence on the coastal physical oceanography outside the Pearl River Estuary. Continental Shelf Research 24: 1745-1760.

31. Jing ZY, Qi YQ, Hua ZL, Zhang H (2009) Numerical study on the summer upwelling system in the northern continental shelf of the South China Sea. Continental Shelf Research 29: 467-478.

32. Strickland JDH, Parsons TR (1972) A Practical Handbook of Seawater Analysis, Bulletin 167, second ed. Fisheries Research Board of Canada, Ottawa.

33. Wong GTF, Zhang LS (1992b) Chemical removal of oxygen with sulfite for the polarographic or voltammetric determination of iodate or iodide in seawater. Marine Chemistry 38: 109-116.

34. Herring JR, Liss PS (1974) A new method for the determination of iodine species in seawater. Deep Sea Research and Oceanographic Abstracts 21: 777-783.

35. Luther Ш GW, Swartz CB, Ullman WJ (1988) Direct determination of iodide in seawater by cathodic stripping square wave voltammetry. Analitical Chemistry 60: 1721-1724.

36. Wong GTF, Tseng CM, Wen LS, Chung SW (2007) Nutrient dynamics and N-anomaly at the SEATS station. Deep Sea Research Part II: Topical Studies in Oceanography 54: 1528-1545.

37. Codispoti LA, Christensen JP (1985) Nitrification, denitrification and nitrous oxide cycling in the Eastern Tropical South Pacific Ocean. Marine Chemistry 16: 277-300.

38. Ward BB (1987) Nitrogen transformation in the Southern California bight. Deep-Sea Research 34: 785-805.

39. Gong GC, Liu KK, Liu CT, Pai SC (1992) The chemical hydrography of the South China Sea west of Luzon and a comparison with the West Philippine Sea. Terrestrial, Atmospheric and Oceanic Sciences 3: 587-602.

40. Shaw PT (1991) The seasonal variation of the intrusion of the Philippine Sea water into the South China Sea. Journal of Geophysical Research 96: 821-827.

41. Shaw PT, Chao SY (1994) Surface circulation in the South China Sea. Deep Sea Research Part I: Oceanographic Research Papers 41: 1663-1683.

42. Wong GTF, Takayanagi K, Todd JF (1985) Dissolved iodine in waters overlying and in the Orca Basin, Gulf of Mexico. Marine Chemistry 17: 177-183.

43. Ullman WJ, Luther GW, De Lange GJ, Woittiez JR (1990) Iodine chemistry in deep anoxic basins and overlying waters of the Mediterranean Sea. Marine Chemistry 31: 153-170.

44. Truesdale VW, Danielssen DS, Waite TJ (2003) Summer and winter distributions of dissolved iodine in the Skagerrak. Estuarine, Coastal and Shelf Science 57: 701-713.

45. Moisan TA, Dunstan WM, Udomkit A, Wong GTF (1994) The uptake of iodate by marine phytoplankton. Journal of Phycology 30: 580-587.

46. Tian RC, Marty JC, Nicolas E, Chiavérini J, Ruiz-Ping D, et al. (1996) Iodine speciation: a potential indicator to evaluate new production versus regenerated production. Deep Sea Research Part I: Oceanographic Research Papers 43: 723-738.

47. ZicV, Truesdale VW, Cukrov N (2008) The distribution of iodide and iodate in anchialine cave waters --Evidence for sustained localised oxidation of iodide to iodate in marine water. Marine Chemistry 112: 168-178.

48. Spokes LJ, Liss PS (1996) Photochemically induced redox reactions in seawater, II. Nitrogen and iodine. Marine Chemistry 54: 1-10.

49. Truesdale VW (2007) On the feasibility of some photochemical reactions of iodide in seawater. Marine Chemistry 104: 266-281.

50. Lu ZM, Gan JP, Dai MH, Cheung AYY (2010) The influence of coastal upwelling and a river plume on the subsurface chlorophyll maximum over the shelf of the northeastern South China Sea. Journal of Marine Systems 82: 35-46.

51. Song XY, Liu HX, Huang LM, Tan YH, Ke ZX, et al. (2010) Distribution characteristics of basic biological production and its influencing factors in the northern South China Sea in summer. Acta Ecologica Sinica 30: 6409-6417.

52. Truesdale VW, Watts SF, Rendell AR (2001b) On the possibility of iodide oxidation in the near-surface of the Black Sea and its implications to iodine in the general ocean. Deep-Sea Research Part I: Oceanographic Research Papers 48: 2397-2412.

53. Zic V, Branica M (2006) The distributions of iodate and iodide in Rogoznica Lake (East Adriatic Coast). Estuarine Coastal and Shelf Science 66: 55-66.

54. Yool A, Martin AP, Fernandez C, Clark DR (2007) The significance of nitrification for oceanic new production. Nature 447: 999-1002.

55. Karl DM, Lukas R (1996) The Hawaii Ocean Time-series (HOT) program: background, rationale and field implementation. Deep-Sea Research Part II: Topical Studies in Oceanography 43: 129-156.

56. Wyrtki K (1961) Scientific Results of Marine Investigations of the South China Sea and the Gulf of Thailand. Physical oceanography of the Southeast Asian waters 2: 195.

PERMISSIONS

All chapters in this book were first published in JMSRD, by OMICS International; hereby published with permission under the Creative Commons Attribution License or equivalent. Every chapter published in this book has been scrutinized by our experts. Their significance has been extensively debated. The topics covered herein carry significant findings which will fuel the growth of the discipline. They may even be implemented as practical applications or may be referred to as a beginning point for another development.

The contributors of this book come from diverse backgrounds, making this book a truly international effort. This book will bring forth new frontiers with its revolutionizing research information and detailed analysis of the nascent developments around the world.

We would like to thank all the contributing authors for lending their expertise to make the book truly unique. They have played a crucial role in the development of this book. Without their invaluable contributions this book wouldn't have been possible. They have made vital efforts to compile up to date information on the varied aspects of this subject to make this book a valuable addition to the collection of many professionals and students.

This book was conceptualized with the vision of imparting up-to-date information and advanced data in this field. To ensure the same, a matchless editorial board was set up. Every individual on the board went through rigorous rounds of assessment to prove their worth. After which they invested a large part of their time researching and compiling the most relevant data for our readers.

The editorial board has been involved in producing this book since its inception. They have spent rigorous hours researching and exploring the diverse topics which have resulted in the successful publishing of this book. They have passed on their knowledge of decades through this book. To expedite this challenging task, the publisher supported the team at every step. A small team of assistant editors was also appointed to further simplify the editing procedure and attain best results for the readers.

Apart from the editorial board, the designing team has also invested a significant amount of their time in understanding the subject and creating the most relevant covers. They scrutinized every image to scout for the most suitable representation of the subject and create an appropriate cover for the book.

The publishing team has been an ardent support to the editorial, designing and production team. Their endless efforts to recruit the best for this project, has resulted in the accomplishment of this book. They are a veteran in the field of academics and their pool of knowledge is as vast as their experience in printing. Their expertise and guidance has proved useful at every step. Their uncompromising quality standards have made this book an exceptional effort. Their encouragement from time to time has been an inspiration for everyone.

The publisher and the editorial board hope that this book will prove to be a valuable piece of knowledge for researchers, students, practitioners and scholars across the globe.

LIST OF CONTRIBUTORS

Henry M Manik
Department of Marine Science and Technology, Faculty of Fisheries and Marine Sciences, Bogor Agricultural University (IPB) Kampus IPB, Darmaga, Bogor-16680, Indonesia

Durogbitan A Abimbola
Petrfac IES, Research and Development, Manchester, UK

Babu A, Varadharajan D, Vengadesh Perumal N, Thilagavathi B, Manikandarajan T, Sampathkumar P and Balasubramanian T
Faculty of Marine Sciences, Annamalai University, Parangipettai -608 502, Tamil Nadu, India

José Luis Bonilla-Gómez
Department of Marine Sciences and Limnology, National Autonomous University of Mexico, Mexico D. F., Mexico

Maribel Badillo, Alfredo Gallardo, Carmen Galindo, Daniel Arceo and Xavier Chiappa- Carrara
UNAM-Sisal, Faculty of Science, National University of Mexico, Sisal, Yucatan, Mexico

Korynthia López
UNAM-Sisal, Faculty of Chemistry, National Autonomous University of Mexico, Sisal, Yucatan, Mexico

Awaleh MO, Hoch Farhan Bouraleh, Bahga HO and Samaleh AI
Chemistry laboratory, Institute of Earth Sciences, Centre for studies and Research in Djibouti (CERD), Airport Road, PO Box 486, Djibouti - city, Republic of Djibouti

Soubaneh YD
Department of Biology, Chemistry and Geography, University of Quebec at Rimouski, 300, Allée des Ursulines, Rimouski, QC, G5L 3A1, Canada

Badran M
Regional Organization for the Conservation of Environment of the Red Sea and Gulf of Aden "PERSGA", PO Box: 53662 Jeddah 21583 Saudi Arabia

Gaspar Banfalvi
Department of Biotechnology and Micro-biology, University of Debrecen, Hungary

Ekowati Chasanah
Indonesia Research Center for Marine and Fisheries Product Processing and Biotechnology, Jl. KS Tubun Petamburan VI, Jakarta 10260, Indonesia

Hedi Indra Januar
Indonesia Research Center for Marine and Fisheries Product Processing and Biotechnology, Jl. KS Tubun Petamburan VI, Jakarta 10260, Indonesia
Magister of Environmental Science, Diponegoro University at Indonesia, Jl. Imam Bardjo, SH No. 5 Semarang, Indonesia

Boedi Hendrarto
Magister of Environmental Science, Diponegoro University at Indonesia, Jl. Imam Bardjo, SH No. 5 Semarang, Indonesia

Anthony D. Wright
University of Hawaii, College of Pharmacy, 34 Rainbow Drive, Hilo, Hawaii 96720, USA

Kevin C Owen
Höegh LNG AS, 15469 Osprey Glen Drive, Lithia, FL, USA

Jenna L Owen
Tindale-Oliver and Associates, Tampa, FL, USA

Daniel P Owen
Department of Biological Sciences, Florida State University, Tallahassee, FL, USA

Tien Anh Tran
School of Energy and Power Engineering, Wuhan University of Technology, Wuhan 430063, P.R. China
Faculty of Marine Engineering, Vietnam Maritime University, Haiphong, Vietnam

Ghoneim SA
Department of Environmental Planning, Cairo University, Egypt

Ibrahim H
Regional Development Department, Cairo University, Egypt

Durogbitan A Abimbola
Petrfac IES, Research and Development, Manchester, UK

Manikandan R and Vijayakumar R
Research Department of Microbiology, Bharathidasan University Constituent College, Kurumbalur, Perambalur District, Tamil Nadu, India

Ravinesh Ram and Paul C. Southgate
Centre for Sustainable Tropical Fisheries and Aquaculture, College of Marine & Environmental Sciences, James Cook University, Townsville, Queensland 4811, Australia

Roveena Vandana Chand
School of Biological and Chemical Sciences, Faculty of Science Technology and Environment, University of the South Pacific, Suva, Fiji Islands

Ding Hao, Song Bao-wei and Cao Yong-hui
College of Marine Engineering, Northwestern Polytechnical University, Xi'an 710072, China

Sunanda Kadiri and Siddaiah Vidavalur
Department of Organic Chemistry and FDW, Andhra University, Visakhapatnam, 530 003, India

Nagendra Sastry Yarla
Department of Biochemistry and Bioinformatics, School of Life Sciences, Institute of Science, GITAM University, Visakhapatnam, 530045, India

Abdolah Raeisi Sarasiab
Department of Marine Ecology, Farhangian Science University of Iran, Iran

Zohreh Mirsalari
Department of Biodiversity, Science and Research Branch, Islamic Azad University, Ahvaz, Iran

Mehdi Hosseini
Department of Marine Biology, Faculty of Biological Science, Shahid Beheshti University, Tehran, Iran

Elżbieta Niemirycz, Lucyna Falkowska, Maria Witt, Andrzej R Reindl and Marta E Kobusińska
University of Gdansk, Institute of Oceanography, Department of Marine Chemistry and Environmental Protection, 46 Marszalka Pilsudskiego Str., Gdynia 81-378, Poland

Joanna Szlinder-Richert
National Marine Fisheries Research Institute, Department of Food and Environmental Chemistry, Kollataja Str, Gdynia 81-332, Poland

Ott Roots
Estonian Environmental Research Centre, Marja 4D, 10 617 Tallinn, Estonia
University of Tartu, Estonian Marine Institute, Mäealuse 10A, 12 618 Tallinn, Estonia

Ksenia Pazdro and Agata Zaborska
Institute of Oceanology of the Polish Academy of Sciences, 55 Powstancow Warszawy Str, Sopot 81-712, Poland

Matti Verta
Finnish Environment Institute (SYKE), P.O. Box 140 Helsinki, Finland

Grażyna Sapota
Maritime Institute in Gdansk, Department of Environmental Protection, 41/42 Dlugi Targ Str, Gdansk 80-830, Poland

Youssouf MO
Centre for Study and Research in Djibouti Airport Road, PO Box 486, Djibouti
UMR6539 UBO/CNRS/IRD/Ifremer European University Institute of the Sea (IUEM) Technopole Brest Iroise, 29280 Plouzane, France

Laurent M
UMR6539 UBO/CNRS/IRD/Ifremer European University Institute of the Sea (IUEM) Technopole Brest Iroise, 29280 Plouzane, France

Xavier C
University of Western Brittany (UBO) UMR 6523 CNRS/IFREMER/UBO Physics Laboratory of the Ocean, IUEM, Plouzané, France

Stamatopoulos C
Consultant of Fisheries Statistics, Department of Fisheries, Ministry of Environment, Doha, State of Qatar

Abdallah M
Consultant of Fisheries Resources, Department of Fisheries, Ministry of Environment, Doha, State of Qatar

Durogbitan A Abimbola
Petrfac IES, Research and Development, Manchester, UK

MWS Lau
School of Mechanical and Systems Engineering, Faculty of Science, Agriculture and Engineering, University of Newcastle, Newcastle (UK), International Singapore

SC Fok
Department of Mechanical Engineering, The Petroleum Institute, P.O. Box 2533, Abu Dhabi, UAE

GL Seet
School of Mechanical and Aerospace Engineering, Nanyang Technological University, 50 Nanyang Avenue, Singapore

E Low
Engineering Accreditation Board, Institution of Engineers, 70 Bukit Tinggi Road, Singapore

Rahaman SMB, Hasanuzzaman AFM, Huq KA, Islam SS and Bir J
Fisheries and Marine Research Technology Discipline, Khulna University, Khulna 9208, Bangladesh

Golder J and Begum S
Environmental Science Discipline, Khulna University, Khulna 9208, Bangladesh

Rahaman MS
Department of Chemistry, Comilla University, Comilla 3500, Bangladesh

Long A and Xiao H
State Key Laboratory of Tropic Oceanography, South China Sea Institute of Oceanology, Chinese Academy of Sciences, 164 West Xingang Road, Guangzhou 510301, P.R China

Dang A
State Key Laboratory of Tropic Oceanography, South China Sea Institute of Oceanology, Chinese Academy of Sciences, 164 West Xingang Road, Guangzhou 510301, P.R China
South China Sea Branch of the State Oceanic Administration, 353 Middle Xingang Road, Guangzhou, 510301, P.R China

Yu X
State Key Laboratory of Tropic Oceanography, South China Sea Institute of Oceanology, Chinese Academy of Sciences, 164 West Xingang Road, Guangzhou 510301, P.R China
University of Chinese Academy of Sciences, Beijing, 100039, P.R China

Index

O

P

R

S

T

U

V

W